Madhumangal Pal · Sovan Samanta ·
Ganesh Ghorai

Modern Trends in Fuzzy
Graph Theory

 Springer

Madhumangal Pal
Department of Applied Mathematics
Vidyasagar University
Midnapore, West Bengal, India

Sovan Samanta
Department of Mathematics
Tamralipta Mahavidyalaya
Tamluk, West Bengal, India

Ganesh Ghorai
Department of Applied Mathematics
Vidyasagar University
Midnapore, West Bengal, India

ISBN 978-981-15-8805-1   ISBN 978-981-15-8803-7   (eBook)
https://doi.org/10.1007/978-981-15-8803-7

Mathematics Subject Classification: 05C72, 05C10, 05C38, 05C62, 05C76, 05C78, 05C85, 94C15, 94D05

This Springer imprint is published by the registered company Springer Nature Singapore Pte Ltd.
The registered company address is: 152 Beach Road, #21-01/04 Gateway East, Singapore 189721, Singapore

Modern Trends in Fuzzy Graph Theory

Preface

Graph Theory and Fuzzy Graph Theory

Graph theory started its journey from the famous "Königsberg bridge problem". This problem is often said to have been the birth of graph theory. In 1736, Euler finally solved this problem with the help of graphs. Though graph theory is a relatively old subject, its growing applications in operations research, chemistry, genetics, electrical engineering, architecture, linguistics, geography, sociology, traffic planning, etc., have kept it young. These days graph theory is being used in communication system (mobile, internet, etc.), social networks, computer design, etc.

A subject of much interest, graph theory is used to model a wide range of problems. It is studied by mathematicians, computer scientists, engineers, and many others. This subject is basically studied in two directions: *algebraic* graph theory and *algorithmic* graph theory. In algebraic graph theory, people try to find out the bounds of the graph parameters and their properties. In order to study algebraic graph theory, the knowledge of set and matrix theories is sufficient. While in algorithmic theory, people need to design an algorithm to find out the exact value of the parameters for small as well as large graphs. To further study algorithmic graph theory, the knowledge of data structure, computer memory organization, operating system, etc., is needed.

In graph theory, it is assumed that the vertices, edges, weights, etc., are certain. That is, there is no doubt regarding the existence of these objects. But the real world is full of uncertainties. This means, in some situations, it is seen that the vertices, edges, and weights may or may not be certain. For example, the vehicle travel time or vehicle capacity on a road network may not be known exactly. To represents such graphs, A. Rosenfeld introduced the "fuzzy graph" in 1975. Like set theory, the background of fuzzy graph theory is the fuzzy set theory developed by L. A. Zadeh in 1965. Rosenfeld considered fuzzy relations on fuzzy sets and developed the structure of fuzzy graph. The invention of fuzzy graph theory has opened a new research area in mathematics, communication systems, economics, and others. Social networks, such as Facebook, ResearchGate, Twitter, LinkedIn, etc., are the appropriate examples of fuzzy graphs.

Description of Chapters

Designed into ten chapters, this book is written mainly based on our research. We also have referred to research by other scientists. This is a very useful handbook for beginners as well as expert fuzzy graph theoreticians. It will also be useful to mathematicians, computer scientists, engineers, and others. This book is written for researchers as well as for postgraduate students. Ample materials are presented in the book so that instructors will be able to select topics appropriate to their needs.

In Chap. 1, some fundamental terms about fuzzy graph are defined and explained with very simple examples. Since this book is designed primarily for researchers, we assume the reader is familiar with basic notions of fuzzy set theory. Basic results on paths, connectivity and different types of edges are presented. Other basic concepts include operations on fuzzy graphs, fuzzy cliques, fuzzy independent sets, eigenvalues, and energy of fuzzy graph, as well as some topological indices. Then results on complement, dominations, regular and irregular, automorphism and isomorphism, connectivity index, Wiener index and Zargeb indices of fuzzy graphs are presented. This chapter is concluded with the development of certain types of generalized fuzzy graphs such as intuitionistic fuzzy graphs, Pythagorean fuzzy graphs, q-Rung orthopair fuzzy graphs, bipolar fuzzy graphs, m-polar fuzzy graphs, picture fuzzy graphs, neutroshopic fuzzy graphs, spherical fuzzy graphs and T-spherical fuzzy graphs, interval-valued fuzzy graphs, cubic graphs, vague graphs, interval-valued intuitionistic fuzzy graphs, soft graphs, Dombi fuzzy graphs and Pythagorean Dombi fuzzy graphs.

Chapter 2 is devoted to the concept of fuzzy planar graph. This graph is defined in a very interesting way. As per our definition, every fuzzy graph is a fuzzy planar graph with a certain degree of planarity. Based on this term, k-fuzzy planar graph is defined, where k is the degree of planarity. For 1-fuzzy planar graph, the fuzzy dual graph is defined and investigated lots of properties. Isomorphism of fuzzy planar graph is also studied in this chapter.

In Chap. 3, a detailed discussion on fuzzy trees including forests, bridges, cut vertices, and end vertices of a fuzzy graph. Attention is then turned to multimin and locamin for a FG. Some characterizations of multimin and fuzzy trees are explained.

Chapter 4 deals with fuzzy competition graphs. Different kinds of fuzzy competition graphs, namely fuzzy k-competition graphs, m-step fuzzy competition graphs, m-step fuzzy neighborhood graphs, fuzzy economic competition graphs, and fuzzy competition number are investigated. Lots of properties and their interrelationship are also presented. The chapter concludes with a study of isomorphism in fuzzy competition graph.

Chapter 5 starts by considering fuzzy threshold graphs. The notion of fuzzy threshold dimension, fuzzy threshold partition number, and fuzzy Ferrers diagraphs are introduced. Some of its important properties are investigated.

Chapter 6 is devoted to fuzzy tolerance graph. This graph is a subclass of fuzzy interval graph and has many interesting properties. The fuzzy bounded tolerance graphs and fuzzy interval containment graphs are the topics of this chapter. The fuzzy φ-tolerance competition graph is defined and some particular cases for different values of φ are also investigated.

Chapter 7 gives new concepts on the coloring of fuzzy graph. Two types of colorings are mainly discussed: vertex/node coloring and edge coloring. Both the methods of coloring are described along with examples. A new method of coloring, namely fuzzy fractional coloring of a fuzzy graph, is introduced along with a fuzzy fractional chromatic number.

In Chap. 8, the work on m-polar fuzzy graphs is presented that is mostly due to Ghorai and Pal. This work includes results on operations of m-polar fuzzy graphs. Results concerning isomorphism of m-polar fuzzy graphs as well as results concerning strong, self-complementary and weak self-complementary m-polar fuzzy graphs are provided. The chapter concludes with the concepts of density of m-polar fuzzy graph and balanced of m-polar fuzzy graph.

Chapter 9 is devoted to the work on intuitionistic fuzzy graphs. This graph is defined based on the Attanosov's intuitionistic fuzzy sets developed in 1983. It includes results concerning different types of operations of intuitionistic fuzzy graphs. Other results deal with degrees, independent sets and coverings, connectivity, strong arc, partial cut node, bridge and block. In addition, intuitionistic fuzzy competition graphs, intuitionistic fuzzy tolerance graph, intuitionistic fuzzy planar graph, and genus of intuitionistic fuzzy graph are studied.

In Chap. 10, some applications of fuzzy graphs are provided. Some problems are modeled as fuzzy graphs, namely representation of ecological problem, representation of social network, representation of telecommunication system, link prediction in fuzzy social networks, competition in manufacturing industries, patrolling of bus network, image contraction, installation of cell phone tower, traffic signalling, job selection, etc.

With adequate examples and diagrams provided for better understanding, this book gives an interconnected presentation of fundamental ideas, concepts, and results. We welcome comments, critiques, suggestions, corrections, etc. from the readers of this book, so that next edition may benefit from experience with the first.

Acknowledgements

In writing this book, we have taken help from several books, research articles, and some websites mentioned in the bibliography. We acknowledge the leading fuzzy graph theoreticians: J. N. Mordeson, P. S. Nair, D. S. Malik, K. R. Bhutani, M. Akram, T. Al-Hawary, A. N. Gani, M. S. Sunitha, S. Mathew, M. G. Karunambigai, H. Rashmanlou, K. K. Krishnan, T. Atanassov, W. L. Craine, R. H. Goetschel, C. M. Klein, Koczy, K. M. Lee, K. C. Lin, M. S. Chern, M. L. N. McAllister, C. S. Peng, S. M. A. Nayeem, A. Pal, S. Okada, R. Parvathi, A. Somasundaram, S. Somasundaram, and many others, for their huge and fundamental contributions to fuzzy graph theory.

The authors would like to acknowledge the assistance of their research group of fuzzy graph theory: S. Sahoo, T. Pramanik, S. K. Amanathulla, M. Bhowmik, S. Mondal, C. Jana, S. Raut, S. Bera, A. Bhattachariya, R. Islam, T. Mahapatra, S. Poulik, S. Das, N. Patra, S. Mandal, U. Mandal, R. Mahapatra, and S. Dogra. We are indebted to Miss. Sreenanda Raut for scrutinizing the original draft copy of the book. We would also like to acknowledge our colleagues for their encouragement. We highly acknowledged our parent institutions—Vidyasagar University and Tamralipta University in India—for providing infrastructural and computational supports.

We feel a great reverence for our family members for their continuous support and encouragements during the preparation of the manuscript. We thank Springer Nature for their sincere care in the publication of the book.

Midnapore, India   Madhumangal Pal
Tamluk, India   Sovan Samanta
Midnapore, India   Ganesh Ghorai

Contents

About the Authors

Madhumangal Pal is Professor at the Department of Applied Mathematics, Vidyasagar University, West Bengal, India. He completed his Ph.D. in Mathematics from the Indian Institute of Technology Kharagpur, India. His areas of research include algorithmic and fuzzy graph theory, fuzzy matrices, fuzzy algebra, genetic algorithms, and parallel algorithms. Under his supervision, 34 research scholars have completed their Ph.D. theses. An author of 10 books and over 330 research articles, Prof. Pal is the Editor-in-Chief of the *Annals of Pure and Applied Mathematics* and *Journal of Physical Sciences* and is on the editorial boards of several renowned national and international journals. He has completed three research projects funded by the University Grants Commission, the Department of Science and Technology, India, and is currently working on an ongoing project. He has delivered invited talks at seminars and conferences held in India and abroad, including the UK, Greece, China, Malaysia, Bangladesh, Hong Kong, UAE, and Thailand. Prof. Pal is a member of several academic bodies, viz. American Mathematical Society, Calcutta Mathematical Society, ADMA, Neutrosophic Science International Association, USA, Ramanujan Mathematical Society, etc. He is also a Course Coordinator of a MOOC.

Sovan Samanta is Assistant Professor at the Department of Mathematics, Tamralipta Mahavidyalaya (affiliated to Vidyasagar University), West Bengal, India. Earlier, he served as an Assistant Professor at the Indian Institute of Information Technology Nagpur, India, and was a Postdoc Fellow at Hanyang University, Ansan, South Korea. He completed his Ph.D. in Fuzzy Graph Theory from Vidyasagar University, West Bengal, India, in 2014. His research interests include fuzzy graph theory, social networks, and various types of uncertain systems. His research papers on fuzzy graph theory have been published in journals of repute. He is also a subject Editor for the Journal of Applied Mathematics and Computing (SCIE indexed Springer journal) and is on editorial boards for several journals of repute.

Ganesh Ghorai is Assistant Professor at the Department of Applied Mathematics, Vidyasagar University, India. He received his Ph.D. in Mathematics from Vidyasagar University, India, in 2017, and M.Sc. in Mathematics from Indian Institute of Technology, Bombay, India, in 2011. He has published more than 30 research papers in international peer-reviewed journals on fuzzy graphs and related topics and is currently guiding 3 research scholars on their Ph.D. theses in the areas of graph theory, fuzzy graph theory, and fuzzy algebra. He has completed a research project funded by the University Grants Commission of India on fuzzy graph theory and is Deputy Coordinator of an ongoing project funded by Science and Engineering Research Board, Department of Science and Technology, Government of India (DST-SERB). He is a member of the Calcutta Mathematical Society, Kolkata, India, and several administrative and academic bodies in Vidyasagar University and other institutes.

Chapter 1
Fundamentals of Fuzzy Graphs

The journey of graph theory started with famous Königsberg seven bridges problem and its negative resolution by Leonhard Euler in 1736. So, it is an old subject, even it is very new due to its wide applications in modeling of modern society. The graphs are being used as mathematical tools for modeling a huge number of real world problems. Graphs can be used in situations where objects can be connected by some rules or relations. For instance, tramways where tram depots are connected by roads, circuit design where nodes are connected by wires, etc. Nowadays, graphs are being used in games and recreational mathematics too. But, the demand of real world does not end here. There are a lot of uncertainties in the real world, in other words, vagueness or fuzziness. As a matter of fuzziness researchers have devoted themselves into studying a new branch of mathematics called "Fuzzy graph" (FG). Rosenfeld [129] first considered fuzzy relations on fuzzy sets and developed the theory of FGs in 1975. Many real life applications such as molecular structure of chemical compounds, electrical network, genetic (DNA) studies, registers allocation, scheduling, networking, communication, image segmentation, data mining, airline routines, planning project, and many more have been successfully dealt with fuzzy graph. Recently, FGs are widely used for modeling communication network (internet, mobile, etc.), social networks (Facebook, ResearchGate, WhatsApp, Twitter, LinkedIn, etc.), and many others related fields.

In this chapter, a complete notion about fuzzy graph theory and its key attributes are given in the basic level. A details discussion about connectivity in fuzzy graph and that of different types of edges and paths have been made. Several important operations on fuzzy graph as well as associated fuzzy matrix of it have been talked about. In order to acquire knowledge about fuzzy graph, it is indispensable to know about few basic terminologies involved in fuzzy graph such as fuzzy cliques, fuzzy independent sets, eigenvalues, and energy as well as some topological indices which have been presented in this chapter particularly. Also, some clear ideas on complement, dominations, regular and irregular, automorphism and isomorphism of fuzzy

M. Pal et al., *Modern Trends in Fuzzy Graph Theory*,
https://doi.org/10.1007/978-981-15-8803-7_1

graphs are discussed. Lastly, to get an idea of the wide expanse of fuzzy graph in the preliminary level, different types of fuzzy graphs have been mentioned and discussed elaborately with examples and diagrammatic representation.

1.1   Basic Terminologies

It is very interesting that graph theory as well as FG theory are studied in two directions, viz. algebraic sense and algorithmic sense. To study in algebraic sense, the knowledge of basic algebra is sufficient, whereas to study in algorithmic sense, the knowledge on data structure, memory organization, etc. are required. Mainly, mathematicians approach in algebraic sense, while computer scientists in algorithmic sense.

Before going to the definition of FG , the conventional or classical graph is defined below. In contrast to FG , the classical graphs are called crisp graphs.

A (crisp) **graph** $G = (V, E)$ consists of two sets called the vertices V and the edges E. The elements of $V = \{v_1, v_2, \ldots\}$ are called **vertices** or **nodes** or **points** and the elements of $E = \{e_1, e_2, \ldots\}$ are called **edges** or **arcs** or **lines**. The vertices v_i, v_j associated with an edge, say, e_k are called the **end vertices** of e_k. An edge whose end points are v_i and v_j is denoted by (v_i, v_j) or $v_i v_j$, and the vertices v_i and v_j are said to be **adjacent** in G. From the definition of the graph, it is observed that any subset R of $V \times V$ is a graph whose vertex set is V and edge set is R. The most common representation of a graph is its diagram. In the diagram, the vertices of a graph are represented by points and the edges by means of line/curved segments.

In crisp graphs, it is assumed that all the vertices and edges are fixed, i.e. their existence is certain. But, in some real life cases, it is seen that some vertices and/or edges are not certain. That means some vertices and/or edges may be present in the graph for a particular time duration and may not present for some other time interval. These type of situations occur very often in social network. Since the appearances or presence of the vertices and/or edges in a graph may be uncertain, some numbers are assigned to the said vertices and edges to indicate their degree of appearance or presence or existence in the graph.

1.2   Definitions of Fuzzy Graphs

The formal definition of a FG is given below. Let $G^* = (V, E)$ be a crisp graph.

Definition 1.1 A **fuzzy graph** [129] $\mathscr{G} = (\mathscr{V}, \sigma, \mu)$ corresponding to the crisp graph G^* is a non-empty set $\mathscr{V}$ together with a pair of functions $\sigma : V \to [0, 1]$ and $\mu : \mathscr{V} \times \mathscr{V} \to [0, 1]$ such that for all $a, b \in \mathscr{V}$,

$$\mu(a, b) \leq \min\{\sigma(a), \sigma(b)\}, \tag{1.1}$$

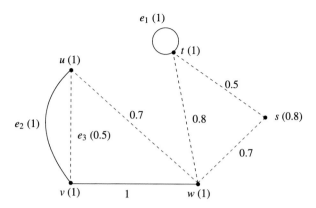

Fig. 1.1 An example of a FG

where $\sigma(a)$ and $\mu(a, b)$ represent the membership values of the vertex a and edge (a, b) in $\mathscr{G}$ respectively.

An edge of a graph is denoted by (a, b) or simply ab.

A FG and a crisp graph are structurally same, the only difference is the membership values assigned to the vertices and edges. If the membership values of all vertices and edges are equal and equal to 1, then the FG becomes a crisp graph. Hence, a crisp graph $G = (V, E)$ may be treated as FG as well. In this case, $\sigma(a) = 1$ for all $a \in \mathscr{V}$, $\mu(a, b) = 1$ if $(a, b) \in E$ and $\mu(a, b) = 0$ otherwise. Note that, if $\sigma(a) = k$, for all $a \in \mathscr{V}, k$ $(0 < k < 1)$ is a constant, then $\mathscr{G}$ cannot be a crisp graph. In this case, all the vertices are uncertain and this type of FG is called a regular FG (discussed later).

There is a restriction on the membership values of the edges (see inequality (1.1)). If '$\leq$' is replaced by '$\geq$' in inequality (1.1) and if the membership value of a vertex or both the vertices are zero, then there may be an edge without end vertices, which is impossible. But, by removing this restriction, a FG is defined called generalized FG.

Both the numbers $\sigma(a)$ and $\mu(a, b)$ lie between 0 and 1. So, in a FG , there are three types of vertices, viz. vertices with membership value 0 (in this case, the vertex a never occurs in the graph), membership value 1 (in this case, the vertex a is certainly present in the graph), and lastly, any real number between 0 and 1 (in this case, the vertex a is present in the graph with some possibility). Similarly, three types of edges are also present in a FG. Edges with membership value 1 (drawn by solid lines), edges with membership value in between 0 and 1 (drawn by dotted lines), and edges with 0 membership values are not mentioned in the graph (see Fig. 1.1). But, normally we draw the edges of a FG by solid lines only except the edges whose membership values are 0. So, we consider only such edges whose membership values are positive.

A FG is depicted in Fig. 1.1.

In Fig. 1.1, u, v, w, s, t are five vertices with membership values 1, 1, 1, 0.8, 1 respectively. There are eight edges $e_1, e_2, (u, v), (u, w), (v, w), (w, s), (w, t)$, and

(s, t) with membership values 1, 1, 0.5, 0.7, 1, 0.7, 0.8, and 0.5 respectively. The edge e_1 is self-loop. The edges e_2 and e_3 are parallel edges. If the membership value of the edge (w, s) is 0.9, then the modified graph is not a valid FG, as it violets the inequality (1.1), though its underlying graph is a valid crisp graph.

If the membership values of the vertices and edges of FG $\mathscr{G} = (\mathscr{V}, \sigma, \mu)$ are all 1, then $\mathscr{G} = (\mathscr{V}, \sigma, \mu)$ becomes a crisp graph. It may be noted that for simple FG the set of edges is $\mathscr{V} \times \mathscr{V}$.

The underlying crisp graph can also be defined as $\mathscr{G}^* = (\mathscr{V}, \sigma^*, \mu^*)$ where $\sigma^* = \{a \in \mathscr{V} : \sigma(a) > 0\}$ and $\mu^* = \{(a, b) \in \mathscr{V} \times \mathscr{V} : \mu(a, b) > 0\}$. Using these notations, it is obvious that $\sigma^* = \mathscr{V}$ and $\mu^* = E$.

The sets σ^* and μ^* are nothing but $supp(\sigma)$ and $supp(\mu)$ respectively.

Depending on the membership values of vertices and edges and their occurrences, Blue et al. [26] have classified FG as follows:

Type I: Crisp vertex set and fuzzy edge set, i.e. certain vertex set and uncertain edge set.

Type II: Crisp vertices and edges with fuzzy connectivity, i.e. both vertices and edges are certain but connectivity is uncertain.

Type III: Fuzzy vertex set and crisp edge set (this classification violates the condition $\mu(a, b) \le \sigma(a) \wedge \sigma(b)$).

Type IV: Crisp graph with fuzzy weights, it is a crisp graph with uncertain vertex and edge weights.

Type V: Fuzzy set of crisp graphs, fuzzy composition of crisp graphs.

Recently, we observed that many real life problems cannot be modeled using crisp graph due to the lack of precise information. Such problems can be modeled properly using FG. For example, interconnection between users of Facebook can be modeled by a FG. All the individuals or organizations in the Facebook are considered as the vertices of Facebook and there is an edge between two vertices if the corresponding individuals or organizations communicate with each other. To have an idea of how it can be done, see Fig. 1.2. The membership values of the individuals or organizations (vertices) are calculated based on the duration of their respective appearance on Facebook. On the other hand, the membership value of an edge is decided by the frequency of interaction between the end vertices.

A FG can be drawn in many different ways by changing the positions of the vertices and drawing the edges by straight lines or curved lines. It is immaterial whether the edges are shown using straight line or curved lines, short or long. The important thing is the relationship between the vertices.

If the vertex set V and edge set E are finite then the graph is called a finite graph. Otherwise, it is called an infinite graph. In general, a graph means a finite graph unless otherwise stated.

Like crisp graph, in a FG, the self-loop, parallel edges may occur. If there is a self-loop at a vertex a, then $\mu(a, a) > 0$. A graph with parallel edges is called **multi-graph**. A graph without self-loops and parallel edges is a **simple graph**. In general, we work with simple FGs, if it is not specified otherwise.

Definition 1.2 The FG $\mathscr{H} = (\mathscr{V}_1, \tau, \nu)$ is called a **partial fuzzy subgraph** of $\mathscr{G} = (\mathscr{V}, \sigma, \mu)$ if $\tau \le \sigma$ and $\nu \le \mu$. In particular, $\mathscr{H}$ is called a **fuzzy subgraph of**

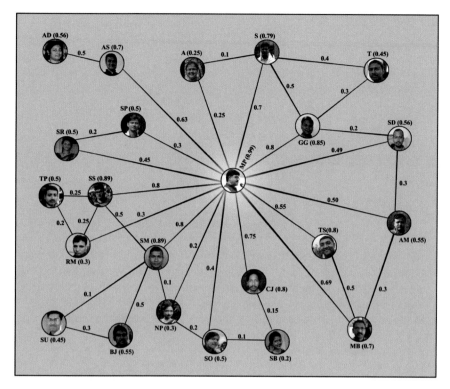

Fig. 1.2 A small part of Facebook

$\mathscr{G} = (\mathscr{V}, \sigma, \mu)$ induced by $\mathscr{V}_1$ if $\mathscr{V}_1 \subseteq \mathscr{V}$, $\tau(a) = \sigma(a)$ for all $a \in \mathscr{V}_1$ and $\nu(a, b) = \mu(a, b)$ for all $a, b \in \mathscr{V}_1$. The FG $\mathscr{G} = (\mathscr{V}, \sigma, \mu)$ is called **trivial** if $|\mathscr{V}| = 1$.

Definition 1.3 Let $\mathscr{G} = (\mathscr{V}, \sigma, \mu)$ be a FG. The **span** of $\mathscr{G}$ is a partial fuzzy subgraph $(\mathscr{V}_1, \tau, \nu)$ if $\sigma = \tau$. In this case, $(\mathscr{V}_1, \tau, \nu)$ is called a spanning fuzzy subgraph of $\mathscr{G} = (\mathscr{V}, \sigma, \mu)$.

Now, we define two very important FGs .

Definition 1.4 A FG $\mathscr{G} = (\mathscr{V}, \sigma, \mu)$ is said to be a **strong FG** if

$$\mu(a, b) = \min\{\sigma(a), \sigma(b)\} \text{ for each edge } (a, b) \text{ in } \mathscr{G}.$$

A strong FG is depicted in Fig. 1.3a. Note that the equality condition is true for edges which are present in the graph.

Definition 1.5 A FG $\mathscr{G} = (\mathscr{V}, \sigma, \mu)$ is said to be **complete fuzzy graph** (CFG) if

$$\mu(a, b) = \min\{\sigma(a), \sigma(b)\} \text{ for all } a, b \in \mathscr{V}.$$

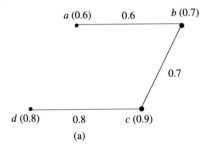 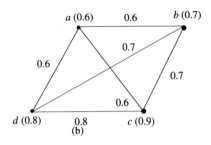

Fig. 1.3 **a** Strong FG, **b** Complete FG

A CFG is shown in Fig. 1.3b.

The equality condition is true for all pairs of vertices of the graph. This implies that there is an edge between every pair of vertices. That is, the underlying crisp graph of the CFG is a complete graph. But, if there is an edge between every pair of vertices in a FG then it need not be a CFG . For example, if the membership value of the edge (a, b) in the graph of Fig. 1.3b is taken as 0.5, then this FG is not a CFG whereas the underling crisp graph is complete.

From Fig. 1.3, it is easy to verify that every CFG is strong, but converse is not true.

Note 1.1 Let $\mathscr{G} = (\mathscr{V}, \sigma, \mu)$ be a CFG containing n vertices $\{1, 2, 3, \ldots, n\}$. Without the loss of generality, it is assumed that the membership value of the vertex i be m_i $(i = 1, 2, \ldots, n)$ and all m_i's are distinct with $m_1 < m_2 < m_3 \cdots < m_n$. Then $\mathscr{G} = (\mathscr{V}, \sigma, \mu)$ has $n - 1$ edges with membership values $m_1, n - 2$ edges with membership value m_2, and so on. Lastly, 1 edge with membership value m_{n-1} and no edge with membership value m_n.

Definition 1.6 A FG $\mathscr{G} = (\mathscr{V}, \sigma, \mu)$ is said to be **bipartite** if the vertex set $\mathscr{V}$ can be partitioned into two non-empty sets $\mathscr{V}_1$ and $\mathscr{V}_2$ such that $\mu(v_1, v_2) = 0$ if $v_1, v_2 \in \mathscr{V}_1$ or $v_1, v_2 \in \mathscr{V}_2$. Further, if $\mu(v_1, v_2) = \min\{\sigma(v_1), \sigma(v_2)\}$ for all $v_1 \in \mathscr{V}_1$ and $v_2 \in \mathscr{V}_2$, then $\mathscr{G}$ is called a **complete bipartite FG** and is denoted by K_{σ_1, σ_2}, where σ_1, σ_2 are respectively the restrictions of σ to $\mathscr{V}_1$ and $\mathscr{V}_2$.

Directed FGs (or simply fuzzy digraphs) are FGs in which the fuzzy relations between edges are not necessarily symmetric. The definition of directed FG is given as follows:

Definition 1.7 ([91]) **Directed fuzzy graph (fuzzy digraph)** $\overrightarrow{\mathscr{G}} = (\mathscr{V}, \sigma, \overrightarrow{\mu})$ is a non-empty set $\mathscr{V}$ together with a pair of functions $\sigma : \mathscr{V} \rightarrow [0, 1]$ and $\overrightarrow{\mu} : \mathscr{V} \times \mathscr{V} \rightarrow [0, 1]$ such that for all $a, b \in \mathscr{V}$, $\overrightarrow{\mu}(a, b) \leq \sigma(a) \wedge \sigma(b)$.

Since $\overrightarrow{\mu}$ is well defined, a fuzzy digraph has at most two directed edges (which must have opposite directions) between any two vertices. Here, $\overrightarrow{\mu}(a, b)$ denotes the membership value of the edge $\overrightarrow{(a, b)}$. The membership value of a loop at a vertex

a is represented by $\vec{\mu}(a, a) \neq 0$. Here, $\vec{\mu}$ need not be symmetric as $\vec{\mu}(a, b)$ and $\vec{\mu}(b, a)$ may have different values. The *underlying crisp graph of directed FG* is the graph similarly obtained except the directed arcs are replaced by undirected edges.

1.3 Paths, Cycles, and Connectedness

A **path** P in a FG $\mathscr{G} = (\mathscr{V}, \sigma, \mu)$ is a sequence of distinct vertices $x_1, x_2, \ldots, x_n$ ($n \geq 2$), $x_1 \neq x_n$, such that $\mu(x_i, x_{i+1}) > 0$, $i = 1, 2, \ldots, (n-1)$. The consecutive pairs are the edges of the path. The number of edges, i.e. $n - 1$ is called the **length** of the path P. A path P is a **cycle** if $x_1 = x_n$ and $n \geq 3$. That is, $P = (\mathscr{V}, \sigma, \mu)$ is a cycle in $\mathscr{G}$ if and only if it is a cycle in the underlying graph $\mathscr{G}^*$. A cycle $P = (\mathscr{V}, \sigma, \mu)$ is a **fuzzy cycle** if it contains more than one weakest edge.

The eccentricity ($ecen(a)$) of a vertex a of a graph $G^* = (V, E)$ is the length of the longest path from a to b, $b \in V$. The diameter $diam(G)$ and radius $\rho(G)$ are respectively defined as $diam(G) = \max\{ecen(a) : a \in V\}$ and $\rho(G) = \min\{ecen(a) : a \in V\}$. For a graph, diameter is not necessarily twice of radius.

The **strength of the path** $P(x_1, x_2, \ldots, x_n)$ is defined as $\min\{\mu(x_i, x_{i+1}) : i = 1, 2, \ldots, n-1\}$. That is, the strength of a path is the membership value of the weakest edge (the edge with least membership in the path).

The maximum strength among all paths between the vertices a and b in $\mathscr{G} = (\mathscr{V}, \sigma, \mu)$ is called the **strength of connectedness** between the vertices a and b and it is denoted as $CONN_{\mathscr{G}}(a, b)$ or $\mu^\infty(a, b)$.

The strength of the strongest path from a to b is $\mu^\infty(a, b)$. An ab path P is called a strongest ab path if its strength equals $CONN_{\mathscr{G}}(a, b)$.

It is easy to verify that if $\mathscr{G}_1 = (\mathscr{V}_1, \tau, \nu)$ is a partial fuzzy subgraph of $\mathscr{G} = (\mathscr{V}, \sigma, \mu)$, then $\nu^\infty(a, b) \leq \mu^\infty(a, b)$. If there is a path between the vertices a, b, then a, b are called connected.

Example 1.1 Let us consider the graph of Fig. 1.4.

In this graph, there are four paths between the vertices x and y, viz. $x - a - b - y$, $x - c - y$, $x - d - e - f - y$, and the edge (x, y). The strength of the path $x - a -$

Fig. 1.4 Four paths between the vertices x and y

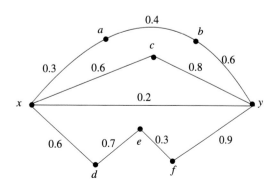

$b - y$ is min$\{0.3, 0.4, 0.6\} = 0.3$. Similarly, the strength of the paths $x - c - y$, $x - d - e - f - y$, and (x, y) are respectively 0.6, 0.3, and 0.2. Thus, the strength of connectedness $CONN_{\mathcal{G}}(x, y)$ is max$\{0.3, 0.6, 0.3, 0.2\} = 0.6$ and the strongest $x - y$ path is $x - c - y$.

1.4 Types of Arcs and Paths

In 2003, the concept of strong edges in FGs was introduced by Bhutani and Rosenfeld. They have published several papers [21–23] on strong edges, fuzzy end vertices, and geodesics in FGs .

A FG $\mathcal{G} = (\mathcal{V}, \sigma, \mu)$ is **connected** if $CONN_{\mathcal{G}}(a, b) > 0$ for every pair of vertices $a, b \in \mathcal{V}$. The FG $\mathcal{G} - (a, b)$ is obtained from the FG $\mathcal{G} = (\mathcal{V}, \sigma, \mu)$ by deleting the edge (a, b) from $\mathcal{G} = (\mathcal{V}, \sigma, \mu)$ or assigning 0 to $\mu(a, b)$.

Definition 1.8 Let $\mathcal{G} = (\mathcal{V}, \sigma, \mu)$ be a FG. An edge (a, b) is called **strong** in $\mathcal{G}$ if $\mu(a, b) > 0$ and $\mu(a, b) \geq CONN_{\mathcal{G}-(a,b)}(a, b)$. A path $P(z_1, z_2, \ldots, z_n)$ from z_1 to z_n is called a **strong path** if (z_i, z_{i+1}) is strong for all $1 \leq i \leq n - 1$.

The vertex b is said to be strong neighbor of a if the arc (a, b) is strong. The set of all strong neighbors of a is denoted by $N_s(a)$.

In a FG, an edge with maximum membership value is always strong. But the converse is not true. To justify it, let us consider the FG $\mathcal{G} = (\mathcal{V}, \sigma, \mu)$ whose underlying crisp graph is K_3. Let $\mathcal{V} = \{a, b, c\}$ and membership values of all vertices be 1, $\mu(a, b) = \mu(a, c) = 0.6$, $\mu(b, c) = 0.4$. Then the edges $(a, b), (a, c)$ are strong edges.

Lemma 1.1 *If $\mu(a, b) = \sigma(a) \wedge \sigma(b)$, then the edge (a, b) is strong.*

Proof Let P be a path from the vertex a to the vertex b. Then P must contain edges (a, c) and (d, b) in $\mathcal{G} - (a, b)$, for some $c \neq b$ and $d \neq a$. Therefore, the strength of the path P is at most $\mu(a, c) \wedge \mu(d, b) \leq \wedge\{\sigma(a), \sigma(c), \sigma(d), \sigma(b)\} \leq \sigma(a) \wedge \sigma(b) = \mu(a, b)$. $\qquad\qquad\square$

Definition 1.9 An edge (a, b) of a FG $\mathcal{G} = (\mathcal{V}, \sigma, \mu)$ is called an **effective edge** if $\mu(a, b) = \sigma(a) \wedge \sigma(b)$. The vertices a and b are called effective neighbors. The set of all effective neighbors of a is called effective neighborhood of a and is denoted by $EN(a)$.

If all the edges of a FG are effective, then the FG becomes CFG. A pendant vertex in a FG is defined as a vertex of an effective incident degree one. A fuzzy edge is called a **fuzzy pendant edge** [133] if one end vertex is fuzzy pendant vertex. The membership value of the pendant edge is the minimum among the membership values of the end vertices.

Mathew and Sunitha [84] have classified the arcs/edges of a FG $\mathcal{G} = (\mathcal{V}, \sigma, \mu)$ into three different classes.

Definition 1.10 The arc (a, b) in $\mathscr{G} = (\mathscr{V}, \sigma, \mu)$ is said to be
(i) α-strong if $\mu(a, b) > CONN_{\mathscr{G}-(a,b)}(a, b)$,
(ii) β-strong if $\mu(a, b) = CONN_{\mathscr{G}-(a,b)}(a, b)$,
(iii) δ-arc if $\mu(a, b) < CONN_{\mathscr{G}-(a,b)}(a, b)$.

The above classification of arcs depends on the given FG. For example, most of the edges in a CFG are β-strong, whereas a fuzzy tree consists of α-strong edges only. Thus, the classification of strong arcs into α and β help us to identify the structure of a FG properly.

Definition 1.11 A δ-arc (a, b) is called a δ^*-arc if $\mu(a, b) > \mu(c, d)$, where (c, d) is the weakest edge of $\mathscr{G}$.

Example 1.2 Let $\mathscr{G} = (\mathscr{V}, \sigma, \mu)$ be a FG, and let $\mathscr{V} = \{u_1, u_2, u_3, u_4\}$, $\sigma(u_i) = 1$ for $i = 1, 2, 3, 4$ and $\mu(u_1, u_2) = 0.5$, $\mu(u_2, u_3) = 0.9$, $\mu(u_3, u_4) = 0.9$, $\mu(u_1, u_4) = 0.5$, $\mu(u_2, u_4) = 0.6$. Here, $\mu(u_2, u_3) = 0.9 > 0.6 = CONN_{\mathscr{G}-(u_2,u_3)}$ and $\mu(u_3, u_4) = 0.9 > 0.6 = CONN_{\mathscr{G}-(u_3,u_4)}$. So, (u_2, u_3) and (u_3, u_4) are α-strong arcs. Again, $\mu(u_1, u_2) = \mu(u_1, u_4) = 0.5 = CONN_{\mathscr{G}-(u_1,u_2)} = CONN_{\mathscr{G}-(u_1,u_4)}$. So, (u_1, u_2) and (u_1, u_4) are β-strong arcs. Also, $\mu(u_2, u_4) = 0.6 < 0.9 = CONN_{\mathscr{G}-(u_2,u_4)}$ and so (u_2, u_4) is a δ-arc.

Definition 1.12 A path in a FG $\mathscr{G} = (\mathscr{V}, \sigma, \mu)$ is called an α-strong path if all its arcs are α-strong and is called a β-strong path if all its arcs are β-strong.

Example 1.3 Let $\mathscr{G} = (\mathscr{V}, \sigma, \mu)$ be a FG where $\mathscr{V} = \{a, b, c, d\}$ such that $\sigma(a) = 1, \sigma(b) = 1, \sigma(c) = 0.6, \sigma(d) = 0.9, \mu(a, b) = 0.9, \mu(a, d) = 0.8, \mu(b, c) = 0.4, \mu(c, d) = \mu(b, d) = 0.6$. Here, the path $b - a - d$ is an α-strong path, whereas path $b - c - d$ is a β-strong path.

Theorem 1.1 ([84]) *A strong path P from the vertex a to the vertex b is called the strongest $a - b$ path in the following cases.*
(i) P contains only α-strong arcs.
(ii) P is the unique strong $a - b$ path.
(iii) All $x - y$ paths in $\mathscr{G}$ are of equal strength.

Lemma 1.2 *Let $\mathscr{G} = (\mathscr{V}, \sigma, \mu)$ be a CFG. Then $\mathscr{G}$ does not contain any δ-arc.*

Proof Let $\mathscr{G} = (\mathscr{V}, \sigma, \mu)$ be a CFG. If possible, let $\mathscr{G}$ has a δ-arc (a, b). Then by definition $\mu(a, b) < CONN_{\mathscr{G}-ab}(a, b)$.

This indicates that there exists a stronger $a - b$ path P other than the edge (a, b) in $\mathscr{G}$. Let the strength of the path P be s_P. Then $\mu(a, b) < s_P$. Let c be the immediate next vertex of a in P. Thus,

$$\mu(a, c) > \mu(a, b). \tag{1.2}$$

Similarly, let d be the last but one vertex in P. Thus,

$$\mu(d, b) > \mu(a, b). \tag{1.3}$$

Since, $\mathscr{G}$ is CFG, $\mu(a, b) = \sigma(a) \wedge \sigma(b)$, i.e. $\mu(a, b)$ is either $\sigma(a)$ or $\sigma(b)$. Now, $\mathscr{G}$ being a CFG, $\sigma(a) = \mu(a, b)$, contradicts (1.3) and $\sigma(b) = \mu(a, b)$ contradicts (1.3). This completes the proof.   □

The following result is obviously true.

Lemma 1.3 *Let $\mathscr{G} = (\mathscr{V}, \sigma, \mu)$ be a CFG. Then $\mathscr{G}$ contains at most one α-strong edge.*

The Lemmas 1.2 and 1.3 lead the following theorems.

Theorem 1.2 *Let $\mathscr{G} = (\mathscr{V}, \sigma, \mu)$ be a CFG with n vertices. Then the number of β-strong edges in $\mathscr{G}$ is $\binom{n}{2}$ or $\binom{n}{2} - 1$.*

Theorem 1.3 *Let $\mathscr{G} = (\mathscr{V}, \sigma, \mu)$ be a CFG. Then there exists a β-strong path between any two vertices of $\mathscr{G}$.*

To find the shortest distance between a pair of vertices with different types of arc weights many methods are available in literature among them the work of Nayeem and Pal [99] is very simple and flexible.

1.5 Degrees, Order, and Size of Fuzzy Graphs

In a crisp graph, the degree of a vertex is the number of edges which are adjacent to the vertex. But, in FG, different types of degrees are defined and they are very interesting.

Definition 1.13 ([93, 151]) The **degree of a vertex** a of the FG $\mathscr{G}$ is defined as $deg_{\mathscr{G}}(a) = \sum_{b \neq a} \mu(a, b)$. The minimum degree ($\delta(\mathscr{G})$) and maximum degree ($\Delta(\mathscr{G})$) of a FG $\mathscr{G}$ are defined as follows:

$$\delta(\mathscr{G}) = \min\{deg_{\mathscr{G}}(a) : a \in \mathscr{V}\} \quad \text{and} \quad \Delta(\mathscr{G}) = \max\{deg_{\mathscr{G}}(a) : a \in \mathscr{V}\}.$$

The upper bound of the degree of a vertex in a simple FG $\mathscr{G} = (\mathscr{V}, \sigma, \mu)$ is n, where n is the number of vertices in $\mathscr{G}$.

In a crisp graph $G = (V, E)$, the size and order of G are $|V|$ and $|E|$. The similar terms are also defined for a FG as well.

Definition 1.14 Let $\mathscr{G} = (\mathscr{V}, \sigma, \mu)$ be a FG. The **order** ($O(\mathscr{G})$) and **size** ($S(\mathscr{G})$) of $\mathscr{G}$ are defined by

$$O(\mathscr{G}) = \sum_{a \in \mathscr{V}} \sigma(a) \text{ and } S(\mathscr{G}) = \sum_{(a,b) \in V \times V} \mu(a, b).$$

Note that $O(\mathscr{G}) \leq |V|$ and $S(\mathscr{G}) \leq |E|$. If $S \subseteq \mathscr{V}$, then the scalar cardinality of S is given by $\sum_{a \in S} \sigma(a)$ and it is denoted by $|S|_s$. $O(\mathscr{G})$ is the scalar cardinality of $\mathscr{V}$ and it is the order of $\mathscr{G}$.

Definition 1.15 Let $\mathcal{G} = (\mathcal{V}, \sigma, \mu)$ be a FG. The **total degree** of a vertex ($td_{\mathcal{G}}(a)$) $a \in \mathcal{V}$ is the sum of the membership values of all edges incident to a and the membership value of the vertex a, i.e. $td_{\mathcal{G}}(a) = \sum_{a \neq b} \mu(a, b) + \sigma(a) = deg_{\mathcal{G}}(a) + \sigma(a)$.

Definition 1.16 ([138]) The **effective degree** (denoted by $deg_E(a)$) of a vertex a is the sum of the membership values of the effective arcs incident at a. The minimum ($\delta_E(\mathcal{G})$) and maximum ($\Delta_E(\mathcal{G})$) effective degrees of a FG $\mathcal{G} = (\mathcal{V}, \sigma, \mu)$ are defined as follows:

$$\delta_E(\mathcal{G}) = \min\{deg_E(a) : a \in \mathcal{V}\} \text{ and } \Delta_E(\mathcal{G}) = \max\{deg_E(a) : a \in \mathcal{V}\}.$$

Definition 1.17 ([69]) The **strong degree** (denoted by $deg_s(a)$) of a vertex $a \in \mathcal{V}$ is defined as the sum of membership values of all strong edges adjacent to a. The minimum ($\delta_s(\mathcal{G})$) and maximum ($\Delta_s(\mathcal{G})$) strong degrees of the FG $\mathcal{G} = (\mathcal{V}, \sigma, \mu)$ are defined as follows:

$$\delta_s(\mathcal{G}) = \min\{deg_s(a) : a \in \mathcal{V}\} \text{ and } \Delta_s(\mathcal{G}) = \max\{deg_s(a) : a \in \mathcal{V}\}.$$

Definition 1.18 The **neighborhood degree** of a vertex a is denoted by $deg_N(a)$ and is defined by $deg_N(a) = \sum_{b \in N(a)} \sigma(b)$. The minimum ($\delta_N(\mathcal{G})$) and maximum ($\Delta_N(\mathcal{G})$) neighborhood degrees of the FG $\mathcal{G} = (\mathcal{V}, \sigma, \mu)$ are given by

$$\delta_N(\mathcal{G}) = \min\{deg_N(a) : a \in \mathcal{V}\} \text{ and } \Delta_N(\mathcal{G}) = \max\{deg_N(a) : a \in \mathcal{V}\}.$$

Definition 1.19 ([138]) The **effective neighborhood degree** of a vertex a is denoted by $deg_{EN}(a)$ and is defined as $deg_{EN}(a) = \sum_{b \in EN(a)} \sigma(b)$. The minimum ($\delta_{EN}(G)$) and maximum ($\Delta_{EN}(G)$) effective neighborhood degrees of the FG $\mathcal{G} = (\mathcal{V}, \sigma, \mu)$ are defined below:

$$\delta_{EN}(G) = \min\{deg_{EN}(a) : a \in \mathcal{V}\} \text{ and } \Delta_{EN}(G) = \max\{deg_{EN}(a) : a \in \mathcal{V}\}.$$

Definition 1.20 ([95]) The **strong neighborhood degree** of a vertex a is denoted $deg_{SN}(a)$ and is defined by $deg_{SN}(a) = \sum_{b \in NS(a)} \sigma(b)$. The minimum ($\delta_{SN}(\mathcal{G})$) and maximum ($\Delta_{SN}(\mathcal{G})$) strong neighborhood degrees of the FG $\mathcal{G} = (\mathcal{V}, \sigma, \mu)$ are defined below:

$$\delta_{SN}(\mathcal{G}) = \min\{deg_{SN}(a) : a \in \mathcal{V}\} \text{ and } \Delta_{SN}(\mathcal{G}) = \max\{deg_{SN}(a) : a \in \mathcal{V}\}.$$

From the definition of effective arc and strong arc, one can conclude that each effective arc is strong. But, the converse is not true, i.e. a strong arc need not be effective.

The following results follow from the definitions.

Lemma 1.4 ([79]) *Let $\mathscr{G} = (\mathscr{V}, \sigma, \mu)$ be a FG, then*
(i) $deg_E(a) \le deg_s(a) \le deg(a)$ for all $a \in \mathscr{V}$.
(ii) $deg_{EN}(a) \le deg_{SN}(a) \le deg_N(a)$ for all $a \in \mathscr{V}$.
(iii) $deg_E(a) \le deg_{EN}(a)$ for all $a \in \mathscr{V}$.
(iv) $deg(a) \le deg_N(a)$ for all $a \in \mathscr{V}$.
(v) $deg_s(a) \le deg_{SN}(a)$ for all $a \in \mathscr{V}$.

Like the degree of a vertex in a FG, one can define the degree of an edge in a FG.

Definition 1.21 Let $\mathscr{G} = (\mathscr{V}, \sigma, \mu)$ be a FG on $G^* = (V, E)$. The **degree of an edge** $(a, b) \in E$ is

$$deg_{\mathscr{G}}(a, b) = \sum_{c \ne a} \mu(a, c) + \sum_{c \ne b} \mu(c, b) - 2\mu(a, b).$$

Definition 1.22 Let $\mathscr{G} = (\mathscr{V}, \sigma, \mu)$ be a FG on $G^* = (V, E)$. The **total degree of an edge** $(a, b) \in E$ is

$$tdeg_{\mathscr{G}}(a, b) = \sum_{c \ne a} \mu(a, c) + \sum_{c \ne b} \mu(c, b) - \mu(a, b) = deg_{\mathscr{G}}(a, b) + \mu(a, b).$$

Example 1.4 Let $\mathscr{G} = (\mathscr{V}, \sigma, \mu)$ be a FG with $\mathscr{V} = \{a, b, c, d, e\}$. Suppose $\sigma(a) = 0.7, \sigma(b) = 0.6, \sigma(c) = 0.5, \sigma(d) = 0.7, \sigma(e) = 0.8, \quad \mu(a, b) = 0.3, \mu(b, c) = 0.4, \mu(c, d) = 0.5, \mu(d, a) = 0.5, \mu(a, e) = 0.7$.

Then $deg(a, b) = 1.6, deg(b, c) = 0.8, deg(c, d) = 0.9, deg(a, d) = 1.5, deg(a, e) = 0.8$, and $tdeg(a, b) = 1.9, tdeg(b, c) = 1.2, tdeg(c, d) = 1.4, tdeg(a, d) = 2.0, deg(a, e) = 1.5$.

Like the famous handshaking theorem in the crisp graph, two similar results are also available in FGs. They are stated below:

Theorem 1.4 ([69] Handshaking theorem for fuzzy graph) *The sum of degrees of all vertices in a FG $\mathscr{G} = (\mathscr{V}, \sigma, \mu)$ is equal to twice the sum of membership values of all edges in $\mathscr{G}$.*

Lemma 1.5 ([69]) *The sum of strong degrees of all vertices in a FG $\mathscr{G} = (\mathscr{V}, \sigma, \mu)$ is equal to twice the sum of membership values of all strong edges in $\mathscr{G}$.*

Lemma 1.6 ([93]) *The sum of effective degrees of all vertices in a FG $\mathscr{G} = (\mathscr{V}, \sigma, \mu)$ is equal to twice the sum of membership values of all effective edges in $\mathscr{G}$.*

Lemma 1.7 ([79]) *In a CFG $\mathscr{G} = (\mathscr{V}, \sigma, \mu)$ all edges are strong as well as effective. Thus,*
(i) $deg_E(a) = deg_s(a) = deg(a)$ for all $a \in \mathscr{V}$,
(ii) $deg_{EN}(a) = deg_{SN}(a) = deg_N(a)$ for all $a \in \mathscr{V}$.

1.6 Automorphism and Isomorphism of Fuzzy Graphs

Here, the notion of a weak isomorphism, co-weak isomorphism, and isomorphisms between FGs are introduced. Some examples are given to explain the definitions. Finally, it is shown that, interestingly, every fuzzy group can be embedded in a FG of the group of automorphisms of some FG.

First, we recall the definition of fuzzy group as follows:

Definition 1.23 (*Fuzzy Group*) A fuzzy group of a group G is a mapping $\phi : G \to [0, 1]$ satisfying

$$\phi(ab) \geq \phi(a) \wedge \phi(b), \; a, b \in G \tag{1.4}$$
$$\phi(a^{-1}) = \phi(a), \; a \in G \tag{1.5}$$

Definition 1.24 (*Homomorphism* [20]) Let $\mathscr{G}_1 = (\mathscr{V}_1, \sigma_1, \mu_1)$ and $\mathscr{G}_2 = (\mathscr{V}_2, \sigma_2, \mu_2)$ be two FGs of the graphs $G_1^* = (\mathscr{V}_1, E_1)$ and $G_2^* = (\mathscr{V}_2, E_2)$ respectively. A homomorphism between $\mathscr{G}_1$ and $\mathscr{G}_2$ is a mapping $\psi : \mathscr{V}_1 \to \mathscr{V}_2$ satisfying the following:

(i) $\sigma_1(a) \leq \sigma_2(\psi(a))$ for all $a \in \mathscr{V}_1$,
(ii) $\mu_1(ab) \leq \mu_2(\psi(a)\psi(b))$ for all $a, b \in \mathscr{V}_1$.

Definition 1.25 (*Isomorphism* [20]) Let $\mathscr{G}_1 = (\mathscr{V}_1, \sigma_1, \mu_1)$ and $\mathscr{G}_2 = (\mathscr{V}_2, \sigma_2, \mu_2)$ be two FGs of the graphs $G_1^* = (\mathscr{V}_1, E_1)$ and $G_2^* = (\mathscr{V}_2, E_2)$ respectively. An isomorphism between $\mathscr{G}_1$ and $\mathscr{G}_2$ is a bijective mapping $\psi : \mathscr{V}_1 \to \mathscr{V}_2$ such that

(i) $\sigma_1(a) = \sigma_2(\psi(a))$ for all $a \in \mathscr{V}_1$,
(ii) $\mu_1(ab) = \mu_2(\psi(a)\psi(b))$ for all $a, b \in \mathscr{V}_1$.
In this case, we write $\mathscr{G}_1 \cong \mathscr{G}_2$.

Definition 1.26 (*Weak Isomorphism* [20]) A weak isomorphism between $\mathscr{G}_1 = (\mathscr{V}_1, \sigma_1, \mu_1)$ and $\mathscr{G}_2 = (\mathscr{V}_2, \sigma_2, \mu_2)$ is a homomorphism $\psi : \mathscr{V}_1 \to \mathscr{V}_2$ such that ψ is bijective and satisfies

$$\sigma_1(a) = \sigma_2(\psi(a)) \text{ for all } a \in \mathscr{V}_1.$$

Hence, the weight of the nodes is preserved by a weak isomorphism. But the weights of the arcs is not necessarily preserved.

Example 1.5 Consider the FGs $\mathscr{G}_1 = (\mathscr{V}_1, \sigma_1, \mu_1)$ and $\mathscr{G}_2 = (\mathscr{V}_2, \sigma_2, \mu_2)$ where $\mathscr{V}_1 = \{u_1, v_1\}$, $\mathscr{V}_2 = \{u_2, v_2\}$, $\sigma_1(u_1) = 0.2, \sigma_1(v_1) = 0.3, \mu_1(u_1v_1) = \mu_1(v_1u_1) = 0.1, \sigma_2(u_2) = 0.3, \sigma_2(v_2) = 0.2$, and $\mu_2(u_2v_2) = \mu_2(v_2u_2) = 0.2$.

Let us now define a mapping $\psi : \mathscr{V}_1 \to \mathscr{V}_2$ such that $\psi(u_1) = v_2$ and $\psi(v_1) = u_2$.

Then we have $\sigma_1(u_1) = 0.2 = \sigma_2(v_2) = \sigma_2(\psi(u_1))$, $\sigma_1(v_1) = 0.3 = \sigma_2(u_2) = \sigma_2(\psi(v_1))$ and $\mu_1(u_1v_1) = 0.1 < 0.2 = \mu_2(v_2u_2) = \mu_2(\psi(u_1)\psi(v_1))$.

Hence, ψ is a bijective homomorphism. Moreover, ψ is a weak isomorphism but it is not an isomorphism.

Definition 1.27 (*Co-Weak Isomorphism* [20]) A co-weak isomorphism between $\mathscr{G}_1 = (\mathscr{V}_1, \sigma_1, \mu_1)$ and $\mathscr{G}_2 = (\mathscr{V}_2, \sigma_2, \mu_2)$ is a homomorphism $\psi : \mathscr{V}_1 \to \mathscr{V}_2$ such that ψ is bijective and satisfies

$$\mu_1(ab) = \mu_2(\psi(a)\psi(b)) \text{ for all } a, b \in \mathscr{V}_1.$$

Hence, the weight of the arcs is preserved by a weak isomorphism. But the weight of the nodes is not necessarily preserved.

Example 1.6 Consider the FGs $\mathscr{G}_1 = (\mathscr{V}_1, \sigma_1, \mu_1)$ and $\mathscr{G}_2 = (\mathscr{V}_2, \sigma_2, \mu_2)$ where $\mathscr{V}_1 = \{u_1, v_1\}$, $\mathscr{V}_2 = \{u_2, v_2\}$, $\sigma_1(u_1) = 0.2, \sigma_1(v_1) = 0.3$, $\mu_1(u_1 v_1) = \mu_1(v_1 u_1) = 0.1, \sigma_2(u_2) = 0.4, \sigma_2(v_2) = 0.3$, and $\mu_2(u_2 v_2) = \mu_2(v_2 u_2) = 0.1$.

Let us now define a mapping $\psi : \mathscr{V}_1 \to \mathscr{V}_2$ such that $\psi(u_1) = v_2$ and $\psi(v_1) = u_2$.

Then we have $\sigma_1(u_1) = 0.2 < 0.3 = \sigma_2(v_2) = \sigma_2(\psi(u_1)), \sigma_1(v_1) = 0.3 < 0.4 = \sigma_2(u_2) = \sigma_2(\psi(v_1))$, and $\mu_1(u_1 v_1) = 0.1 = \mu_2(v_2 u_2) = \mu_2(\psi(u_1)\psi(v_1))$.

Hence, ψ is a bijective homomorphism. Moreover, ψ is a co-weak isomorphism but it is not an isomorphism.

Remark 1.1 (i) If $\mathscr{G}_1 = \mathscr{G}_2 = \mathscr{G}$, then the homomorphism $\psi : \mathscr{V}_1 \to \mathscr{V}_1$ is called an endomorphism. An isomorphism $\psi : \mathscr{V}_1 \to \mathscr{V}_1$ is called an automorphism.

(ii) If $\mathscr{G}_1 = \mathscr{G}_2$, then the weak and co-weak isomorphisms actually become isomorphic.

(iii) Let $\mathscr{G} = (\mathscr{V}, \sigma, \mu)$ be a FG and $Aut(\mathscr{G})$ be the set of all automorphisms of $\mathscr{G}$. Let $\psi : \mathscr{V} \to \mathscr{V}$ be the identity mapping $\psi(x) = x$ for all $x \in \mathscr{V}$. Then Clearly, $\psi \in Aut(\mathscr{G})$.

1.6.1 Some Related Results

Theorem 1.5 *Let $\mathscr{G} = (\mathscr{V}, \sigma, \mu)$ be a FG. Then $(Aut(\mathscr{G}), \circ)$ forms a group where $\circ$ is the composition of two mappings.*

Proof Let $\psi_1, \psi_2 \in Aut(\mathscr{G})$ and $a, b \in \mathscr{V}$. Then $\psi_1 \circ \psi_2 \in Aut(\mathscr{G})$ since

$$\sigma((\psi_1 \circ \psi_2)(a)(\psi_1 \circ \psi_2)(b)) = \sigma(\psi_1(\psi_2(a))\psi_1(\psi_2(b))) = \sigma(\psi_2(a)\psi_2(b)) = \sigma(ab),$$

$$\sigma((\psi_1 \circ \psi_2)(a)) = \sigma(\psi_1(\psi_2(a))) = \sigma(\psi_2(a)) = \sigma(a).$$

Clearly, the operation $\circ$ is associative on $Aut(\mathscr{G})$.

Let $I : \mathscr{G} \to \mathscr{G}$ be the identity function. Then $I \in Aut(\mathscr{G})$ and $\psi \circ I = I \circ \psi = \psi$ for all $\psi \in Aut(\mathscr{G})$.

So, I is the identity element of $Aut(\mathscr{G})$.

Now for any $\psi \in Aut(\mathscr{G})$, we have $\psi^{-1} \in Aut(\mathscr{G})$ and also $\psi \circ \psi^{-1} = \psi^{-1} \circ \psi = I$.

Hence, ψ^{-1} is the inverse of ψ in $Aut(\mathscr{G})$.

Therefore, $(Aut(\mathscr{G}), \circ)$ is a group. □

The next theorem is due to Bhattacharyya [19]. This theorem describes how to construct a fuzzy group from a given FG.

Theorem 1.6 *Let $\mathscr{G} = (\mathscr{V}, \sigma, \mu)$ be a FG and $Aut(\mathscr{G})$ be the set of all automorphisms of $\mathscr{G}$. Define a mapping $\tau : Aut(\mathscr{G}) \to [0, 1]$ by*

$$\tau(\psi) = \sup\{\mu(\psi(a)\psi(b)) : a, b \in \mathscr{V}\}$$

for all $\psi \in Aut(\mathscr{G})$. Then τ is a fuzzy group on $Aut(\mathscr{G})$.

Definition 1.28 (*Embedding of fuzzy group*) A fuzzy group $\sigma_1 : G_1 \to [0, 1]$ has an embedding into a fuzzy group $\sigma_2 : G_2 \to [0, 1]$ if there exists a mapping $\phi : G_1 \to G_2$ which is one-one and

$$\sigma_1(a) \leq \sigma_2(\phi(a)) \text{ for all } a \in G_1.$$

Note 1.2 Let (G, σ) be a fuzzy group. Define $\mu_\sigma(a, b) = \sigma(a) \wedge \sigma(b)$ for all $a, b \in G$. Then (G, σ, μ_σ) is a FG with the set of vertices as G. Let $Aut(G)$ be the set of all automorphisms of the fuzzy graph G. Then $Aut(G)$ is a group where the degree of membership value of each $\psi \in Aut(G)$ as given by the fuzzy group $\tau : Aut(G) \to [0, 1]$ is exactly equal to $\sigma(e)$, since

$$\begin{aligned}
\tau(\psi) &= \vee \mu_\sigma(\psi(x), \psi(y)) \\
&= \vee \mu_\sigma(x, y) \\
&= \vee \sigma(x) \wedge \sigma(y) \\
&= \sigma(e).
\end{aligned}$$

Some important results are stated below:

Theorem 1.7 *Every fuzzy group can be embedded into the fuzzy group of the group of all automorphisms of some FG.*

Proof (G, σ) be a fuzzy group and (G, σ, μ_σ) be the corresponding FG of G. Let $(Aut(G), \tau)$ be the fuzzy group associated with $Aut(G)$.

In order to show that there exists an embedding of (G, σ) into $(Aut(G), \tau)$, we need to show that there exists a one-one mapping $\phi : G \to Aut(G)$ such that

$$\sigma(a) \leq \tau(\phi(a)) \text{ for all } a \in G.$$

Define $\phi(a)$ to be an automorphism of G given by $\phi(a) : G \to G, a \to a^{-1}x$.

Then by Note 1.2, we have $\tau(\phi(a)) = \sigma(e)$.

In a group G, we have $\sigma(e) \geq \sigma(a)$ for all $a \in G$. Hence, we have $\sigma(a) \leq \tau(\phi(a))$.

This completes the proof. □

Theorem 1.8 *The relation of isomorphism is an equivalence relation on the collection of all FGs.*

Proof Clearly, the reflexive property holds.

For symmetry, let us assume that $\mathscr{G}_1 = (\mathscr{V}_1, \sigma_1, \mu_1)$ and $\mathscr{G}_2 = (\mathscr{V}_2, \sigma_2, \mu_2)$ be two isomorphic FGs, i.e. $\mathscr{G}_1 \cong \mathscr{G}_2$. Then there exists an isomorphism $\psi : \mathscr{V}_1 \to \mathscr{V}_2$.

Then it is obvious that $\psi^{-1} : \mathscr{V}_2 \to \mathscr{V}_1$ is an isomorphism between G_2 onto $\mathscr{G}_1$. So $G_2 \, cong \, G_1$.

To show the transitivity, let $\mathscr{G}_1 = (\mathscr{V}_1, \sigma_1, \mu_1)$, $\mathscr{G}_2 = (\mathscr{V}_2, \sigma_2, \mu_2)$ and $\mathscr{G}_3 = (\mathscr{V}_3, \sigma_3, \mu_3)$ be FGs such that $G_1 \cong G_2$ and $G_2 \cong G_3$.

Then there exist isomorphisms $\psi_1 : \mathscr{V}_1 \to \mathscr{V}_2$ and $\psi_2 : \mathscr{V}_2 \to \mathscr{V}_3$ of $\mathscr{G}_1$ onto $\mathscr{G}_2$ and $\mathscr{G}_2$ onto $\mathscr{G}_3$ respectively.

Therefore, $\psi_2 \circ \psi_1 : \mathscr{V}_1 \to \mathscr{V}_3$ is an onto mapping.

Also,

$$\sigma_3((\psi_2 \circ \psi_1)(a)) = \sigma_3((\psi_2(\psi_1)(a)))$$
$$= \sigma_2(\psi_1)(a))$$
$$= \sigma_1(a)$$

and

$$\mu_3((\psi_2 \circ \psi_1)(a)(\psi_2 \circ \psi_1)(b)) = \mu_3((\psi_2(\psi_1)(a))(\psi_2(\psi_1)(b)))$$
$$= \mu_2(\psi_1)(a)\psi_1)(b))$$
$$= \mu_1(ab).$$

This shows that $\psi_2 \circ \psi_1$ is the required isomorphism between G_1 and G_3.

Hence, the result. $\square$

1.7 Complement of Fuzzy Graph

In [92], Mordeson and Nair defined the **complement** of a FG $\mathscr{G} = (\mathscr{V}, \sigma, \mu)$. The complement of a FG $\mathscr{G}$ is denoted by $\mathscr{G}^c = (\mathscr{V}, \sigma^c, \mu^c)$, where $\sigma^c(a) = \sigma(a)$ for all $a \in \mathscr{V}$ and

$$\mu^c(a, b) = \begin{cases} 0, & \text{if } \mu(a, b) > 0, \\ \sigma(a) \wedge \sigma(b), & \text{otherwise.} \end{cases}$$

This definition is very close to the complement of a crisp graph $G = (V, E)$. That is, if $(a, b) \in E$, then $(a, b) \notin E^c$ and if there is no edge between the vertices c and d, then there must be an edge (c, d) in G^c.

But, this definition does not satisfy the very common property of complement of a FG, viz. $(\mathscr{G} = (\mathscr{V}, \sigma, \mu)^c)^c = \mathscr{G}$. This result is valid if and only if $\mathscr{G}$ is strong. Sunitha and Kumar [140] modified this definition in such a way that the property $(\mathscr{G} = (\mathscr{V}, \sigma, \mu)^c)^c = \mathscr{G}$ is true for any FG $\mathscr{G}$.

The complement of a FG $\mathscr{G} = (\mathscr{V}, \sigma, \mu)$ is another FG $\mathscr{G} = (\mathscr{V}, \sigma, \mu)^c = (\mathscr{V}, \sigma^c, \mu^c)$, where $\sigma^c(a) = \sigma(a)$ for all $a \in \mathscr{V}$ and $\mu^c(a, b) = \sigma(a) \wedge \sigma(b) - \mu(a, b)$ for all $a, b \in \mathscr{V}$.

Note that in both the definitions, the membership values of vertices remain same. The only change in the definition occurs in the membership values of edges. But, the graph undergoes a major change structurally due to this change in membership values of edges. As per this definition if (a, b) is an edge in $\mathscr{G}$ then this may be an edge in $\mathscr{G}^c$ also. The complement of a FG is illustrated in the following example.

Example 1.7 Let $\mathscr{G} = (\mathscr{V}, \sigma, \mu)$ be a FG with $\mathscr{V} = \{a, b, c, d\}$ Also, let $\sigma(a) = 0.9, \sigma(b) = 0.7, \sigma(c) = 0.6, \sigma(d) = 0.4, \mu(a, b) = 0.6, \mu(b, c) = 0.5, \mu(c, d) = 0.4$. And, $\mu(a, c) = 0, \mu(a, d) = 0, \mu(b, d) = 0$ since there is no edge in $\mathscr{G}$.

As per definition, the membership values of vertices remain unchanged. The membership values of edges are calculated below:
$\mu^c(u, v) = \sigma(u) \wedge \sigma(v) - \mu(u, v)$.
$\mu^c(a, b) = 0.7 - 0.6 = 0.1, \mu^c(a, c) = 0.6 - 0 = 0.6, \mu^c(a, d) = 0.4, \mu^c(b, c) = 0.1, \mu^c(b, d) = 0.4, \mu^c(c, d) = 0$. The (big) dotted edges are newly added in the graph. Note the (small) dotted edge (c, d) whose membership value is 0. According to convention this type of edge does not appear in the graph.

The diagrammatic representation is shown in Fig. 1.5.

Note that all the edges (except (c, d)) in $\mathscr{G}$ are also present in $\mathscr{G}^c$ with different membership values. For an edge (a, b), if $\mu(a, b) = \sigma(a) \wedge \sigma(b)$, then the edge (a, b) does not appear in $\mathscr{G}^c$.

Like crisp complete graph, the complement of CFG is a null graph and vice versa. Let $a_i, i = 1, 2, \ldots, n$ be the vertices of a fuzzy null graph $(O(\mathscr{G})_n)$ and the membership value of the vertex a_i be m_i. Then the membership value of the edge (a_i, a_j) in $O(\mathscr{G})_n^c$ is $\min\{m_i, m_j\}$ for all $i, j = 1, 2, \ldots, n$. That is, $\mu(a_i, a_j) = \min\{m_i, m_j\}$, so $O(\mathscr{G})_n^c$ is a CFG.

Theorem 1.9 *Let* $\mathscr{G}_1 = (\mathscr{V}_1, \sigma_1, \mu_1)$ *and* $\mathscr{G}_2 = (\mathscr{V}_2, \sigma_2, \mu_2)$ *be strong FGs. Then* $\mathscr{G}_1 \cong \mathscr{G}_2$ *if and only if* $\mathscr{G}_1^c \cong \mathscr{G}_2^c$.

Proof Assume that, $\mathscr{G}_1 \cong \mathscr{G}_2$. Then there exists an isomorphism $\psi : \mathscr{V}_1 \to \mathscr{V}_2$ which is bijective and satisfies

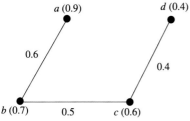
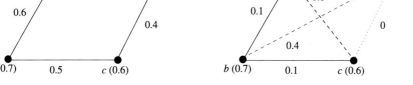

Fig. 1.5 Illustration of complement of FG

$$\sigma_1(a) = \sigma_2(\psi(a)) \text{ for all } a \in \mathcal{V}_1,$$

$$\mu_1(ab) = \mu_2(\psi(a)\psi(b)) \text{ for all } a, b \in \mathcal{V}_1.$$

If $\mu_1(a, b) = 0$, then

$$\begin{aligned}
\mu_1^c(a, b) &= \min\{\sigma_1(a), \sigma_1(b)\} \\
&= \min\{\sigma_2(\psi(a)), \sigma_2(\psi(b))\} \\
&= \mu_2^c(\psi(a)\psi(b)).
\end{aligned}$$

If $0 < \mu_1(a, b) \le 1$, then $0 < \mu_2(\psi(a), \psi(b)) \le 1$.
Therefore, $\mu_1^c(a, b) = 0 = \mu_2^c(\psi(a), \psi(b))$.
So, $\mathcal{G}_1^c \cong \mathcal{G}_2^c$.
Conversely, let $\mathcal{G}_1^c \cong \mathcal{G}_2^c$.
Then there exists a bijective mapping $\psi : \mathcal{V}_1 \to \mathcal{V}_2$ satisfying

$$\sigma_1^c(a) = \sigma_2^c(\psi(a)) \text{ for all } a \in \mathcal{V}_1 \text{ and}$$

$$\mu_1^c(a, b) = \mu_2^c(\psi(a), \psi(b)) \text{ for all } a, b \in \mathcal{V}_1.$$

If $\mu_1(a, b) = 0$, then

$$\begin{aligned}
\mu_2^c(\psi(a), \psi(b)) &= \mu_1^c(a, b) \\
&= \min\{\sigma_1(a), \sigma_1(b)\} \\
&= \min\{\sigma_1^c(a), \sigma_1^c(b)\} \\
&= \min\{\sigma_2^c(\psi(a)), \sigma_2^c(\psi(b))\} \\
&= \min\{\sigma_2(\psi(a)), \sigma_2(\psi(b))\}.
\end{aligned}$$

Again, $\mu_2^c(\psi(a), \psi(b)) = \min\{\sigma_2(\psi(a)), \sigma_2(\psi(b))\} - \mu_2(\psi(a), \psi(b))$
So, $\mu_2(\psi(a), \psi(b)) = 0 = \mu_1(a, b)$.
If $0 < \sigma_1(a, b) \le 1$, then $\mu_2^c(\psi(a), \psi(b)) = \mu_1^c(\psi(a), \psi(b)) = 0$.
Thus, we have,

$$\begin{aligned}
\mu_2(\psi(a), \psi(b)) &= \min\{\sigma_2(\psi(a)), \sigma_2(\psi(b))\} - 0 \\
&= \min\{\sigma_2^c(\psi(a)), \sigma_2^c(\psi(b))\} \\
&= \min\{\sigma_1^c(a), \sigma_1^c(b)\} \\
&= \mu_1(a, b).
\end{aligned}$$

Hence, $\mathcal{G}_1 \cong \mathcal{G}_2$.   □

Theorem 1.10 *Let $\mathcal{G} = (\mathcal{V}, \sigma, \mu)$ be a FG. Then the automorphism groups of $\mathcal{G}$ and $\mathcal{G}^c$ are identical.*

Proof Let $\psi \in (Aut(\mathcal{G}), \circ)$.

Then $\psi : \mathcal{V} \to \mathcal{V}$ is an isomorphism.

Now $\sigma^c(\psi(a)) = \sigma(\psi(a)) = \sigma(a) = \sigma^c(a)$ and

$$\mu^c(\psi(a), \psi(b)) = \sigma(\psi(a)) \wedge \sigma(\psi(b)) - \mu(\psi(a), \psi(b))$$
$$= \sigma(a) \wedge \sigma(b) - \mu(a, b)$$
$$= \mu^c(a, b).$$

Hence, the result. □

Definition 1.29 A FG $\mathcal{G} = (\mathcal{V}, \sigma, \mu)$ is said to be self-complementary if $\mathcal{G} \cong \mathcal{G}^c$.

A necessary condition for a FG to be self-complementary is given as follows.

Theorem 1.11 *Let $\mathcal{G} = (\mathcal{V}, \sigma, \mu)$ be a self-complementary FG. Then*

$$\sum_{a \neq b} \mu(a, b) = \frac{1}{2} \sum_{a \neq b} [\sigma(a) \wedge \sigma(b)].$$

Proof Assumed that $\mathcal{G} = (\mathcal{V}, \sigma, \mu)$ is a self-complementary FG. Then there exists an isomorphism $f : \mathcal{V} \to \mathcal{V}$, such that $\mu^c(f(a), f(b)) = \mu(a, b)$ for all $a, b \in \mathcal{V}$ and $\sigma^c f(a) = \sigma(a)$ for all $a \in \mathcal{V}$.

Now, $\mu^c(f(a), f(b)) = \sigma^c(f(a)) \wedge \sigma^c(f(b)) - \mu(f(a), f(b))$
This gives, $\mu(a, b) = \sigma(a) \wedge \sigma(b) - \mu(f(a), f(b))$
Summing up over all distinct vertices,

$$\sum_{a \neq b} \mu(a, b) + \sum_{a \neq b} \mu(f(a), f(b)) = \sum_{a \neq b} \sigma(a) \wedge \sigma(b)$$

or, $\quad 2 \sum_{a \neq b} \mu(a, b) = \sum_{a \neq b} \sigma(a) \wedge \sigma(b)$

or, $\quad \sum_{a \neq b} \mu(a, b) = \frac{1}{2} \sum_{a \neq b} \sigma(a) \wedge \sigma(b).$

The converse of this theorem is not true.

The condition of Theorem 1.11 is not sufficient. Now, we present the following theorem whose condition is sufficient for a FG to be self-complementary.

Theorem 1.12 *If $\mu(a, b) = \frac{1}{2}(\sigma(a) \wedge \sigma(b))$ for all $a, b \in \mathcal{V}$ in a FG $\mathcal{G} = (\mathcal{V}, \sigma, \mu)$, then $\mathcal{G}$ is self-complementary.*

Proof If $\mathcal{G} = (\mathcal{V}, \sigma, \mu)$ is a FG satisfying $\mu(a, b) = \frac{1}{2}(\sigma(a) \wedge \sigma(b))$ for all $a, b \in \mathcal{V}$, then the identity mapping $I : \mathcal{V} \to \mathcal{V}$ is an isomorphism from $\mathcal{G}$ to $\mathcal{G}^c$.

Clearly, I satisfies the first condition for isomorphism, i.e. $\sigma(a) = \sigma^c(I(a))$ for all $a \in \mathcal{V}$ and

$$\mu^c(I(a), I(b)) = \mu^c(a, b)$$
$$= \sigma(a) \wedge \sigma(b) - \mu(a, b)$$
$$= \sigma(a) \wedge \sigma(b) - \frac{1}{2}(\sigma(a) \wedge \sigma(b))$$
$$= \frac{1}{2}(\sigma(a) \wedge \sigma(b)) = \mu(a, b).$$

i.e. $\mu^c(I(a), I(b)) = \mu(a, b)$ for all $a, b \in \mathcal{V}$. Hence, $\mathcal{G} \cong \mathcal{G}^c$, i.e. $\mathcal{G}$ is self-complementary. $\qquad\qquad\square$

Example 1.8 Let $\mathcal{G} = (\mathcal{V}, \sigma, \mu)$ be a FG where $\mathcal{V} = \{a, b, c, d\}$, $\sigma(a) = 0.1$ for all $a \in \mathcal{V}$ and $\mu(a, b) = \mu(a, c) = \mu(c, d) = 0.1$. Then G is self-complementary.

Let $\mathcal{G}^c = (\mathcal{V}, \sigma^c, \mu^c)$ be the complement of $\mathcal{G}$, where $\sigma^c = \sigma$, $\mu^c(a, b) = \mu^c(a, c) = \mu^c(c, d) = 0$, $\mu^c(a, d) = \mu^c(b, d) = \mu^c(b, c) = 0.1$.

Let us now define a mapping $\psi : \mathcal{V} \to \mathcal{V}$ by $\psi(a) = b$, $\psi(b) = c$, $\psi(c) = d$, $\psi(d) = a$. Then clearly, ψ is a bijective mapping and
$\sigma(a) = \sigma(\psi(a))$ for all $a \in \mathcal{V}$.

Also $\mu(a, b) = 0.1 = \mu^c(\psi(a), \psi(b)) = \mu^c(b, c)$, $\mu(a, c) = 0.1 = \mu^c(\psi(a), \psi(c)) = \mu^c(b, d)$,
$\mu(c, d) = 0.1 = \mu^c(\psi(c), \psi(d)) = \mu^c(a, d)$, $\mu(b, c) = 0 = \mu^c(\psi(b), \psi(c)) = \mu^c(c, d)$,
$\mu(b, d) = 0 = \mu^c(\psi(b), \psi(d)) = \mu^c(a, c)$, $\mu(a, d) = 0 = \mu^c(\psi(a), \psi(d)) = \mu^c(a, b)$.

Hence, ψ is an isomorphism from $\mathcal{G}$ onto $\mathcal{G}^c$, i.e. $\mathcal{G} \cong \mathcal{G}^c$, i.e. $\mathcal{G}$ is self-complementary.

Lemma 1.8 For any FG $\mathcal{G} = (\mathcal{V}, \sigma, \mu)$, $size(\mathcal{G}) + size(\mathcal{G}^c) \leq \sum_{a,b \in \mathcal{V}} 2[\sigma(a) \wedge \sigma(b)]$.

Proof For any FG $\mathcal{G}$, $\mu(a, b) \leq \sigma(a) \wedge \sigma(b)$ and also $\mu^c(a, b) \leq \sigma(a) \wedge \sigma(b)$.
Thus, $\mu(a, b) + \mu^c(a, b) \leq 2[\sigma(a) \wedge \sigma(b)]$.
Summing over all pairs of vertices, $\sum_{a,b \in \mathcal{V}} \mu(a, b) + \mu^c(a, b) \leq 2 \sum_{a,b \in \mathcal{V}} [\sigma(a) \wedge \sigma(b)]$.
Thus, $size(\mathcal{G}) + size(\mathcal{G}^c) \leq 2 \sum_{a,b \in \mathcal{V}} [\sigma(a) \wedge \sigma(b)]$. $\qquad\qquad\square$

A strong edge in $\mathcal{G}$ is not necessarily a strong edge in $\mathcal{G}^c$ and vice versa.
For example, let $\mathcal{G} = (\mathcal{V}, \sigma, \mu)$ be a FG with $\mathcal{V} = \{a, b, c, d\}$. Let $\sigma(a) = 0.8$, $\sigma(b) = 1$, $\sigma(c) = 0.6$, $\sigma(d) = 0.4$, $\mu(a, b) = 0.3$, $\mu(a, c) = 0.1$, $\mu(a, d) = 0.2$, $\mu(b, c) = 0.2$, $\mu(c, d) = 0.1$, $\mu(a, d) = 0.2$.
Let the complement of $\mathcal{G}$ be $\mathcal{G}^c$. Then
$\mu^c(a, b) = 0.5$, $\mu^c(a, c) = 0.5$, $\mu^c(a, d) = 0.2$, $\mu^c(b, c) = 0.4$, $\mu^c(b, d) = 0.4$, $\mu^c(c, d) = 0.3$.

Here, $\mu(a, d) = 0.2$, $CONN_{\mathscr{G}}(a, d) = 0.2$. So (a, d) is a strong edge in $\mathscr{G}$. But $\mu^c(a, d) = 0.2$, $CONN_{\mathscr{G}^c}(a, d) = 0.4$, and hence, (a, d) is not a strong edge in $\mathscr{G}^c$. Again, $\mu^c(a, c) = 0.5$, $CONN_{\mathscr{G}^c}(a, c) = 0.5$, i.e. (a, c) is a strong edge in $\mathscr{G}$, but (a, c) is not a strong edge in $\mathscr{G}$.

We have defined the complement of a FG $\mathscr{G}$ whose underlying crisp graph is G^*. The complement of $\mathscr{G}$ is denoted by $\mathscr{G}^c$. In this definition of complement, the underlying crisp graph of $\mathscr{G}^c$ is not isomorphic to the complement of G^*. This means, if (a, b) is an edge in $\mathscr{G}$, then it may be an edge in $\mathscr{G}^c$, but this does not happen in crisp graph. To maintain parity with the complement of a crisp graph, a new type of complement is defined known as μ-complement and it is denoted by $\mathscr{G}^\mu$.

Definition 1.30 The **μ-complement** of a FG $\mathscr{G} = (\mathscr{V}, \sigma, \mu)$ with underlying graph $G^* = (V, E)$ is denoted by $\mathscr{G}^\mu = (\mathscr{V}^\mu, \sigma^\mu, \mu^\mu)$ and is defined by

$$\mathscr{V}^\mu = \mathscr{V}$$
$$\sigma^\mu(a) = \sigma(a) \quad \text{for all } a \in \mathscr{V}$$
$$\mu^\mu(a, b) = \begin{cases} \sigma(a) \wedge \sigma(a) - \mu(a, b) & \text{for all } (a, b) \in E \\ 0 & \text{for all } (a, b) \notin E \end{cases}$$

Remark 1.2 If G^* is complete, then $\mathscr{G}^c$ and $\mathscr{G}^\mu$ both are null graphs.
For complete graph, $\mu(a, b) = \sigma(a) \wedge \sigma(b)$ for all $a, b \in \mathscr{V}$.
Thus, $\mu^c(a, b) = \sigma(a) \wedge \sigma(b) - \mu(a, b) = 0$ for all $a, b \in \mathscr{V}$.
Also, $\mu^\mu(a, b) = \sigma(a) \wedge \sigma(b) - \mu(a, b) = 0$ for all $a, b \in \mathscr{V}$.
In both cases, the complement graph is null.

Note 1.3 The number of edges of $\mathscr{G}^\mu$ is less than or equal to the number of edges in $\mathscr{G}$. But, the number of edges in $\mathscr{G}^c$ may be more than that of $\mathscr{G}$.

Theorem 1.13 *Let $\mathscr{G} = (\mathscr{V}, \sigma, \mu)$ be a FG and $G^* = (V, E)$ be its underlying crisp graph. The FG $\mathscr{G}^\mu$ is null graph if and only if $\mathscr{G}$ is strong FG.*

Proof Let $\mathscr{G}$ be strong. Then $\mu(a, b) = \sigma(a) \wedge \sigma(b) - \mu(a, b)$ for all $(a, b) \in E$ and $\mu(a, b) = 0$ otherwise.

By definition of $\mathscr{G}^\mu$, $\mu^\mu(a, b) = 0$ for all $(a, b) \in E$ and $(a, b) \notin E$., i.e. for all $(a, b) \in V \times V$.S Thus, $\mathscr{G}\mu$ is a null graph.

Conversely, let $\mathscr{G}^\mu$ be a null graph.
Then $\mu^\mu(a, b) = 0$ for all $(a, b) \in V \times V$.
That is, $\mu^\mu(a, b) = 0$ for all $(a, b) \in E$ and $(a, b) \notin E$.

By definition of $\mathscr{G}^\mu$, if $(a, b) \in E$, $\mu^\mu(a, b) = \sigma(a) \wedge \sigma(b) - \mu(a, b)$.

Then $0 = \sigma(a) \wedge \sigma(b) - \mu(a, b)$, or $\mu(a, b) = \sigma(a) \wedge \sigma(b)$ for $(a, b) \in E$.
Hence, $\mathscr{G}$ is a strong FG. $\qquad\square$

The above result is also true for CFG as it is a strong FG.

Theorem 1.14 *If $\mathscr{G}_1$ and $\mathscr{G}_2$ are isomorphic, then $\mathscr{G}_1^\mu$ and $\mathscr{G}_2^\mu$ are also isomorphic.*

1.8   Regular and Irregular Fuzzy Graphs

In this section, regular and irregular FGs are discussed most of whose results were developed in [68, 94, 96].

Definition 1.31 ([94]) A FG $\mathcal{G} = (\mathcal{V}, \sigma, \mu)$ is said to be **regular** (RFG), if for a positive real number k, $deg(a) = k$, for all $a \in \mathcal{V}$. In this case, $\mathcal{G}$ is called k-regular FG (k-RFG).

If each vertex of $\mathcal{G}$ has same total degree, say k, then $\mathcal{G}$ is said to be a **totally regular FG** (TRFG) of total degree k or a k-**totally regular FG** (k-**TRFG**).

Example 1.9 Let $\mathcal{G} = (\mathcal{V}, \sigma, \mu)$ be a FG and $\mathcal{V} = \{a, b, c, d\}$. Let $\sigma(a) = 0.5$, $\sigma(b) = 0.7, \sigma(c) = 0.5, \sigma(d) = 0.6$ and $\mu(a, b) = 0.4, \mu(b, c) = 0.3, \mu(c, d) = 0.4, \mu(d, a) = 0.3$. Then $\mathcal{G}$ is regular.

Every CFG may not be RFG. This is illustrated through the following example.

Example 1.10 Let as assume a FG $\mathcal{G} = (\mathcal{V}, \sigma, \mu)$ having four nodes such that $\sigma(a) = 0.5, \sigma(b) = 0.7, \sigma(c) = 0.4, \sigma(e) = 0.6$ and $\mu(a, b) = 0.5, \mu(b, c) = 0.4, \mu(c, e) = 0.4, \mu(a, e) = 0.5, \mu(c, a) = 0.4, \mu(b, e) = 0.6$. Here, $deg(a) = 1.4$, $deg(b) = 1.5, deg(c) = 1.2, deg(e) = 1.5$. Hence, $\mathcal{G}$ is not a RFG.

Theorem 1.15 ([68]) *Let $\mathcal{G} = (\mathcal{V}, \sigma, \mu)$ be a FG and σ be a c-constant function. Then $\mathcal{G}$ is regular if and only if $\mathcal{G}$ is a TRFG.*

Proof Let $\mathcal{G} = (\mathcal{V}, \sigma, \mu)$ be a RFG. Therefore, $deg(a) = k$, for all $a \in \mathcal{V}$, where k is constant. Now, we are to show that $tdeg(a) = constant$, for all $a \in \mathcal{V}$. Now, $tdeg(a) = deg(a) + \sigma(a) = k + c$, since $\sigma(a) = c$, for all $a \in \mathcal{V}$.

This is true for every $a \in \mathcal{V}$. Hence, $\mathcal{G}$ is a TRFG.

Conversely, let $\mathcal{G}$ be a TRFG. Then, $tdeg(a)$ is constant and let it be p, for all $a \in \mathcal{V}$. We need to prove $deg(a) = constant$, for all $a \in \mathcal{V}$.
Now, $tdeg(a) = deg(a) + \sigma(a)$, i.e. $p = deg(a) + c$ or, $deg(a) = p - c$, a constant, for all $a \in \mathcal{V}$.

Therefore, $\mathcal{G}$ is a RFG.   □

Theorem 1.16 ([94]) *If $\mathcal{G} = (\mathcal{V}, \sigma, \mu)$ is a RFG as well as TRFG, then σ is a constant function.*

Proof Since, $\mathcal{G}$ is a RFG, therefore, $deg(a) = p_1$ (say) for all $a \in \mathcal{V}$. Again, $\mathcal{G}$ is a TRFG, therefore, $tdeg(a) = p_2$ (say), for all $a \in \mathcal{V}$.

Now, $tdeg(a) = deg(a) + \sigma(a)$, i.e. $p_2 = p_1 + \sigma(a)$ for all $a \in \mathcal{V}$
Then $\sigma(a) = p_2 - p_1$, a constant.

This is true for all $a \in \mathcal{V}$. Hence, σ is a constant function.   □

The converse of the above theorem is not necessarily true. This is justified by an example.

Example 1.11 Let $\mathcal{G} = (\mathcal{V}, \sigma, \mu)$ be a FG with three vertices a, b, c, and membership value of each vertex be 0.7. Again, let $\mu(a, b) = 0.4, \mu(b, c) = 0.5$. Here, $deg(a) = 0.4, deg(b) = 0.9, deg(c) = 0.5$ and $tdeg(a) = 1.1, tdeg(b) = 1.6, tdeg(c) = 1.2$. This shows that, $\mathcal{G}$ is neither RFG nor TRFG, but σ is constant.

Theorem 1.17 ([68]) *Let $\mathcal{G} = (\mathcal{V}, \sigma, \mu)$ be a FG and its underlying crisp graph be $G^* = (V, E)$, where G^* is an odd cycle. Then $\mathcal{G}$ is regular if and only if μ is a constant function.*

Proof Let $\mathcal{G} = (\mathcal{V}, \sigma, \mu)$ be a c-RFG. Let $e_1, e_2, \ldots, e_{2k+1}$ be the edges in G^*. Let $\mu(e_1) = p_1$. Then, $\mu(e_2) = c - p_1, \mu(e_3) = p_1, \mu(e_4) = c - p_1, \mu(e_5) = p_1$, and so on.

In general,

$$\mu(e_j) = \begin{cases} p_1, & \text{if } j \text{ is odd} \\ c - p_1, & \text{if } j \text{ is even} \end{cases}$$

Hence, $\mu(e_1) = \mu(e_{2k+1}) = p_1$. If two edges e_1, e_{2k+1} are incident in a common vertex, say a, then $\mu(e_1) + \mu(e_{2k+1}) = deg(a) = c$, i.e. $p_1 + p_1 = c$. Thus, $p_1 = \frac{c}{2}$.

Also, $c - p_1 = \frac{c}{2}$. Hence, $\mu(e_j) = \frac{c}{2}$ for all edges e_j, i.e. μ is a constant function.

Conversely, let $\mathcal{G} = (\mathcal{V}, \sigma, \mu)$ be a FG such that μ is a constant function, say $\mu(a, b) = s$, for all $(a, b) \in E$. Then $deg(a) = 2s$, for all $a \in \mathcal{V}$. Hence, $\mathcal{G}$ is a RFG. $\qquad\square$

Theorem 1.18 ([68]) *The size of a p-RFG is $\frac{pn}{2}$, where n is the number of vertices in the graph.*

Proof Let $\mathcal{G} = (\mathcal{V}, \sigma, \mu)$ be a p-RFG and its underlying crisp graph be $G^* = (V, E)$. The size of $\mathcal{G}$ is $S(\mathcal{G}) = \sum\limits_{(a,b) \in E} \mu(a, b)$. Since $\mathcal{G}$ is a p-RFG, therefore, $deg(a) = p$, for all $a \in \mathcal{V}$.

Again, by handshaking theorem for FG,

$$\sum_{a \in \mathcal{V}} deg(a) = 2 \sum_{(a,b) \in E} \mu(a, b) = 2S(\mathcal{G}).$$

Hence,

$$2S(\mathcal{G}) = \sum_{a \in \mathcal{V}} deg(a) = \sum_{a \in \mathcal{V}} p = pn.$$

Hence, $S(\mathcal{G}) = \frac{pn}{2}$. $\qquad\square$

Theorem 1.19 ([94]) *If $\mathcal{G} = (\mathcal{V}, \sigma, \mu)$ is a t-TRFG, then $2S(\mathcal{G}) + O(\mathcal{G}) = tn$, where n is the number of vertices in $\mathcal{G}$.*

Proof Since $\mathcal{G}$ is t-TRFG. Therefore,

$$t = tdeg(a) = deg(a) + \sigma(a), \, for \, all \, a \in \mathcal{V}.$$

So, $\sum_{a \in \mathcal{V}} t = \sum_{a \in \mathcal{V}} tdeg(a) = \sum_{a \in \mathcal{V}} deg(a) + \sum_{a \in \mathcal{V}} \sigma(a)$. That is, $tn = 2S(\mathcal{G}) + O(\mathcal{G})$. $\square$

Theorem 1.20 *If $\mathcal{G} = (\mathcal{V}, \sigma, \mu)$ be a p-regular as well as t-TRFG, then $O(\mathcal{G}) = n(t - p)$, where n is the total number of vertices of $\mathcal{G}$.*

Proof From Theorem 1.18, $S(\mathcal{G}) = \frac{pn}{2}$. Therefore, $2S(\mathcal{G}) = pn$. Again, from Theorem 1.19, $2S(\mathcal{G}) + O(\mathcal{G}) = tn$.

Hence, $2S(\mathcal{G}) + O(\mathcal{G}) = tn$, or, $pn + O(\mathcal{G}) = tn$, i.e. $O(\mathcal{G}) = n(t - p)$. $\square$

Definition 1.32 ([96]) A FG is said to be **irregular**, if there is a vertex which is adjacent to vertices with distinct degrees. A FG is called **neighborly irregular FG** (NIFG) if every two adjacent vertices of the graph have different degrees. A FG is said to be **totally irregular FG** (TIFG) if there is a vertex which is adjacent to vertices with distinct total degrees. If every two adjacent vertices have distinct total degrees of a FG then it is called **neighborly total irregular FG** (NTIFG). A FG is called **highly irregular FG** (HIFG) if every vertex of G is adjacent to vertices with distinct degrees.

Example 1.12 Let $\mathcal{G} = (\mathcal{V}, \sigma, \mu)$ be a FG and $\mathcal{V} = \{a, b, c, d, e\}$. Let $\sigma(a) = 0.5, \sigma(b) = 0.7, \sigma(c) = 0.5, \sigma(d) = 0.6, \sigma(e) = 0.6$ and $\mu(a, b) = 0.4, \mu(b, c) = 0.4, \mu(c, d) = 0.3, \mu(d, e) = 0.3, \mu(e, a) = 0.4, \mu(c, e) = 0.1$. This FG, $\mathcal{G}$ is irregular.

Remark 1.3 If $\mathcal{G} = (\mathcal{V}, \sigma, \mu)$ is NIFG, then every fuzzy subgraph of $\mathcal{G}$ may not be NIFG. This is illustrated in the following example.

Example 1.13 Let us consider a FG $\mathcal{G} = (\mathcal{V}, \sigma, \mu)$ having four nodes a, b, c, d. Let $\sigma(a) = 0.6, \sigma(b) = 0.7, \sigma(c) = 0.8, \sigma(d) = 0.9$ and $\mu(a, b) = 0.6, \mu(b, c) = 0.5, \mu(c, d) = 0.5, \mu(a, d) = 0.4$.

Let us consider a fuzzy subgraph $\mathcal{H} = (\mathcal{V}, \sigma', \mu')$ of $\mathcal{G}$ as follows: $\sigma'(a) = 0.5, \sigma'(b) = 0.6, \sigma'(c) = 0.6, \sigma'(d) = 0.7$ and $\mu(a, b) = 0.3, \mu(b, c) = 0.4, \mu(c, d) = 0.35, \mu(a, d) = 0.2$. Clearly, $\mathcal{H}$ is not a NIFG.

Remark 1.4 A CFG may not be NIFG.

The following results hold trivially.

Lemma 1.9 ([96]) *(i) A HIFG $\mathcal{G} = (\mathcal{V}, \sigma, \mu)$ is not necessarily a NIFG.*
(ii) A NIFG $\mathcal{G} = (\mathcal{V}, \sigma, \mu)$ is not necessarily a HIFG.

Theorem 1.21 ([96]) *Let $\mathcal{G} = (\mathcal{V}, \sigma, \mu)$ be a FG. Then $|G$ is HIFG and NIFG if and only if the degrees of all vertices of $\mathcal{G}$ are distinct.*

Proof Let $\mathscr{G} = (\mathscr{V}, \sigma, \mu)$ be a HIFG and NIFG having n vertices. Let $a_1 \in \mathscr{V}$ and let the vertices $a_2, a_3, \ldots, a_n$ be adjacent to a_1. Suppose the degree of $a_2, a_3, \ldots, a_n$ be $p_2, p_3, \ldots, p_n$ respectively, in which $p_i \neq p_j$, for $i, j = 2, 3, \ldots, n$ as $\mathscr{G}$ is HIFG. Now, $deg(a_1)$ is not equal to any p_j, for $j = 2, 3, \ldots, n$ as $\mathscr{G}$ is NIFG. Hence, all nodes have distinct degrees.

Conversely, let all the vertices have distinct degrees. Then, it is obvious that $\mathscr{G}$ is HIFG and NIFG. □

Lemma 1.10 *The complement of a NIFG is not neighborly irregular.*

Theorem 1.22 ([96]) *Let $\mathscr{G} = (\mathscr{V}, \sigma, \mu)$ be a FG, where σ is a constant function. Then $\mathscr{G}$ is neighborly irregular if and only if $\mathscr{G}$ is a NTIFG.*

Proof Let $\mathscr{G}$ is NIFG and v_1, v_2 be two adjacent vertices in $\mathscr{G}$. Since $\mathscr{G}$ is a NIFG, therefore, the degree of v_1, v_2 are distinct. Suppose p_1, p_2 $(p_1 \neq p_2)$ be the degrees of v_1, v_2 respectively. Now, we have to show that the total degree of these two nodes $tdeg(v_1), tdeg(v_2)$ are distinct.

If possible, let $tdeg(v_1) = tdeg(v_2)$.

Then,

$tdeg(v_1) = tdeg(v_2)$, i.e. $deg(v_1) + \sigma(v_1) = deg(v_2) + \sigma(v_2)$

$p_1 + c = p_2 + c$ [since σ is constant, therefore $\sigma(v_1) = \sigma(v_2) = c$, say]

$p_1 - p_2 = c - c$. Thus, $p_1 = p_2$.

This contradicts that p_1 and p_2 are distinct. Hence, $\mathscr{G}$ is a NTIFG.

Conversely, let $\mathscr{G}$ be a NTIFG. Let u_1, u_2 be two adjacent vertices in $\mathscr{G}$. Since $\mathscr{G}$ is a NTIFG, therefore, $tdeg(v_1) \neq tdeg(v_2)$. Suppose p_1, p_2 be the degrees of u_1, u_2 respectively.

Now,

$tdeg(v_1) \neq tdeg(v_2)$, i.e. $deg(v_1) + \sigma(v_1) \neq deg(v_1) + \sigma(v_1)$

So, $p_1 + c \neq p_2 + c$ [since σ is constant, so $\sigma(v_1) = \sigma(v_2) = c$, say]

or, $p_1 - p_2 \neq c - c$. Therefore, $p_1 \neq p_2$.

Hence, $\mathscr{G}$ is NIFG. □

Definition 1.33 Let $\mathscr{G} = (\mathscr{V}, \sigma, \mu)$ be a FG and its underlying crisp graph be $G^* = (V, E)$. If the degrees of all edges in $\mathscr{G}$ are same and it is k, then $\mathscr{G}$ is called **edge regular FG** or k**-edge regular FG** .

Similarly, if the total degrees of all the edges in $\mathscr{G}$ are equal and equal to k, then $\mathscr{G}$ is called **totally edge regular FG** or k**-totally edge regular FG** .

It can be verified by examples that if $\mathscr{G}$ is an edge regular FG then $\mathscr{G}^c$ is neither edge regular nor totally edge regular FG.

The following theorem gives the condition for which complement of an edge regular FG is also edge regular.

Theorem 1.23 ([114]) *Let $\mathscr{G} = (\mathscr{V}, \sigma, \mu)$ be a FG such that $\mu(a, b) = \frac{1}{2}[\sigma(a) \wedge \sigma(b)]$ for all $a, b \in \mathscr{V}$. Then $\mathscr{G}$ is a k-edge regular FG if and only if $\mathscr{G}^c$ is k-edge regular FG.*

Proof It is given that $\mu(a, b) = \frac{1}{2}[\sigma(a) \wedge \sigma(b)]$ for all $a, b \in \mathcal{V}$.

According to the definition,

$\mu^c(a, b) = \{\sigma(a) \wedge \sigma(b)\} - \mu(a, b)$ for all $a, b \in \mathcal{V}$.

Therefore, $\mu^c(a, b) = [\sigma(a) \wedge \sigma(b)]/2$ for all $a, b \in \mathcal{V}$.

Hence,

$$\mu^c(a, b) = \mu(a, b) \text{ for all } a, b \in \mathcal{V}. \tag{1.6}$$

That is, the membership value of an edge (a, b) in $\mathcal{G}$ is same as in $\mathcal{G}^c$.

Now,

$$\begin{aligned}
deg_{\mathcal{G}^c}(a, b) &= \sum_{c \neq a} \mu^c(a, c) + \sum_{c \neq b} \mu^c(c, b) - 2\mu^c(a, b) \\
&= \sum_{c \neq a} \mu(a, c) + \sum_{c \neq b} \mu(c, b) - 2\mu(a, b) \quad \text{[by 1.6]} \\
&= deg_{\mathcal{G}}(a, b) \quad [\text{ for all } a, b \in \mathcal{V}.
\end{aligned}$$

Hence, if $\mathcal{G}$ is a k-edge regular FG then so is $\mathcal{G}^c$ and vice-versa. $\square$

The converse of the above theorem is not true. It can be verified by an example. The Theorem 1.23 can be extended for totally edge regular FG which is stated below.

Theorem 1.24 ([114]) *Let $\mathcal{G} = (\mathcal{V}, \sigma, \mu)$ be a FG such that $\mu(a, b) = \frac{1}{2}[\sigma(a) \wedge \sigma(b)]$ for all $a, b \in \mathcal{V}$. Then $\mathcal{G}$ is a k-totally edge regular FG if and only if $\mathcal{G}^c$ is a k-totally edge regular FG.*

Theorem 1.25 *Let $\mathcal{G} = (\mathcal{V}, \sigma, \mu)$ be a FG and its underlying crisp graph ($G^* = (V, E)$) be a regular graph. Also, let μ is a constant function. Then $\mathcal{G}$ is edge regular FG.*

Proof Let G^* be k-regular. Then $deg_{G^*}(a) = k$ (say). Since μ is constant, $\mu(a, b) = c$ for all edge $(a, b) \in E$.

Now,

$$\begin{aligned}
deg_{\mathcal{G}}(a, b) &= \sum_{z \neq a} \mu(a, z) + \sum_{z \neq b} \mu(z, b) - 2\mu(a, b) \\
&= \sum_{z \neq a} c + \sum_{z \neq b} c - 2c \\
&= c \, deg_{G^*}(a) + c \, deg_{G^*}(b) - 2c \\
&= kc + kc - 2c = 2c(k - 1).
\end{aligned}$$

This shows that $\mathcal{G} = (\mathcal{V}, \sigma, \mu)$ is a $2c(k - 1)$-edge regular FG. $\square$

Example 1.14 Let $\mathscr{G} = (\mathscr{V}, \sigma, \mu)$ be a FG on a regular graph $G^* = (V, E)$. Let $\mathscr{V} = \{a, b, c, d\}$ with $\sigma(a) = 0.5, \sigma(b) = 0.6, \sigma(c) = 0.8, \sigma(d) = 0.7, \mu(a, b) = \mu(b, c) = \mu(c, d) = \mu(d, a) = 0.3$. That is, G^* is 2-regular and μ is constant.

Its complement $\mathscr{G}^c$ is
$\sigma(a) = 0.5, \sigma(b) = 0.6, \sigma(c) = 0.8, \sigma(d) = 0.7, \mu^c(a, b) = 0.2, \mu^c(b, c) = 0.3,$
$\mu^c(c, d) = 0.4, \mu^c(d, a) = 0.2, \mu^c(a, c) = 0.5, \mu^c(b, d) = 0.6..$

Now, $deg_{\mathscr{G}^c}(a, b) = 0.5, deg_{\mathscr{G}^c}(b, c) = 0.6.$

The result shows that $\mathscr{G}^c$ is not edge regular.

Now, we give the definition of another kind of RFG.

Definition 1.34 ([35]) A **perfectly regular FG** is a FG that is both regular and totally regular.

Lemma 1.11 ([35]) *Let $\mathscr{G} = (\mathscr{V}, \sigma, \mu)$ be a perfectly regular FG. Then σ is a constant function.*

Proof Since $\mathscr{G} = (\mathscr{V}, \sigma, \mu)$ is perfectly regular, $\mathscr{G}$ is both c_1-regular and c_2-totally regular.

Let a, b be any two vertices of $\mathscr{V}$. Then
$deg(a) = deg(b) = c_1$ and $tdeg(a) = tdeg(b) = c_2$.

Now, $tdeg(a) = deg(a) + \sigma(a) = c_2 = tdeg(b) = deg(b) + \sigma(b)$.

That is, $\sigma(a) = \sigma(b)$ for all $a, b \in \mathscr{V}$. Hence, σ is constant. $\qquad\square$

Lemma 1.12 ([35]) *Let $\mathscr{G} = (\mathscr{V}, \sigma, \mu)$ be a perfectly regular FG on $G^* = (V, E)$ with n number of vertices.*
(i) If σ is c-constant then $O(\mathscr{G}) = c\ n$.
(ii) If $deg(a) = k$ for all $a \in \mathscr{V}$, then $S(\mathscr{G}) = kn/2$.

Proof (i) By definition, $O(\mathscr{G}) = \sum_{a \in \mathscr{V}} \sigma(a) = \sum_{a \in \mathscr{V}} c = c\ n$.
(ii) Since $deg(a) = \sum_{b \neq a} \mu(a, b)$,

$$\sum_{a \in \mathscr{V}} deg(a) = \sum_{a \in \mathscr{V}} \sum_{b \neq a} \mu(a, b) = 2 \sum_{(a,b) \in E} \mu(a, b) = 2S(\mathscr{G}).$$

Thus, $2S(\mathscr{G}) = \sum_{a \in \mathscr{V}} deg(a) = \sum_{a \in \mathscr{V}} k = nk$.

Hence, $S(\mathscr{G}) = nk/2$. $\qquad\square$

1.9 Product of Fuzzy Graphs

In 1994, Mordeson and Peng [90] investigated different kinds of operations on FGs viz. Cartesian product, composition, union and join of FGs . In 2002, Sunitha and Vijayakumar [140] defined complement of FGs and investigated many of its properties based on these operations. The necessary and sufficient conditions for an arbitrary partial fuzzy subgraph of $\mathscr{G} = (\mathscr{G}' * \mathscr{G}'')$ to be also formed by the same operation from partial fuzzy subgraphs of $\mathcal{G}'$ and $\mathcal{G}''$, where $*$ is any one of the above four operations are provided in [90].

Recently, some other products, viz. direct product, semi-strong product, strong product, tensor product, normal product, modular product, homomorphic product, etc. on FG are available in the literature.

In this section, let $\mathcal{G} = (\mathcal{V}, \sigma, \mu)$, $\mathcal{G}' = (\mathcal{V}', \sigma', \mu')$, and $\mathcal{G}'' = (\mathcal{V}'', \sigma'', \mu'')$ be the FGs and their corresponding underlying graphs be $G = (V, E)$, $G' = (V', E')$, and $G'' = (V'', E'')$.

1.9.1 Union of Fuzzy Graphs

Definition 1.35 Let $\mathcal{G}' = (\mathcal{V}', \sigma', \mu')$ and $\mathcal{G}'' = (\mathcal{V}'', \sigma'', \mu'')$ be two FGs . The union between them is defined as $\mathcal{G} = \mathcal{G}' \cup \mathcal{G}'' = (\mathcal{V}, \sigma' \cup \sigma'', \mu' \cup \mu'')$, where the sets of vertices and edges are given by
$\mathcal{V} = \mathcal{V}' \cup \mathcal{V}''$ and $E = E' \cup E''$ with

$$(\sigma' \cup \sigma'')(u) = \begin{cases} \sigma'(u), & \text{if } u \in \mathcal{V}' - \mathcal{V}'' \\ \sigma''(u), & \text{if } u \in \mathcal{V}'' - \mathcal{V}' \\ \sigma'(u) \vee \sigma''(u), & \text{if } u \in \mathcal{V}' \cap \mathcal{V}'' \end{cases}$$

and

$$(\mu' \cup \mu'')(u, v) = \begin{cases} \mu'(u, v), & \text{if } (u, v) \in E' - E'' \\ \mu''(u, v), & \text{if } (u, v) \in E'' - E' \\ \mu'(u, v) \vee \mu''(u, v), & \text{if } (u, v) \in E' \cap E'' \end{cases}$$

Definition 1.36 Let $\mathcal{G} = \mathcal{G}' \cup \mathcal{G}'' = (\mathcal{V}, \sigma' \cup \sigma'', \mu' \cup \mu'')$ be the union of two FGs $\mathcal{G}' = (\mathcal{V}', \sigma', \mu')$ and $\mathcal{G}'' = (\mathcal{V}'', \sigma'', \mu'')$. Then the degree of the vertex (u) in $\mathcal{V}$ is denoted by $deg_{\mathcal{G}' \cup \mathcal{G}''}(u)$ and is determined by

$$deg_{\mathcal{G}' \cup \mathcal{G}''}(u) = \sum_{(v,u) \in E} (\mu' \cup \mu'')(u, v)$$

$$= \sum_{(u,v) \in E' - E''} \mu'(u, v) + \sum_{(u,v) \in E'' - E'} \mu''(u, v) + \sum_{(u,v) \in E' \cap E''} \mu'(u, v) \vee \mu''(u, v).$$

Theorem 1.26 Let $\mathcal{G}' = (\mathcal{V}', \sigma', \mu')$ and $\mathcal{G}'' = (\mathcal{V}'', \sigma'', \mu'')$ be two FGs . Then

$$deg_{\mathcal{G}' \cup \mathcal{G}''}(u) = deg_{\mathcal{G}'}(u) + deg_{\mathcal{G}''}(u) - \sum_{(u,v) \in E' \cap E''} \mu'(u, v) \wedge \mu''(u, v).$$

Proof From definition, we have

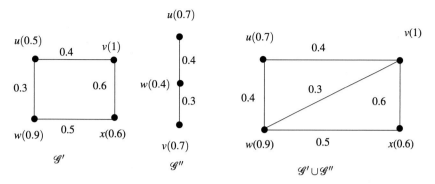

Fig. 1.6 Union of two FGs

$$deg_{\mathscr{G}'\cup\mathscr{G}''}(u) = \sum_{(v,u)\in E} (\mu' \cup \mu'')(u,v)$$

$$= \sum_{(u,v)\in E'-E''} \mu'(u,v) + \sum_{(u,v)\in E''-E'} \mu''(u,v) + \sum_{(u,v)\in E'\cap E''} \mu'(u,v) \vee \mu''(u,v)$$

$$= \sum_{(u,v)\in E'-E''} \mu'(u,v) + \sum_{(u,v)\in E''-E'} \mu''(u,v) + \sum_{(u,v)\in E'\cap E''} \mu'(u,v) \vee \mu''(u,v)$$

$$+ \sum_{(u,v)\in E'\cap E''} \mu'(u,v) \wedge \mu''(u,v) - \sum_{(u,v)\in E'\cap E''} \mu'(u,v) \wedge \mu''(u,v)$$

$$= \sum_{(u,v)\in E'} \mu'(u,v) + \sum_{(u,v)\in E''} \mu''(u,v) - \sum_{(u,v)\in E'\cap E''} \mu'(u,v) \wedge \mu''(u,v)$$

$$= deg_{\mathscr{G}'}(u) + deg_{\mathscr{G}''}(u) - \sum_{(u,v)\in E'\cap E''} \mu'(u,v) \wedge \mu''(u,v).$$

Example 1.15 Consider two FGs $\mathscr{G}'$ and $\mathscr{G}''$. The corresponding union of these two FGs is shown in Fig. 1.6. Here, $deg_{\mathscr{G}'\cup\mathscr{G}''}(u) = deg_{\mathscr{G}'}(u) + deg_{\mathscr{G}''}(u) - \sum_{(u,v)\in E'\cap E''} \mu'(u,v) \wedge \mu''(u,v) = (0.3 + 0.4) + 0.4 - (0.3 \wedge 0.4) = 0.8$. Similarly, we can find out the degree of all other vertices.

1.9.2 Join of Fuzzy Graphs

Definition 1.37 Let $\mathscr{G}' = (\mathscr{V}', \sigma', \mu')$ and $\mathscr{G}'' = (\mathscr{V}'', \sigma'', \mu'')$ be two FGs . The join between them is defined as $\mathscr{G} = \mathscr{G}' + \mathscr{G}'' = (\mathscr{V}, \sigma' + \sigma'', \mu' + \mu'')$, where the set of vertices and edges are given by $\mathscr{V} = \mathscr{V}' \cup \mathscr{V}''$ and $E = E' \cup E'' \cup E^\star$, where $E^\star$ is the set of all edges joining vertices of $\mathscr{V}'$ with vertices of $\mathscr{V}''$ such such

$$(\sigma' + \sigma'')(u) = (\sigma' \cup \sigma'')(u)$$

and

$$(\mu' + \mu'')(u, v) = \begin{cases} (\mu' \cup \mu'')(u, v), & \text{if } (u, v) \in E' \cup E'' \\ \sigma'(u) \wedge \sigma''(v), & \text{if } (u, v) \in E^{\star} \end{cases}$$

Definition 1.38 Let $\mathscr{G} = \mathscr{G}' + \mathscr{G}'' = (\mathscr{V}, \sigma' + \sigma'', \mu' + \mu'')$ be the join of two FGs $\mathscr{G}' = (\mathscr{V}', \sigma', \mu')$ and $\mathscr{G}'' = (\mathscr{V}'', \sigma'', \mu'')$. Then the degree of the vertex (u) in $\mathscr{V}$ is denoted by $deg_{\mathscr{G}'+\mathscr{G}''}(u)$ and it is determined by

$$deg_{\mathscr{G}'+\mathscr{G}''}(u) = \sum_{(v,u)\in E} (\mu' + \mu'')(u, v)$$

$$= \sum_{(u,v)\in E'\cup E''} (\mu' \cup \mu'')(u, v) + \sum_{(u,v)\in E^{\star}} \sigma'(u) \wedge \sigma''(v).$$

Theorem 1.27 Let $\mathscr{G}' = (\mathscr{V}', \sigma', \mu')$ and $\mathscr{G}'' = (\mathscr{V}'', \sigma'', \mu'')$ be two FGs .
1. If $\sigma' \geq \sigma''$, then

$$deg_{\mathscr{G}'+\mathscr{G}''}(u) = \begin{cases} deg_{\mathscr{G}'}(u) + o(\mathscr{G}''), & \text{if } u \in \mathscr{V}' \\ deg_{\mathscr{G}''}(u) + p_1\sigma''(u), & \text{if } u \in \mathscr{V}'' \end{cases}$$

2. If $\sigma' \leq \sigma''$, then

$$deg_{\mathscr{G}'+\mathscr{G}''}(u) = \begin{cases} deg_{\mathscr{G}'}(u) + p_2\sigma'(u), & \text{if } u \in \mathscr{V}' \\ deg_{\mathscr{G}''}(u) + o(\mathscr{G}'), & \text{if } u \in \mathscr{V}'' \end{cases}$$

Proof 1. If $\sigma' \geq \sigma''$ and $u \in \mathscr{V}'$, then from definition

$$deg_{\mathscr{G}'+\mathscr{G}''}(u) = \sum_{(v,u)\in E} (\mu' + \mu'')(u, v)$$

$$= \sum_{(u,v)\in E'\cup E''} (\mu' \cup \mu'')(u, v) + \sum_{(u,v)\in E^{\star}} \sigma'(u) \wedge \sigma''(v).$$

$$= \sum_{(u,v)\in E'} \mu'(u, v) + \sum_{v\in \mathscr{V}''} (\sigma'(u) \wedge \sigma'')(v)$$

$$= deg_{\mathscr{G}'}(u) + \sum_{v\in \mathscr{V}''} \sigma''(v)$$

$$= deg_{\mathscr{G}'}(u) + o(\mathscr{G}'').$$

Again, if $\sigma' \geq \sigma''$ and $u \in \mathscr{V}''$, then from definition

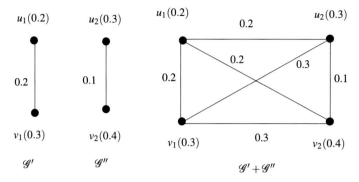

Fig. 1.7 Join of two FGs

$$deg_{\mathscr{G}'+\mathscr{G}''}(u) = \sum_{(v,u)\in E} (\mu' + \mu'')(u, v)$$

$$= \sum_{(u,v)\in E'\cup E''} (\mu' \cup \mu'')(u, v) + \sum_{(u,v)\in E^*} \sigma'(u) \wedge \sigma''(v).$$

$$= \sum_{(u,v)\in E''} \mu''(u, v) + \sum_{v\in \mathscr{V}'} (\sigma'(u) \wedge \sigma'')(v)$$

$$= deg_{\mathscr{G}''}(u) + \sum_{v\in \mathscr{V}'} \sigma'(u)$$

$$= deg_{\mathscr{G}''}(u) + p_1\sigma''(u).$$

Hence, if $\sigma' \geq \sigma''$, then

$$deg_{\mathscr{G}'+\mathscr{G}''}(u) = \begin{cases} deg_{\mathscr{G}'}(u) + o(\mathscr{G}''), & \text{if } u \in \mathscr{V}' \\ deg_{\mathscr{G}''}(u) + p_1\sigma''(u), & \text{if } u \in \mathscr{V}'' \end{cases}$$

Similarly, we can prove that, if $\sigma' \leq \sigma''$, then

$$deg_{\mathscr{G}'+\mathscr{G}''}(u) = \begin{cases} deg_{\mathscr{G}'}(u) + p_2\sigma'(u), & \text{if } u \in \mathscr{V}' \\ deg_{\mathscr{G}''}(u) + o(\mathscr{G}'), & \text{if } u \in \mathscr{V}'' \end{cases}$$

Example 1.16 Consider two FGs $\mathscr{G}'$ and $\mathscr{G}''$. The corresponding join of these two FGs is shown in Fig. 1.7. Here, $\sigma' \leq \sigma''$, so by Theorem 1.27 we have

$$deg_{\mathscr{G}'+\mathscr{G}''}(u_1) = deg_{\mathscr{G}'}(u_1) + p_2\sigma'(u_1) = 0.2 + 2 \times 0.2 = 0.6$$

$$deg_{\mathscr{G}'+\mathscr{G}''}(u_2) = deg_{\mathscr{G}''}(u_2) + o(\mathscr{G}') = 0.1 + (0.2 + 0.3) = 0.6.$$

Similarly, we can calculate the degree of all others vertices.

1.9.3 Composition of Fuzzy Graphs

The composition of two FGs stand different from their union or join. It can be viewed as a composition of two functions obeying fuzzy relations. Here, we define the composition of two FGs as well as determine the degree of each vertex of the resultant FG.

Definition 1.39 Let $\mathcal{G}' = (\mathcal{V}', \sigma', \mu')$ and $\mathcal{G}'' = (\mathcal{V}'', \sigma'', \mu'')$ be two FGs . The composition between them is defined as $\mathcal{G} = \mathcal{G}'[\mathcal{G}''] = (\mathcal{V}, \sigma' \circ \sigma'', \mu' \circ \mu'')$, where the set of vertices and edges are given by
$\mathcal{V} = \mathcal{V}' \times \mathcal{V}''$ and $E = \{((a_1, b_1), (a_2, b_2)) \mid a_1 = a_2, (b_1, b_2) \in E''$ or $b_1 = b_2, (a_1, a_2) \in E'\}$
$\cup \{((a_1, b_1), (a_2, b_2)) \mid b_1 \neq b_2, (a_1, a_2) \in E'\}$ with

$$(\sigma' \circ \sigma'')(x, y) = \sigma'(x) \wedge \sigma''(y)$$

for all $(a_1, b_1) \in \mathcal{V}' \times \mathcal{V}''$ and

$$(\mu' \circ \mu'')((a_1, b_1), (a_2, b_2)) = \begin{cases} \sigma'(a_1) \wedge \mu''(b_1, b_2), & \text{if } a_1 = a_2, (b_1, b_2) \in E'' \\ \mu'(a_1, a_2) \wedge \sigma''(b_1), & \text{if } b_1 = b_2, (a_1, a_2) \in E' \\ \mu'(a_1, a_2) \wedge \sigma'(b_1) \wedge \sigma''(b_2), & \text{if } b_1 \neq b_2, (a_1, a_2) \in E' \end{cases}$$

Definition 1.40 Let $\mathcal{G} = \mathcal{G}'[\mathcal{G}''] = (\mathcal{V}, \sigma' \circ \sigma'', \mu' \circ \mu'')$ be the composition of two FGs $\mathcal{G}' = (\mathcal{V}', \sigma', \mu')$ and $\mathcal{G}'' = (\mathcal{V}'', \sigma'', \mu'')$. Then the degree of the vertex (a_1, b_1) in $\mathcal{V}$ is denoted by $deg_{\mathcal{G}'[\mathcal{G}'']}(a_1, b_1)$ and it is determined by

$$deg_{\mathcal{G}'[\mathcal{G}'']}(a_1, b_1) = \sum_{((a_1, b_1)(a_2, b_2)) \in E} (\mu' \circ \mu'')((a_1, b_1)(a_2, b_2))$$

$$= \sum_{a_1 = a_2, (b_1, b_2) \in E''} \sigma'(a_1) \wedge \mu''(b_1, b_2) + \sum_{b_1 = b_2, (a_1, a_2) \in E'} \mu'(a_1, a_2) \wedge \sigma''(b_1)$$

$$+ \sum_{b_1 \neq b_2, (a_1, a_2) \in E'} \mu'(a_1, a_2) \wedge \sigma'(b_1) \wedge \sigma''(b_2).$$

Theorem 1.28 Let $\mathcal{G}' = (\mathcal{V}', \sigma', \mu')$ and $\mathcal{G}'' = (\mathcal{V}'', \sigma'', \mu'')$ be two FGs . If $\sigma' \geq \mu''$ and $\sigma'' \geq \mu'$, then $deg_{\mathcal{G}'[\mathcal{G}'']}(a_1, b_1) = |\mathcal{V}''| deg_{\mathcal{G}'}(a_1) + deg_{\mathcal{G}''}(b_1)$.

Proof By definition

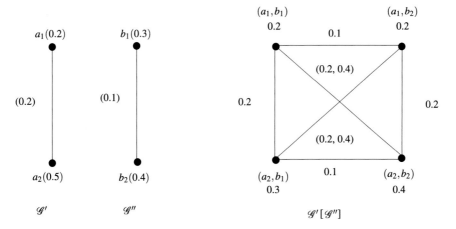

Fig. 1.8 Composition of two FGs

$$deg_{\mathscr{G}'[\mathscr{G}'']}(a_1, b_1) = \sum_{((a_1,b_1)(a_2,b_2))\in E} (\mu' \circ \mu'')((a_1, b_1)(a_2, b_2))$$

$$= \sum_{a_1=a_2,(b_1,b_2)\in E''} \sigma'(a_1) \wedge \mu''(b_1, b_2) + \sum_{b_1=b_2,(a_1,a_2)\in E'} \mu'(a_1, a_2) \wedge \sigma''(b_1)$$

$$+ \sum_{b_1\neq b_2,(a_1,a_2)\in E'} \mu'(a_1, a_2) \wedge \sigma'(b_1) \wedge \sigma''(b_2)$$

$$= \sum_{(b_1,b_2)\in E''} \mu''(b_1, b_2) + \sum_{b_1=b_2,(a_1,a_2)\in E'} \mu'(a_1, a_2) + \sum_{b_1\neq b_2,(a_1,a_2)\in E'} \mu'(a_1, a_2)$$

$$= \sum_{b_1=b_2,(a_1,a_2)\in E'} \mu'(a_1, a_2) + \sum_{b_1\neq b_2,(a_1,a_2)\in E'} \mu'(a_1, a_2) + \sum_{(b_1,b_2)\in E''} \mu''(b_1, b_2)$$

$$= |\mathscr{V}''|deg_{1\mathscr{G}'}(a_1) + deg_{1\mathscr{G}''}(b_1).$$

Hence, $deg_{\mathscr{G}'[\mathscr{G}'']}(a_1, b_1) = |\mathscr{V}''|deg_{\mathscr{G}'}(a_1) + deg_{\mathscr{G}''}(b_1)$ (Fig. 1.8). $\qquad\square$

1.9.4 Direct Product of Fuzzy Graphs

The direct product of two FGs is defined below:

Definition 1.41 Let $\mathscr{G}' = (\mathscr{V}', \sigma', \mu')$ and $\mathscr{G}'' = (\mathscr{V}'', \sigma'', \mu'')$ be two FGs such that $\mathscr{V}' \cap \mathscr{V}'' = \phi$. The direct product of $\mathscr{G}'$ and $\mathscr{G}''$ is defined as FG $\mathscr{G}' \sqcap \mathscr{G}'' = (\mathscr{V}, \sigma' \sqcap \sigma'', \mu' \sqcap \mu'')$ where $\mathscr{V} = \mathscr{V}' \times \mathscr{V}''$, $E = \{((a_1, b_1), (a_2, b_2)) \mid (a_1, a_2) \in (b_1, b_2) \in E''\}$. The membership value of the vertex (x, y) in the resultant graph $\mathscr{G}' \sqcap \mathscr{G}''$ is given by

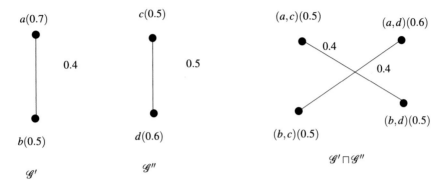

Fig. 1.9 Direct product of two FGs

$$(\sigma' \sqcap \sigma'')(x, y) = \sigma'(x) \wedge \sigma''(y).$$

And the membership value of the edge $((a_1, b_1), (a_2, b_2))$ in $\mathscr{G}' \sqcap \mathscr{G}''$ is determined by

$$(\mu' \sqcap \mu'')((a_1, b_1), (a_2, b_2)) = \mu'(a_1, a_2) \wedge \mu''(b_1, b_2).$$

Example 1.17 Let $\mathscr{G}' = (\mathscr{V}', \sigma', \mu')$ and $\mathscr{G}'' = (\mathscr{V}'', \sigma'', \mu'')$ be two FGs such that $\mathscr{V}' = \{a, b\}$, $\mathscr{V}'' = \{c, d\}$, $E' = \{(a, b)\}$, $E'' = \{(c, d)\}$, $\sigma'(a) = 0.7$, $\sigma'(b) = 0.5$, $\mu'(a, b) = 0.4$ and $\sigma''(c) = 0.5$, $\sigma''(d) = 0.6$, $\mu''(c, d) = 0.5$. The FGs $\mathscr{G}'$, $\mathscr{G}''$ and $\mathscr{G}' \sqcap \mathscr{G}''$ are shown in Fig. 1.9.

Theorem 1.29 *If $\mathscr{G}'$ and $\mathscr{G}''$ are strong FGs , then $\mathscr{G}' \sqcap \mathscr{G}''$ is also strong.*

Proof Since $\mathscr{G}' = (\mathscr{V}', \sigma', \mu')$ and $\mathscr{G}'' = (\mathscr{V}'', \sigma'', \mu'')$ are strong FGs , then $\mu'(a_1, a_2) = \sigma'(a_1) \wedge \sigma''(a_2)$ and $\mu''(b_1, b_2) = \sigma''(b_1) \wedge \sigma''(b_2)$ for all $(a_1, a_2) \in E'$ and $(b_1, b_2) \in E''$.
Now,

$$
\begin{aligned}
(\mu' \sqcap \mu'')((a_1, b_1), (a_2, b_2)) &= \mu'(a_1, a_2) \wedge \mu''(b_1, b_2) \\
&= [\sigma'(a_1) \wedge \sigma'(a_2)] \wedge [\sigma''(b_1) \wedge \sigma''(b_2)] \\
&= [\sigma'(a_1) \wedge \sigma''(b_1)] \wedge [\sigma'(a_2) \wedge \sigma''(b_2)] \\
&= (\sigma' \sqcap \sigma'')(a_1, b_1) \wedge (\sigma' \sqcap \sigma'')(a_2, b_2)
\end{aligned}
$$

Hence, $\mathscr{G}' \sqcap \mathscr{G}''$ is a strong FG. $\square$

The converse of Theorem 1.29 is partially true, if not always completely.

Theorem 1.30 *Let $\mathscr{G}'$ and $\mathscr{G}''$ be two FGs . If $\mathscr{G}' \sqcap \mathscr{G}''$ is strong, then at least one of $\mathscr{G}'$ and $\mathscr{G}''$ is strong.*

Proof If possible, let $\mathcal{G}'$ and $\mathcal{G}''$ be not strong. Then there exist at least one edge $(a_1, a_2) \in E'$, $(b_1, b_2) \in E''$ satisfies the conditions $\mu'(a_1, a_2) < \sigma'(a_1) \wedge \sigma'(a_2)$ and $\mu''(b_1, b_2) < \sigma''(b_1) \wedge \sigma''(b_2)$. Now,

$$
\begin{aligned}
(\mu' \sqcap \mu'')((a_1, b_1), (a_2, b_2)) &= \mu'(a_1, a_2) \wedge \mu''(b_1, b_2) \\
&< [\sigma'(a_1) \wedge \sigma'(a_2)] \wedge [\sigma''(b_1) \wedge \sigma''(b_2)] \\
&= [\sigma'(a_1) \wedge \sigma''(b_1)] \wedge [\sigma'(a_2) \wedge \sigma''(b_2)] \\
&= (\sigma' \sqcap \sigma'')(a_1, b_1) \wedge (\sigma' \sqcap \sigma'')(a_2, b_2)
\end{aligned}
$$

Therefore,

$$
(\mu' \sqcap \mu'')((a_1, b_1), (a_2, b_2)) < (\sigma' \sqcap \sigma'')(a_1, b_1) \wedge (\sigma' \sqcap \sigma'')(a_2, b_2)
$$

This shows that $\mathcal{G}' \sqcap \mathcal{G}''$ is not strong, a contradiction. Hence, at least one of $\mathcal{G}'$ and $\mathcal{G}''$ is strong. $\qquad\square$

1.9.5 Semi-Strong Product of Fuzzy Graphs

Based on different kinds of fuzzy operations, we classify the product of two FGs . To start with we define semi-strong product of two FGs , followed by their strong product as well as their Cartesian, tensor, normal, modular, and homomorphic products.

Definition 1.42 Let $\mathcal{G}' = (\mathcal{V}', \sigma', \mu')$ and $\mathcal{G}'' = (\mathcal{V}'', \sigma'', \mu'')$ be two FGs such that $\mathcal{V}' \cap \mathcal{V}'' = \phi$.

Then the semi-strong product of two FGs is defined as FG $\mathcal{G}' \bullet \mathcal{G}'' = (\mathcal{V}, \sigma' \bullet \sigma'', \mu' \bullet \mu'')$, where $V = \mathcal{V}' \times \mathcal{V}''$, $E = \{((x, b_1), (x, b_2)) \mid x \in \mathcal{V}', (b_1, b_2) \in E''\}$ $\cup \{((a_1, b_1), (a_2, b_2)) \mid (a_1, a_2) \in E', (b_1, b_2) \in E''\}$. The membership value of the vertex (x, y) in $\mathcal{G}' \bullet \mathcal{G}''$ is given by

$$
(\sigma' \bullet \sigma'')(x, y) = \sigma'(x) \wedge \sigma''(y).
$$

The membership value of the edge in $\mathcal{G}' \bullet \mathcal{G}''$ is computed as

$$
\begin{cases}
(\mu' \bullet \mu'')((x, b_1), (x, b_2)) = \sigma'(x) \wedge \mu''(b_1, b_2) \\
(\mu' \bullet \mu'')((a_1, b_1), (a_2, b_2)) = \mu'(a_1, a_2) \wedge \mu''(b_1, b_2)
\end{cases}
$$

Example 1.18 Let $\mathcal{G}' = (\mathcal{V}', \sigma', \mu')$ and $\mathcal{G}'' = (\mathcal{V}'', \sigma'', \mu'')$ be two FGs , where $\mathcal{V}' = \{a, b\}$, $\mathcal{V}'' = \{c, d\}$, $E' = \{(a, b)\}$, $E'' = \{(c, d)\}$. Let $\sigma'(a) = 0.7$, $\sigma'(b) = 0.5$, $\mu'(a, b) = 0.4$ and $\sigma''(c) = 0.5$, $\sigma''(d) = 0.6$, $\mu''(c, d) = 0.5$. The FGs $\mathcal{G}'$, $\mathcal{G}''$ and $\mathcal{G}' \bullet \mathcal{G}''$ are depicted in Fig. 1.10.

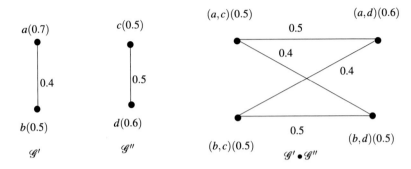

Fig. 1.10 Semi-strong product of two FGs

Theorem 1.31 *The semi-strong product of two strong FGs is strong.*

Proof If $((x, b_1), (x, b_2)) \in E$, then

$$
\begin{aligned}
(\mu' \bullet \mu'')((x, b_1), (x, b_2)) &= \sigma'(x) \wedge \mu''(b_1, b_2)) \\
&= \sigma'(x) \wedge [\sigma''(b_1) \wedge \sigma''(b_2)] \\
&= [\sigma'(x) \wedge \sigma''(b_1)] \wedge [\sigma'(x) \wedge \sigma''(b_2)] \\
&= (\sigma' \bullet \sigma'')(x, b_1) \wedge (\sigma' \bullet \sigma'')(x, b_2).
\end{aligned}
$$

Again, if $((a_1, b_1), (a_1, b_2)) \in E$, then

$$
\begin{aligned}
(\mu' \bullet \mu'')((a_1, b_1), (a_2, b_2)) &= \mu'(a_1, a_2) \wedge \mu''(b_1, b_2) \\
&= [\sigma'(a_1) \wedge \sigma'(a_2)] \wedge [\sigma''(b_1) \wedge \sigma''(b_2)] \\
&= [\sigma'(a_1) \wedge \sigma''(b_1)] \wedge [\sigma'(a_2) \wedge \sigma''(b_2)] \\
&= (\sigma' \bullet \sigma'')(a_1, b_1) \wedge (\sigma' \bullet \sigma'')(a_2, b_2).
\end{aligned}
$$

Therefore, $\mathscr{G}' \bullet \mathscr{G}''$ is also strong FG. $\qquad\square$

Theorem 1.32 *If $\mathscr{G}' \bullet \mathscr{G}''$ is strong, then at least one of $\mathscr{G}'$ and $\mathscr{G}''$ is strong.*

The proof of this theorem is quite straight forward and is left for the readers.

1.9.6 Strong Product of Fuzzy Graphs

Now, we can define strong product of two FGs .

Definition 1.43 Let $\mathscr{G}' = (\mathscr{V}', \sigma', \mu')$ and $\mathscr{G}'' = (\mathscr{V}'', \sigma'', \mu'')$ be two FGs such that $\mathscr{V}' \cap \mathscr{V}'' = \phi$. The strong product between them is defined as $\mathscr{G}' \otimes \mathscr{G}'' = (\mathscr{V}, \sigma' \otimes \sigma'', \mu' \otimes \mu'')$, where $\mathscr{V} = \mathscr{V}' \times \mathscr{V}''$, $E = \{((x, b_1), (x, b_2)) \mid x \in \mathscr{V}', (b_1, b_2) \in E''\} \cup \{((a_1, y), (a_2, y)) \mid (a_1, a_2) \in E', y \in \mathscr{V}''\} \cup \{((a_1, b_1), (a_2, b_2)) \mid (a_1, a_2) \in$

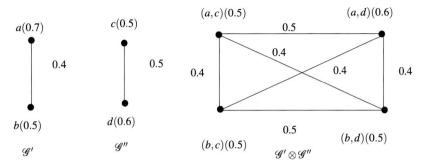

Fig. 1.11 Strong product of two FGs

E', $(b_1, b_2) \in E''$. The membership value of the vertex (x, y) in $\mathscr{G}' \otimes \mathscr{G}''$ is given by

$$(\sigma' \otimes \sigma'')(x, y) = \sigma'(x) \wedge \sigma''(y).$$

And the membership value of the edge in $\mathscr{G}' \otimes \mathscr{G}''$ is given by

$$\begin{cases} (\mu' \otimes \mu'')((x, b_1), (x, b_2)) = \sigma'(x) \wedge \mu''(b_1, b_2) \\ (\mu' \otimes \mu'')((a_1, y), (a_2, y)) = \mu'(a_1, a_2) \wedge \sigma''(y) \\ (\mu' \otimes \mu'')((a_1, b_1), (a_2, b_2)) = \mu'(a_1, a_2) \wedge \mu''(b_1, b_2) \end{cases}$$

Example 1.19 Let $\mathscr{G}' = (\mathscr{V}', \sigma', \mu')$ and $\mathscr{G}'' = (\mathscr{V}'', \sigma'', \mu'')$ be two FGs and $\mathscr{V}' = \{a, b\}$, $\mathscr{V}'' = \{c, d\}$. Also, let $E' = \{(a, b)\}$, $E'' = \{(c, d)\}$, $\sigma'(a) = 0.7$, $\sigma'(b) = 0.5$, $\mu'(a, b) = 0.4$ and $\sigma''(c) = 0.5$, $\sigma''(d) = 0.6$, $\mu''(c, d) = 0.5$. The FGs $\mathscr{G}'$, $\mathscr{G}''$ and $\mathscr{G}' \otimes \mathscr{G}''$ are drawn in Fig. 1.11.

Theorem 1.33 *If $\mathscr{G}'$ and $\mathscr{G}''$ are CFGs, $\mathscr{G}' \otimes \mathscr{G}''$ is also a CFG.*

Proof We know that strong product of FGs is a FG and every pair of vertices are adjacent. Now if $((x, b_1), (x, b_2)) \in E$, then

$$\begin{aligned} (\mu' \otimes \mu'')((x, b_1), (x, b_2)) &= \sigma'(x) \wedge \mu''(b_1, b_2)) \\ &= \sigma'(x) \wedge [\sigma''(b_1) \wedge \sigma''(b_2)] \\ &= [\sigma'(x) \wedge \sigma''(b_1)] \wedge [\sigma'(x) \wedge \sigma''(b_2)] \\ &= (\sigma' \otimes \sigma'')(x, b_1) \wedge (\sigma' \otimes \sigma'')(x, b_2). \end{aligned}$$

and

$$\begin{aligned} (\mu'_2 \otimes \mu''_2)((x, b_1), (x, b_2)) &= \sigma'_2(x) \vee \mu''_2(b_1, b_2)) \\ &= \sigma'_2(x) \vee [\sigma''_2(b_1) \vee \sigma''_2(b_2)] \\ &= [\sigma'_2(x) \vee \sigma''_2(b_1)] \vee [\sigma'_2(x) \vee \sigma''_2(b_2)] \\ &= (\sigma'_2 \otimes \sigma''_2)(x, b_1) \vee (\sigma'_2 \otimes \sigma''_2)(x, b_2). \end{aligned}$$

If $((a_1, y), (a_1, y)) \in E$, then

$$\begin{aligned}
(\mu' \otimes \mu'')((a_1, y), (a_2, y)) &= \mu'(a_1, a_2) \wedge \sigma''(y)) \\
&= [\sigma'(a_1) \wedge \sigma'(a_2)] \wedge \sigma''(y)] \\
&= [\sigma'(a_1) \wedge \sigma''(y)] \wedge [\sigma'(a_2) \wedge \sigma''(y)] \\
&= (\sigma' \otimes \sigma'')(a_1, y) \wedge (\sigma' \otimes \sigma'')(a_2, y).
\end{aligned}$$

Again, if $((a_1, b_1), (a_1, b_2)) \in E$, then

$$\begin{aligned}
(\mu' \otimes \mu'')((a_1, b_1), (a_2, b_2)) &= \mu'(a_1, a_2) \wedge \mu''(b_1, b_2) \\
&= [\sigma'(a_1) \wedge \sigma'(a_2)] \wedge [\sigma''(b_1) \wedge \sigma''(b_2)] \\
&= [\sigma'(a_1) \wedge \sigma''(b_1)] \wedge [\sigma'(a_2) \wedge \sigma''(b_2)] \\
&= (\sigma' \otimes \sigma'')(a_1, b_1) \wedge (\sigma' \otimes \sigma'')(a_2, b_2).
\end{aligned}$$

Hence, $\mathscr{G}' \otimes \mathscr{G}''$ is complete. $\square$

The following theorem follows from the above result.

Theorem 1.34 *If* $\mathscr{G}' = (\mathscr{V}', \sigma', \mu')$ *and* $\mathscr{G}'' = (\mathscr{V}'', \sigma'', \mu'')$ *are two FGs such that* $\mathscr{G}' \otimes \mathscr{G}''$ *is strong, then at least one of* $\mathscr{G}'$ *or* $\mathscr{G}''$ *must be strong.*

1.9.7   Cartesian Product of Fuzzy Graphs

Definition 1.44 The Cartesian product of two FGs $\mathscr{G}' = (\mathscr{V}', \sigma', \mu')$ and $\mathscr{G}'' = (\mathscr{V}'', \sigma'', \mu'')$ is a FG and it is defined as $\mathscr{G} = \mathscr{G}' \times \mathscr{G}'' = (\mathscr{V}, \sigma' \times \sigma'', \mu' \times \mu'')$, where $\mathscr{V} = \mathscr{V}' \times \mathscr{V}''$ and $E = \{((a_1, b_1), (a_2, b_2)) \mid a_1 = a_2, (b_1, b_2) \in E'' \text{ or } b_1 = b_2, (a_1, a_2) \in E'\}$ with

$$(\sigma' \times \sigma'')(x, y) = \sigma'(x) \wedge \sigma''(y)$$

for all $(a_1, b_1) \in \mathscr{V}' \times \mathscr{V}''$ and

$$(\mu' \times \mu'')((a_1, b_1), (a_2, b_2)) = \begin{cases} \sigma'(a_1) \wedge \mu''(b_1, b_2), & \text{if } a_1 = a_2, (b_1, b_2) \in E'' \\ \mu'(a_1, a_2) \wedge \sigma''(b_1), & \text{if } b_1 = b_2, (a_1, a_2) \in E' \end{cases}$$

Definition 1.45 Let $\mathscr{G} = \mathscr{G}' \times \mathscr{G}'' = (\mathscr{V}, \sigma' \times \sigma'', \mu' \times \mu'')$ be the Cartesian product of $\mathscr{G}'$ and $\mathscr{G}''$. Then the degree of the vertex (a_1, b_1) in $\mathscr{V}$ is denoted by $deg_{\mathscr{G}' \times \mathscr{G}''}(a_1, b_1)$ and is defined by

$$deg_{\mathscr{G}' \times \mathscr{G}''}(a_1, b_1) = \sum_{((a_1,b_1)(a_2,b_2)) \in E} (\mu' \times \mu'')((a_1, b_1)(a_2, b_2))$$

$$= \sum_{a_1=a_2,(b_1,b_2) \in E''} \sigma'(a_1) \wedge \mu''(b_1, b_2) + \sum_{b_1=b_2,(a_1,a_2) \in E'} \mu'(a_1, a_2) \wedge \sigma''(b_1)$$

Following theorem gives the relation on degrees of the vertices of the original graphs and that of the product graph.

Theorem 1.35 *Let $\mathscr{G}'$ and $\mathscr{G}''$ be two FGs . If $\sigma' \geq \mu''$ and $\sigma'' \geq \mu'$, then $deg_{\mathscr{G}' \times \mathscr{G}''}(a_1, b_1) = deg_{\mathscr{G}'}(a_1) + deg_{\mathscr{G}''}(b_1)$, where $a_1 \in \mathscr{V}'$ and $b_1 \in \mathscr{V}''$.*

Proof From the definition of degree of a vertex in Cartesian product, one can write

$$deg_{\mathscr{G}' \times \mathscr{G}''}(a_1, b_1) = \sum_{((a_1,b_1)(a_2,b_2)) \in E} (\mu' \times \mu'')((a_1, b_1)(a_2, b_2))$$

$$= \sum_{a_1=a_2,(b_1,b_2) \in E''} \sigma'(a_1) \wedge \mu''(b_1, b_2) + \sum_{b_1=b_2,(a_1,a_2) \in E'} \mu'(a_1, a_2) \wedge \sigma''(b_1)$$

$$= \sum_{(b_1,b_2) \in E''} \mu''(b_1, b_2) + \sum_{(a_1,a_2) \in E'} \mu'(a_1, a_2) \qquad [\text{since } \sigma' \geq \mu'' \text{ and } \sigma'' \geq \mu']$$

$$= \sum_{(a_1,a_2) \in E'} \mu'(a_1, a_2) + \sum_{(b_1,b_2) \in E''} \mu''(b_1, b_2)$$

$$= deg_{\mathscr{G}'}(a_1) + deg_{\mathscr{G}''}(b_1).$$

Hence, $deg_{\mathscr{G}' \times \mathscr{G}''}(a_1, b_1) = deg_{\mathscr{G}'}(a_1) + deg_{\mathscr{G}''}(b_1)$. □

1.9.8   Tensor Product of Fuzzy Graphs

The conjunction or tensor product between two FGs is defined here.

Definition 1.46 Let $\mathscr{G}'$ and $\mathscr{G}''$ be two FGs . The tensor product between them is defined as
$\mathscr{G} = \mathscr{G}' \otimes \mathscr{G}'' = (\mathscr{V}, \sigma' \otimes \sigma'', \mu' \otimes \mu'')$, where $\mathscr{V} = \mathscr{V}' \times \mathscr{V}''$ and $E = \{((a_1, b_1), (a_2, b_2)) \mid (a_1, a_2) \in E', (b_1, b_2) \in E''\}$ with

$$(\sigma' \otimes \sigma'')(x, y) = \sigma'(x) \wedge \sigma''(y)$$

for all $(a_1, b_1) \in \mathscr{V}' \times \mathscr{V}''$ and

$$(\mu' \otimes \mu'')((a_1, b_1), (a_2, b_2)) = \mu'(a_1, a_2) \wedge \mu''(b_1, b_2).$$

Definition 1.47 Let $\mathscr{G} = \mathscr{G}' \otimes \mathscr{G}'' = (\mathscr{V}, \sigma' \otimes \sigma'', \mu' \otimes \mu'')$ be the tensor product of two FGs $\mathscr{G}'$ and $\mathscr{G}''$. Then the degree of the vertex (a_1, b_1) in $\mathscr{V}$ is denoted by $deg_{\mathscr{G}' \otimes \mathscr{G}''}(a_1, b_1)$ and is defined by

$$deg_{\mathscr{G}' \otimes \mathscr{G}''}(a_1, b_1) = \sum_{((a_1,b_1)(a_2,b_2)) \in E} (\mu' \otimes \mu'')((a_1, b_1)(a_2, b_2))$$

$$= \sum_{(a_1,a_2) \in E'} \mu'(a_1, a_2) \wedge \mu''(b_1, b_2),$$

Theorem 1.36 Let $\mathscr{G}' = (\mathscr{V}', \sigma', \mu')$ and $\mathscr{G}'' = (\mathscr{V}'', \sigma'', \mu'')$ be two FGs. If $\mu'' \geq \mu'$, then $deg_{\mathscr{G}' \otimes \mathscr{G}''}(a_1, b_1) = deg_{\mathscr{G}'}(a_1)$ and if $\mu' \geq \mu''$, then $deg_{\mathscr{G}' \otimes \mathscr{G}''}(a_1, b_1) = deg_{\mathscr{G}''}(b_1)$.

Proof Let $\mu'' \geq \mu'$, then

$$deg_{\mathscr{G}' \otimes \mathscr{G}''}(a_1, b_1) = \sum_{((a_1,b_1)(a_2,b_2)) \in E} (\mu' \otimes \mu'')((a_1, b_1)(a_2, b_2))$$

$$= \sum_{(a_1,a_2) \in E'} \mu'(a_1, a_2) \wedge \mu''(b_1, b_2)$$

$$= \sum \mu'(a_1, a_2) = deg_{\mathscr{G}'}(a_1).$$

Hence, $deg_{\mathscr{G}' \otimes \mathscr{G}''}(a_1, b_1) = deg_{\mathscr{G}'}(a_1)$.
Similarly, if $\mu' \geq \mu''$, then $deg_{\mathscr{G}' \otimes \mathscr{G}''}(a_1, b_1) = deg_{\mathscr{G}''}(b_1)$. $\square$

Example 1.20 The degree of the vertices of $\mathscr{G}' \otimes \mathscr{G}''$ can be calculated using the above theorem.

Consider the FGs $\mathscr{G}'$, $\mathscr{G}''$ and $\mathscr{G}'[\mathscr{G}'']$ depicted in Fig. 1.12. Here, $\mu'' \geq \mu'$, then $deg_{\mathscr{G}' \otimes \mathscr{G}''}(a_1, b_1) = deg_{\mathscr{G}'}(a_1) = 0.2$. Hence, $deg_{\mathscr{G}' \otimes \mathscr{G}''}(a_1, b_1) = 0.2$.

1.9.9 Normal Product of Fuzzy Graphs

Another new kind of product of FGs called normal product is defined below.

Definition 1.48 The normal product of two FGs $\mathscr{G}' = (\mathscr{V}', \sigma', \mu')$ and $\mathscr{G}'' = (\mathscr{V}'', \sigma'', \mu'')$ is defined as a FG $\mathscr{G} = \mathscr{G}' \bullet \mathscr{G}'' = (\mathscr{V}, \sigma' \bullet \sigma'', \mu' \bullet \mu'')$ where $\mathscr{V} = \mathscr{V}' \times \mathscr{V}''$ and $E = \{((a_1, b_1), (a_2, b_2)) \mid a_1 = a_2, (b_1, b_2) \in E''$ or $b_1 = b_2, (a_1, a_2) \in E'\} \cup \{((a_1, b_1), (a_2, b_2)) \mid (a_1, a_2) \in (b_1, b_2) \in E''\}$ with

$$(\sigma' \bullet \sigma'')(x, y) = \sigma'(x) \wedge \sigma''(y)$$

for all $(a_1, b_1) \in \mathscr{V}' \times \mathscr{V}''$ and

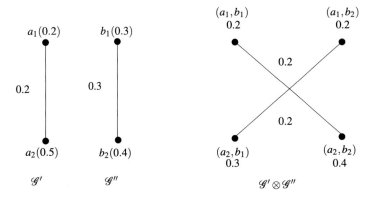

Fig. 1.12 Tensor product of two FGs

$$(\mu' \bullet \mu'')((a_1, b_1), (a_2, b_2)) = \begin{cases} \sigma'(a_1) \wedge \mu''(b_1, b_2), & \text{if } a_1 = a_2, (b_1, b_2) \in E'' \\ \mu'(a_1, a_2) \wedge \sigma''(b_1), & \text{if } b_1 = b_2, (a_1, a_2) \in E' \\ \mu'(a_1, a_2) \wedge \mu''(b_1, b_2), & \text{if } (a_1, a_2) \in E', (b_1, b_2) \in E'' \end{cases}$$

Definition 1.49 Let $\mathscr{G} = \mathscr{G}' \bullet \mathscr{G}'' = (\mathscr{V}, \sigma' \bullet \sigma'', \mu' \bullet \mu'')$ be the normal product of two FGs $\mathscr{G}'$ and $\mathscr{G}''$. Then the degree of the vertex (a_1, b_1) in $\mathscr{V}$ is defined by

$$deg_{\mathscr{G}' \bullet \mathscr{G}''}(a_1, b_1) = \sum_{((a_1, b_1)(a_2, b_2)) \in E} (\mu' \bullet \mu'')((a_1, b_1)(a_2, b_2))$$

$$= \sum_{a_1 = a_2, (b_1, b_2) \in E''} \sigma'(a_1) \wedge \mu''(b_1, b_2) + \sum_{b_1 = b_2, (a_1, a_2) \in E'} \mu'(a_1, a_2) \wedge \sigma''(b_1)$$

$$+ \sum_{(a_1, a_2) \in (b_1, b_2) \in E''} \mu'(a_1, a_2) \wedge \mu''(b_1, b_2)$$

Theorem 1.37 Let $\mathscr{G}' = (\mathscr{V}, \sigma', \mu')$ and $\mathscr{G}'' = (\mathscr{V}'', \sigma'', \mu'')$ be two FGs. If $\sigma' \geq \mu''$, $\sigma'' \geq \mu'$ and $\mu' \leq \mu''$, then $deg_{\mathscr{G}' \bullet \mathscr{G}''}(a_1, b_1) = |\mathscr{V}''|deg_{\mathscr{G}'}(a_1) + deg_{\mathscr{G}''}(b_1)$, where $a_1 \in \mathscr{V}'$ and $b_1 \in \mathscr{V}''$.

Proof Let $\sigma' \geq \mu''$, $\sigma'' \geq \mu'$ and $\mu' \leq \mu''$, then by definition

$$deg_{\mathscr{G}' \bullet \mathscr{G}''}(a_1, b_1) = \sum_{((a_1, b_1)(a_2, b_2)) \in E} (\mu' \bullet \mu'')((a_1, b_1)(a_2, b_2))$$

$$= \sum_{a_1 = a_2, (b_1, b_2) \in E''} \sigma'(a_1) \wedge \mu''(b_1, b_2) + \sum_{b_1 = b_2, (a_1, a_2) \in E'} \mu'(a_1, a_2) \wedge \sigma''(b_1)$$

$$+ \sum_{(a_1, a_2) \in E', (b_1, b_2) \in E''} \mu'(a_1, a_2) \wedge \mu''(b_1, b_2)$$

$$= \sum_{(b_1, b_2) \in E''} \mu''(b_1, b_2) + \sum_{b_1 = b_2, (a_1, a_2) \in E'} \mu'(a_1, a_2) + \sum_{(a_1, a_2) \in E'} \mu'(a_1, a_2)$$

$$= \sum_{b_1=b_2,(a_1,a_2)\in E'} \mu'(a_1,a_2) + \sum_{(a_1,a_2)\in E'} \mu'(a_1,a_2) + \sum_{(b_1,b_2)\in E''} \mu''(b_1,b_2)$$

$$= |\mathscr{V}''|deg_{1\mathscr{G}'}(a_1) + deg_{1\mathscr{G}''}(b_1).$$

Hence, $deg_{\mathscr{G}'\bullet\mathscr{G}''}(a_1,b_1) = |\mathscr{V}''|deg_{\mathscr{G}'}(a_1) + deg_{\mathscr{G}''}(b_1)$. □

1.9.10 Modular Product of Fuzzy Graphs

Definition 1.50 Let $\mathscr{G}'$ and $\mathscr{G}''$ be two FGs . The modular product of $\mathscr{G}'$ and $\mathscr{G}''$ is defined as $\mathscr{G} = \mathscr{G}' \circledcirc \mathscr{G}'' = (\mathscr{V}, \sigma' \circledcirc \sigma'', \mu' \circledcirc \mu'')$, where $\mathscr{V} = \mathscr{V}' \circledcirc \mathscr{V}'' = \{(a_1,b_1) \mid a_1 \in \mathscr{V}' \text{ and } b_1 \in \mathscr{V}''\}$ and $E = E' \circledcirc E'' = \{((a_1,b_1),(a_2,b_2)) \mid (a_1,a_2) \in E', (b_1,b_2) \in E'' \text{ or } (a_1,a_2) \notin E', (b_1,b_2) \notin E''\}$ along with

$$(\sigma' \circledcirc \sigma'')(a_1,b_1) = \sigma'(a_1) \wedge \sigma''(b_1)$$

for all $(a_1,b_1) \in V$ and

$$(\mu' \circledcirc \mu_1'')((a_1,b_1),(a_2,b_2)) = \begin{cases} \mu'(a_1,a_2) \wedge \mu''(b_1,b_2), & \text{if } (a_1,a_2) \in E', (b_1,b_2) \in E'' \\ \sigma'(a_1) \wedge \sigma'(a_2) \wedge \sigma''(b_1) \wedge \sigma''(b_2), & \text{if } (a_1,a_2) \notin E', (b_1,b_2) \notin E'' \end{cases}$$

Definition 1.51 Let $\mathscr{G} = \mathscr{G}' \circledcirc \mathscr{G}'' = (\mathscr{V}, \sigma' \circledcirc \sigma'', \mu' \circledcirc \mu'')$ be the modular product of two FG $\mathscr{G}' = (\mathscr{V}', \sigma', \mu')$ and $\mathscr{G}'' = (\mathscr{V}'', \sigma'', \mu'')$. Then the degree of the vertex (a_1,b_1) in $\mathscr{V}$ is denoted by $deg_{\mathscr{G}'\circledcirc\mathscr{G}''}(a_1,b_1)$ and defined by

$$deg_{\mathscr{G}'\circledcirc\mathscr{G}''}(a_1,b_1) = \sum_{((a_1,b_1)(a_2,b_2))\in E} (\mu' \circledcirc \mu'')((a_1,b_1)(a_2,b_2))$$

$$= \sum_{(a_1,a_2)\in E',(b_1,b_2)\in E''} \mu'(a_1,a_2) \wedge \mu''(b_1,b_2) + \sum_{(a_1,a_2)\notin E',(b_1,b_2)\notin E''} \sigma'(a_1) \wedge \sigma'(a_2) \wedge \sigma''(b_1) \wedge \sigma''(b_2)$$

Theorem 1.38 *If* $\mathscr{G}' = (\mathscr{V}', \sigma', \mu')$ *and* $\mathscr{G}'' = (\mathscr{V}'', \sigma'', \mu'')$ *are CFGs and* $\mu' \leq \mu''$, *then* $deg_{\mathscr{G}'\circledcirc\mathscr{G}''}(a_1,b_1) = deg_{\mathscr{G}'}(a_1)$.

Proof From the definition of degree of a vertex in the modular product, we have

$$deg_{\mathscr{G}'\circledcirc\mathscr{G}''}(a_1,b_1) = \sum_{((a_1,b_1)(a_2,b_2))\in E} (\mu' \circledcirc \mu'')((a_1,b_1)(a_2,b_2))$$

$$= \sum_{(a_1,a_2)\in E',(b_1,b_2)\in E''} \mu'(a_1,a_2) \wedge \mu''(b_1,b_2) + \sum_{(a_1,a_2)\notin E',(b_1,b_2)\notin E''} \sigma'(a_1) \wedge \sigma'(a_2) \wedge \sigma''(b_1) \wedge \sigma''(b_2)$$

$$= \sum_{(a_1,a_2)\in E',(b_1,b_2)\in E''} \mu'(a_1,a_2) \wedge \mu''(b_1,b_2) \text{ [Since both FGs are complete]}$$

$$= \sum_{(a_1,a_2)\in E'} \mu'(a_1,a_2) \text{ [Since } \mu' \leq \mu'']$$

$$= deg_{\mathscr{G}'}(a_1).$$

Hence, $deg_{\mathscr{G}'\circledcirc\mathscr{G}''}(a_1,b_1) = deg_{\mathscr{G}'}(a_1)$. □

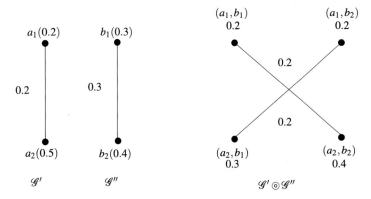

Fig. 1.13 Modular product of two FGs

Let us consider the following example to illustrate the degree of modular product of two FGs .

Example 1.21 Let us consider the FGs shown in Fig. 1.13.

Here, the FGs $\mathcal{G}'$, $\mathcal{G}''$ are complete and $\mu' \leq \mu''$.

Now, $deg_{\mathcal{G}' \odot \mathcal{G}''}(a_1, b_1) = deg_{\mathcal{G}'}(a_1) = 0.2$.

Hence, $deg_{\mathcal{G}' \odot \mathcal{G}''}(a_1, b_1) = 0.2$.

1.9.11 Homomorphic Product of Fuzzy Graphs

This is the last kind of product on FGs that we discuss in this section.

Definition 1.52 Let $\mathcal{G}' = (\mathcal{V}', \sigma', \mu')$ and $\mathcal{G}'' = (\mathcal{V}'', \sigma'', \mu'')$ be two FGs . Then the homomorphic product is defined as $\mathcal{G} = \mathcal{G}' \diamond \mathcal{G}'' = (\mathcal{V}, \sigma' \diamond \sigma'', \mu' \diamond \mu'')$, where $\mathcal{V} = \mathcal{V}' \diamond \mathcal{V}'' = \{(a_1, b_1) \mid a_1 \in \mathcal{V}' \text{and } b_1 \in \mathcal{V}''\}$ and $E = E' \diamond E'' = \{((a_1, b_1), (a_2, b_2)) \mid a_1 = a_2, (b_1, b_2) \in E'' \text{ or } (a_1, a_2) \in E', (b_1, b_2) \notin E''\}$ with

$$(\sigma' \diamond \sigma'')(a_1, b_1) = \sigma'(a_1) \wedge \sigma''(b_1)$$

for all $(a_1, b_1) \in V$ and

$$(\mu' \diamond \mu_1'')((a_1, b_1), (a_2, b_2)) = \begin{cases} \sigma'(a_1) \wedge \mu''(b_1, b_2), & \text{if } a_1 = a_2, (b_1, b_2) \in E'' \\ \mu'(a_1, a_2) \wedge \sigma''(b_1) \wedge \sigma''(b_2), & \text{if } (a_1, a_2) \in E', (b_1, b_2) \notin E'' \end{cases}$$

Definition 1.53 Let $\mathcal{G} = \mathcal{G}' \diamond \mathcal{G}'' = (\mathcal{V}, \sigma' \diamond \sigma'', \mu' \diamond \mu'')$ be the homomorphic product of two FGs $\mathcal{G}' = (\mathcal{V}', \sigma', \mu')$ and $\mathcal{G}'' = (\mathcal{V}'', \sigma'', \mu'')$. Then the degree of the vertex (a_1, b_1) in $\mathcal{V}$ is denoted by $deg_{\mathcal{G}' \diamond \mathcal{G}''}(a_1, b_1)$ and defined by

$$deg_{\mathscr{G}'\diamond\mathscr{G}''}(a_1,b_1)= \sum_{((a_1,b_1)(a_2,b_2))\in E} (\mu' \diamond \mu'')((a_1,b_1)(a_2,b_2))$$

$$= \sum_{a_1=a_2,(b_1,b_2)\in E''} \sigma'(a_1)\wedge\mu''(b_1,b_2)+ \sum_{(a_1,a_2)\in E',(b_1,b_2)\notin E''} \mu'(a_1,a_2)\wedge\sigma''(b_1)\wedge\sigma''(b_2).$$

Theorem 1.39 *Let $\mathscr{G}' = (\mathscr{V}',\sigma',\mu')$ and $\mathscr{G}'' = (\mathscr{V}'',\sigma'',\mu'')$ be two FGs. If $\mathscr{G}''$ is CFG and $\sigma' \le \mu''$, then $deg_{\mathscr{G}'\diamond\mathscr{G}''}(a_1,b_1) = (|\,\mathscr{V}''\,|-1)\sigma'(a_1)$.*

Proof From the definition of the degree of a vertex in homomorphic product, one can write

$$deg_{\mathscr{G}'\diamond\mathscr{G}''}(a_1,b_1)= \sum_{((a_1,b_1)(a_2,b_2))\in E} (\mu' \diamond \mu'')((a_1,b_1)(a_2,b_2))$$

$$= \sum_{a_1=a_2,(b_1,b_2)\in E''} \sigma'(a_1)\wedge\mu''(b_1,b_2)+ \sum_{(a_1,a_2)\in E',(b_1,b_2)\notin E''} \mu'(a_1,a_2)\wedge\sigma''(b_1)\wedge\sigma''(b_2)$$

$$= \sum_{a_1=a_2,(b_1,b_2)\in E''} \sigma'(a_1)\wedge\mu''(b_1,b_2) \quad [\text{Since } \mathscr{G}'' \text{ is a CFG }]$$

$$= \sum_{a_1=a_2} \sigma'(a_1) \quad [\text{Since } \sigma' \le \mu'']$$

$$= (|\,\mathscr{V}''\,|-1)\sigma'(a_1).$$

Hence, $deg_{\mathscr{G}'\diamond\mathscr{G}''}(a_1,b_1) = (|\,\mathscr{V}''\,|-1)\sigma'(a_1)$. $\square$

1.10 Matrix Representation of Fuzzy Graphs

A very common representation of a graph is its pictorial form. But this form is not quite useful in computer applications. A graph can also be represented in many other ways. Among them, matrix representation seems befitting for computer processing and also to design efficient algorithms. Different types of matrix representations are available in literature, such as adjacent matrix, incident matrix, etc.

The **adjacent matrix** of an undirected (crisp) graph G with n vertices and no parallel edges is an $n \times n$ symmetric matrix $A = [a_{ij}]$, where

$$a_{ij} = \begin{cases} 1, & \text{if there is an edge between } i\text{th and } j\text{th vertices,} \\ 0, & \text{if there is no edge between them} \end{cases}$$

If there is a self-loop at the vertex i, then $a_{ii} = 1$.

The adjacent matrix for a weighted graph is defined below.

Let G be a graph and if each edge be associated to a real number called weight, then G is called **weighted graph**. The adjacent matrix for the graph G is $A(G) = [a_{ij}]$, where

$$a_{ij} = \begin{cases} w_{ij}, & \text{if the vertices } i \text{ and } j \text{ are connected by an edge, where } w_{ij} \\ & \text{is the weight of the edge } (i,j), \\ 0, & \text{otherwise} \end{cases}$$

The adjacent matrix of a weighted graph carries more information. So, we will use this concept to obtain the matrix form of a FG.

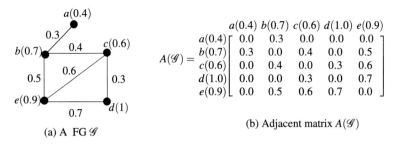

Fig. 1.14 A FG $\mathscr{G}$ and its adjacent matrix $A(\mathscr{G})$

In crisp graph, all the vertices are certain, that is, the membership values of all vertices are 1. But, in FG , the membership values of the vertices are different. So, to incorporate this additional information to the matrix, we have to update the adjacent matrix. Let $\mathscr{G} = (\mathscr{V}, \sigma, \mu)$ be a FG with n vertices, where $\mathscr{V} = \{v_1, v_2, \ldots, v_n\}$ be the set of vertices. The adjacent matrix for the FG $\mathscr{G} = (\mathscr{V}, \sigma, \mu)$ is denoted by

$$A(\mathscr{G}) = [r_A(i)][a_{ij}]_{n \times n},$$

where
$$r_A(i) = \sigma(v_i) \text{ and } a_{ij} = \mu(v_i, v_j) \text{ for all } i, j = 1, 2, \ldots, n.$$

Let us consider the FG $\mathscr{G} = (\mathscr{V}, \sigma, \mu)$ of Fig. 1.14a. The number associated with each vertex and each edge represent membership values. The adjacent matrix $A(\mathscr{G})$ is shown in Fig. 1.14b.

The adjacent matrix depends on the numbering or ordering of the vertices. Different ordering of vertices gives different adjacent matrices. Since n vertices can be ordered in $n!$ different ways, the number of different adjacent matrices for a given graph with n vertices is $n!$.

Some observations on the adjacent matrix of a FG are listed below.

(i) The adjacent matrix is symmetric for undirected graphs, i.e. $a_{ij} = a_{ji}$ for all i, j.
(ii) For all $i, j = 1, 2, \ldots, n$, $a_{ij} \leq r_A(i) \wedge r_A(j)$.
(iii) The adjacent matrix for a digraph is not symmetric.
(iv) The entries of $A(\mathscr{G})$ are real numbers in $[0, 1]$.
(v) The sum of the entries of a row is the degree of the corresponding vertex.
(vi) For a simple FG, the sum of all entries is the twice the sum of membership values of all edges.
(vii) For a simple FG, the diagonal entries are all 0's.
(viii) If $\mathscr{G}$ is a complete graph then the diagonal elements of $A(\mathscr{G})$ are 0 and all other elements are non-zero numbers in $[0, 1]$.

(ix) The number of elements in an adjacent matrix for a FG with n vertices is n^2, and number of rows and columns is $2n$. Therefore, to store a FG using adjacent matrix, $n^2 + 2n$ memory units are required in a computer.

(x) A FG $\mathcal{G}$ is disconnected and it has two components $\mathcal{G}_1, \mathcal{G}_2$ if and only if its adjacent matrix $A(\mathcal{G})$ can be represented as

$$A(\mathcal{G}) = \begin{bmatrix} A(\mathcal{G}_1) & \vdots & 0 \\ \cdots & \cdots & \cdots \\ 0 & \vdots & A(\mathcal{G}_2) \end{bmatrix},$$

where $A(\mathcal{G}_1)$ and $A(\mathcal{G}_2)$ are the adjacent matrices of the components $\mathcal{G}_1$ and $\mathcal{G}_2$ respectively.

If $\mathcal{G}$ has three components $\mathcal{G}_1, \mathcal{G}_2, \mathcal{G}_3$, then $A(\mathcal{G})$ looks like

$$A(\mathcal{G}) = \begin{bmatrix} A(\mathcal{G}_1) & 0 & 0 \\ 0 & A(\mathcal{G}_2) & 0 \\ 0 & 0 & A(\mathcal{G}_3) \end{bmatrix}.$$

Thus, the connectedness of a FG can easily be identified from its adjacent matrix.

(xi) For a given symmetric matrix $A = [r_A(i)][a_{ij}]_n$ of order n with entries from $[0, 1]$, one can construct a FG $\mathcal{G}$ of n vertices such that A is the adjacent matrix of $\mathcal{G}$.

(xii) Using permutations of rows and of the corresponding columns one can reorder the vertices of a FG. If two or more rows are interchanged in $A(\mathcal{G})$, the corresponding columns should also be interchanged. By suitable permutations of rows and columns of $A(\mathcal{G})$, one can test the graph isomorphism as follows. Two FGs $\mathcal{G}_1$ and $\mathcal{G}_2$ with no parallel edges are isomorphic if and only if their adjacent matrices $A(\mathcal{G}_1)$ and $A(\mathcal{G}_2)$ are related as

$$A(\mathcal{G}_2) = P^{-1} A(\mathcal{G}_1) P,$$

where P is a permutation matrix.

The adjacent matrix of a FG is not a fuzzy matrix, due to restriction $a_{ij} \leq r_A(i) \wedge r_A(j)$, for all $i, j = 1, 2, \ldots, n$. In fuzzy matrix, a_{ij} is any number between 0 and 1 including 0 and 1. This particular type of fuzzy matrix is known as "fuzzy matrix with fuzzy row and fuzzy column (FMFRC)" and has been studied in [105] and one of its variants in [104]. This special type of fuzzy matrix is referred to as to g-FM. This is a particular case of FMFRC.

It is discussed that from a FG one can construct the g-FM. Conversely, from a g-FM, we can draw the corresponding FG. If the given matrix is a g-FM, then one can draw the FG directly. A FG whether it is g-FM or not can be tested by the formula: $a_{ij} \leq r_A(i) \wedge r_A(j)$ for all elements a_{ij} of the matrix.

For each row or column draw a vertex with specified membership value. Now, draw an edge between the vertices i and j if $a_{ij} \neq 0$ with membership value a_{ij}. If the diagonal element a_{ii} for any i, is non-zero then draw a self-loop on the vertex i with a_{ii} as the membership value.

1.10.1 Null FMFRC

Depending on the membership values of rows, columns, and elements, two types of null g-FM are defined.

Type I: If $r_A(i) = 0$ and $a_{ij} = 0$ for all i and j, then g-FM A is called p-null, denoted by 0_p. For example

$$
\begin{array}{c}
0\ 0\ 0 \\
\begin{array}{c} 0 \\ 0 \\ 0 \end{array}
\begin{bmatrix}
0\ 0\ 0 \\
0\ 0\ 0 \\
0\ 0\ 0
\end{bmatrix}
\end{array}
$$

is a 3×3 order p-null g-FM.

The graph corresponding to this type of g-FM does not exist in reality. It exists hypothetically.

Type II: If $a_{ij} = 0$ for all i and j, whatever may be the values of $r_A(i)$, then the g-FM A is called e-null, and it is denoted by 0_e.

In this case, the FG has some vertices with specified membership values, but there is no edge, that is null graph.

Like null g-FM, different types of identity g-FMs are also defined in [105].

1.10.2 Operations on g-FMs

Many operators are defined on fuzzy matrix. Here, only two operators, viz. $\vee$ and $\wedge$ are discussed on g-FM. We denote

$$a \vee b = max\{a, b\} \text{ and } a \wedge b = min\{a, b\}.$$

1.10.3 $\vee$ Operator

Let $A = [r_A(i)][a_{ij}]_n$ and $B = [r_B(i)][b_{ij}]_n$ be two g-FMs. Then

$$A \vee B = D = [r_D(i)][d_{ij}]_n,$$

where
$r_D(i) = r_A(i) \vee r_B(i)$,
$d_{ij} = a_{ij} \vee b_{ij}$ for all i, j.

1.10.4 ∧ *Operator*

The $\wedge$ operation is similar to $\vee$ operation.
Let $A = [r_A(i)][a_{ij}]_n$ and
$B = [r_B(i)][b_{ij}]_n$ be two g-FMs. Then

$$A \wedge B = D = [r_D(i)][d_{ij}]_n,$$

where
$r_D(i) = r_A(i) \wedge r_B(i)$,
$d_{ij} = a_{ij} \wedge b_{ij}$ for all i, j.

Theorem 1.40 *If A and B be two g-FMs of same order, then $A \vee B$ and $A \wedge B$ both are g-FMs.*

Proof Let $A = [r_A(i)][a_{ij}]_n$ and $B = [r_B(i)][b_{ij}]_n$ be two g-FMs. Also, let $C = A \vee B$. Then,
$a_{ij} \leq r_A(i) \vee r_A(j)$ and $b_{ij} \leq r_B(i) \vee r_B(j)$ for all i, j.
Now, $r_C(i) \leq r_A(i) \vee r_B(i)$ for all i and $r_C(j) \leq r_A(j) \vee r_B(j)$ for all j.
$c_{ij} = a_{ij} \vee b_{ij} \leq \{r_A(i) \vee r_A(j)\} \vee \{r_B(i) \vee r_B(j)\}$
$\quad = \{r_A(i) \vee r_B(i)\} \vee \{r_A(j) \vee r_B(j)\} = r_C(i) \vee r_C(j)$

Definition 1.54 Let $A = [r_A(i)][a_{ij}]_{m \times n}$ be a g-FM. The **density** of A is denoted as $\mathscr{D}(A)$ and is defined as

$$\mathscr{D}(A) = \frac{\sum\limits_{i,j} a_{ij}}{\sum\limits_{i,j} r_A(i) \wedge c_A(j)},$$

provided $\sum\limits_{i,j} r_A(i) \wedge c_A(j) \neq 0$.

For a g-FM, $a_{ij} \geq 0$ for all i, j. Thus, $\mathscr{D}(A)$ is non-negative for any FMFRC A. $\mathscr{D}(A)$ is zero only when all $a_{ij} = 0$. Higher value of $\mathscr{D}(A)$, indicates the matrix is denser. If the value of $D(A)$ is close to zero, then the g-FM is called a sparse g-FM with a degree of sparsity $\mathscr{D}(A)$.

For any g-FM A, $\quad 0 \leq \mathscr{D}(A) \leq 1$. [105]. The left and right equalities hold when A is null g-FM and complete g-FM respectively.

Definition 1.55 A g-FM A is called balanced if $\mathscr{D}(S) \leq \mathscr{D}(A)$ for all sub g-FMs S of A.

Now, we define a particular type of balanced g-FM.

Definition 1.56 A g-FM A is called strictly balanced if $\mathscr{D}(S) = \mathscr{D}(A)$ for all sub g-FMs S of A.

$$\text{Example 1.22} \quad \text{Let } A = \begin{array}{c} 0.3 \quad 0.2 \\ 0.3\begin{bmatrix} 0.225 & 0.150 \\ 0.2\begin{bmatrix} 0.150 & 0.150 \end{bmatrix} \end{array}.$$

Then $\mathscr{D}(A) = 0.75$.

Let $S_1 = \begin{array}{c} 0.3 \\ 0.3[0.225] \end{array}$, $S_2 = \begin{array}{c} 0.3 \\ 0.2[0.15] \end{array}$, $S_3 = \begin{array}{c} 0.2 \\ 0.3[0.15] \end{array}$, $S_4 = \begin{array}{c} 0.2 \\ 0.2[0.15] \end{array}$.

Therefore, $\mathscr{D}(S_1) = 0.75$, $\mathscr{D}(S_2) = 0.75$, $\mathscr{D}(S_3) = 0.75$, $\mathscr{D}(S_4) = 0.75$.

Thus, $\mathscr{D}(S_i) = \mathscr{D}(A) = 0.75$ for all i. Hence, A is a strictly balanced g-FM.

1.11 Balanced Fuzzy Graphs

The density of a crisp graph $G = (V, E)$ is defined by

$$\mathscr{D}(G) = \frac{|E|}{|V|(|V| - 1)/2}. \tag{1.7}$$

For a simple graph, the maximum number of edges of a graph $G = (V, E)$ is $|V|(|V| - 1)/2$ (for complete graph). Thus, the $\mathscr{D}(G)$ represents the ratio of the edges with respect to the complete graph. $\mathscr{D}(G)$ is non-negative for any graph G and its maximum value (for simple graph) is 1, when G is complete. Thus, for simple graph $0 \leq \mathscr{D}(G) \leq 1$. If $\mathscr{D}(G)$ is high the graph is dense, i.e. higher the value of $\mathscr{D}(G)$, more the number of edges. For null graph, $\mathscr{D}(G)$ is 0. For multi-graph $\mathscr{D}(G)$ may be greater than 1.

However, for a FG $\mathscr{G} = (\mathscr{V}, \sigma, \mu)$, the density is defined as follows:

$$\mathscr{D}(\mathscr{G}) = \frac{2 \sum\limits_{a,b \in \mathscr{V}} \mu(a, b)}{\sum\limits_{a,b \in \mathscr{V}} \sigma(a) \wedge \sigma(b)}. \tag{1.8}$$

The term $\sum\limits_{a,b \in \mathscr{V}} \mu(a, b)$ finds the sum of membership values of all edges, i.e. the sum is taken over all edges, while in the denominator ($\sum\limits_{a,b \in \mathscr{V}} \sigma(a) \wedge \sigma(b)$) the sum is taken over all pairs of vertices.

Like crisp graph, the lower bound of $\mathscr{D}(\mathscr{G})$ is 0 when $\mu(a, b) = 0$ for all edges (a, b). For the CFG, the density is maximum and it is 2. That is, for the simple FG $\mathscr{G}, 0 \leq \mathscr{D}(\mathscr{G}) \leq 2$.

Definition 1.57 A FG $\mathscr{G} = (\mathscr{V}, \sigma, \mu)$ is **balanced** if $\mathscr{D}(\mathscr{H}) \leq \mathscr{D}(\mathscr{G})$, for all subgraphs $\mathscr{H}$ of $\mathscr{G}$.

Example 1.23 Let $\mathscr{G} = (\mathscr{V}, \sigma, \mu)$ be a FG, $\mathscr{V} = \{v_1, v_2, v_3, v_4\}$ and set of edges be $\{(v_1, v_2), (v_2, v_3), (v_3, v_4), (v_1, v_4), (v_2, v_4), (v_1, v_3)\}$. Let $\sigma(v_1) = 0.4, \sigma(v_2) = 0.3, \sigma(v_3) = 0.3, \sigma(v_4) = 0.2$, and $\mu(v_1, v_2) = 0.24, \mu(v_2, v_3) = 0.24, \mu(v_3, v_4) = 0.16, \mu(v_1, v_4) = 0.16, \mu(v_2, v_4) = 0.16, \mu(v_1, v_3) = 0.24$.

Then, $\mathscr{D}(\mathscr{G}) = \frac{2(0.24+0.24+0.16+0.16+0.16+0.24)}{0.3+0.3+0.2+0.2+0.2+0.3} = 1.6$.

The density of an isolated vertex is 0, because it has no edge. The all possible subgraphs $\mathscr{H}_i = (\mathscr{V}_i, \sigma, \mu)$ of $\mathscr{G}$ are
$\mathscr{H}_1 = \{v_1\}, \mathscr{H}_2 = \{v_2\}, \mathscr{H}_3 = \{v_3\}, \mathscr{H}_4 = \{v_4\}, \quad \mathscr{H}_5 = \{v_1, v_2\}, \mathscr{H}_6 = \{v_1, v_3\},$
$\mathscr{H}_7 = \{v_1, v_4\}, \mathscr{H}_8 = \{v_2, v_3\}, \mathscr{H}_9 = \{v_2, v_4\}, \mathscr{H}_{10} = \{v_3, v_4\}, \mathscr{H}_{11} = \{v_1, v_2, v_3\},$
$\mathscr{H}_{12} = \{v_1, v_3, v_4\}, \mathscr{H}_{13} = \{v_1, v_2, v_4\}, \mathscr{H}_{14} = \{v_2, v_3, v_4\}, \mathscr{H}_{15} = \{v_1, v_2, v_3, v_4\}$
be the non-empty subgraphs of $\mathscr{G}$.

The densities of these subgraphs are
$\mathscr{D}(\mathscr{H}_1) = 0, \mathscr{D}(\mathscr{H}_2) = 0, \mathscr{D}(\mathscr{H}_3) = 0, \mathscr{D}(\mathscr{H}_4) = 0, \mathscr{D}(\mathscr{H}_5) = 1.6, \mathscr{D}(\mathscr{H}_6) = 1.6,$
$\mathscr{D}(\mathscr{H}_7) = 1.6, \mathscr{D}(\mathscr{H}_8) = 1.6, \mathscr{D}(\mathscr{H}_9) = 1.6, \mathscr{D}(\mathscr{H}_{10}) = 1.6, \mathscr{D}(\mathscr{H}_{11}) = 1.6,$
$\mathscr{D}(\mathscr{H}_{12}) = 1.6, \mathscr{D}(\mathscr{H}_{13}) = 1.6, \mathscr{D}(\mathscr{H}_{14}) = 1.6, \mathscr{D}(\mathscr{H}_{15}) = 1.6.$

Therefore, $\mathscr{D}(\mathscr{H}_i) \leq \mathscr{D}(\mathscr{G})$ for all subgraphs $\mathscr{H}_i$ of $\mathscr{G}$. Hence, $\mathscr{G}$ is a balanced FG.

Definition 1.58 A FG $\mathscr{G} = (\mathscr{V}, \sigma, \mu)$ is strictly balanced if for every non-null subgraph $\mathscr{H}$, $\mathscr{D}(\mathscr{H}) = \mathscr{D}(\mathscr{G})$.

Lemma 1.13 *Every CFG is balanced.*

Proof It is easy to verify that density of CFG is 2. Again, every subgraph of CFG is CFG. Hence, $\mathscr{D}(\mathscr{H}) = 2$ for every $\mathscr{H} \subseteq \mathscr{G}$. Thus, $\mathscr{G}$ is balanced. $\qquad\square$

But, the converse of above result need not be true, that is every balanced FG is not necessarily complete.

Theorem 1.41 *Let $\mathscr{G} = (\mathscr{V}, \sigma, \mu)$ be a FG and $\mathscr{G}^c = (\mathscr{V}, \sigma^c, \mu^c)$ be its complement. Then $\mathscr{D}(\mathscr{G}) + \mathscr{D}(\mathscr{G}^c) = 2$.*

Proof Let $\mathscr{G} = (\mathscr{V}, \sigma, \mu)$ be a FG with $\mathscr{V} = \{a_1, a_2, \ldots, a_n\}$ and let $G^* = (V, E)$ be the underlying graph. Since, $\mathscr{G}^c$ is the complement of $\mathscr{G}$, therefore, $\mu^c(a_i, a_j) = \sigma(a_i) \wedge \sigma(a_j) - \mu(a_i, a_j), \sigma^c(a_i) = \sigma(a_i), \sigma^c(a_j) = \sigma(a_j)$ for all $a_i, a_j \in \mathscr{V}$.

Now,

$$\mathscr{D}(\mathscr{G}^c) = \frac{\sum\limits_{a_i, a_j \in \mathscr{V}} 2\mu^c(a_i, a_j)}{\sum\limits_{a_i, a_j \in \mathscr{V}} \sigma^c(a_i) \wedge \sigma^c(a_j)}$$

$$= \frac{2 \sum\limits_{a_i, a_j \in \mathscr{V}} [\sigma(a_i) \wedge \sigma(a_j) - \mu(a_i, a_j)]}{\sum\limits_{a_i, a_j \in \mathscr{V}} \sigma(a_i) \wedge \sigma(a_j)}$$

$$= 2 - \frac{2 \sum\limits_{a_i, a_j \in \mathcal{V}} \mu(a_i, a_j)}{\sum\limits_{a_i, a_j \in \mathcal{V}} \sigma(a_i) \wedge \sigma(a_j)}$$

$$= 2 - \mathcal{D}(\mathcal{G}).$$

Hence, $\mathcal{D}(\mathcal{G}) + \mathcal{D}(\mathcal{G}^c) = 2$. □

Theorem 1.42 ([62]) *Let $\mathcal{G}_1$ and $\mathcal{G}_2$ be two isomorphic FGs. If $\mathcal{G}_1$ is balanced, then $\mathcal{G}_2$ is also balanced.*

Definition 1.59 Let $\mathcal{G} = (\mathcal{V}, \sigma, \mu)$ be a FG. Then σ is said to be c-constant function if $\sigma(a) = c$ for all $a \in \mathcal{V}$ and similarly μ is called a d-constant function if $\mu(a, b) = d$ for all edges (a, b) in $\mathcal{G}$.

Theorem 1.43 *Let $\mathcal{G} = (\mathcal{V}, \sigma, \mu)$ be a k-regular FG with n vertices. Then*

$$\mathcal{D}(\mathcal{G}) = \frac{nk}{\sum \sigma(a) \wedge \sigma(b)}, \quad a, b \in \mathcal{V}.$$

Proof Since $\mathcal{G}$ is k-regular, $deg_{\mathcal{G}}(a) = k$ for all $a \in \mathcal{V}$. Now, by handshaking theorem (Theorem 1.4) of FG,
$$2 \sum_{a,b \in \mathcal{V}} \mu(a, b) = \sum_{a \in \mathcal{V}} deg_{\mathcal{G}}(a) = kn.$$
Thus, $\mathcal{D}(\mathcal{G}) = \frac{nk}{\sum \sigma(a) \wedge \sigma(b)}$. □

Now, we define another type of density called ***-density** of a FG $\mathcal{G} = (\mathcal{V}, \sigma, \mu)$.

Definition 1.60 ([62]) Let $\mathcal{G} = (\mathcal{V}, \sigma, \mu)$ be a FG. The *-density of $\mathcal{G}$ is denoted by $\mathcal{D}^*(\mathcal{G})$ and is defined as

$$\mathcal{D}^*(\mathcal{G}) = \frac{2 \sum_{a,b \in \mathcal{V}} \mu(a, b)}{\sum_{a \in \mathcal{V}} \sigma(a)}.$$

Definition 1.61 ([62]) Let $\mathcal{G} = (\mathcal{V}, \sigma, \mu)$ be a FG. The FG $\mathcal{G}$ is said to be ***-balanced** if $\mathcal{D}^*(\mathcal{H}) \leq \mathcal{D}^*(\mathcal{G})$ for all non-empty subgraphs $\mathcal{H}$ of $\mathcal{G}$.

Lemma 1.14 *Let $\mathcal{G} = (\mathcal{V}, \sigma, \mu)$ be a k-regular FG and σ is c-constant function. Then $\mathcal{D}^*(\mathcal{G}) = \frac{k}{c}$.*

Lemma 1.15 *Let σ and μ be c- and d-constant functions. Then*
(i) $\mathcal{D}^*(\mathcal{K}_n) = n - 1$, $\mathcal{K}_n$ is CFG,
(ii) $\mathcal{D}^*(\mathcal{C}_n) \leq 2$, $\mathcal{C}_n$ is fuzzy cycle whose underlying graph is C_n.

Proof (i) Since $\mathcal{K}_n$ is CFG, $\mu(a, b) = \sigma(a) \wedge \sigma(b) = c$.

$$\mathscr{D}^*(\mathscr{K}_n) = 2\frac{\sum\limits_{a,b\in\mathscr{K}_n}\mu(a,b)}{\sum\limits_{a\in\mathscr{K}_n}\sigma(a)} = 2\frac{cn(n-1)/2}{cn} = n-1.$$

(ii) For $\mathscr{C}_n$, $\mu(a,b) \leq \sigma(a) \wedge \sigma(b)$, i.e. $d \leq c$.

$$\mathscr{D}^*(\mathscr{C}_n) = 2\frac{\sum\limits_{a,b\in\mathscr{C}_n}\mu(a,b)}{\sum\limits_{a\in\mathscr{C}_n}\sigma(a)} = 2\frac{dn}{cn} = 2\frac{d}{c} \leq 2.$$

$\square$

Theorem 1.44 *If σ is c-constant function, then CFG $\mathscr{K}_n$ is *-balanced.*

Proof From Lemma 1.15 $\mathscr{D}^*(\mathscr{K}_n) = n - 1$.

There are two types of subgraphs $(\mathscr{H} = (\mathscr{V}(\mathscr{H}), \sigma_{\mathscr{H}}, \mu_{\mathscr{H}}))$ of $\mathscr{K}_n$.
(i) subgraph with same set of vertices with less number of edges,
(ii) subgraph with less number of vertices and less number of edges.

For case (i), it is clear that $\mathscr{D}^*(\mathscr{H}) \leq \mathscr{D}^*(\mathscr{K}_n)$.

For case (ii), it is assumed that $\mathscr{H}$ has $n - p$, $p \geq 1$ number of vertices. Then the number of edges in $\mathscr{H}$ is

$$\frac{n(n-1)}{2} - [(n-1) + (n-2) + \cdots + (n-p)]$$
$$= \frac{n(n-1)}{2} - \left[pn - \frac{p(p+1)}{2}\right]$$
$$= \frac{n(n-1) - 2pn + p(p+1)}{2}.$$

Hence,

$$\mathscr{D}^*(\mathscr{H}) = \frac{2c[n(n-1) - 2pn + p(p+1)]/2}{c(n-p)} = n - (p+1).$$

Thus, $\mathscr{D}^*(\mathscr{H}) \leq \mathscr{D}^*(\mathscr{G})$ and hence $\mathscr{K}_n$ is *-balanced. $\square$

Theorem 1.45 ([62]) *Let σ is a c-constant function.*
(i) If $\mathscr{C}_n$ is strong for $n > 3$, then $\mathscr{C}_n$ *-balanced.
(ii) If $\mathscr{K}_{n,n}$ is strong, then $\mathscr{K}_{n,n}$ is *-balanced.
(ii) If Petersen FG $\mathscr{P}_{10}$ is strong, then $\mathscr{P}_{10}$ is *-balanced.

1.12 Cliques in Fuzzy Graphs

Cliques in graphs play a very important role as they are used extensively in many areas of graph theory. The clique of a graph $G = (V, E)$ is a set of vertices such that every vertex is adjacent to all other vertices, i.e. the subgraph induced by a clique is complete. Nair and Cheng for first time [98] defined clique for FG, but, as per their definition fuzzy subgraph induced by a fuzzy clique is not complete [141].

Also, their definition of fuzzy clique fails to draw any connection between fuzzy clique and complete fuzzy subgraphs, unlike a crisp clique. To overcome this, in 2016, sun et al. [141] modified the definition of fuzzy clique.

Definition 1.62 (Nair and Cheng [98]) A fuzzy subgraph $H = (\mathcal{V}', \sigma', \mu')$ of a FG $\mathcal{G} = (\mathcal{V}, \sigma, \mu)$ is a **fuzzy clique** if H^* (underlying crisp graph of H) is a clique and every cycle in H is a fuzzy cycle.

According to this definition, a fuzzy clique is a fuzzy subgraph with the following characterization.

Lemma 1.16 (Nair and Cheng [98]) A fuzzy subgraph $H = (\mathcal{V}', \sigma', \mu')$ of a FG $\mathcal{G} = (\mathcal{V}, \sigma, \mu)$ is a fuzzy clique if and only if every cycle of length 3 in H is a fuzzy cycle.

According to the Definition 1.62, the fuzzy clique is not necessarily complete. To make parity with crisp clique Sun et al. [141] modified the definition of fuzzy clique as follows.

The contribution of this part is mostly from [141].

Definition 1.63 ([141]) Let $\mathcal{G} = (\mathcal{V}, \sigma, \mu)$ be a FG. A subset σ' of σ is called a **fuzzy clique** if the fuzzy subgraph induced by σ' is complete.

To the best of our knowledge, these are the two types of fuzzy cliques that are available in literature. To avoid confusion, the clique defined in Definition 1.62 is referred to as f-**clique** and the that in Definition 1.63 as 'fuzzy clique'.

The relationship between fuzzy clique, complete fuzzy subgraph, and f-clique has been established in the following theorem.

Theorem 1.46 Let $H = (\mathcal{V}', \rho, \delta)$ be a fuzzy subgraph of $\mathcal{G} = (\mathcal{V}, \sigma, \mu)$. If H is complete then H is f-clique.

Proof Let $H = (\mathcal{V}', \rho, \delta)$ be a fuzzy subgraph of $\mathcal{G}$ such that H is complete. Since H is complete, therefore its underlying crisp graph is also complete. Let (v'_1, v'_2, v'_3) be a cycle of length 3 in H [v'_1, v'_2, v'_3 belong to the vertex set of H^*]. Without loss of generality, let $\rho(v'_1) \leq \rho(v'_2) \leq \rho(v'_3)$ such that $\delta(v'_1, v'_2) = \rho(v'_1) = \delta(v'_1, v'_3)$ and $\delta(v'_3, v'_2) = \rho(v'_2)$. This shows that (v'_1, v'_2, v'_3) is a cycle in H as there does not exist any weakest edge in (v'_1, v'_2, v'_3).

As the cycle (v'_1, v'_2, v'_3) was chosen arbitrarily, it proves that H is an f-clique. □

Corollary 1.1 *If H is a fuzzy subgraph induced by a fuzzy clique in $\mathscr{G}$, then H is an f-clique.*

The converse of the Theorem 1.46 or Corollary 1.1 is not true in general.

Theorem 1.47 *If $H = (\mathscr{V}', \rho, \delta)$ is an f-clique in a FG $\mathscr{G} = (\mathscr{V}, \sigma, \mu)$ such that H is strong, then H is complete.*

Proof Let $H = (\mathscr{V}', \rho, \delta)$ be an f-clique in a FG $\mathscr{G} = (\mathscr{V}, \sigma, \mu)$ such that H is strong. Since H is an f-clique, its underlying crisp graph H^* is a clique that is, it is complete in crisp sense.

Again, as H is strong we have, $\delta(v_i', v_j') = \rho(v_i') \wedge \rho(v_j')$, for all $i, j = 1, 2, \ldots, n$ where $v_i', v_j' \in H^*$. This proves that H is complete. □

The minimum cardinality of a fuzzy clique cover of $\mathscr{G}$ is called the fuzzy clique cover number of $\mathscr{G}$, denoted by $cc(\mathscr{G})$. A minimum fuzzy clique cover is a fuzzy clique cover FC such that $|FC| = cc(\mathscr{G})$.

The following results are proved in [141].

Theorem 1.48 *(i) Each complete fuzzy subgraph is an f-clique.*
(ii) The fuzzy subgraph induced by a fuzzy clique is an f-clique.
(iii) Each strong f-clique is complete.
(iv) A strong f-clique is a fuzzy clique.

Theorem 1.49 *([141]) Let $\mathscr{G} = (\mathscr{V}, \sigma, \mu)$ be a FG. A subset σ' of σ is a fuzzy clique if and only if $\sigma'(a) \wedge \sigma'(b) \leq \mu(a, b)$ for all $a \neq b$ and $a, b \in \mathscr{V}$.*

Proof Let σ' be a fuzzy clique of the FG $\mathscr{G} = (\mathscr{V}, \sigma, \mu)$. Also, let $H = (\mathscr{V}', \sigma', \mu')$ be the fuzzy subgraph induced by the fuzzy clique σ'. By definition of fuzzy clique, H is complete. Now, by the definition of fuzzy subgraph and CFG, $\sigma'(a) \wedge \sigma'(b) = \mu'(a, b) = \sigma'(a) \wedge \sigma'(b) \wedge \mu(a, b) \leq \mu(a, b)$ for all $a, b \ (\neq a) \in \mathscr{V}'$.

Conversely, let σ' be a subset of σ and $H = (\mathscr{V}', \sigma', \mu')$ be the fuzzy subgraph induced by σ'. Thus, $\sigma'(a) \wedge \sigma'(b) \leq \mu(a, b)$ for all $a, b \in \mathscr{V}' \ (a \neq b)$.

By definition of fuzzy subgraph $\mu'(a, b) = \sigma'(a) \wedge \sigma'(b) \wedge \mu(a, b)$ for all $a, b \in \mathscr{V}'$. Now, by the assumption $\sigma'(a) \wedge \sigma'(b) \leq \mu(a, b)$, so $\mu'(a, b) = \sigma'(a) \wedge \sigma'(b)$. Thus, H is complete and hence σ' is a fuzzy clique. □

The following results follow from this theorem.

Corollary 1.2 *[141] Every non-empty subset of a fuzzy clique of a FG $\mathscr{G}$ is a fuzzy clique in $\mathscr{G}$.*

Corollary 1.3 *[141] Any α-cut of a fuzzy clique of a FG $\mathscr{G}$ is a clique in $\mathscr{G}_\alpha$, where $\mathscr{G}_\alpha$ is the α-cut of the FG $\mathscr{G}$.*

Definition 1.64 Let $\mathscr{G} = (\mathscr{V}, \sigma, \mu)$ be a FG and ρ' be a fuzzy clique. ρ' is called maximal if there is no fuzzy clique ρ such that $\rho' < \rho$. A maximal fuzzy clique ρ' is maximum if it contains the maximum number of vertices.

Theorem 1.50 *Let ρ' be a maximal fuzzy clique in a FG $\mathscr{G} = (\mathscr{V}, \sigma, \mu)$. Then there is at least one vertex a in $\mathscr{V}$ such that $\rho'(a) = \sigma(a)$.*

Like crisp graph, one can define clique cover for a FG. A **clique cover** is a set of cliques of a graph $G = (V, E)$ such that all the edges are covered by the clique cover. The definition of fuzzy clique cover is given below.

Definition 1.65 Let $\mathscr{C}$ be a collection of fuzzy cliques of a FG $\mathscr{G}$. $\mathscr{C}$ is called **fuzzy clique cover** of $\mathscr{G}$ if $\mathscr{G}$ can be decomposed as the union of all the fuzzy subgraphs induced by the fuzzy cliques in $\mathscr{C}$.

The minimum cardinality of a fuzzy clique cover of $\mathscr{G}$ is called the fuzzy clique cover number.

A minimum fuzzy clique cover is a fuzzy clique cover $\mathscr{C}$ whose cardinality is minimum among all other clique covers.

Example 1.24 Let $\mathscr{G} = (\mathscr{V}, \sigma, \mu)$ be a FG where $\mathscr{V} = \{v_1, v_2, \ldots, v_5\}$ such that $\sigma(v_1) = 0.8, \sigma(v_2) = 0.6, \sigma(v_3) = 0.3, \sigma(v_4) = 0.2, \sigma(v_5) = 0.5$ and $\mu(v_1, v_2) = 0.4, \mu(v_1, v_3) = 0.3, \mu(v_1, v_4) = 0.2, \mu(v_3, v_2) = 0.3, \mu(v_4, v_2) = 0.2, \mu(v_5, v_2) = 0.5, \mu(v_3, v_4) = 0.2, \mu(v_3, v_5) = 0.3, \mu(v_5, v_4) = 0.2$.

In this FG $\mathscr{G}$, the fuzzy subgraph induced by (v_2, v_3, v_4, v_5) is a fuzzy clique, hence an f-clique and is also a maximal fuzzy clique.

On the other hand, (v_1, v_2, v_3, v_4) is a maximal clique in crisp sense as well as a maximal f-clique. But, it is not a fuzzy clique because the fuzzy subgraph so induced is not complete as $\mu(v_1, v_2) = 0.4 \neq \sigma(v_1) \wedge \sigma(v_2)$.

Consider two fuzzy subgraphs H' and H'' of $\mathscr{G}$ such that $H' = \{\sigma'(v_1) = 0.8, \sigma'(v_2) = 0.4, \sigma'(v_3) = 0.3, \sigma'(v_4) = 0.2, \mu'(v_1, v_2) = 0.4, \mu'(v_1, v_3) = 0.3, \mu'(v_1, v_4) = 0.2, \mu'(v_3, v_2) = 0.3, \mu'(v_4, v_2) = 0.2, \mu'(v_3, v_4) = 0.2\}$ and $H'' = \{\sigma''(v_2) = 0.6, \sigma''(v_3) = 0.3, \sigma''(v_4) = 0.2, \sigma''(v_5) = 0.5, \mu''(v_3, v_2) = 0.3, \mu''(v_2, v_4) = 0.2, \mu''(v_2, v_5) = 0.5, \mu''(v_3, v_4) = 0.2, \mu''(v_3, v_5) = 0.2, \mu''(v_5, v_4) = 0.2\}$.

Clearly, H' and H'' are two fuzzy cliques in $\mathscr{G}$ such that their union is the FG $\mathscr{G}$. Therefore, $\{H', H''\}$ is a fuzzy clique cover and it is also a maximum fuzzy clique cover of $\mathscr{G}$ such that $|FC| = cc(\mathscr{G}) = 2$.

1.13 Independent Sets in Fuzzy Graphs

In crisp graph $G = (V, E)$, a set of vertices S $(\subseteq V)$ is said to be an **independent set** if there is no edge between every pair of vertices of S. The definition of an independent set for a FG is given below.

Definition 1.66 (*Fuzzy independent set*) Let $\mathscr{G} = (\mathscr{V}, \sigma, \mu)$ be a FG. Two vertices in a FG $\mathscr{G}$ are said to be **fuzzy independent** if there is no strong edge between them. A subset S of $\mathscr{V}$ is said to be **fuzzy independent set** for $\mathscr{G}$ if every two vertices of S are fuzzy independent.

Definition 1.67 (*Maximal fuzzy independent set*) Let $\mathcal{G} = (\mathcal{V}, \sigma, \mu)$ be a FG. A fuzzy independent set S of $\mathcal{G}$ is said to be **maximal fuzzy independent set** if there is no fuzzy independent set whose cardinality is larger than the cardinality of S. The maximum cardinality among all maximal fuzzy independent sets is called **fuzzy independence number** of $\mathcal{G}$.

A new concept in fuzzy independent set for FG is introduced below. In the new concept, the degree of independence (DOI) of a FG is defined. If the DOI of a graph is 1 then the FG is fully independent, i.e. the underlying crisp graph is independent and if it is 0, then the underlying graph is complete.

The independence strength of an edge (a, b) in $\mathcal{G} = (\mathcal{V}, \sigma, \mu)$ is denoted by $I_{\mathcal{G}}(a, b)$ or simply, $I(a, b)$ (where there is no ambiguity) and is defined by

$$I_{\mathcal{G}}(a, b) = \frac{\mu(a, b)}{\min\{\sigma(a), \sigma(b)\}}.$$

In crisp sense, an independent set is a set of vertices so that every pair of vertices are not connected by an edge. But, in fuzzy sense, any subset of vertices is an independent set with some degree of independence.

Thus, every set of vertices is an independent set with some degree of independence and hence every FG has some degree of independence. Let $S \subseteq \mathcal{V}$ be the set of p number of vertices of a FG $\mathcal{G} = (\mathcal{V}, \sigma, \mu)$. Then the degree of independence $DOI_p(S)$ [113] of S is defined as

$$DOI_p(S) = 1 - \frac{\displaystyle\sum_{a,b \in S} I_S(a, b)}{\mathcal{N}(p)},$$

where $\mathcal{N}(p) = \frac{p(p-1)}{2}$ is the total number of possible positive edges among p vertices.

Note 1.4 In a null FG $\mathcal{G}$ with p vertices, $\displaystyle\sum_{a,b \in S} I_S(a, b) = 0$ for every pair of vertices. Then $DOI_p(\mathcal{G}) = 1 - 0 = 1$. This leads to the notion of crisp independent set. If $DOI_p(\mathcal{G}) = 1$, then the set is nothing but an independent set in crisp sense.

Note 1.5 For a CFG $\mathcal{G} = (\mathcal{V}, \sigma, \mu)$ with n vertices, $\displaystyle\sum_{a,b \in \mathcal{V}} I_{\mathcal{G}}(a, b) = \mathcal{N}(n)$. Thus, $DOI_n(\mathcal{G}) = 0$.

Example 1.25 Let us consider two FGs $\mathcal{G}_1$ and $\mathcal{G}_2$ having four vertices each, shown in Fig. 1.15.

For these graphs, $\displaystyle\sum_{a,b \in \mathcal{V}} I_{\mathcal{G}_1}(a, b) = \frac{0.5}{0.6} + \frac{0.1}{0.7} + \frac{0.3}{0.5} = 0.83 + 0.14 + 0.6 = 1.57$

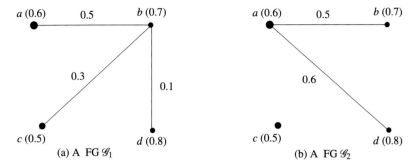

Fig. 1.15   The relation between the degree of independence and the number of vertices

and $\displaystyle\sum_{a,b\in\mathcal{V}} I_{\mathcal{G}_2}(a,b) = \frac{0.5}{0.6} + \frac{0.6}{0.6} = 0.83 + 1 = 1.83.$

Thus,   for   Fig. 1.15a, $DOI_4(\mathcal{G}_1) = 1 - \frac{1.57}{\mathcal{N}(4)} = 1 - \frac{1.57}{6} \approx 0.74$ and   for
Fig. 1.15b, $DOI_4(\mathcal{G}_2) = 1 - \frac{1.83}{\mathcal{N}(4)} = 1 - \frac{1.83}{6} \approx 0.69.$

Note 1.6   Although the FG in Fig. 1.15a has more number of edges than the one in
Fig. 1.15b, the vertices in Fig. 1.15a are more independent in nature than those in
Fig. 1.15b which is clearly evident as the edges in Fig. 1.15a are less strong than in
Fig. 1.15b.

This inspection leads to propose the next important theorem given followed by a
definition.

Definition 1.68   In a FG, an edge is said to be an independent strong edge if $I(a, b) \geq$
0.5.

Theorem 1.51   *Let $\mathcal{G} = (\mathcal{V}, \sigma, \mu)$ be a FG. If there is an independent strong edge
between every two vertices of $\mathcal{G}$, then $DOI_n(\mathcal{G}) \leq 0.5$.*

Proof   Given G is a FG with n vertices and between every pair of vertices there is
an independent strong edge, i.e. $\forall a, b \in \mathcal{V}$, $I(a, b) \geq 0.5$, that is,

$$\sum_{a,b\in\mathcal{V}} I(a, b) \geq \mathcal{N}(n) \times 0.5 \quad \text{or} \quad 1 - \frac{\displaystyle\sum_{a,b\in\mathcal{V}} I(a, b)}{\mathcal{N}(n)} \leq 1 - 0.5 = 0.5.$$

Hence, $DOI_n(\mathcal{G}) \leq 0.5$.   □

The following result follows from above theorem.

Corollary 1.4   *If there is no independent strong edge in a FG $\mathcal{G} = (\mathcal{V}, \sigma, \mu)$ with
n vertices, then $DOI_n(\mathcal{G}) > 0.5$.*

Theorem 1.52 *Let $H = (V_H, \sigma_H, \mu_H)$ be a fuzzy subgraph of a FG $\mathscr{G} = (\mathscr{V}, \sigma, \mu)$ induced by $V_H (\subseteq \mathscr{V})$,*

    Then

$$DOI_m(H) = DOI_n(\mathscr{G}) - \frac{\mathscr{N}(n) - \mathscr{N}(m)}{\mathscr{N}(m) \cdot \mathscr{N}(n)} \cdot \mathscr{I} + \frac{\Delta\mathscr{I}}{\mathscr{N}(m)}$$

where, $m = |V_H|$ and $n = |\mathscr{V}|$, $\mathscr{I} = \sum_{a,b \in \mathscr{V}} I(a,b)$ and $\Delta\mathscr{I} = \sum_{a,b \in \mathscr{V}} I(a,b) - \sum_{a,b \in V_H} I(a,b)$.

Proof By definition,

$$DOI_n(\mathscr{G}) = 1 - \frac{\sum_{a,b \in \mathscr{V}} I(a,b)}{\mathscr{N}(n)} \tag{1.9}$$

$$\text{and } DOI_m(H) = 1 - \frac{\sum_{a,b \in V_H} I(a,b)}{\mathscr{N}(m)} \tag{1.10}$$

In terms of $\mathscr{I}$, $\Delta\mathscr{I}$, the relations (1.9) and (1.10) become

$$DOI_n(\mathscr{G}) = 1 - \frac{\mathscr{I}}{\mathscr{N}(n)} \tag{1.11}$$

$$\text{and } DOI_m(H) = 1 - \frac{\mathscr{I} - \Delta\mathscr{I}}{\mathscr{N}(m)} \tag{1.12}$$

Now,

$$DOI_m(H) - DOI_n(\mathscr{G}) = 1 - \frac{\mathscr{I} - \Delta\mathscr{I}}{\mathscr{N}(m)} - 1 + \frac{\mathscr{I}}{\mathscr{N}(n)}$$

$$= \frac{\mathscr{I}}{\mathscr{N}(n)} - \frac{\mathscr{I}}{\mathscr{N}(m)} + \frac{\Delta\mathscr{I}}{\mathscr{N}(m)}$$

Hence,

$$DOI_m(H) = DOI_n(\mathscr{G}) - \frac{\mathscr{N}(n) - \mathscr{N}(m)}{\mathscr{N}(m) \cdot \mathscr{N}(n)} \cdot \mathscr{I} + \frac{\Delta\mathscr{I}}{\mathscr{N}(m)}.$$

$\square$

    This result is very useful for a FG with a huge number of vertices as one needs to just work on a fuzzy subgraph without inspecting the what FG itself. The case is shown for a smaller FG in Example 1.26.

Example 1.26 Consider a FG which is shown in Fig. 1.15a. For this FG $\mathcal{G}_1$, the degree of independence $DOI_4(\mathcal{G}_1)$ is 0.74. Now the degree of independence of the fuzzy subgraph H induced by the fuzzy subset $\{a(0.6), b(0.7)\}$ of the fuzzy vertex set $\mathcal{V}$ of $\mathcal{G}_1$ is calculated by the formula obtained in Theorem 1.52 as follows.

$$DOI_2(H) = DOI_4(G_1) - \frac{\mathcal{N}(4) - \mathcal{N}(2)}{\mathcal{N}(4) \cdot \mathcal{N}(2)} \cdot 1.57 + \frac{0.83}{\mathcal{N}(2)}$$

$$= 0.74 - \frac{6-1}{6} \cdot 1.57 + \frac{0.83}{1}$$

$$\simeq 0.26$$

Here, we have seen that the degree of independence decreases for a subset of a vertex set. But, this is not in general true. The case is shown in the following corollaries.

Corollary 1.5 Let $\mathcal{G} = (\mathcal{V}, \sigma, \mu)$ be a FG with n vertices. Suppose a vertex is deleted from $\mathcal{G}$ and the induced subgraph be $\mathcal{G}'$. Then

$$DOI_{n-1}(\mathcal{G}') = DOI_n(\mathcal{G}) - \frac{4}{n(n-1)(n-2)} \cdot \mathcal{I} + \frac{24\mathcal{I}}{(n-1)(n-2)}.$$

Proof Let the vertex x be deleted from $\mathcal{G}$. So, the number of vertices of $\mathcal{G}'$ is $n-1$. Then, by using Theorem 1.52,

$$DOI_{n-1}(\mathcal{G}') = DOI_n(\mathcal{G}) - \frac{\mathcal{N}(n) - \mathcal{N}(n-1)}{\mathcal{N}(n-1) \cdot \mathcal{N}(n)} \cdot \mathcal{I} + \frac{\Delta\mathcal{I}}{\mathcal{N}(n-1)}$$

$$= DOI_n(\mathcal{G}) - \frac{4}{n(n-1)(n-2)} \cdot \mathcal{I} + \frac{24\mathcal{I}}{(n-1)(n-2)}.$$

$\square$

The following theorem is left to the reader.

Theorem 1.53 Let H be a fuzzy subgraph of the FG $\mathcal{G}$ and the number of vertices of H and $\mathcal{G}$ are m, n respectively. Then, $DOI_m(H) \leq DOI_n(\mathcal{G})$ if and only if $\Delta\mathcal{I} \leq \frac{\mathcal{N}(n) - \mathcal{N}(m)}{\mathcal{N}(n)} \cdot \mathcal{I}$.

Example 1.27 Let us consider the FG $\mathcal{G}$ shown in Fig. 1.15a. After usual computation $DOI_4(\mathcal{G}) = 0.74$ and $\mathcal{I} = 0.9$.

Now, we want to determine a maximal fuzzy subgraph of this FG so that $DOI_m(H) \geq 0.75$, $[m \leq 4]$.

Now, $DOI_3(H) > DOI_4(\mathcal{G})$ if $\Delta\mathcal{I} > \frac{2}{4}\mathcal{I} = \frac{0.9}{2} = 0.45$.

This shows that deletion of the vertex c or d does not affect its degree of independence. So, if one of a or b is deleted then the DOI may increase.

If b is deleted, $DOI_3(H) = 1$. If a is deleted, then by Corollary 1.5

$$DOI_3(H) = 0.74 - \frac{4}{4 \times 3 \times 2} \times 0.9 + \frac{0.5 \times 2}{3 \times 2}$$

$$= 0.74 + \frac{1}{60} \approx 0.75.$$

Therefore, the maximal fuzzy subgraph with $DOI > 0.75$ is $\{a, c, d\}$.

1.14 Domination in Fuzzy Graphs

Chess is a famous game played all over the world. From the game of chess, the idea of domination in graph theory came into play in approximately 1850. The problem began determining the minimum number of queens in a condition that all the squares are either involved or attacked by a queen. In 1962, Ore [102] and Berge [18] represented the idea of domination in graph theory and it was further discussed by Cockayne and Hedetniemi [39, 40]. Somasundaram [138, 139] explained domination in FGs with the help of effective edges. Nagoorgani et al. [48] explored domination using strong arcs in graph. Various types of domination in FG have been discussed elaborately in [80, 81, 85].

Herein, the concept of domination in FG has been introduced and demonstrated with examples. In this section, various terms related to domination have been discussed. Also, minimal domination, maximal domination, and total domination have been studied with appropriate examples.

Definition 1.69 A vertex a dominates other vertex b in a FG $\mathscr{G} = (\mathscr{V}, \sigma, \mu)$ if $\mu(a, b) = \min\{\sigma(a), \sigma(b)\}$ for $a, b \in \mathscr{V}$. A set $\mathbb{D} \subseteq \mathscr{V}$ is said to be a **dominating set** in $\mathscr{G}$ if for every $b \in \mathscr{V} - \mathbb{D}$, there exists $a \in \mathbb{D}$ such that a dominates b.

A dominating set with minimum cardinality (i.e. sum of membership values is minimum) is called the **domination number** of $\mathscr{G}$ and is denoted by $\gamma(\mathscr{G})$.

Definition 1.70 A dominating set $\mathbb{D}$ of a FG $\mathscr{G}$ is called minimal if there is no proper subset of $\mathbb{D}$ which is dominating.

Example 1.28 Let $\mathscr{G} = (\mathscr{V}, \sigma, \mu)$ be a FG with set of vertices $\mathscr{V} = \{a, b, c, d\}$, shown in Fig. 1.16. Also, let the membership values of vertices and edges be $\sigma(a) = 0.6$, $\sigma(b) = 0.4$, $\sigma(c) = 0.5$, $\sigma(d) = 0.3$, $\mu(a, b) = 0.4$, $\mu(b, c) = 0.4$, $\mu(b, d) = 0.3$, $\mu(c, a) = 0.4$, $\mu(a, d) = 0.1$.

For this graph, the dominating sets are $\{b\}$, $\{a, b\}$, $\{b, c\}$, $\{a, c, d\}$, etc. By calculation, $\{b\}$ is a minimal dominating set. Therefore, $\gamma(\mathscr{G}) = 0.4$.

Normally, a vertex a of a FG $\mathscr{G} = (\mathscr{V}, \sigma, \mu)$ is said to be an isolated vertex if $\mu(a, b) = 0$ for all $b \in \mathscr{V}$. But, in the sense of fuzzy dominating set, the definition of isolated vertex is different, which is given below:

Definition 1.71 A vertex a of a FG is said to be an **isolated vertex** if $\mu(ab) < \min\{\sigma(a), \sigma(b)\}$ for all $b \in \mathscr{V} \setminus \{a\}$.

Fig. 1.16 Illustration of
Ex. 1.28

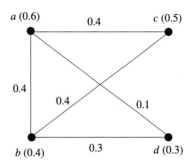

Fig. 1.17 Illustration of
Example 1.29

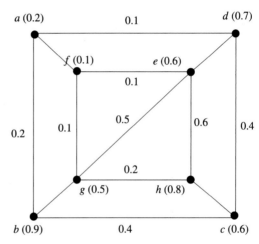

Remark 1.5 1. As per definition, no isolated vertex dominates any other vertex.
2. According to the definition, if a dominates b, b dominates a. Hence, domination is a symmetric relation on $\mathcal{V}$.
3. $\mu(ab) < \min\{\sigma(a), \sigma(b)\}$ for all $a, b \in \mathcal{V}$, then $\mathcal{V}$ is the only dominating set of $\mathcal{G}$.

Example 1.29 Let us consider a FG, $\mathcal{G} = (\mathcal{V}, \sigma, \mu)$ of Fig. 1.17, where $\mathcal{V} = \{a, b, c, d, e, f, g, h\}$ and $\sigma(a) = 0.2$, $\sigma(b) = 0.9$, $\sigma(c) = 0.6$, $\sigma(d) = 0.9, \sigma(e) = 0.6$, $\sigma(f) = 0.1, \sigma(g) = 0.5$, $\sigma(h) = 0.8$, $\mu(a, b) = 0.2$, $\mu(a, f) = 0.1$, $\mu(a, d) = 0.1$, $\mu(b, c) = 0.4$, $\mu(b, g) = 0.2$, $\mu(c, d) = 0.4$, $\mu(c, h) = 0.5$, $\mu(d, e) = 0.3, \mu(e, f) = 0.1$, $\mu(e, g) = 0.5 \, \mu(f, g) = 0.1$, $\mu(g, h) = 0.2$, $\mu(h, e) = 0.1$.

In this FG $\mathcal{G}$, the vertex 'd' is not dominated by any other vertex in $\mathcal{V} \smallsetminus \{d\}$. Hence, '$d$' is an isolated vertex in $\mathcal{G}$.

Definition 1.72 A set $\mathbb{D}$ in a FG $\mathscr{G} = (\mathscr{V}, \sigma, \mu)$ without isolated vertices is a total dominating fuzzy set if for every vertex $b \in \mathscr{V}$, there exists a vertex $a \in \mathbb{D}$ so that a dominates b.

Definition 1.73 The minimum (maximum) cardinality of a total dominating fuzzy set is said to be lower (upper) total domination number of $\mathscr{G}$ and is denoted by $T_l(\mathscr{G})$ $(T_u(\mathscr{G}))$.

Example 1.30 From Example 1.28, it is clear that there is no isolated vertex in the FG shown in Fig. 1.16. Hence, $\{a, b\}$, $\{b, c\}$, etc. are total dominating sets.

$\mathbb{D} = \{b\}$ is not a total dominating set as there does not exist any vertex in $\mathbb{D}$ which dominates $b \in \mathscr{V}$.

Again, $\mathbb{D} = \{a, c, d\}$ is not a total dominating as there does not exist any vertex in $\mathbb{D}$ which dominates a or c or d in $\mathscr{V}$.

Theorem 1.54 *Let $\mathscr{G} = (\mathscr{V}, \sigma, \mu)$ be a FG without isolated vertices. Then $\mathscr{V} \setminus \mathbb{D}$ is a dominating set, if $\mathbb{D}$ is a minimal dominating set.*

Proof Since $\mathscr{G} = (\mathscr{V}, \sigma, \mu)$ is a FG without isolated nodes so each vertex in $\mathscr{G} = (\mathscr{V}, \sigma, \mu)$ is dominated by at least one other vertex in $\mathscr{G}$. Let $\mathscr{G}$ has a minimal dominating set $\mathbb{D}$, then every vertex in $\mathscr{V} \setminus \mathbb{D}$ is dominated by any vertex in $\mathbb{D}$. That means conversely every vertex in $\mathbb{D}$ is dominated by any vertex in $\mathscr{V} \setminus \mathbb{D}$. Hence, $\mathscr{V} \setminus \mathbb{D}$ is also a dominating set. $\square$

Theorem 1.55 *Let $\mathscr{G} = (\mathscr{V}, \sigma, \mu)$ be a FG. Then $T_l(\mathscr{G}) + \overline{T_l(\mathscr{G})} \leq 2O(\mathscr{G})$, and the equality holds if and only if $0 < \mu(a, b) < \min\{\sigma(a), \sigma(b)\}$ for all $a, b \in \mathscr{V}$.*

Proof $T_l(\mathscr{G}) = O(\mathscr{G})$ if and only if $\mu(a, b) = \min\{\sigma(a), \sigma(b)\}$ for all $a, b \in \mathscr{V}$. Also, since $\overline{T_l(\mathscr{G})} = O(\mathscr{G})$ if and only if $\overline{\mu(a, b)} = \min\{\sigma(a), \sigma(b)\} - \mu(a, b)$ for all $a, b \in \mathscr{V}$. Therefore, $T_l(\mathscr{G}) + \overline{T_l(\mathscr{G})} = 2O(\mathscr{G})$ if and only if $0 < \mu(a, b) < \min\{\sigma(a), \sigma(b)\}$ for all $a, b \in \mathscr{V}$. $\square$

Definition 1.74 A set $I_\mathbb{D} \subseteq \mathscr{V}$ is said to be an **independent set** of a FG $\mathscr{G} = (\mathscr{V}, \sigma, \mu)$ if $\mu(a, b) < \min\{\sigma(a), \sigma(b)\}$, for all $a, b \in I_\mathbb{D}$.

That is, the collection of all isolated vertices forms an independent set.

Definition 1.75 An independent set $I_\mathbb{D}$ of a FG $\mathscr{G} = (\mathscr{V}, \sigma, \mu)$ is said to be a maximal independent set of $\mathscr{G}$ if for every vertex $a \in \mathscr{V} - I_\mathbb{D}$, the set $I_\mathbb{D} \cup \{a\}$ is not independent.

Example 1.31 From Example 1.29, it is clear that there are several independent sets in Fig. 1.17. If we consider the independent set $I_\mathbb{D} = \{a, c, d, e\}$ then it is maximal independent set in $\mathscr{G} = (\mathscr{V}, \sigma, \mu)$ since for every vertex in $\mathscr{V} - I_\mathbb{D} = \{b, f, g, h\}$ the set is not independent.

Definition 1.76 The lower and upper independence numbers of a FG $\mathcal{G} = (\mathcal{V}, \sigma, \mu)$ denoted by $i_{\mathbb{D}}(\mathcal{G})$ and $I_{\mathbb{D}}(\mathcal{G})$ respectively and defined as the minimum cardinality and maximum cardinality among all maximal independent sets of $\mathcal{G} = (\mathcal{V}, \sigma, \mu)$ respectively.

Definition 1.77 Let $\mathcal{G} = (\mathcal{V}, \sigma, \mu)$ be a FG. For any two vertices $a, b \in \mathcal{V}$, a strongly dominates b in $\mathcal{G}$ if $\mu(a, b) = \min\{\sigma(a), \sigma(b)\}$ and $deg_{\mathcal{G}}(a) \geq deg_{\mathcal{G}}(b)$, where $deg_{\mathcal{G}}(a) = \sum_{b \in \mathcal{V}} \mu(a, b)$ for $a \neq b$. Similarly, a weakly dominates b in $\mathcal{G}$ if $\mu(a, b) = \min\{\sigma(a), \sigma(b)\}$ and $deg_{\mathcal{G}}(a) \leq deg_{\mathcal{G}}(b)$.

Example 1.32 Let us consider Example 1.28. Clearly, vertex 'b' strongly dominates the other vertices in $\mathcal{V}$ since $\mu(a, b) = \min\{\sigma(a), \sigma(b)\}$ and $deg_{\mathcal{G}}(b) \geq deg_{\mathcal{G}}(a)$ and $\mu(b, c) = \min\{\sigma(b), \sigma(c)\}$ and $deg_{\mathcal{G}}(b) \geq deg_{\mathcal{G}}(c)$ and $\mu(b, d) = \min\{\sigma(b), \sigma(d)\}$ and $deg_{\mathcal{G}}(b) \geq deg_{\mathcal{G}}(d)$. Similarly, vertex '$d$' weakly dominates the other vertices in $\mathcal{V}$.

Definition 1.78 A set $\mathbb{D} \subseteq \mathcal{V}$ is called a **strong dominating set** of $\mathcal{G}$ if for every vertex $b \notin \mathbb{D}$, there exists at least a vertex $a \in \mathbb{D}$ such that a strongly dominates b.

Definition 1.79 A set $\mathbb{D} \subseteq \mathcal{V}$ is called a weak dominating set of a FG $\mathcal{G}$ if for every vertex $b \notin \mathbb{D}$, there is a vertex $a \in \mathbb{D}$ such that a weakly dominates b.

Example 1.33 Let us consider Example 1.28. If we consider the set $\{d\}$ in $\mathcal{V}$ as a dominating set then it will be weakly dominating set since vertex 'd' weakly dominates all the other vertices in $\mathcal{V}$.

Definition 1.80 The **strong domination number** of a FG $\mathcal{G} = (\mathcal{V}, \sigma, \mu)$ is the minimum cardinality of a strong dominating set of $\mathcal{G}$ and is denoted by $\gamma_s(\mathcal{G})$. Similarly, the **weak domination number** of $\mathcal{G}$ is the minimum cardinality of a weak dominating set of $\mathcal{G}$ and is denoted by $\gamma_w(\mathcal{G})$.

Example 1.34 Let us consider Example 1.28. Clearly, the set $\{b\}$ is a strong dominating fuzzy set in $\mathcal{G} = (\mathcal{V}, \sigma, \mu)$ and the set $\{d\}$ is a weakly dominating set. Therefore, $\gamma_s(\mathcal{G}) = 0.4$ and $\gamma_w(\mathcal{G}) = 0.3$

Definition 1.81 A dominating set $\mathbb{D}$ of a FG is called a **strong minimal** if there is no proper subset of $\mathbb{D}$ which is a strong dominating set.

Example 1.35 Let us consider Example 1.28. Clearly, by routine calculation, vertex 'b' strongly dominates the other vertices in $\mathcal{V}$. Therefore, if we consider the set $\{b\}$ as dominating set then it will be clearly a strong dominating fuzzy set in $\mathcal{G}$.

Theorem 1.56 *In any CFG $\mathcal{G}$, $\gamma_w(\mathcal{G}) \leq \gamma_s(\mathcal{G})$.*

Proof Let $\mathcal{G} = (\mathcal{V}, \sigma, \mu)$ is a CFG .

Case(i)   Suppose, $\forall x \in \mathcal{V}, \sigma(x)$ is equal to any other vertex in $\mathcal{G}$. Hence, $\mu(x, y) = \sigma(x) \wedge \sigma(y) = \sigma(x)$ since $\mathcal{G}$ is complete.

Thus, $\gamma_w(\mathcal{G}) = \gamma_s(\mathcal{G}) = \sigma(x)$.

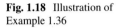

Fig. 1.18 Illustration of
Example 1.36

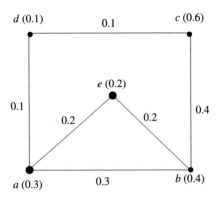

Case(ii)   Suppose, $\forall x, y \in \mathcal{V}, \sigma(x) \neq \sigma(y)$. For a CFG , any vertex dominates all
other remaining vertices. If the degree of this vertex is least among the remaining
then, the dominating set formed with that vertex will be a weak dominating set.
Thus, the minimum cardinality of that set that is the weak domination number
$\gamma_w(\mathcal{G}) = \wedge\{\sigma(x), \sigma(y)\}$. Obviously, the strong dominating set has a vertex other
than the least value of the vertex set. Hence, strong domination number will be
strictly greater than weak domination number. That means $\gamma_w(\mathcal{G}) < \gamma_s(\mathcal{G})$.

Therefore, from Case(i) and Case(ii), $\gamma_w(\mathcal{G}) \leq \gamma_s(\mathcal{G})$.   □

Definition 1.82 A subset $\mathbb{D} \subseteq \mathcal{V}$ is called a **secure dominating set** in $\mathcal{G} =$
$(\mathcal{V}, \sigma, \mu)$, if for each vertex x in $\mathcal{V} - \mathbb{D}$ is adjacent to a vertex $y \in \mathbb{D}$ such that
$(\mathbb{D} - \{y\}) \cup \{x\}$ is a dominating set.
    The parameter secure domination number is denoted by $\gamma_{sc}(\mathcal{G})$ and it is the min-
imum cardinality of all secure dominating sets of $\mathcal{G}$.

Example 1.36 Let us consider a FG $\mathcal{G} = (\mathcal{V}, \sigma, \mu)$ of Fig. 1.18. For this graph, the
vertex set is $\mathcal{V} = \{a, b, c, d, e\}$. Let the membership values of vertices and edges be
$\sigma(a) = 0.3$, $\sigma(b) = 0.4$, $\sigma(c) = 0.6$, $\sigma(d) = 0.1$, $\sigma(e) = 0.2$, $\mu(a, b) = 0.3$,
$\mu(b, c) = 0.4$, $\mu(c, d) = 0.1$, $\mu(d, a) = 0.1$, $\mu(a, e) = 0.2$, $\mu(b, e) = 0.2$.
    The secure dominating sets for this graph are $\{a, e, c\}, \{a, b, e\}, \{a, e, d\}$, etc.

Definition 1.83 ([97]) A subset $\mathbb{D} \subseteq \mathcal{V}$ is said to be a **2-dominating set** in $\mathcal{G} =$
$(\mathcal{V}, \sigma, \mu)$, if every vertex of $\mathcal{V} - \mathbb{D}$ has at least two neighbors in $\mathbb{D}$. The 2-dominating
number of a FG $\mathcal{G}$ is denoted by $\gamma_2(\mathcal{G})$ and it is the minimum cardinality over all
2-dominating sets of $\mathcal{G}$.

Example 1.37 From Example 1.36, Fig. 1.18, if we consider the dominating set
$\{a, b, d\}$ then every vertex from $\{c, e\}$ has at least two neighbor in $\{a, b, d\}$. Hence,
$\{a, b, d\}$ is a 2-dominating set.

Definition 1.84 ([71]) If $\mathbb{D}$ is 2-dominating set and the subgraph induced by $\mathbb{D}$ has
no isolated vertices, then $\mathbb{D} \subseteq \mathcal{V}$ is called a **2-total dominating set** in $\mathcal{G}$. The 2-total

domination number of $\mathscr{G}$ is denoted by $\gamma_{2t}(\mathscr{G})$ and it is the minimum cardinality over all 2-total dominating sets of $\mathscr{G}$.

Example 1.38 From Example 1.36, Fig. 1.18, if we consider the dominating set $\{a, b, d\}$ then every vertex from $\{c, e\}$ has at least two neighbors in $\{a, b, d\}$. Hence, $\{a, b, d\}$ is a 2-dominating set and also $\mathscr{G}$ has no isolated vertices. Therefore, the set $\{a, b, d\}$ is a 2-total dominating set in $\mathscr{G}$.

Definition 1.85 ([71]) A 2-dominating set $\mathbb{D}$ of $\mathscr{G} = (\mathscr{V}, \sigma, \mu)$ is a secure 2-dominating set, if for every vertex $x \in \mathscr{V} - \mathbb{D}$ is adjacent to a vertex $y \in \mathbb{D}$ such that $(\mathbb{D} - \{y\}) \cup \{x\}$ is 2-dominating set. Minimum fuzzy cardinality of all 2-secure dominating sets of $\mathscr{G}$ is 2-secure domination number of $\mathscr{G}$ and is denoted by $\gamma_{2sc}(\mathscr{G})$.

Example 1.39 Let us consider Example 1.36 and $\{a, c, e\}$ be one of the dominating sets of this graph. For this graph, every vertex from $\{b, d\}$ has at least two neighbors in $\{a, c, e\}$. Hence, $\{a, c, e\}$ is a 2-dominating set. Again, if we consider for the vertex 'b' from the set $\{b, d\}$, then 'b' is adjacent to the vertex 'a'. Therefore, $\{a, c, e\}$ is a secure 2-dominating set.

Definition 1.86 ([71]) Suppose $\mathscr{G} = (\mathscr{V}, \sigma, \mu)$ be a FG without isolated vertices. A 2- secure dominating set $\mathbb{D}$ of $\mathscr{G}$ is called a 2-secure total dominating set if the subgraph $< \mathbb{D} >$ induced by $\mathbb{D}$ has no isolated vertices.

The 2-secure total domination number of $\mathscr{G}$ is denoted by $\gamma_{2set}(\mathscr{G})$, which is the minimum among the cardinalities of all 2-secure total dominating sets of $\mathscr{G}$.

Theorem 1.57 *If $\mathbb{D}$ is a minimal dominating set in a CFG $\mathscr{G} = (\mathscr{V}, \sigma, \mu)$, then*

1. *$\mathbb{D}$ is a secure dominating set.*
2. *$\mathbb{D}$ is not a secure total dominating set.*

Proof Since $\mathbb{D}$ is a minimal dominating set of a CFG $\mathscr{G} = (\mathscr{V}, \sigma, \mu)$, all edges of $\mathscr{G}$ are strong edges. Then minimal dominating set $\mathbb{D}$ contains only one vertex x (say), i.e. $\mathbb{D} = \{x\}$. Now for any vertex $y \in \mathscr{V} - \mathbb{D}$ and y is adjacent to x. Hence, $(\mathbb{D} - \{x\}) \cup \{y\} = \{y\}$ is a dominating set. Therefore, $\mathbb{D}$ is secure dominating set. Since any secure dominating set of a CFG contains a vertex y, by definition of a total dominating set, $\mathbb{D}$ is not secure total dominating set. □

Theorem 1.58 *If $\mathbb{D}$ is a minimal dominating set in a CFG $\mathscr{G} = (\mathscr{V}, \sigma, \mu)$, then*

1. *$\mathbb{D}$ is not a 2-dominating set.*
2. *$\mathbb{D}$ is not a 2-total dominating set.*

Proof If $\mathbb{D}$ is a minimal dominating set in a CFG $\mathscr{G} = (\mathscr{V}, \sigma, \mu)$, then $\mathbb{D}$ contains a vertex of minimum cardinality but we know 2-dominating set to contain at least two vertices. Therefore, $\mathbb{D}$ is not a 2-dominating set. Similarly, we can say, $\mathbb{D}$ is not a 2-total dominating set. □

Theorem 1.59 *For a CFG, $\gamma_{sc}(\mathscr{G}) = \gamma(\mathscr{G})$.*

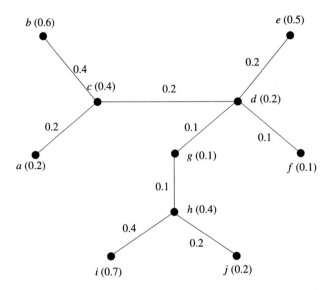

Fig. 1.19 Illustration of Example 1.40

Proof If $\mathbb{D}$ is a minimal dominating set in a CFG $\mathscr{G} = (\mathscr{V}, \sigma, \mu)$, then $\mathbb{D}$ contains a vertex $\{x\}$ i.e. $\mathbb{D} = \{x\}$. The minimum cardinality of $\mathbb{D}$ is $\gamma(\mathscr{G})$. Again since $\mathbb{D}$ is a secure dominating set, minimum cardinality is denoted by $\gamma_{sc}(\mathscr{G})$. Thus, $\gamma_{sc}(\mathscr{G}) = \gamma(\mathscr{G})$. $\square$

Definition 1.87 ([128]) A dominating set $\mathbb{D}$ of a FG $\mathscr{G} = (\mathscr{V}, \sigma, \mu)$ is a perfect dominating set if for each vertex $x \notin \mathbb{D}$, x is adjacent to exactly one vertex $y \in \mathbb{D}$. Minimum fuzzy cardinality of all perfect dominating sets of $\mathscr{G}$ is perfect domination number of $\mathscr{G}$ and is denoted by $\gamma_{pf}(\mathscr{G})$.

Example 1.40 Let us consider a FG $\mathscr{G} = (\mathscr{V}, \sigma, \mu)$ depicted in Fig. 1.19. For this graph, the set of vertices is $\mathscr{V} = \{a, b, c, d, e\}$. The membership values of the vertices are $\sigma(a) = 0.2$, $\sigma(b) = 0.6$, $\sigma(c) = 0.4$, $\sigma(d) = 0.2$, $\sigma(e) = 0.5$, $\sigma(f) = 0.1$, $\sigma(g) = 0.1$, $\sigma(h) = 0.4$, $\sigma(i) = 0.7$, $\sigma(j) = 0.2$. and that of edges are $\mu(a, c) = 0.2$, $\mu(b, c) = 0.4$, $\mu(c, d) = 0.2$, $\mu(d, e) = 0.2$, $\mu(d, f) = 0.1$, $\mu(d, g) = 0.1$, $\mu(g, h) = 0.1$, $\mu(h, i) = 0.4$, $\mu(h, j) = 0.2$.

For this graph, one perfect dominating set is $\{c, d, h\}$.

Note 1.7 Theories have been developed by researchers working in this field following Definition 1.69. But, strangely this definition does not justify properly the word "domination". More precisely, in which condition can we say that given two vertices $a, b \in \mathscr{V}$, a dominates b while b does not dominate a?

Till date, no research has been carried on keeping these facts in mind. Interested readers may consider it as an important area of research.

1.15 Eigenvalues and Energy of Fuzzy Graphs

Given a (crisp) graph $G = (V, E)$, its energy has many mathematical properties. A concept that is related to the spectrum of a graph is its energy. To find out the energy of a graph we need the adjacency matrix $A = [a_{ij}]$ of the graph whose elements a_{ij} represent the number of edges between vertices v_i and v_j which become 0 ($i = j$) and 1 ($i \neq j$ and there exists an edge) when the graph is multi-graph. The same idea has been extended to FGs . Many works are available on eigenvalues and energy of FGs , see [9, 101, 108].

Definition 1.88 ([89]) Let $\mathscr{G} = (\mathscr{V}, \sigma, \mu)$ be a FG. The **adjacency matrix** A of $\mathscr{G}$ is an $n \times n$ matrix defined as

$$A = [a_{ij}] \text{ where } a_{ij} = \begin{cases} \mu(v_i, v_j); & i \neq j \\ 0; & i = j \end{cases}$$

Definition 1.89 (*Eigenvalues of a fuzzy graph* [89]) The eigenvalues of the adjacency matrix A of a FG $\mathscr{G}$ are the eigenvalues of $\mathscr{G}$.

Definition 1.90 (*Energy of a fuzzy graph* [89]) Given a FG $\mathscr{G}$, its energy is defined as the sum of absolute values of eigen values of $\mathscr{G}$.

Example 1.41 Consider a FG $\mathscr{G} = (\mathscr{V}, \sigma, \mu)$, where $\mathscr{V} = \{v_1, v_2, v_3\}$ with $\sigma : \mathscr{V} \to [0, 1]$ such that $\sigma(v_1) = 0.7$, $\sigma(v_2) = 0.5$, $\sigma(v_3) = 0.6$ and $\mu : \mathscr{V} \times \mathscr{V} \to [0, 1]$ such that $\mu(v_1, v_2) = 0.3$, $\sigma(v_1, v_3) = 0.4$ and $\sigma(v_2, v_3) = 0.5$.

The adjacency matrix A for $\mathscr{G}$ is

$$A = \begin{bmatrix} 0 & 0.3 & 0.4 \\ 0.3 & 0 & 0.5 \\ 0.4 & 0.5 & 0 \end{bmatrix}$$

The characteristic polynomial of A is $\lambda^3 - 0.50\lambda - 0.12 = 0$. For this graph $\mathscr{G}$, the spectrum is $\{0.80558, -0.51803, -0.28755\}$ and hence the energy $(E(\mathscr{G}))$ of $\mathscr{G}$ is $E(\mathscr{G}) = 0.80558 + 0.51803 + 0.28755 = 1.61116$

Note 1.8 The energy of a FG is not in general greater than 1, unlike a nontrivial simple crisp graph [89].

Theorem 1.60 *Let $\mathscr{G} = (\mathscr{V}, \sigma, \mu)$ be a FG whose underlying crisp graph be $G^* = (V, E)$. Also, let the number of vertices of $\mathscr{G}$ be n and it's set of edges be $E = \{e_1, e_2, \cdots, e_p\}$. Then,*

$$E(\mathscr{G}) \leq 2 \sum_{i=1}^{p} \mu(e_i).$$

Proposition 1.1 *Let $\mathscr{G} = (\mathscr{V}, \sigma, \mu)$ be a FG with underlying crisp graph $G^* = (V, E)$. If $|V| = n$ and $E = \{e_1, e_2, \cdots, e_p\}$, then*

$$E(\mathcal{G}) \leq (n-1) \sum_{i=1}^{p} \sigma(v_i).$$

Proof Using the result of Theorem 1.60,

$$E(\mathcal{G}) \leq 2 \sum_{i=1}^{p} \mu(e_i) = 2 \sum_{i=1}^{\frac{n(n-1)}{2}} \mu(e_i)$$

where $\mu(e_i) = 0$ for all $i > p$.

For all v_i, $v_j \in \mathcal{V}$, $\mu(e_i) \leq \min\{\sigma(v_i), \sigma(v_j)\}$.

Therefore,

$$E(\mathcal{G}) \leq 2 \sum_{i=1}^{\frac{n(n-1)}{2}} \mu(e_i) = \sum_{i=1}^{\frac{n(n-1)}{2}} [\mu(e_i) + \mu(e_i)]$$

$$\leq \sum_{1 \leq i \leq j \leq n} [\sigma(v_i) + \sigma(v_j)] = (n-1) \sum_{i=1}^{n} \sigma(v_i).$$

$\square$

1.15.1 Connected Energy of Fuzzy Graphs

Definition 1.91 (*Connectedness strength matrix of fuzzy graphs* [72]) Given a FG $\mathcal{G} = (\mathcal{V}, \sigma, \mu)$, its connectedness strength matrix, $C = [c_{ij}]$ where,

$$c_{ij} = \begin{cases} CONNG(v_i, v_j) & ; i \neq j \\ 0 & ; i = j \end{cases}$$

Example 1.42 Consider the same FG as given in Example 1.41. Its connectedness strength matrix is given by

$$C = \begin{bmatrix} 0 & 0.4 & 0.4 \\ 0.4 & 0 & 0.5 \\ 0.4 & 0.5 & 0 \end{bmatrix}$$

The characteristic polynomial of C is $\lambda^3 - 0.57\lambda - 0.16 = 0$, giving connectedness eigen values of C (also of $\mathcal{G}$) as $\lambda_1 = 0.86847$, $\lambda_2 = -0.5$, $\lambda_3 = -0.36847$.

Therefore, connectedness strength energy for $\mathcal{G}$ is $0.86847 + 0.5 + 0.36847 = 1.73694$

Theorem 1.61 *Let $\mathcal{G} = (\mathcal{V}, \sigma, \mu)$ be a FG. Then, the connectedness energy $CE(\mathcal{G})$ is greater or equal to its energy $E(\mathcal{G})$.*

Proof Given a FG $\mathcal{G} = (\mathcal{V}, \sigma, \mu)$, its energy $E(\mathcal{G})$ and connectedness energy $CE(\mathcal{G})$ are respectively defined as sum of the absolute values of eigenvalues of the adjacency matrix A and connectedness strength matrix C of $\mathcal{G}$.

Now,

$$A = [a_{ij}] = \begin{cases} \mu(v_i, v_j) & ; i \neq j \\ 0 & ; i = j \end{cases} \tag{1.13}$$

$$C = [c_{ij}] = \begin{cases} CONNG(v_i, v_j) & ; i \neq j \\ 0 & ; i = j \end{cases} \tag{1.14}$$

As $CONNG(v_i, v_j) \geq \mu(v_i, v_j)$, from Eqs. (1.13) and (1.14), we can conclude that sum of absolute values of eigen values of C is greater than that of A.

Thus, $CE(\mathcal{G}) > E(\mathcal{G})$.

Equality holds when the FG $\mathcal{G}$ is connected and matrices C and A are equal. $\square$

Note 1.9 While dealing with fuzzy sets and fuzzy relations, all values are expected to lie within $[0, 1]$. As evident from above the eigenvalues as well as connectedness eigenvalues of a FG may fail to lie within $[0, 1]$ which is not quite acceptable for a FG or equivalently a fuzzy matrix. A possible reason for getting negative eigenvalues in the above examples is that normal arithmetic operations have been used to find out the eigenvalues as well as the energy of a FG instead of fuzzy arithmetics.

Not much work has been done in this area considering fuzzy arithmetics. Interested readers may take up this aspect as an interesting area of research.

1.16 Topological Indices on Fuzzy Graphs

Topological indices are molecular descriptors widely used in areas like mathematical chemistry, molecular topology, and chemical graph theory. Such numerical parameters are representatives of molecular compounds that characterize the topology of the corresponding molecular graph. Molecular properties are correlated with the chemical structure of the compound. In a molecular graph, vertices and edges are representatives of atoms and their bonding respectively. The first investigation into Wiener index was made by Harold Wiener in 1947 [148], during a study on the boiling point of paraffin. It was chemists who used Wiener index decades before it captivated the attention of mathematicians. Innovative results connected to the Wiener index were reported during the middle of the 1970s and this gradually reaped great esteem. Zagreb index is one such topological indices which is a degree-based topological index and was introduced by Gutman et al. in [58]. This topological index is used to calculate π-electron energy of a conjugate system [59]. One can see for more topological indices like Harmonic index, connectivity index, Randic index, Estrada index, Padmakar-Ivan index, Szeged index, etc. in [14, 36, 45, 46, 63, 64, 73, 117]. Recently, Binu et al. studied connectivity index and Wiener index of FGs in [24, 25] and gave an application on human trafficking and illegal immigration networks. In [70], Kalathian et al. also studied some topological indices on FG.

In this section, some topological indices are defined on FGs and some important results are presented.

Definition 1.92 ([89]) Let $\mathscr{G} = (\mathscr{V}, \sigma, \mu)$ be a FG and $a, b \in \mathscr{V}$. The geodesic (GDS) between a and b is the shortest strong path from a to b. Sum of membership values of all edges in the GDS is called weight of the GDS.

The first Zagreb index of a graph (crisp) is defined as follows:

Definition 1.93 ([44]) Let $G = (V, E)$ be a (crisp) graph. Then the first Zagreb index of the graph G is denoted by $M_1(G)$ and is defined by

$$M_1(G) = \sum_{a \in V} deg^2(a).$$

In 2019, Kalathian et al. [70] defined Zagreb index for a FG as follows:

Definition 1.94 ([70]) Let $\mathscr{G} = (\mathscr{V}, \sigma, \mu)$ be a FG. Then the first Zagreb index of the FG $\mathscr{G}$ is denoted by $ZF_1(\mathscr{G})$ and is defined by

$$ZF_1(\mathscr{G}) = \sum_{a \in \mathscr{V}} \sigma(a) deg^2(a).$$

The second Zagreb index of a (crisp) graph is defined as follows:

Definition 1.95 ([44]) The second Zagreb index of the graph $G = (V, E)$ is denoted by $M_2(G)$ and is defined by

$$M_2(G) = \sum_{(a,b) \in E} deg(a) deg(b).$$

In 2019, Kalathian et al. [70] defined second Zagreb index for a FG as follows:

Definition 1.96 ([70]) Let $\mathscr{G} = (\mathscr{V}, \sigma, \mu)$ be a FG. The second Zagreb index of the FG $\mathscr{G}$ is denoted by $ZF_2(\mathscr{G})$ and is defined as follows:

$$ZF_2(\mathscr{G}) = \sum_{(a,b) \in E} \sigma(a) \sigma(a) deg(a) deg(b).$$

In 2019, Binu et al. [24] defined connectivity index for a FG as follows:

Definition 1.97 ([24]) Let $\mathscr{G} = (\mathscr{V}, \sigma, \mu)$ be a FG. The connectivity index of $\mathscr{G}$ is denoted by $CIF(\mathscr{G})$ and defined by

$$CIF(\mathscr{G}) = \sum_{a,b \in \mathscr{V}} \sigma(a) \sigma(b) CONN_{\mathscr{G}}(a, b).$$

Fig. 1.20 A FG with (i)
$ZF_1(\mathcal{G}) = 2.848$, (ii)
$ZF_2(\mathcal{G}) = 1.8252$, (iii)
$CIF(\mathcal{G}) = 1.601$, and (iv)
$WIF(\mathcal{G}) = 3.054$

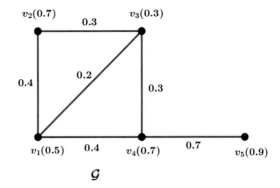

In 1947, Wiener [148] defined Wiener index for a crisp graph as follows:

Definition 1.98 ([148]) Let $G = (V, E)$ be a (crisp) graph. The **Wiener index** of the graph G is denoted by $WI(G)$ and is defined by

$$WI(G) = \sum_{a,b \in V} d(a, b),$$

where $d(a, b)$ denotes distance between a and b in G.

In 2020, Binu et al. [25] defined Wiener index for a FG as follows:

Definition 1.99 ([25]) Let $\mathcal{G} = (\mathcal{V}, \sigma, \mu)$ be a FG. The **Wiener index** of $\mathcal{G}$ is denoted by $WIF(\mathcal{G})$ and defined by

$$WIF(\mathcal{G}) = \sum_{a,b \in \mathcal{V}} \sigma(a)\sigma(b)dist_S(a, b),$$

where $dist_S(a, b)$ is minimum sum of weights of GDS between a and b and is called strong distance.

Example 1.43 Let $\mathcal{G} = (\mathcal{V}, \sigma, \mu)$ be a FG shown in Fig. 1.20 with vertex set $\mathcal{V} = \{v_1, v_2, \cdots, v_5\}$ and $\sigma(v_1) = 0.5, \sigma(v_2) = 0.7, \sigma(v_3) = 0.3, \sigma(v_4) = 0.7, \sigma(v_5) = 0.9, \mu(v_1, v_2) = 0.4, \mu(v_1, v_3) = 0.2, \mu(v_1, v_4) = 0.4, \mu(v_2, v_3) = 0.3, \mu(v_3, v_4) = 0.3, \mu(v_4, v_5) = 0.7$.

Then $deg(v_1) = 1.0, deg(v_2) = 0.7, deg(v_3) = 0.8, deg(v_4) = 1.4, deg(v_5) = 0.7$. Therefore,

$$ZF_1(\mathcal{G}) = \sum_{a \in V} \sigma(a)deg^2(a)$$

$$= 0.5 \times 1.0^2 + 0.7 \times 0.7^2 + 0.3 \times 0.8^2 + 0.7 \times 1.4^2 + 0.9 \times 0.7^2$$

$$= 2.848.$$

$$ZF_2(\mathcal{G}) = \sum_{(a,b) \in E} \sigma(a)\sigma(b)deg(a)deg(b) = 1.8252.$$

Now connectedness between pairs of vertices are $CONN(v_1, v_2) = 0.4$, $CONN(v_1, v_3) = 0.3$, $CONN(v_1, v_4) = 0.4$, $CONN(v_1, v_5) = 0.4$, $CONN(v_2, v_3) = 0.3$, $CONN(v_2, v_4) = 0.4$, $CONN(v_2, v_5) = 0.4$, $CONN(v_3, v_4) = 0.3$, $CONN(v_3, v_5) = 0.3$, $CONN(v_4, v_5) = 0.7$. Therefore,

$$CIF(\mathcal{G}) = \sum_{a,b \in V} \sigma(a)\sigma(b)CONN(a, b) = 1.601.$$

Now strong distance between pairs of vertices are $dist_S(v_1, v_2) = 0.4$, $dist_S(v_1, v_3) = 0.7$, $dist_S(v_1, v_4) = 0.4$, $dist_S(v_1, v_5) = 1.1$, $dist_S(v_2, v_3) = 0.3$, $dist_S(v_2, v_4) = 0.8$, $dist_S(v_2, v_5) = 1.5$, $dist_S(v_3, v_4) = 0.3$, $dist_S(v_3, v_5) = 1.0$, $dist_S(v_4, v_5) = 0.7$. Therefore,

$$WIF(\mathcal{G}) = \sum_{a,b \in V} \sigma(a)\sigma(b)dist_S(a, b) = 3.054.$$

Theorem 1.62 Let $\mathcal{H} = (\mathcal{V}', \pi, \omega)$ be a partial fuzzy subgraph of the FG $\mathcal{G} = (\mathcal{V}, \sigma, \mu)$, then (i) $ZF_1(\mathcal{H}) \leq ZF_1(\mathcal{G})$, (ii) $ZF_2(\mathcal{H}) \leq ZF_2(\mathcal{G})$, (iii) $CIF(\mathcal{H}) \leq CIF(\mathcal{G})$.

Proof As $\mathcal{H}$ is a PFSG of $\mathcal{G}$ then for any $a, b \in \mathcal{V}'$, $\sigma'(a) \leq \sigma(a)$ and $\mu'(a, b) \leq \mu(a, b)$. So,

$$deg_{\mathcal{H}}(a) = \sum_{a \in \mathcal{V}'} \mu'(a, b) \leq \sum_{b \in \mathcal{V}'} \mu(a, b) \leq \sum_{b \in \mathcal{V}} \mu(a, b) = deg_{\mathcal{G}}(a).$$

Therefore,

$$ZF_1(\mathcal{H}) = \sum_{a \in \mathcal{V}'} \sigma'(a)deg_{\mathcal{H}}^2(a) \leq \sum_{a \in V} \sigma(a)deg_{\mathcal{G}}^2(a) = ZF_1(\mathcal{G}).$$

Hence, $ZF_1(\mathcal{H}) \leq ZF_1(\mathcal{G})$. Similarly, one can prove other inequalities. □

Theorem 1.63 (Theorem 4.2 of [24] and Theorem 3.4 of [25]) Let $\mathcal{G} = (V, \sigma, \mu)$ be a CFG with $\mathcal{G} = (V, \sigma, \mu)$ and $V = \{a_1, a_2, \cdots, a_n\}$ such that $p_1 \leq p_2 \leq \cdots \leq p_n$, where $p_i = \sigma(a_i)$ for $1 \leq i \leq n$. Then

$$(i)\ CIF(\mathcal{G}) = \sum_{i=1}^{n-1} p_i^2 \sum_{j=i+1}^{n} p_j, \qquad (ii)\ WIF(\mathcal{G}) = \sum_{i=1}^{n-1} p_i^2 \sum_{j=i+1}^{n} p_j.$$

Theorem 1.64 *Let $\mathcal{G}_1 = (\mathcal{V}_1, \sigma_1, \mu_1)$ and $\mathcal{G}_2 = (\mathcal{V}, \sigma_2, \mu_2)$ be isomorphic FGs . Then (i) $ZF_1(\mathcal{G}_1) = ZF_1(\mathcal{G}_2)$, (ii) $ZF_2(\mathcal{G}_1) = ZF_2(\mathcal{G}_2)$, (iii) $CIF(\mathcal{G}_1) = CIF(\mathcal{G}_2)$, (iv) $WIF(\mathcal{G}_1) = WIF(\mathcal{G}_2)$.*

Proof As $\mathcal{G}_1$ and $\mathcal{G}_2$ are two isomorphic FGs , there exist an isomorphism ϕ between $\mathcal{G}_1$ and $\mathcal{G}_2$, i.e. $\phi : \mathcal{V}_1 \to \mathcal{V}_2$ is a bijection such that for all $a, b \in \mathcal{V}_1$, $\sigma_1(b) = \sigma_2(\phi(b))$ and $\mu_1(a, b) = \mu_2(\phi(a), \phi(b))$. Then

$$deg_{\mathcal{G}_1}(b) = \sum_{a \in \mathcal{V}_1} \mu_1(a, b) = \sum_{a \in \mathcal{V}_1} \mu_2(\phi(a), \phi(b)) = \sum_{\phi(a) \in \mathcal{V}_2} \mu_2(\phi(a), \phi(b)) = deg_{\mathcal{G}_2}(\phi(b)).$$

Therefore,
$$ZF_1(\mathcal{G}_1) = \sum_{b \in \mathcal{V}_1} \sigma_1(b) deg_{\mathcal{G}_1}^2(b) = \sum_{b \in \mathcal{V}_1} \sigma_2(\phi(b)) deg_{\mathcal{G}_2}^2(\phi(b))$$
$$= \sum_{\phi(b) \in \mathcal{V}_2} \sigma_2(\phi(b)) deg_{\mathcal{G}_2}^2(\phi(b)) = ZF_1(\mathcal{G}_2).$$

Similarly, one can prove for other indices. □

Theorem 1.65 *Let $\mathcal{G} = (\mathcal{V}, \sigma, \mu)$ be a FG and its number of vertices be n. Let $\mathcal{G}_1 = (\mathcal{V}, \sigma_1, \mu_1)$ be CFG spanned by $\mathcal{V}$. Then (i) $ZF_1(\mathcal{G}) \leq ZF_1(\mathcal{G}_1)$, (ii) $ZF_2(\mathcal{G}) \leq ZF_2(\mathcal{G}_1)$, (iii) $CIF(\mathcal{G}) \leq CIF(\mathcal{G}_1)$, (iv) $WIF(\mathcal{G}) \leq WIF(\mathcal{G}_1)$.*

Proof It is easy to show that $\mathcal{G}$ is a partial fuzzy subgraph of $\mathcal{G}_1$. Then the required inequality follows by Theorem 1.11. □

Theorem 1.66 *Let $\mathcal{G} = (\mathcal{V}, \sigma, \mu)$ be a connected FG with n vertices and $\sigma(a) = 1$, $\forall\, a \in \mathcal{V}$, $\mu(e) = c$, $\forall$ edge e in $\mathcal{G}$, $0 < c \leq 1$. Then $CIF(\mathcal{G}) = \frac{n(n-1)}{2} c$.*

Proof Since $\mu(e) = c$, $\forall\, e \in E$, $0 < c \leq 1$, then each edge is strong. Therefore, $CONF_G(u, v) = c$, $\forall\, a, b \in \mathcal{V}$. Then connectivity index of $\mathcal{G}$ is

$$CIF(\mathcal{G}) = \sum_{a,b \in \mathcal{V}} \sigma(a)\sigma(b) CONF_{\mathcal{G}}(a, b) = \sum_{a,b \in \mathcal{V}} c = \frac{n(n-1)}{2} c.$$

□

Theorem 1.67 *Let S be the maximal spanning tree of a fuzzy tree $\mathcal{G} = (\mathcal{V}, \sigma, \mu)$. Then $WIF(\mathcal{G}) = WIF(S)$.*

Proof Given $\mathcal{G} = (\mathcal{V}, \sigma, \mu)$ be a fuzzy tree and S be the maximal spanning tree of $\mathcal{G}$. Then those edges not in S are δ-edges. Also δ-edges do not belong to any geodesic in G. Hence, the result. □

Theorem 1.68 *Let* $\mathscr{G} = (\mathscr{V}, \sigma, \mu)$ *be an n-vertex saturated fuzzy cycle. Let strength of the each* α*-st edge be* κ *and each* β*-st edge be* η*, then*

$$WI(\mathscr{G}) = \begin{cases} \frac{n^3}{16}(\kappa + \eta) & \text{if } n \equiv 0 \ (mod \ 4) \\ \frac{n}{2}\eta + \frac{n(n^2-4)}{16}(\kappa + \eta) & \text{if } n \equiv 2 \ (mod \ 4). \end{cases}$$

Proof Let $\mathscr{G} = (\sigma, \mu)$ be a saturated fuzzy cycle. It is clear that alternate edges of $\mathscr{G}$ have membership values κ and η and all edges of the fuzzy cycle $\mathscr{G}$ are strong. We know that n is an even number by Theorem 2.3 of [24]. So, geodesic in $\mathscr{G}$ can be $1, 2, \cdots, \frac{n}{2}$.

For $1 \leq l' \leq \frac{n}{2}$, define $P_{l'} = \{(a', b') \in \theta^* \times \theta^* | \text{ length of the geodesic between } a' \text{ and } b' \text{ is } l'\}$. Then the cardinality of $P_{\frac{n}{2}}$ is $\frac{n}{2}$. Now for any $(a', b') \in P_{\frac{n}{2}}$,

$$d_s(a', b') = \begin{cases} \frac{n}{4}(\kappa + \eta) & \text{if } n \equiv 0 \ (mod \ 4) \\ \eta + \frac{n-2}{4}(\kappa + \eta) & \text{if } n \equiv 2 \ (mod \ 4). \end{cases}$$

For any vertex $a' \in \theta^*$, there are exactly two vertices at distance l' from a'. Number of such pairs of vertices is $2n$. Avoiding repetition, we get n number of distinct pairs $(u', v') \in \theta^* \times \theta^*$ where length of the geodesic between u' and v' is l'.

For $1 \leq l' < \frac{n}{2}$, l' even and let $(a', b') \in P_{l'}$ then $d_s(a', b') = \frac{l'}{2}(\kappa + \eta)$.

Let $1 \leq l' < \frac{n}{2}$, l' odd and $a' \in \theta^*$. Let $b', c' \in \theta^*$ such that distance from a' to b' and c' is l'. Then one of the geodesics of length l' from a' contains $\frac{l'+1}{2}$ α-strong, $\frac{l'-1}{2}$ β-strong edges and other geodesics contain $\frac{l'-1}{2}$ α-strong, $\frac{l'+1}{2}$ β-strong edges.

Then for $n \equiv 0 \ (mod \ 4)$:

$$WI(\mathscr{G}) = \sum_{(a',b') \in P_{\frac{n}{2}}} [d_s(a', b')] + \sum_{l'=2,4,\cdots}^{\frac{n}{2}-2} \left(\sum_{(a',b') \in P_{l'}} [d_s(a', b')] \right) + \sum_{l'=1,3,\cdots}^{\frac{n}{2}-1} \left(\sum_{(a',b') \in P_{l'}} [d_s(a', b')] \right)$$

$$= \frac{n}{2}[\frac{n}{4}(\kappa + \eta)] + \sum_{l'=2,4,\cdots}^{\frac{n}{2}-2} n[\frac{l'}{2}(\kappa + \eta)] + \sum_{l'=1,3,\cdots}^{\frac{n}{2}-1} \frac{n}{2}[(\frac{l'+1}{2}\eta + \frac{l'-1}{2}\kappa) + (\frac{l'-1}{2}\eta + \frac{l'+1}{2}\kappa)]$$

$$= \frac{n}{2}[\frac{n}{4}(\kappa + \eta)] + \sum_{l'=2,4,\cdots}^{\frac{n}{2}-2} n[\frac{l'}{2}(\kappa + \eta)] + \sum_{l'=1,3,\cdots}^{\frac{n}{2}-1} \frac{n}{2}[l'(\kappa + \eta)]$$

$$= \frac{n^2}{8}(\kappa + \eta) + \frac{n^2(n - 4)}{32}(\kappa + \eta) + \frac{n^3}{32}(\kappa + \eta)$$

$$= \frac{n^3}{16}(\kappa + \eta).$$

For $n \equiv 2 \ (mod \ 4)$:

$$WI(G) = \sum_{(a',b') \in P_{\frac{n}{2}}} [d_s(a', b')] + \sum_{l'=2,4,\cdots}^{\frac{n}{2}-1} \left(\sum_{(a',b') \in P_{l'}} [d_s(a', b')] \right) + \sum_{l'=1,3,\cdots}^{\frac{n}{2}-2} \left(\sum_{(a',b') \in P_{l'}} [d_s(a', b')] \right)$$

$$= \frac{n}{2}\left[\mu + \frac{n-2}{4}(\kappa + \eta)\right] + \sum_{l'=2,4,\cdots}^{\frac{n}{2}-1} n[\frac{l'}{2}(\kappa + \eta)] + \sum_{l'=1,3,\cdots}^{\frac{n}{2}-2} \frac{n}{2}\left[\frac{l'+1}{2}\eta + \frac{l'-1}{2}\kappa + \frac{l'-1}{2}\eta + \frac{l'+1}{2}\kappa\right]$$

$$= \frac{n}{2}\left[\eta + \frac{n-2}{4}(\kappa + \eta)\right] + \frac{n(n^2 - 4)}{32}(\kappa + \eta) + \frac{n(n-2)^2}{32}(\kappa + \eta)$$

$$= \frac{n}{2}\eta + \frac{n(n^2 - 4)}{16}(\kappa + \eta)$$

This completes the proof. □

Theorem 1.69 ([25]) *Let* $\mathscr{G} = (\mathscr{V}, \sigma, \mu)$ *be a connected FG which satisfies the following conditions:*

(i) Each edge in $\mathscr{G}$ *is either* α*-st or* β*-st,*

(ii) There is an edge (a, b) *for any vertices* $a, b \in \mathscr{V}$.

Then $CIF(\mathscr{G}) = WIF(\mathscr{G})$.

1.17 Extension of Fuzzy Graphs

We have discussed different features and types of FGs. In a FG all vertices and edges have membership values which may represent their degree of existence or appearance or occurrence, etc. in the graph. But, the degree of non-existence or non-appearance or non-occurrence, etc. are not included in the theory of FGs. This section will shed light upon the existence as well as the non-exitance of vertices and/or edges in a FG.

1.17.1 Intuitionistic Fuzzy Graphs

In 1983, Atanasov [16] defined a new type of fuzzy set called intuitionistic fuzzy set (IFS) which considers both the parameters, viz. membership and non-membership of all elements of the set. The IFS is defined below:

Definition 1.100 An **intuitionistic fuzzy set** [16] A defined on the universal set U is characterized as follows $A = \{(x, \mu_A(x), \nu_A(x)) : x \in U\}$, where the membership function $\mu_A : U \rightarrow [0, 1]$, and non-membership function $\nu_A : U \rightarrow [0, 1]$ satisfy the condition $0 \leq \mu_A(x) + \nu_A(x) \leq 1$, for all $x \in U$.

Observe that if $\mu_A(x) + \nu_A(x) = 1$, i.e. $\mu_A(x) = 1 - \nu_A(x)$ for all $x \in U$, then IFS becomes fuzzy set (FS) because only one parameter $\mu_A(x)$ characterizes the FS. So, for IFS, $\mu_A(x) + \nu_A(x) < 1$ for at least one $x \in U$. For IFS, a function $\pi_A : U \rightarrow [0, 1]$ is defined as $\pi_A(x) = 1 - \mu_A(x) - \nu_A(x)$ for all $x \in U$.

Based on the concept of IFS, Parvathi and Karunambigai [107] defined **intuitionistic fuzzy graph** (IFG).

Definition 1.101 ([107]) An IFG is of structure $G = (V, \sigma, \mu)$ where $\sigma = (\sigma_1, \sigma_2)$, $\mu = (\mu_1, \mu_2)$ and

(i) $V = \{v_0, v_1, \ldots, v_n\}$ such that $\sigma_1 : V \rightarrow [0, 1]$ and $\sigma_2 : V \rightarrow [0, 1]$, denote the degree of membership and non-membership values of the node $v_i \in V$ respectively and $0 \leq \sigma_1(v_i) + \sigma_2(v_i) \leq 1$ for all $v_i \in V$ for all $i = 1, 2, n$.

Fig. 1.21 An intuitionistic
fuzzy graph

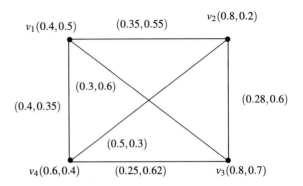

(ii) $\mu_1 : V \times V \rightarrow [0, 1]$ and $\mu_2 : V \times V \rightarrow [0, 1]$, where $\mu_1(v_i, v_j)$ and $\mu_2(v_i, v_j)$ denote the degree of membership and non-membership value of the arc (v_i, v_j) respectively such that $\mu_1(v_i, v_j) \leq \min\{\sigma_1(v_i), \sigma_1(v_j)\}$ and $\mu_2(v_i, v_j) \leq \max\{\sigma_2(v_i), \sigma_2(v_j)\}, 0 \leq \mu_1(v_i, v_j) + \mu_2(v_i, v_j) \leq 1$ for all (v_i, v_j).

That is in IFG, all vertices and edges are associated with two numbers: one is membership value and other is non-membership value. The membership value may represent the degree of existence/presence of the vertex/edge in the graph and the non-membership value represents the degree of non-existence/absence of the same.

An example of an IFG is shown in Fig. 1.21.

Many works have been published on IFGs with applications. In this book, detailed discussions on IFGs have been made in Chap. 10.

1.17.2 Pythagorean Fuzzy Graphs

Recall that the sum of membership and non-membership values of each vertex (also for edge) is always lies between 0 and 1 including 0 and 1.

In IFS membership value (μ), non-membership value (ν) and hesitation margin (π) are restricted by the relations $0 \leq \mu + \nu \leq 1$ and $\mu + \nu + \pi = 1$. But, there are some cases where $\mu + \nu \geq 1$ happens. Such type of cases can be handled by the Pythagorean fuzzy set (PyFS) with one more restriction.

Definition 1.102 ([149]) Let μ, ν be membership and nonmembership functions defined on universal set U. The **Pythagorean fuzzy set** A is defined as follows:
$A = \{x, \mu_A(x), \nu_A(x) : x \in X\}$ where $\mu : X \rightarrow [0, 1]$ and $\nu : X \rightarrow [0, 1]$ are such that $0 \leq (\mu_A(x))^2 + (\nu_A(x))^2 \leq 1$ and for all $x \in X$.

Let there be a function $\pi_A(x)$ such that $(\mu_A(x))^2 + (\nu_A(x))^2 + (\pi_A(x))^2 = 1$, i.e. $\pi_A(x) = \sqrt{1 - [(\mu_A(x))^2 + (\nu_A(x))^2]}$. Obviously, $\pi_A(x) \in [0, 1)$ and the function $\pi_A : X \rightarrow [0, 1]$ is called the hesitation margin.

Now, we define **Pythagorean fuzzy relation** (PFR).

Definition 1.103 A PyFS on $U \times U$ is said to be a PFR on U, denoted by $R = \{\langle ab, \mu_R(ab), \nu_R(ab)\rangle : a, b \in U\}$, where $\mu_R : U \times U \to [0, 1]$ and $\nu_R : U \times U \to [0, 1]$ represent the membership and non-membership of values of ab in R respectively such that $0 \leq \mu_R^2(ab) + \nu_R^2(ab) \leq 1$ for all $a, b \in U$.

Using this concept, Naz et al. [101] defined the Pythagorean fuzzy graphs (PyFGs). Verma et al. [145] defined the concept of strong PyFGs and complement of PyFg. Energy of PyFG is discussed by Akram and Naz [9]. More results on PyFGs are available in [10–12].

The formal definition of PyFG is given below.

Definition 1.104 ([101]) A PyFG on a universe of discourse U is denoted by $\mathscr{G} = (\mathscr{V}, A, B)$, where A is a PyFS on U and B is a PyFR on U which satisfies the following conditions:
$\mu_B(a, b) \leq \mu_A(a) \wedge \mu_A(b), \nu_B(a, b) \leq \nu_A(a) \vee \nu_A(b)$ and $0 \leq \mu_B^2(a, b) + \nu_B^2(a, b) \leq 1$ for all $a, b \in U$, also $\mu_B : U \times U \to [0, 1]$ and $\nu_B : U \times U \to [0, 1]$ represent the membership and non-membership values of B, respectively.

Here, A and B are the vertex set and edge set respectively.

Example 1.44 Let $\mathscr{G} = (\mathscr{V}, A, B)$, where $\mathscr{V} = \{a, b, c, d\}$ with the membership and non-membership values of the vertices a, b, c, d are respectively $\langle 0.2, 0.7\rangle$, $\langle 0.3, 0.6\rangle$, $\langle 0.3, 0.5\rangle$, $\langle 0.4, 0.6\rangle$. The edges set is $\{(a, b), (b, c), (c, d)\}$ and membership and non-membership values of the edges are $\langle 0.2, 0.65\rangle$, $\langle 0.25, 0.4\rangle$, $\langle 0.2, 0.55\rangle$ respectively. This is an example of a PyFG.

1.17.3 q-Rung Orthopair Fuzzy Graphs

Yager [150] introduced the concept of q-rung orthopair fuzzy set (q-ROFS) in 2017. After that lots of works have been done in decision-making problem, graph theory, etc. The q-ROFS is a generalization of IFS and PyFS. The *q***-rung orthopair fuzzy graph** (q-ROFG) is an extension of IFG and PyFG. In 2019, Habib et al. [60] introduced the concept of q-ROFG based on the idea of the q-ROFS. Also, Habib et al. [61] studied different types of q-rung orthopair fuzzy competition graphs. Yin et al. [152] defined the degree and total degree of a vertex on q-ROFG and defined some common products between q-ROFGs.

Definition 1.105 A **q-ROFS** A in a finite universe of discourse U is given $A = \{\langle x, \mu_A(x), \nu_A(x)\rangle : x \in X\}$, where $\mu_A : X \to [0, 1]$ denotes the degree of membership and $\nu_A : X \to [0, 1]$ denotes the degree of non-membership of the element $x \in X$ to the set A respectively with the condition that $0 \leq (\mu_A(x))^q + (\nu_A(x))^q \leq 1$, $(q \geq 1)$. The degree of indeterminacy is given as $\pi_A(x) = [(\mu_A(x))^q + (\nu_A(x))^q - (\mu_A(x))^q(\nu_A(x))^q]^{\frac{1}{q}}$.

Fig. 1.22 Representation of
IFS, PyFS, and q-ROFS

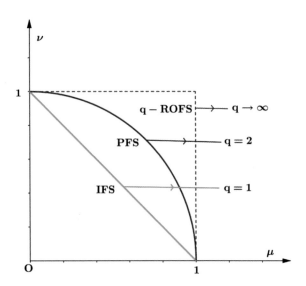

Definition 1.106 ([61]) A q-ROFS R on $U \times U$ is said to be a **q-rung orthopair fuzzy relation** (q-ROFR) on U, denoted by

$$R = \{\langle ab, \mu_R(ab), \nu_R(ab)\rangle : ab \in U \times U\},$$

where $\mu_R : U \times U \to [0, 1]$ and $\nu_R : U \times U \to [0, 1]$ represent the membership and nonmembership functions of R respectively and satisfy the conditions $0 \leq \mu_R^q(ab) + \nu_R^q(ab) \leq 1$ for all $ab \in U \times U$ and $q \geq 1$.

Now, we can give the definition of q-ROFG.

Definition 1.107 ([61]) A q-ROFG on a universal set U is a pair $\mathscr{G} = (A, B)$, where A is a q-ROFS on U and B is a q-ROFR on $U \times U$ such that $\mu_B^q(a, b) \leq \mu_A(a) \wedge \mu_A(b)$, $\nu_B^q(a, b) \geq \nu_A(a) \vee \nu_A(b)$ and $0 \leq \mu_B^q(a, b) + \nu_B^q(a, b) \leq 1$ for all $a, b \in U$ and $q \geq 1$.

Here, A and B are respectively the vertex set and edge set of $\mathscr{G}$ respectively.

A diagrammatic representation of IFS, PyFS and q-ROFS is shown in Fig. 1.22.
– when $q = 1$, q-ROFG becomes IFG
– when $q = 2$, q-ROFG becomes PyFG
Thus, q-ROFG carries more information than IFG and PyFG.

1.17.4 Bipolar Fuzzy Graphs

In 1994, Zhang [154, 155] introduced the concept of bipolar fuzzy sets (BFSs) as a generalization of fuzzy sets. BFSs are extension of FSs whose membership degree range is $[-1, 1]$. In a BFS, the membership degree $(0, 1]$ of an element indicates that the element somewhat satisfies the property, the membership degree $[-1, 0)$ of an element indicates that the element somewhat satisfies the implicit counter-property, the membership degree 0 of an element means that the element is irrelevant to the corresponding property. Apparently, BFSs and IFSs have similar look, but they are different sets with certain differences. Interested readers may go through [76] for more detailed discussion.

A wide variety of human decision-making is based on double-sided or bipolar judgmental thinking on a positive side and a negative side. For example, cooperation and competition, friendship and hostility, common interests and conflict of interests, effect and side effect, likelihood and unlikelihood, feedforward and feedback, etc. are often the two sides in decision and coordination.

Definition 1.108 Let U be the universal set. A **bipolar fuzzy set** A on U is a structure having the form

$$A = \{(x, \mu_A^P(x), \mu_A^N(x)): x \in U\},$$

where $\mu_A^P: U \to [0, 1]$ is the positive membership function and $\mu_A^N: U \to [-1, 0]$ is the negative membership function.

For simplicity, the symbol $A = (\mu_A^P, \mu_A^N)$ is used to represent the BFS $A = \{(x, \mu_A^P(x), \mu_A^N(x)): x \in U\}$.

Definition 1.109 ([154]) Let $A = (\mu_A^P, \mu_A^N)$ and $B = (\mu_B^P, \mu_B^N)$ be two BFSs on the universal set U. If $A = (\mu_A^P, \mu_A^N)$ is a **bipolar fuzzy relation** (BFR) on U, then $A = (\mu_A^P, \mu_A^N)$ is said to be a BFR on $B = (\mu_B^P, \mu_B^N)$ if $\mu_A^P(a, b) \leq \mu_B^P(a) \wedge \mu_B^P(b)$ and $\mu_A^N(a, b) \geq \mu_B^N(a) \vee \mu_B^N(b)$ for all $a, b \in U$.

If $\mu_A^P(a, b) = \mu_A^P(b, a)$ and $\mu_A^N(a, b) = \mu_A^N(b, a)$ for all $a, b \in U$, then A is called symmetric BFR.

Definition 1.110 ([1]) Let $G^* = (V, E)$ be a crisp graph. A **bipolar fuzzy graph** (BFG) on G^* is defined as a pair $\mathscr{G} = (A, B)$, where $A = (\mu_A^P, \mu_A^N)$ is BFS in V and $B = (\mu_B^P, \mu_B^N)$ is a BFR in $E \subseteq V \times V$, such that

$$\mu_B^P(a, b) \leq \mu_A^P(a) \wedge \mu_A^P(b), \quad \mu_B^N(a, b) \geq \mu_A^N(a) \vee \mu_A^N(b),$$

for all $(a, b) \in E$.

A BFG is shown in Fig. 1.23.
For further work on BFGs see [7, 56, 75, 87, 112, 122–124, 131, 132, 134].

1.17.5 m-Polar Fuzzy Graphs

In BFS, only two parameters are considered, one represents positive side and the other negative side. In 2014, Chen et al. [38] extended BFS by introducing multiple

Fig. 1.23 A BFG

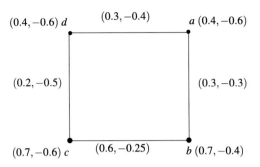

parameters called m-polar fuzzy sets. In an m-polar fuzzy set, the membership value of an element belongs to $[0, 1]^m$ which represents all the m different properties of an element. This is a more generalised structure to solve many real world problems.

Here, $[0, 1]^m$ (m-power of $[0, 1]$) is taken as a poset with point-wise order '$\leq$', where m is a natural number. $\leq$ is defined by $a \leq b \Leftrightarrow p_i(a) \leq p_i(b)$ for all $i = 1, 2, \ldots, m$, where $a, b \in [0, 1]^m$ and $p_i : [0, 1]^m \to [0, 1]$ is the ith projection mapping.

Definition 1.111 ([5]) An m-**polar fuzzy set** (m-PFS) (or a $[0, 1]^m$-set) on U is a mapping $A : U \to [0, 1]^m$.

Definition 1.112 ([8]) Let A be a m-PFS on U. An m-**polar fuzzy relation** (m-PFR) on A is an m-PFS R of $U \times U$, such that $R(a, b) \leq \min\{A(a), A(b)\}$ for all $a, b \in U$, i.e. $p_i \cdot R(a, b) \leq \min\{p_i \cdot A(a), p_i \cdot A(b)\}$ for all $i = 1, 2, \ldots, m$.

An m-PFR R on U is said to be symmetric if $R(a, b) = R(b, a)$ for all $a, b \in U$. Motivated from m-PFS, Ghorai and Pal [49] defined m-polar fuzzy graphs.

Definition 1.113 ([49]) An m-**polar fuzzy graph** (m-PFG) of the crisp graph $G^* = (V, E)$ is a pair $\mathscr{G} = (A, B)$, where $A : V \to [0, 1]^m$ is an m-PFS in V and $B : \tilde{V}^2 \to [0, 1]^m$ is an m-PFS in $\tilde{V}^2$, such

$$p_i \cdot B(a, b) \leq \min\{p_i \cdot A(a), p_i \cdot A(b)\}$$

for all $i = 1, 2, \ldots, m$ and $(a, b) \in \tilde{V}^2$, and $B(a, b) = \mathbf{0}$ for all $(a, b) \in \tilde{V}^2 - E$. $\mathbf{0} = (0, 0, \ldots, 0)$ is the smallest element in $[0, 1]^m$.

A is called the m-polar fuzzy vertex set of $\mathscr{G}$ and B is called the m-polar fuzzy edge set of $\mathscr{G}$.

A 4-polar fuzzy graph is shown in Fig. 1.24.

Ghorai and Pal, Akram et al., Ramprasad et al. did many works on m-PFGs [8, 49–52, 54, 55, 78, 82, 83, 116].

Fig. 1.24 A 4-polar fuzzy
graph

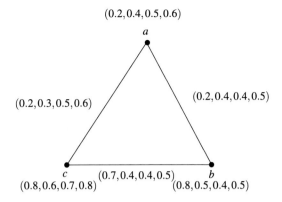

1.17.6 Picture Fuzzy Graphs

FS deals with only one parameter called membership value, IFS deals with two parameters, viz. membership and non-membership values. But, some times it may happen that some decision-makers may not give their opinion about a statement in acceptance or rejection, they may just stay neutral. Incorporating this case (neutral) Cuong [41] defined picture fuzzy sets (PFSs).

Definition 1.114 A **picture fuzzy set** (PFS) A on a universe U is an object of the form
$A = \{\langle x, \mu_A(x), \eta_A(x), \nu_A(x)\rangle : x \in U\}$, where
$\mu_A(x) : U \to [0, 1]$ is called the degree of positive membership of x in A, $\eta_A(x) :$ $U \to [0, 1]$ is called the degree of neutral membership of x in A, $\nu_A(x) : U \to [0, 1]$ is called the degree of non-membership of x in A.

The numbers $\mu_A(x)$, $\eta_A(x)$, $\nu_A(x)$ must satisfy the condition $0 \leq \mu_A(x) + \eta_A(x) + \nu_A(x) \leq 1$ for all $x \in U$.

All three numbers $\mu_A(x)$, $\eta_A(x)$, $\nu_A(x)$ lie between 0 and 1 including them, but they are not independent.

The term $1 - (\mu_A(x) + \eta_A(x) + \nu_A(x))$ for all $x \in U$ is called the degree of refusal membership of x in A.

By changing the dependence relation among $\mu_A(x)$, $\eta_A(x)$, $\nu_A(x)$, another set called spherical fuzzy set is defined in [77].

Definition 1.115 Let $G^* = (V, E)$ be a crisp graph. $G = (\mathcal{V}, A, B)$ is said to be a **picture fuzzy graph** on G^*, where $A = \langle \mu_A(x), \eta_A(x), \nu_A(x)\rangle$ is a picture fuzzy set on $\mathcal{V}$ and $B = \langle \mu_B(x, y), \eta_B(x, y), \nu_B(x, y)\rangle$ is a picture fuzzy set on $E \subseteq \mathcal{V} \times \mathcal{V}$ such that for each edge (x, y)
$\mu_B(x, y) \leq \min\{\mu_A(x), \mu_A(y)\}$, $\eta_B(x, y) \leq \min\{\eta_A(x), \eta_A(y)\}$, $\nu_B(x, y) \geq \max \{\nu_A(x), \nu_A(y)\}$.

Also, $0 \leq \mu_A(x) + \eta_A(x) + \nu_A(x) \leq 1$ for all $x \in \mathcal{V}$ and $0 \leq \mu_B(x, y) + \eta_B (x, y) + \nu_B(x, y) \leq 1$ for all $(x, y) \in \mathcal{V} \times \mathcal{V}$.

1.17.7   Neutroshopic Fuzzy Graphs

In 1998, Smarandache introduced neutrosophic sets (NSs) [136]. This is a general-ization of FSs and IFSs, and it is a powerful tool to express incomplete, indeterminate and inconsistent information that occur in many real world problems. NSs are char-acterized by three quantities, viz. truth-membership function (T), indeterminacy-membership function (I), and falsity-membership function (F). In single-valued neutrosophic set (SVNS) the truth-membership degree, indeterminacy-membership degree, and falsity-membership degree are independent and lie between 0 and 1 including them.

Definition 1.116 Let U be the universe of discourse. A NS A in U is defined as

$$A = \{\langle x, (T_A(x), I_A(x), F_A(x)) \rangle : x \in U, T_A(x), I_A(x), F_A(x) \in]^-0, 1^+[\},$$

where $T_A(x), I_A(x)$ and $F_A(x)$ are the truth-membership, indeterminacy-membership, and falsity-membership functions respectively. They are real standard or non-standard subsets of $]^0, 1^+[$. $]^0, 1^+[$ represents non-standard hyperreal unit interval.

The three values $T_A(x), I_A(x)$ and $F_A(x)$ are independent and hence their sum lies between $^-0$ and 3^+, i.e.

$$^-0 \leq T_A(x) + I_A(x) + F_A(x) \leq 3^+$$

for all $x \in U$.

One limitation in NS is the use of non-standard interval $]^0, 1^+[$. So instead of $]^0, 1^+[$ Wang et al. [147] considered the interval $[0, 1]$ for technical applications, because it is a bit complicated to use $]^0, 1^+[$ in real life applications. Hence, they defined **single-valued neutrosophic set** (SVNS) as follows.

Definition 1.117 Let U be a universe of discourse, i.e. a space of points (objects). A SVNS A in U is characterized by a truth-membership function $T_A(x)$, an indeterminacy-membership function $I_A(x)$ and a falsity membership function $F_A(x)$, for each element $x \in U$, where $T_A(x), I_A(x), F_A(x) \in [0, 1]$.

A SVSN A can be written as follows:

$$A = \{\langle x, (T_A(x), I_A(x), F_A(x)) \rangle : x \in U\}.$$

After the development of NS, a huge number of papers have been published in different fields of mathematics, physics, philosophy, etc.

In 2015, Smarandache [137] defined four basic categories of neutrosophic graphs (NGs),
(i) first two are based on literal indeterminacy (I), i.e. I-vertex NG and I-edge NG [143, 144].

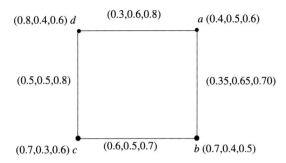

Fig. 1.25 A SVNG

(ii) other two are based on (T, I, F) components, i.e. (T, I, F)-edge NG and the (T, I, F)- vertex NG.

In 2016, Broumi et al. [30] proposed another type of NG combining both (T, I, F)-edge and (T, I, F)-vertex NGs. This NG is called **single-valued neutrosophic graph** (SVNG). The SVNG is the generalization of FG and IFG .

Definition 1.118 ([30]) A **single-valued neutrosophic graph** (SVNG) on the underlying graph $G^* = (V, E)$ is defined as $\mathscr{G} = (A, B)$, where
(i) The functions T_A, T_I, T_F are mapped from V to $[0, 1]$ and they denote the degree of truth-membership, indeterminacy-membership, and falsity-membership of the element $x \in V$ such that
$0 \le T_A(x) + I_A(x) + F_A(x) \le 3$ for all $x \in V$.
(ii) The functions T_B, I_B, F_B are mapped from $E \subseteq V \times V$ to $[0, 1]$ such that

$$0 \le T_B(x, y) + I_B(x, y) + F_B(x, y) \le 3 \quad \text{for all } (x, y) \in E$$

and satisfy the conditions
$T_B(x, y) \le T_A(x) \wedge T_A(y),\ \ I_B(x, y) \ge I_A(x) \vee I_A(y)$ and $F_B(x, y) \ge F_A(x) \vee F_A(y)$ for all $x, y \in V$.
$T_B(x, y), I_B(x, y), F_B(x, y)$ denote the degree of truth-membership, indeterminacy-membership, and falsity-membership of the edge $(x, y) \in E$.

The set A is called the **single-valued neutrosophic vertex** set of V and B the **single-valued neutrosophic edge set** of E respectively.

A SVNG is depicted in Fig. 1.25.
Ample amount of works on SVNGs have been done by many researchers in [30–34, 100, 130].

1.17.8 Spherical Fuzzy Graphs and T-Spherical Fuzzy Graphs

By making changes in the condition of IFS, the definition of PyFG has been obtained. Similar idea being used to construct spherical and T-spherical fuzzy set whose study initiated in [77].

Definition 1.119 ([77]) A **spherical fuzzy set** (SFS) S in a universe of discourse U is defined as follows:

$$S = \{\langle x, \mu_S(x), \eta_S(x), \nu_S(x)\rangle : x \in U\},$$

where $\mu_S : U \to [0, 1]$, $\eta_S : U \to [0, 1]$ and $\nu_S : U \to [0, 1]$ denote the degree of membership, degree of neutral membership (abstain) and degree of non-membership respectively and

$$\mu_S^2(x) + \eta_S^2(x) + \nu_S^2(x) \leq 1 \quad \text{for all } x \in U. \tag{1.15}$$

The degree of refusal for SFS S for any $x \in U$ is given by

$$\pi_S(x) = \sqrt{1 - (\mu_S^2(x) + \eta_S^2(x) + \nu_S^2(x))}.$$

By changing the Eq. (1.15) of SFS, another fuzzy set called T-sphere fuzzy set (T-SFS) has been defined in [77].

Definition 1.120 ([77]) A T-**spherical fuzzy set** (T-SFS) S in a universe of discourse U is defined as

$$S = \{\langle x, \mu_S(x), \eta_S(x), \nu_S(x)\rangle : x \in U\},$$

where $\mu_S : U \to [0, 1]$, $\eta_S : U \to [0, 1]$ and $\nu_S : U \to [0, 1]$ denote the degree of membership, degree of neutral membership (abstain), and degree of non-membership respectively, and

$$\mu_S^n(x) + \eta_S^n(x) + \nu_S^n(x) \leq 1 \quad \text{for all } x \in U. \tag{1.16}$$

The degree of refusal for T-SFS S for any $x \in U$ is given by

$$\pi_S(x) = \sqrt[n]{1 - (\mu_S^n(x) + \eta_S^n(x) + \nu_S^n(x))}.$$

Note 1.10 As evident from the definitions, T-SFS is the most generalised form of fuzzy set because
– when $n = 2$, T-SFS becomes SFS.
– when $n = 1$, T-SFS becomes PFS.
– when $n = 2$ and $\eta_s(x) = 0$ for all $x \in U$, T-SFS becomes PyFS.
– when $n = 1$ and $\eta_s(x) = 0$ for all $x \in U$, T-SFS becomes IFS.

Motivated from T-SFS, Guleria and Bajaj [57] introduced T-spherical fuzzy graph (T-SFG) along with operations like product, composition, union, join, and complement. Also, using the concept of T-SFG, a decision-making problem on supply chain management has been solved.

Definition 1.121 Let U be a universal set. A **T-spherical fuzzy relation** (T-SFR) in U is a T-SFS R in $U \times U$, given by $R = \{\langle (a, b), \mu_R(a, b), \eta_R(a, b), \nu_R(a, b) \rangle :$ $(a, b) \in U \times U\}$, where $\mu_R : U \times U \to [0, 1]$, $\eta_R : U \times U \to [0, 1]$, $\nu_R : U \times U \to [0, 1]$ represent the degree of membership, degree of neutral membership (abstain), and degree of non-membership respectively, satisfying the condition

$$\mu_R^n(a, b) + \eta_R^n(a, b) + \nu_R^n(a, b) \leq 1; \quad \text{for all } (a, b) \in U \times U.$$

Definition 1.122 Let U be a universal set. A **T-spherical fuzzy graph** on U is denoted by $\mathscr{G} = (A, B)$, where A, the T-SFS on U satisfies
$\mu_A^n(a) + \eta_A^n(a) + \nu_A^n(a) \leq 1$ for all $a \in U$ and
B, a T-SFR defined on $U \times U$, such that
$\mu_B(a, b) \leq \mu_A(a) \wedge \mu_A(b)$
$\eta_B(a, b) \leq \eta_A(a) \wedge \eta_A(b)$
$\nu_B(a, b) \leq \nu_A(a) \wedge \nu_A(b)$
satisfying the condition
$\mu_B^n(a, b) + \eta_B^n(a, b) + \nu_B^n(a, b) \leq 1$ for all $a, b \in U$.
   The sets A and B are called set of vertices and set of edges respectively of a T-SFG $\mathscr{G}$.

Note 1.11
– If B is not symmetry, then T-SFG is a directed one.
– If $n = 2$, then the T-SFG becomes SFG.
– If $n = 1$, then the T-SFG becomes IFG.
– If $n = 2$ and B is not symmetry, then the T-SFG becomes directed SFG.

   A geometrical comparison of the sets PyFS, NS, IFS, PFS, SFS is depicted in Fig. 1.26 [57].

1.17.9   Interval-Valued Fuzzy Graphs

In 1975, Zadeh [153] generalized the idea of fuzzy set into an interval-valued fuzzy set (IVFSs). IVFS expresses the uncertainty in assigning membership value to an element in belongingness to a fuzzy set. Membership value of an element in belongingness to a fuzzy set is expressed as a closed interval. Small interval indicates less amount uncertainty and large interval indicates more uncertainty in belongingness in IVFS. As the membership value of an element is expressed in an interval, hence it has been named so.

   Let $D[0, 1]$ be the set of all subsets of the closed interval $[0, 1]$. Like FS, $[\mu_A^-(x), \mu_A^+(x)] \subseteq D[0, 1]$ represents the membership value of an element x in the IVFS A. That is, membership value of an element is a sub-interval of the closed interval $[0, 1]$ instead of a single number in $[0, 1]$.

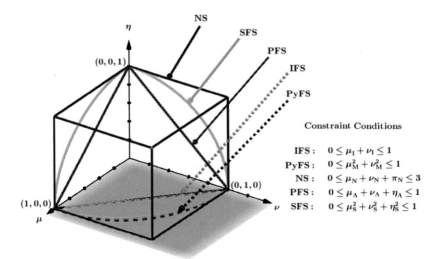

Fig. 1.26 Geometrical comparison of some fuzzy sets

Definition 1.123 Let A be an IVFS defined on the universal set U. Then $A = \{(x, [\mu_A^-(x), \mu_A^+(x)]) : x \in U\}$, where $\mu_A^- : X \to [0, 1]$ and $\mu_A^+ : X \to [0, 1]$ are the lower and upper limits of membership values. Because it is an interval, so $\mu_A^-(x) \leq \mu_A^+(x)$ for all $x \in U$.

Akram and Dudek [2] extended this idea on FG and defined interval-valued fuzzy graph (IVFG).

Definition 1.124 Let $G^* = (V, E)$ be a crisp graph. An **interval-valued fuzzy graph** on G^* is a graph $\mathscr{G} = (A, B)$, where $A = [\mu_A^-, \mu_A^+]$ is an IVFS on V and $B = [\mu_B^-, \mu_B^+]$ is an IVFS on E such that
$\mu_B^-(a, b) \leq \mu_A^-(a) \wedge \mu_A^-(b)$,
$\mu_B^+(a, b) \leq \mu_A^+(a) \wedge \mu_A^+(b)$, for all $(a, b) \in E$.

Many works have been done on IVFGs. Interested readers may refer to [3, 4, 37, 106, 109–111, 118–121, 125].

1.17.10 Cubic Graphs

Using a fuzzy set and IVFS, Jun et al. [67] introduced a new notion on sets called a (internal, external) cubic set. After that many works have been made in different directions.

Let $A = [A^-, A^+]$ and $B = [B^-, B^+]$ be two interval numbers, i.e. $A, B \in D[0, 1]$. Then,

Fig. 1.27 A cubic graph

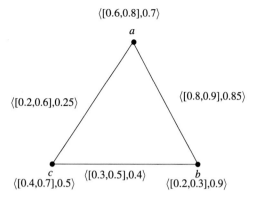

$$\langle[0.6,0.8],0.7\rangle$$
$$a$$
$$\langle[0.2,0.6],0.25\rangle \qquad \langle[0.8,0.9],0.85\rangle$$
$$c \qquad \langle[0.3,0.5],0.4\rangle \qquad b$$
$$\langle[0.4,0.7],0.5\rangle \qquad \langle[0.2,0.3],0.9\rangle$$

$\text{rmin}\{A, B\} = [\min\{A^-, B^-\}, \min\{A^+, B^+\}]$

$A \succeq B$ if and only if $A^- \geq B^-$ and $A^+ \geq B^+$.

Similarly, $A \preceq B$ and $A = B$ are defined.

Definition 1.125 ([67]) Let U be a non-empty set. A **cubic set** in U is a structure

$$\mathscr{A} = \{\langle x, A(x), \mu(x)\rangle : x \in U\},$$

where A is an IVFS in U and μ is a fuzzy set in U.

Definition 1.126 ([67]) Let U be a non-empty set. A cubic set $\mathscr{A} = \langle A, \mu \rangle$ in U is said to be
(i) an **internal cubic set** if $A^-(x) \leq \mu(x) \leq A^+(x)$ for all $x \in U$.
(ii) an **external cubic set** if $\mu(x) < A^-(x)$ or $\mu(x) > A^+(x)$ for all $x \in U$.

Exploring the concept of cubic set, Rashid et al. [126] defined cubic fuzzy graph.

Definition 1.127 ([126]) Let $G^* = (V, E)$ be a graph. A **cubic graph** on G^* is a structure $\mathscr{G} = (\mathscr{P}, \mathscr{Q})$, where $\mathscr{P} = \langle A_\mathscr{P}, \mu_\mathscr{P}\rangle$ is a cubic set on the vertex set V and $\mathscr{Q} = \langle A_\mathscr{Q}, \mu_\mathscr{Q}\rangle$ is the cubic set on the edge set E, i.e.
$A_\mathscr{P} : V \to D[0, 1], \qquad \mu_\mathscr{P} : V \to [0, 1],$
$A_\mathscr{Q} : E \to D[0, 1], \qquad \mu_\mathscr{Q} : E \to [0, 1],$
such that
$A_\mathscr{Q}(a, b) \preceq \text{rmin}\{A_\mathscr{P}(a), A_\mathscr{P}(b)\}, \qquad \mu_\mathscr{Q}(a, b) \leq \max\{\mu_P(a), \mu_P(b)\}$
for all $(a, b) \in E$.

A (internal) cubic graph is shown in Fig. 1.27.

Jan et al. [66] defined and investigated cubic bipolar fuzzy graph. For other works on cubic fuzzy graphs, readers may refer to [74, 127, 146].

1.17.11 Vague Graphs

Gau and Buehrer [47] proposed the concept of a vague set, which is a generalization of the fuzzy set.

Definition 1.128 ([47]) A **vague set** A in an ordinary finite non-empty set U is a pair (t_A, f_A), where $t_A : U \rightarrow [0, 1]$, $f_A : U \rightarrow [0, 1]$ are true and false membership functions respectively such that $0 \leq t_A(x) + f_A(x) \leq 1$ for all $x \in U$.

The number $t_A(x)$ is the lower bound for the degree of membership of x in A and $f_A(x)$ is the lower bound for negative membership of x in A. Thus, the degree of membership of x in A is characterized by the interval $[t_A(x), 1 - f_A(x)]$. Also, this interval is called the vague value of $x \in U$.

It is noted that the IVFSs are not vague sets. In IVFSs, an Interval-valued membership value is assigned to each element of U considering the evidence for x only, without considering the evidence against x. In vague sets, both are independently decided by the decision-maker. This makes a major difference in the judgment about the grade of membership. A vague relation is a generalization of a fuzzy relation.

Definition 1.129 ([115]) Let A and B be two ordinary finite non-empty sets. A **vague relation** is a vague subset of $A \times B$, that is, vague relation R defined by

$$R = \{\langle (a, b), t_R(a, b), f_R(a, b) \rangle : a \in A, b \in B\},$$

where $t_R : A \times B \rightarrow [0, 1]$, $f_R : A \times B \rightarrow [0, 1]$, which satisfy the condition $0 \leq t_R(a, b) + f_R(a, b) \leq 1$ for all $(a, b) \in A \times B$.

This definition leads to the vague graph (see Fig. 1.28).

Definition 1.130 ([115]) Let $G^* = (V, E)$ be a crisp graph. A graph $\mathscr{G} = (\mathscr{V}, A, B)$ is said to be a **vague graph** on G^*, where $A = (t_A, f_A)$ is a vague set on $\mathscr{V}$ and $B = (t_B, f_B)$ is a vague set on E, such that

$$t_B(a, b) \leq t_A(a) \wedge t_A(b), \quad f_B(a, b) \geq f_A(a) \wedge f_A(b)$$

for each edge $(a, b) \in E$.

For further investigation on vague graph, readers may read [27–29, 42, 43, 53, 135, 142].

1.17.12 Interval-Valued Intuitionistic Fuzzy Graphs

In 1989, Atanassov and Gargov [17] defined interval-valued intuitionistic fuzzy sets (IVIFSs) as a combination of IVFS and IFS.

The definition of IVIFS is given below.

Fig. 1.28 A vague graph

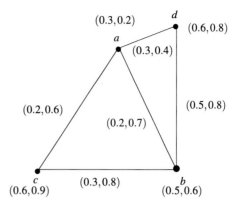

Definition 1.131 An IVIFS A on the universal set U is an algebraic structure

$$A = \{\langle x, M_A(x), N_A(x)\rangle : x \in U\},$$

where $M_A : U \rightarrow D[0, 1]$ and $N_A : U \rightarrow D[0, 1]$ are the membership and non-membership values respectively of the element x in U. Here, the unit interval $M_A(x) = [M_A^-(x), M_A^+(x)]$ and $N_A(x) = [N_A^-(x), N_A^+(x)]$ must satisfy the condition

$$sup\ M_A(x) + sup\ N_A(x) \leq 1,$$

for all $x \in U$.

Using the concept of IVIFS, interval-valued intuitionistic fuzzy graphs (IVIFGs) are defined. This graph is a combination of IVFGs and IFGs. Both the membership and non-membership values are intervals on unit interval $[0, 1]$.

Definition 1.132 Let $G^* = (V, E)$ be a graph. An IVIFG is denoted by $\mathscr{G} = (\mathscr{V}, P, Q)$ which is defined on G^*, where P, Q are subintervals of $[0, 1]$ and
(i) $M_P : \mathscr{V} \rightarrow D[0, 1]$ and $N_P : \mathscr{V} \rightarrow D[0, 1]$ represent the membership and non-membership values of an element $x \in \mathscr{V}$, such that $0 \leq M_P^+(x) + N_P^+(x) \leq 1$ for all $x \in \mathscr{V}$.
(ii) $M_Q : E \rightarrow D[0, 1]$ and $N_Q : E \rightarrow D[0, 1]$ represent the membership and non-membership values of an edge $(x, y) \in E$, such that
$M_Q^-(x, y) \leq M_P^-(x) \wedge M_P^-(y), \quad N_Q^-(x, y) \geq N_P^-(x) \vee N_P^-(y)$
$M_Q^+(x, y) \leq M_P^+(x) \wedge M_P^+(y), \quad N_Q^+(x, y) \geq N_P^+(x) \vee N_P^+(y)$
such that $0 \leq M_Q^+(x, y) + N_Q^+(x, y) \leq 1$ for all $(x, y) \in E$.

An IVIFG is illustrated in Fig. 1.29.
For other works on IVIFGs, see [65, 86, 124].

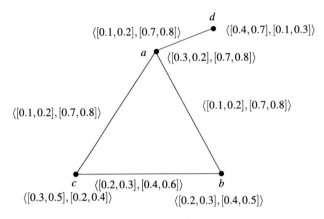

Fig. 1.29 An interval-valued intuitionistic fuzzy graph

1.17.13 Soft Graphs

In 1999, Molodtsov [88] introduced a new kind of set to interpret a particular type of uncertainty. This set is defined for a parametric set on the set of the universe. Let U be the universe of discourse and P be the universe of all possible parameters related to the objects in U. A parameter may be a word or a sentence. Most cases the parameters are considered as characteristics, attributes, properties of objects, etc. in U. The pair (U, P) is also known as a soft universe. The power set of U is denoted by $\mathscr{P}(U)$.

Definition 1.133 (*Soft set* [88]) Let U be an initial universe set and let P be a set of parameters. A pair (F, P) is called a soft set (over U) if and only if F is a mapping of P into the set of all subsets of the set U, i.e. the power set ($\mathscr{P}(U)$) of U.

In other words, the soft set is a parameterized family of subsets of the set U. Every set $F(a)$, $a \in P$, from this family may be considered as the set of a-elements of the soft set (F, P), or as the set of a-approximate elements of the soft set.

Akram and Nawaz [5] defined the soft graphs and presented many results, viz. vertex- and edge-induced soft graphs, etc. In [6] regular soft graphs, irregular soft graphs, highly irregular soft graphs, and neighborly irregular soft graphs have been defined and along with many important results.

Definition 1.134 (*Soft graph* [5]) Let P be the set of parameters. The soft graph over P is denoted by $G = (G^*, F, K, P)$ and it satisfies the following conditions:
(i) $G^* = (V, E)$ is a simple graph,
(ii) P is a non-empty set of parameters,
(iii) (F, P) is a soft set over V,
(iv) (K, P) is a soft set over E,
(v) $(F(x), K(x))$ is a subgraph of G^* for all $x \in P$.

Fig. 1.30 A graph G^*

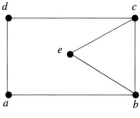

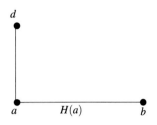

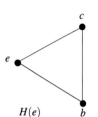

Fig. 1.31 Subgraphs $H(a)$ and $H(e)$

The subgraph $(F(x), K(x))$, $x \in P$ is denoted by $H(x)$. A soft graph can also be represented by

$$G = (F, K, A) = \{H(a) : a \in P\}.$$

Basically, soft graph is a graph on a parameter set. In the following example, a soft graph is illustrated.

Example 1.45 Let us consider a crisp graph $G^* = (V, E)$, where $V = \{a, b, c, d, e\}$ and $E = \{ab, ad, bc, be, ce, cd\}$, shown in Fig. 1.30.

Let the set of parameters be $P = \{a, e\} \subseteq V$ and (F, P) be a soft set with its approximate function $F : P \to \mathscr{P}(V)$ defined by $F(x) = \{y \in V : d(x, y) \leq 1, d(x, y)$ is the distance between x and y $\}$ for all $a \in P$.

Then,
$F(a) = \{a, b, d\}$ and $F(e) = \{b, c, e\}$.

Let (K, P) be a soft set over the set of edges E with its approximate function $K : P \to \mathscr{P}(E)$ and is defined by $K(z) = \{xy \in E : \{x, y\} \subseteq F(z)\}$ for all $z \in P$.

As per this definition,
$K(a) = \{ab, ad\}$ and $K(e) = \{bc, be, ce\}$.

Thus, $H(a) = \{F(a), K(a)\}$ and $H(e) = \{F(e), K(e)\}$ are subgraphs of G^* and they are depicted in Fig. 1.31.

Hence, $G = \{H(a), H(e)\}$ is a soft graph of G^*.

1.17.14 Dombi Fuzzy Graphs and Pythagorean Dombi Fuzzy Graph

In 2018, Ashraf et al. [15] defined Dombi fuzzy graphs based on the Dombi t-norm and t-conorm which are defined below:

$$\text{Dombi } t\text{-norm: } \quad \frac{1}{1 + \left[\left(\frac{1-x}{x}\right)^{\kappa} + \left(\frac{1-y}{y}\right)^{\kappa}\right]^{1/\kappa}}, \quad \kappa > 0$$

$$\text{Dombi } t\text{-conorm: } \quad \frac{1}{1 + \left[\left(\frac{1-x}{x}\right)^{-\kappa} + \left(\frac{1-y}{y}\right)^{-\kappa}\right]^{1/(-\kappa)}}, \quad \kappa > 0$$

$$\text{Negation: } 1 - x$$

The product of Dombi FGs , their homomorphism, complement as well as strong Dombi FG have been investigated in [15].

Definition 1.135 ([15]) Let $G^* = (V, E)$ be a crisp graph. A **Dombi fuzzy graph** on G^* is an ordered pair $\mathscr{G} = (A, B)$, where A and B represent respectively the set of vertices and set of edges of $\mathscr{G}$. The sets A and B are defined as $A : V \to [0, 1]$ and $B : V \times V \to [0, 1]$. B is a symmetric fuzzy relation on V. Also, for all $a, b \in V$ the following conditions must hold:

$$\mu_B(a, b) \leq \frac{\mu_A(a)\mu_A(b)}{\mu_A(a) + \mu_A(b) - \mu_A(a)\mu_A(b)}$$

where $\mu_A(a)$ and $\mu_B(a, b)$ represent the membership values of a and (a, b) in A and B respectively.

It can be proved that

$$\mu_B(a, b) \leq \frac{\mu_A(a)\mu_A(b)}{\mu_A(a) + \mu_A(b) - \mu_A(a)\mu_A(b)} \leq \mu_A(a)\mu_A(b) \leq \mu_A(a) \wedge \mu_A(b).$$

This shows that every Dombi FG is a FG, but converse is not true. Also, the edge membership value in a Dombi FG is comparatively less than FG. That is, more uncertainty is involved on the edges of Dombi FG .

Combining Dombi FG and PyFG, a new type of FG called Pythagorean Dombi FG has been defined by Akram et al. in [13].

Definition 1.136 ([13]) Let $G^* = (V, E)$ be a crisp graph. A **Pythagorean Dombi fuzzy graph** (PDFG) on G^* is $\mathscr{G} = (A, B)$, where $A = (\mu_A, \nu_A)$ is the set of vertices and $B = (\mu_B, \nu_B)$ is the set of edges with $A : V \to [0, 1]$ being a subset of V and $B : V \times V \to [0, 1]$ being a symmetric PyFR on A such that

$$\mu_B(a, b) \leq \frac{\mu_A(a)\mu_A(b)}{\mu_A(a) + \mu_A(b) - \mu_A(a)\mu_A(b)}$$

$$\nu_B(a, b) \leq \frac{\nu_A(a) + \nu_A(b) - 2\nu_A(a)\nu_A(b)}{1 - \nu_A(a)\nu_A(b)}$$

and $0 \leq \mu_B^2(a, b) + \nu_B^2(a, b) \leq 1$ for all $a, b \in V$, where $\mu_A(a)$ and $\nu_A(a)$ represent the membership and non-membership values of a in A respectively.

It is obvious that A and B represent the vertex set and edge set of PDFG $\mathcal{G}$. If B is symmetric, then $\mathcal{G}$ is called undirected PDFG and if B is not symmetric then $\mathcal{G}$ is called Pythagorean Dombi fuzzy digraph.

References

1. M. Akram, Bipolar fuzzy graphs. Inf. Sci. **181**, 5548–5564 (2011)
2. M. Akram, W.A. Dudek, Interval-valued fuzzy graphs. Comput. Math. Appl. **61**(2), 289–299 (2011)
3. M. Akram, N.O. Al-Shehri, W.A. Dudek, Certain types of interval-valued fuzzy graphs. J. Appl. Math. **85** (2013)
4. M. Akram, M.M. Yousaf, W.A. Dudek, Self centered interval-valued fuzzy graphs. Afr. Mat. **26**, 887–898 (2015). https://doi.org/10.1007/s13370-014-0256-9
5. M. Akram, S. Nawaz, Operations on soft graphs. Fuzzy Inf. Eng. **7**(4), 423–449 (2015)
6. M. Akram, S. Nawaz, Certain types of soft graphs. U.P.B. Sci. Bull., Ser. A **78**(4) (2016)
7. M. Akram, S. Samanta, M. Pal, Application of bipolar fuzzy sets in planar graphs. Int. J. Appl. Comput. Math. **3**(2), 773–785 (2017)
8. M. Akram, A. Adeel, m-polar fuzzy graphs and m-polar fuzzy line graphs. J. Discret. Math. Sci. Cryptogr. **20**(8), 1597–1617 (2017)
9. M. Akram, S. Naz, Energy of Pythagorean fuzzy graphs with applications. Mathematics **6**(8), 136 (2018)
10. M. Akram, A. Habib, F. Ilyas, J.M. Dar, Specific types of Pythagorean fyzzy graphs and application to decision-making. Math. Comput. Appl. **23**(42) (2018). https://doi.org/10.3390/mca23030042
11. M. Akram, J.M. Dar, S. Naz, Certain graphs under pythagorean fuzzy environment. Complex Intell. Syst. **5**, 127–144 (2019). https://doi.org/10.1007/s40747-018-0089-5
12. M. Akram, J.M. Dar, A. Farooq, Planar graphs under Pythagorean fuzzy environment. Mathematics **6**(12), 278 (2018). https://doi.org/10.3390/math6120278
13. M. Akram, J.M. Dar, S. Naz, Pythagorean Dombi fuzzy graphs. Complex Intell. Syst. **6**, 29–54 (2020)
14. A.R. Ashrafi, T. Doslic, A. Hamzeh, The Zagreb coindices of graph operations. Discret. Appl. Math. **158**, 1571–1578 (2010)
15. S. Ashraf, S. Naz, E.E. Kerre, Dombi fuzzy graphs. Fuzzy Inf. Eng. **10**(1), 58–79 (2018)
16. K.T. Atanassov, Intuitionistic fuzzy sets. VII ITKR's Sess., Deposed Central Sci.-Tech. Libr. Bulgarian Acad. Sci. **1697**(84) (1983)
17. K.T. Atanassov, G. Gargov, Interval valued intuitionistic fuzzy sets. Fuzzy Sets Syst. **31**(3), 343–349 (1989)
18. C. Berge, *Graphs and Hypergraphs* (1973)
19. P. Bhattacharyya, Some remarks on fuzzy graphs. Pattern Recognit. Lett. **6**(297), 302 (1987)
20. K.R. Bhutani, On automorphisms of fuzzy graphs. Pattern Recognit. Lett. **9**, 159–162 (1989)

21. K.R. Bhutani, A. Rosenfeld, Fuzzy end vertices in fuzzy graphs. Inf. Sci. **152**, 323–326 (2003)
22. K.R. Bhutani, A. Rosenfeld, Geodesics in fuzzy graphs. Electron. Notes Discret. Math. **15**, 51–54 (2003)
23. K.R. Bhutani, A. Rosenfeld, Strong arcs in fuzzy graphs. Inf. Sci. **152**, 319–322 (2003)
24. M. Binu, S. Mathew, J.N. Mordeson, Connectivity index of a fuzzy graph and its application to human trafficking. Fuzzy Set Syst. **360**, 117–136 (2019)
25. M. Binu, S. Mathew, J.N. Mordeson, Wiener index of a fuzzy graph and application to illegal immigration networks. Fuzzy Set Syst. **384**, 132–147 (2020)
26. M. Blue, B. Bush, J. Puckett, Unified approach to fuzzy graph problems. Fuzzy Sets Syst. **125**, 355–368 (2002)
27. R.A. Borzooei, H. Rashmanlou, More results on vague graphs. U.P.B. Sci. Bull. **78**(1) (2016)
28. R.A. Borzooei, H. Rashmanlou, New concepts of vague graphs. Int. J. Mach. Learn. Cyber. **8**, 1081–1092 (2017). https://doi.org/10.1007/s13042-015-0475-x
29. R.A. Borzooei, B.S. Hoseini, Y.B. Jun, Lexicographic product of vague graphs with application. Matematicki Vesnik Matematiqki Vesnik **72**(1), 43–57 (2020)
30. S. Broumi, M. Talea, A. Bakali, F. Smarandache, Single valued neutrosophic graphs. J. New Theory **10**, 86–101 (2016)
31. S. Broumi, M. Talea, A. Bakali, F. Smarandache, On bipolar single valued neutrosophic graphs. J. New Theory **11**, 84–102 (2016)
32. S. Broumi, M. Talea, A. Bakali, F. Smarandache, Interval valued neutrosophic graphs, in *Proceedings of the Annual Symposium of the Institute of Solid Mechanics and Session of the Commission of Acoustics* (2016)
33. S. Broumi, A. Bakali, M. Talea, F. Smarandache, A. Dey, L.H. Son, Spanning tree problem with neutrosophic edge weights. Proc. Comput. Sci. **127**, 190–199 (2018)
34. S. Broumi, A. Bakali, M. Talea, F. Smarandache, P.K. Singh, Properties of interval-valued neutrosophic graphs, in *Fuzzy Multi-criteria Decision-Making Using Neutrosophic Sets*. Studies in Fuzziness and Soft Computing, ed. by C. Kahraman, Ä. Otay, vol. 369 (2019)
35. M. Cary, Perfectly regular and perfectly edge-regular fuzzy graphs. Ann. Pure Appl. Math. **16**(2), 461–469 (2018)
36. G. Cash, S. Klavzar, M. Petkovsek, Three methods for calculation of the hyper-wiener index of molecular graphs. J. Chem. Inf. Comput. Sci. **42**, 571–576 (2002)
37. S.M. Chen, Interval-valued fuzzy hypergraph and fuzzy partition. IEEE Trans. Syst. Man Cybern. B **27**, 725–733 (1997)
38. J. Chen, S. Li, S. Ma, X. Wang, m-Polar fuzzy sets: an extension of bipolar fuzzy sets. Sci. World J. (2014). https://doi.org/10.1155/2014/416530
39. E.J. Cockayne, T.H. Stephen, Towards a theory of domination in graphs. Networks **7**(3), 247–261 (1977)
40. E.J. Cockayne, R.M. Dawes, T.H. Stephen, Total domination in graphs. Networks **10**(3), 211–219 (1980)
41. B.C. Cuong, Picture fuzzy sets first results, in *Part 1, in Preprint of Seminar on Neuro-Fuzzy Systems with Applications* (Institute of Mathematics, Hanoi, 2013)
42. E. Darabian, R.A. Borzooei, H. Rashmanlou, M. Azadi, New concepts of regular and (highly) irregular vague graphs with applications. Fuzzy Inf. Eng. **9**(2), 161–179 (2017)
43. E. Darabian, R.A. Borzooei, Results on vague graphs with applications in human trafficking. New Math. Nat. Comput. **14**(01), 37–52 (2018)
44. S. Das, O. Egecioglu, A.E. Abbadi, Anonymizing weighted social network graphs, in *Proceedings in ICDE Conference* (2010)
45. A.A. Dobrynin, I. Gutman, S. Klavzar, P. Zigert, Wiener index of hexagonal systems. Acta Appl. Math. **72**, 247–294 (2002)
46. E. Estrada, Characterization of $3D$ molecular structure. Chem. Phys. Lett. **319**, 7–13 (2002)
47. W.L. Gau, D.J. Buehrer, Vague sets. IEEE Trans. Syst. Man Cybern. **23**(2), 610–614 (1993)
48. A.N. Gani, M.B. Ahamed, Strong and weak domination in fuzzy graphs. East Asian Math. J. **23**(1), 1–8 (2007)

49. G. Ghorai, M. Pal, A study on m-polar fuzzy planar graphs. Int. J. Comput. Sci. Math. **7**(3), 283–292 (2016)
50. G. Ghorai, M. Pal, Faces and dual of m-polar fuzzy planar graphs. J. Intell. Fuzzy Syst. **31**, 2041–2049 (2016)
51. G. Ghorai, M. Pal, Some isomorphic properties of m-polar fuzzy graphs with applications. SpringerPlus **5**, 1–21 (2016). https://doi.org/10.1186/s40064-016-3783-z
52. G. Ghorai, M. Pal, Some properties of m-polar fuzzy graphs. Pac. Sci. Rev. A: Nat. Sci. Eng. **18**(1), 38–46 (2016)
53. G. Ghorai, M. Pal, Regular product vague graphs and product vague line graphs. Cogent Math. **3**(1), 121–214 (2016). https://doi.org/10.1080/23311835.2016.1213214
54. G. Ghorai, M. Pal, H. Rashmanlou, R.A. Borzooei, New concept of regularity in product m-polar fuzzy graphs. Int. J. Math. Comput. **28**(4), 9–20 (2017)
55. G. Ghorai, M. Pal, Novel concepts of strongly edge irregular m-polar fuzzy graphs. Int. J. Appl. Comput. Math. **3**(4), 3321–3332 (2017)
56. G. Ghorai, M. Pal, A note on regular bipolar fuzzy graphs. Neural Comput. Appl. **21**(1), 197–205 (2012)
57. A. Guleria, R.K. Bajaj, T-spherical fuzzy graphs: operations and applications in various selection processes. Arabian J. Sci. Eng. **45**, 2177–2193 (2020). https://doi.org/10.1007/s13369-019-04107
58. I. Gutman and N. Trinajstic N, Graph theory and molecular orbitals: total π-electron energy of alternant hydrocarbons. Chem. Phys. Lett. **17**, 535–538 (1972)
59. I. Gutman, B. Ruscic, N. Trinajstic, C.F. Wilcox, Graph theory and molecular orbitals. XII. Acyclic polyenes. J. Chem. Phys. **62**, 3399–3405 (1975)
60. A. Habib, M. Akram, A. Farooq, q-Rung orthopair fuzzy competition graphs with application in the soil ecosystem. Mathematics **7**, 91 (2019)
61. A. Habib, M. Akram, A. Farooq, q-Rung orthopair fuzzy competition graphs with application in the soil ecosystem. Mathematics **7**, 91 (2019). https://doi.org/10.3390/math7010091
62. T. Al-Hawary, L. AlMomani, Balanced fuzzy graphs (2018), http://arxiv-export-lb.library.cornell.edu/abs/1804.08677
63. H. Hosoya, Topological index. A newly proposed quantity characterizing the topological nature of structural isomers of saturated hydrocarbons. Bull. Chem. Sco. Jpn. **44**(9), 2332–2339 (1971)
64. S.R. Islam, S. Maity, M. Pal, Comment on "Wiener index of a fuzzy graph and application to illegal immigration networks". Fuzzy Sets Syst. **384**, 148–151 (2020)
65. A.M. Ismayil, A.M. Ali, On strong interval-valued intuitionistic fuzzy graph. Int. J. Fuzzy Math. Syst. **4**(2), 161–168 (2014)
66. N. Jan, L. Zedam, T. Mahmood, K. Ullah, Cubic bipolar fuzzy graphs with applications. J. Intell. Fuzzy Syst. **37**(2), 2289–2307 (2019)
67. Y.B. Jun, C.S. Kim, K.O. Yang, Cubic sets. Ann. Fuzzy Math. Inf. **4**(1), 83–98 (2012)
68. K.K. Kakkad, S. Sharma, New approach on regular fuzzy graph. Glob. J. Pure Appl. Math. **13**(7), 3753–3766 (2017)
69. S. Kalathodi, M.S. Sunitha, *Distance in Fuzzy Graphs* (LAP LAMBERT Academic Publishing, 2012)
70. S. Kalathian, S. Ramalingam, S. Raman, N. Srinivasan, Some topological indices in fuzzy graphs, in *INFUS 2019: Intelligent and Fuzzy Techniques in Big Data analytics and Decision Making* (2019), pp. 73–81
71. M.G. Karunambigai, S. Sivasankar, K. Palanivel, Secure domination in fuzzy graphs and intuitionistic fuzzy graphs. Ann. Fuzzy Math. Inf. **14**(4), 419–431 (2017)
72. K. Kalpana, S. Lavanya, Connectedness energy of fuzzy graph. J. Comput. Math. Sci. **9**(5), 485–492 (2018)
73. M.H. Khalifeh, H.Y. Azari, A.R. Ashrafi, The hyper wiener index of graph operations. Comput. Math. Appl. **56**, 1402–1407 (2008)
74. P.K.K. Kumar, H. Rashmanlou, A.A. Talebi, F. Mofidnakhaei, Regularity of cubic graph with application. J. Indones. Math. Soc. **25**(01), 1–15 (2019)

75. A. Lakdashti, H. Rashmanlou, R.A. Borzooei, S. Samanta, M. Pal, New concepts of bipolar fuzzy graphs. J. Mult.-Valued Logic Soft Comput. **33**, 117–133 (2019)
76. K.M. Lee, Comparison of interval-valued fuzzy sets, intuitionistic fuzzy sets, and bipolar-valued fuzzy sets. J. Fuzzy Logic Intell. Syst. **14**, 125–129 (2004)
77. T. Mahmood, U. Kifayat, Q. Khan, N. Jan, An approach toward decision-making and medical diagnosis problems using the concept of spherical fuzzy sets. Neural Comput. Appl. **31**, 7041–7053 (2019). https://doi.org/10.1007/s00521-018-3521-2
78. T. Mahapatra, M. Pal, Fuzzy colouring of m-polar fuzzy graph and its application. J. Intell. Fuzzy Syst. **35**, 6379–6391 (2018)
79. O.T. Manjusha, M.S. Sunitha, Notes on domination in fuzzy graphs. J. Intell. Fuzzy Syst. **27**, 3205–3212 (2014)
80. O.T. Manjusha, M.S. Sunitha, Total domination in fuzzy graphs using strong arcs. Ann. Pure Appl. Math. **9**(1), 23–33 (2014)
81. O.T. Manjusha, M.S. Sunitha, Strong domination in fuzzy graphs. Fuzzy Inf. Eng. **7**(3), 369–377 (2015)
82. S. Mandal, S. Sahoo, G. Ghorai, M. Pal, Genus value of m-polar fuzzy graphs. J. Intell. Fuzzy Syst. **34**, 1947–1957 (2018)
83. S. Mandal, S. Sahoo, G. Ghorai, M. Pal, Application of strong arcs in m -polar fuzzy graphs. Neural Process. Lett. **50**(1), 771–784 (2019)
84. S. Mathew, M.S. Sunitha, Types of arcs in a fuzzy graph. Inf. Sci. **179**, 1760–1768 (2009)
85. H.B. Merouane, M. Chellali, On secure domination in graphs. Inf. Process. Lett. **115**(10), 786–790 (2015)
86. S.N. Mishra, A. Pal, Regular interval-valued intuitionistic fuzzy graphs. J. Inf. Math. Sci. **9**(3), 609–621 (2017)
87. V. Mohanaselvi, S. Sivamani, N. Revathi, Global domination in bipolar fuzzy graphs. J. Inf. Math. Sci. **9**(3), 815–825 (2017)
88. D.A. Molodtsov, Soft set theory-first results. Comput. Math. Appl. **37**, 19–31 (1999)
89. J.N. Mordeson, S. Mathew, *Advanced Topics in Fuzzy Graph Theory* (Springer, 2019)
90. J.N. Mordeson, C.S. Peng, Operations on fuzzy graphs. Inf. Sci. **79**, 159–170 (1994)
91. J.N. Mordeson, P.S. Nair, Successor and source of (fuzzy) finite state machines and (fuzzy) directed graphs. Inf. Sci. **95**(1–2), 113–124 (1996)
92. J. N. Mordeson, P.S. Nair, *Fuzzy Graphs and Hypergraphs* (Physica Verlag, 2000)
93. A. Nagoorgani, M. Basheer Ahamed, Order and size in fuzzy graph. Bull. Pure Appl. Sci. **22**(1), 145–148 (2003)
94. A. Nagoorgani, K. Radha, On regular fuzzy graphs. J. Phys. Sci. **12**, 33–40 (2008)
95. A. Nagoorgani, P. Vijayalaakshmi, Insentive arc in domination of fuzzy graph. Int. J. Contemp. Math. Sci. **6**(26), 1303–1309 (2011)
96. A. Nagoorgani, A. Latha, On irregular fuzzy graphs. Appl. Math. Sci. **6**(11), 517–523 (2012)
97. A.N. Gani, K.P. Devi, 2-domination in fuzzy graphs. Int. J. Fuzzy Math. Arch. **9**(1), 119–124 (2015)
98. P.S. Nair, S.C. Cheng, Cliques and fuzzy cliques in fuzzy graphs, in *IFSA World Congress and 20th NAFIPS International Conference*, vol. 4 (2001), pp. 2277– 2280
99. S.M.A. Nayeem, M. Pal, Shortest path problem on a network with imprecise edge weight. Fuzzy Optim. Decis. Making **4**(4), 293–312 (2005)
100. S. Naz, H. Rashmanlou, M.A. Malik, Operations on single valued neutrosophic graphs with application. J. Intell. Fuzzy Syst. **32**(3), 2137–2151 (2017)
101. S. Naz, S. Ashraf, F. Karaaslan, Energy of a bipolar fuzzy graph and its application in decision making. Italian J. Pure Appl. Math. **40**, 339–352 (2018)
102. O. Ore, *Theory of Graphs*, vol. 38 (American Mathematical Society, 1965)
103. M. Pal, H. Rashmanlou, Irregular interval-valued fuzzy graphs. Ann. Pure Appl. Math. **3**(1), 56–66 (2013)
104. M. Pal, Interval-valued fuzzy matrices with interval-valued fuzzy rows and columns. Fuzzy Inf. Eng. **7**(3), 335–368 (2015). https://doi.org/10.1016/j.fiae.2015.09.006

105. M. Pal, Fuzzy matrices with fuzzy rows and columns. J. Intell. Fuzzy Syst. **30**(1), 561–573 (2016)
106. M. Pal, An introduction to intersection graphs, Chapter 2, in *An Handbook of Research on Advanced Applications of Graph Theory in Modern Society*, ed. by M. Pal, S. Samanta, A. Pal (IGI Global, USA, 2020), pp. 24–65
107. R. Parvathi, M.G. Karunambigai, Intuitionistic fuzzy graphs. Comput. Intell. Theory Appl. **38**, 139–150 (2000)
108. B. Praba, V.M. Chandrasekaran, G. Deepa, Energy of an intuitionistic fuzzy graphs. Italian J. Pure Appl. Math. **32**, 431–444 (2014)
109. T. Pramanik, S. Samanta, M. Pal, Interval-valued fuzzy planar graphs. Int. J. Mach. Learn. Cyber. **7**(4), 653–664 (2016). https://doi.org/10.1007/s13042-014-0284-7
110. T. Pramanik, M. Pal, S. Mondal, Interval-valued fuzzy threshold graph. Pac. Sci. Rev. A: Nat. Sci. Eng. **18**, 66–71 (2016)
111. T. Pramanik, S. Samanta, M. Pal, S. Mondal, B. Sarkar, Interval-valued fuzzy ϕ-tolerance competition graphs. SpringerPlus **5**, 1981 (2016). https://doi.org/10.1186/s40064-016-3463-z
112. T. Pramanik, S. Mondal, S. Samanta, M. Pal, A study on bipolar fuzzy planar graph and its application in image shrinking. J. Intell. Fuzzy Syst. **34**(3), 1863–1874 (2018)
113. T. Pramanik, G. Muhiuddin, A.M. Alanazi, M. Pal, An extension of fuzzy competition graph and its uses in manufacturing industries. Mathematics **8**, 1008 (2020)
114. K. Radha, N. Kumaravel, Edge regular property of complement and μ-complement of a fuzzy graph and edge adjacency sequence in fuzzy graph. Int. J. Pure Appl. Math. **107**(3), 673–682 (2016)
115. N. Ramakrishna, Vague graphs. Int. J. Comput. Cognit. **7**, 51–58 (2009)
116. C. Ramprasad, P.L.N. Varma, S. Satyanarayana, N. Srinivasarao, Morphism of m-polar fuzzy graph. Adv. Fuzzy Syst. (2017). https://doi.org/10.1155/2017/4715421
117. M. Randic, Novel molecular descriptor for structure-property studies. Chem. Phys. Lett. **211**(10), 478–483 (1993)
118. H. Rashmanlou, M. Pal, Isometry on interval-valued fuzzy graphs. Int. J. Fuzzy Math. Arch. **3**, 28–35 (2013)
119. H. Rashmanlou, M. Pal, Balanced interval-valued fuzzy graphs. J. Phys. Sci. **17**, 43–57 (2013)
120. H. Rashmanlou, M. Pal, Antipodal interval-valued fuzzy graphs. Int. J. Appl. Fuzzy Sets Artif. Intell. **3**, 107–130 (2013)
121. H. Rashmanlou, M. Pal, Some properties of highly irregular interval valued fuzzy graphs. World Appl. Sci. J. **27**(12), 1756–1773 (2013)
122. H. Rashmanlou, S. Samanta, M. Pal, R.A. Borzooei, A study on bipolar fuzzy graphs. J. Intell. Fuzzy Syst. **28**, 571–580 (2015)
123. H. Rashmanlou, S. Samanta, M. Pal, R.A. Borzooei, Bipolar fuzzy graphs with categorical properties. Int. J. Comput. Intell. Syst. **8**(5), 808–818 (2015)
124. H. Rashmanlou, S. Samanta, M. Pal, R.A. Borzooei, Product of bipolar fuzzy graphs and their degree. Int. J. Gen. Syst. **45**(1), 1–14 (2016)
125. H. Rashmanlou, M. Pal, R.A. Borzooei, F. Mofidnakhaei, B. Sarkar, Product of interval-valued fuzzy graphs and degree. J. Intell. Fuzzy Syst. **35**, 6443–6451 (2018). https://doi.org/10.3233/JIFS-181488
126. S. Rashid, N. Yaqoob, M. Akram, M. Gulistan, Cubic graphs with application. Int. J. Anal. Appl. **16**(5), 733–750 (2018)
127. H. Rashmanlou, G. Muhiuddin, S.K. Amanathulla, F. Mofidnakhaei, M. Pal, A study on cubic graphs with novel application. J. Intell. Fuzzy Syst. https://doi.org/10.3233/JIFS-182929
128. S. Revathi, P.J. Jayalakshmi, C.V.R. Harinarayanan, Perfect dominating sets in fuzzy graphs. IOSR J. Math. **8**(3), 43–47 (2013)
129. A. Rosenfeld, Fuzzy graphs, in *Fuzzy Sets and Their Applications*, ed. by L.A. Zadeh, K.S. Fu, M. Shimura (Academic Press, New York, 1975), pp. 77–95
130. M. Åahin, A. Kargin, Single valued neutrosophic quadruple graphs. Asian J. Math. Comput. Res. **26**(4), 243–250 (2019)

131. S. Samanta, M. Pal, Irregular bipolar fuzzy graphs. Int. J. Appl. Fuzzy Sets **2**, 91–102 (2012)
132. S. Samanta, M. Pal, Bipolar fuzzy hypergraphs. Int. J. Fuzzy Logic Syst. **2**(1), 17–28 (2012)
133. S. Samanta, M. Pal, Fuzzy k-competition graphs and p-competition fuzzy graphs. Fuzzy Eng. Inf. **5**(2), 191–204 (2013)
134. S. Samanta, M. Pal, Some more results on bipolar fuzzy sets and bipolar fuzzy intersection graphs. J. Fuzzy Math. **22**(2), 253–262 (2014)
135. S. Samanta, M. Pal, H. Rashmanlou, R.A. Borzooei, Vague graphs and strengths. J. Intell. Fuzzy Syst. **30**(6), 3675–3680 (2016)
136. F. Smarandache, *Neutrosophy: Neutrosophic Probability, Set, and Logic: Analytic Synthesis and Synthetic Analysis* (American Research Press, Rehoboth, 1998)
137. F. Smarandache, *Types of Neutrosophic Graphs and neutrosophic Algebraic Structures together with their Applications in Technology, seminar* (Universitatea Transilvania din Brasov, Facultatea de Design de Produs si Mediu, Brasov, Romania, 2015)
138. A. Somasundaram, S. Somasundaram, Domination in fuzzy graphs-I. Pattern Recognit. Lett. **19**, 787–791 (1998)
139. A. Somasundaram, Domination in products of fuzzy graphs. Int. J. Uncertainty, Fuzziness Knowl.-Based Syst. **13**(2), 195–204 (2005)
140. M.S. Sunitha, A. Vijayakumar, Complement of a fuzzy graph. Indian J. Pure Appl. Math. **33**, 1451–1464 (2002)
141. F. Sun, X.-P. Wang, X.-B. Qu, Cliques and clique covers in fuzzy graphs. J. Intell. Fuzzy Syst. **31**, 1245–1256 (2016)
142. Y. Talebi, H. Rashmanlou, New concepts of domination sets in vague graphs with applications. Int. J. Comput. Sci. Math. **10**(4), 375–389 (2019)
143. W.B. Vasantha Kandasamy, F. Smarandache, *Fuzzy Cognitive Maps and Neutrosophic Cognitive Maps* (Xiquan Phoenix, 2013)
144. W.B. Vasantha Kandasamy, K. Ilanthenral, F. Smarandache, *Neutrosophic Graphs: A New Dimension to Graph Theory*, Kindle Edition (2015)
145. R. Verma, J.M. Merigo, M. Sahni, Pythagorean fuzzy graphs: some results, arXiv:1806.06721v1
146. M. Vijaya, A. Kannan, Number of Hamiltonian fuzzy cycles in cubic fuzzy graphs with vertices n. Int. J. Pure Appl. Math. **117**(6), 91–98 (2017)
147. H. Wang, F. Smarandache, Y.Q. Zhang, R. Sunderraman, *Interval Neutrosophic Sets and Logic: Theory and Applications in Computing* (Hexis, Phoenix, 2005)
148. H. Wiener, Structural determination of paraffin boiling points. J. Am. Chem. Soc. **69**, 17–20 (1947)
149. R.R. Yager, Pythagorean fuzzy subsets, in *Joint IFSA World Congress and NAFIPS Annual Meeting* (2013), pp. 57–61
150. R.R. Yager, Generalized orthopair fuzzy sets. IEEE Trans. Fuzzy Syst. **25**, 1222–1230 (2017)
151. R.T. Yeh, S.Y. Bang, Fuzzy relations, fuzzy graphs and their applications to clustering analysis, in *It Fuzzy Sets and Their Applications*, ed. by L.A. Zadeh, K.S. Fu, M. Shimura (Academic, New Jersey, 1975), pp. 125–149
152. S. Yin, H. Li, Y. Yang, Product operations on q-rung orthopair fuzzy graphs. Symmetry **11**, 588 (2019). https://doi.org/10.3390/sym11040588
153. L.A. Zadeh, The concept of a linguistic and application to approximate reasoning I. Inf. Sci. **8**, 199–249 (1975)
154. W.R. Zhang, Bipolar fuzzy sets and relations: a computational framework for cognitive modeling and multiagent decision analysis, in *Proceedings of IEEE Conference* (1994), pp. 305–309
155. W.R. Zhang, Bipolar fuzzy sets, in *Proceedings of FUZZ-IEEE* (1998), pp. 835–840

Chapter 2
Fuzzy Planar Graphs

In this chapter, we discuss about planarity and isomorphism in FGs. The first four sections deal with fuzzy planar graph (FPG) and fuzzy dual graph, whereas the last section deals with isomorphism. Having started with work of Abdul Jabbar *et al.* [1], the first work on the planarity of FGs, it becomes an interesting as well as highly researched area in FG theory. Nirmala and Dhanabal [12] also made a noteworthy contribution in the said field. In these papers, the authors considered that FPG is a graph whose underlying crisp graph is planar and investigated some important properties. But, these works had one limitation—they did not allow crossing of edges in the FPG. In this chapter a new concept of FPG has been discussed that allows the crossing of edges in the FG. To investigate such type of FPG, a parameter called degree of planarity has been defined which measures how much the graph is planar. Most of the results in the next few sections are mostly based on the works of [3, 7, 8, 14–17, 19].

Based on the paper [19], many related works have been published recently [2, 4–6, 9, 10, 13, 18].

2.1 Concept of Crossing Between Edges

A graph (crisp) G is called a planar graph if it can be traced on a plane without the intersection of its edges for some geometric representation of G. Otherwise, the graph is called non-planar. A tracing of a graph on any surface such that none of its edges intersects for any geometric representation is called an embedding of the graph. Therefore, in order to show that a graph G is non-planar, we need to show that it can not be embedded in the plane for any of its geometric representations of G. Analogously, a geometric graph G is said to be planar if there exists a graph which is isomorphic to G and can be embedded in a plane. A plane representation of G is an embedding of a planar graph G on a plane.

M. Pal et al., *Modern Trends in Fuzzy Graph Theory*, https://doi.org/10.1007/978-981-15-8803-7_2

Fig. 2.1 A FG in which crossing between edges has no significance

Suppose G is a crisp graph such that there is only one crossing between two edges (a_1, a_2) and (a_3, a_4) for a certain geometrical representation. It can be thought that these two edges have membership values equal to 1 in fuzzy sense. If the edge (a_3, a_4) is removed from the graph, then the graph becomes planar. In this case, we say that the membership value of the edge (a_3, a_4) is 0.

Let $\mathscr{G} = (\mathscr{V}, \sigma, \mu)$ be a FG. For simplicity, we assume that there is only one crossing between the edges $((a_4, a_1), \mu(a_4, a_1))$ and $((a_2, a_3), \mu(a_2, a_3))$ for a particular geometric representation. Then we have the following three cases.

Case(i): Suppose $\mu(a_4, a_1) = 1$ and $\mu(a_2, a_3) = 0$ or $\mu(a_4, a_1) = 0$ and $\mu(a_2, a_3) = 0$. In this case, there is no crossing in the graph.

For explanation, we assume that $\mu(a_4, a_1)$ represents the degree of crowdness on the road (a_4, a_1) (see Fig. 2.1).

In Fig. 2.1, both membership values $\mu(a_4, a_1)$ and $\mu(a_2, a_3)$ are tending to zero. So, in this case, we say that there is no crossing between the edges or we can say that the crossing that exists is very very less significance.

Case(ii): If $\mu(a_4, a_1)$ is equal to 1 or close to 1 and the value of $\mu(a_2, a_3)$ is near to 0, in this case also the crossing is not significant for maintaining the traffic.

In Fig. 2.2, the value of $\mu(a_4, a_1)$ is near to 1 and $\mu(a_2, a_3)$ is close to 0. In this case also, one can ignore the crossing.

Case(iii): If the values of both $\mu(a_4, a_1)$ and $\mu(a_2, a_3)$ are close to 1, then we say that the crossing is highly significant for the planarity of the graph.

In Fig. 2.3, the membership values for both the edges are close to 1. So the crossing cannot be ignored at all and has high significance for maintaining the traffic accordingly.

Before defining the FPG, some associated terms need to be introduced.

Fig. 2.2 A FG in which crossing between edges is not significant

Fig. 2.3 A FG in which crossing between edges is highly significant

2.1.1 Intersecting Value in Fuzzy Multi-graph

In a (fuzzy) multi-graph, parallel edges are allowed. If two edges intersect at a point in a fuzzy multi-graph (FMG), a value is assigned to the corresponding point and it is called the intersecting value. Suppose $\mathscr{G} = (\mathscr{V}, \sigma, \mu)$ is a FMGand its underlying graph is G^* and let two edges $((a_1, a_2), (a_1, a_2)_{\mu^k})$ $((a_1, a_2)_{\mu^k}$ represents the membership value of the kth edge $(a_1, a_2))$ and $((a_3, a_4), (a_3, a_4)_{\mu^l})$ intersect at a point P.

The **strength of the edge** (a_1, a_2) is determined by the following formula

$$I_{(a_1, a_2)} = \frac{(a_1, a_2)_{\mu^k}}{\min\{\sigma(a_1), \sigma(a_2)\}}.$$

If $I_{(a_1, a_2)} \geq 0.5$, then the edge is said to be strong. Otherwise, it is called a weak edge.

Based on the strengths on the edges, another parameter called **intersecting value** at the point P is defined below

$$\mathscr{I}_P = \frac{I_{(a_1, a_2)} + I_{(a_3, a_4)}}{2} , \mu(a_1, a_2) > 0 \text{ and } \mu(a_3, a_4) > 0.$$

If there is no intersection between the edges, then the value of intersection is taken as 0.

Motivated from the above discussion, we define FPG in a new way. In our approach, any FG is a FPG whatever may be the number of crossings among the edges in the graph. If there is no crossing among the edges, then the graph is same as a crisp planar graph. If in a graph, there is a crossing among the edges whose membership values are very low, then such graph is also considered as a FPG with certain degree of planarity (DOP). This parameter depends on the number of points of intersection in the graph and it is defined below.

Definition 2.1 For a certain geometrical representation of a FMG $\mathscr{G}$, let $P_1, P_2, \ldots, P_z$ be the points of intersection between the edges in $\mathscr{G}$. The FG $\mathscr{G}$ is called a FPG with DOP $f(\mathscr{G})$, where

$$f(\mathscr{G}) = \frac{1}{1 + \{\mathscr{I}_{P_1} + \mathscr{I}_{P_2} + \cdots + \mathscr{I}_{P_z}\}}.$$

From the definition, it is obvious that the range of $f(\mathscr{G})$ is $0 < f(\mathscr{G}) \leq 1$.

It is clear that if there is no point of intersection between any pair of edges for a particular representation of a FG, then its DOP is 1. In this case, the underlying graph of this FG is nothing but the crisp planar graph. For the FG $\mathscr{G}$, if $f(\mathscr{G})$ decreases, then the number of points of intersection between the edges increases or the intersecting values of the edges increases and hence the nature of planarity decreases. Hence, every FG is a FPG with some DOP.

Fig. 2.4 A FPG with DOP 0.4683

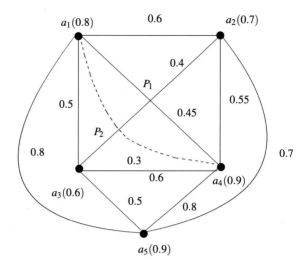

Example 2.1 Let us consider a FG $\mathscr{G} = (\mathscr{V}, \sigma, \mu)$ whose set of vertices be $\mathscr{V} = \{a_1, a_2, a_3, a_4, a_5\}$ and the membership values of the vertices and edges be $\sigma(a_1) = 0.8$, $\sigma(a_2) = 0.7$, $\sigma(a_3) = 0.6$, $\sigma(a_4) = 0.9$, $\sigma(a_5) = 0.9$ and $(a_1, a_2)_{\mu^1} = 0.6$, $(a_1, a_3)_{\mu^1} = 0.5$, $(a_1, a_4)_{\mu^1} = 0.45$, $(a_1, a_4)_{\mu^2} = 0.3$, $(a_2, a_3)_{\mu^1} = 0.4$, $(a_2, a_4)_{\mu^1} = 0.55$, $(a_3, a_4)_{\mu^1} = 0.6$, $(a_1, a_5)_{\mu^1} = 0.8$, $(a_3, a_5)_{\mu^1} = 0.5$, $(a_5, a_4)_{\mu^1} = 0.8$, $(a_5, a_2)_{\mu^1} = 0.7$ (Fig. 2.4). For this graph, there are two points of intersections P_1 and P_2. P_1 is a point of intersection between the edges $((a_1, a_4), 0.45)$ and $((a_2, a_3), 0.4)$ and P_2 is between $((a_1, a_4), 0.3)$ and $((a_2, a_3), 0.4)$.

Now, $I_{(a_1,a_4)} = \frac{0.45}{0.8} = 0.5625$ and $I_{(a_2,a_3)} = \frac{0.4}{0.6} = 0.6667$. Also $\mathscr{I}_{P_1} = 0.6146$ and $\mathscr{I}_{P_2} = 0.5208$. Hence, the DOP of the FMG of Fig. 2.4 is 0.4683.

The DOP of a complete FG is determined in the following theorem.

Theorem 2.1 *For a complete FG $\mathscr{G}$, the DOP $f(\mathscr{G})$ is given by $f(\mathscr{G}) = \frac{1}{1+N}$, where N represents the number of points of intersection between the edges in $\mathscr{G}$.*

Proof Let $\mathscr{G} = (\mathscr{V}, \sigma, \mu)$ be a complete FMG.   □

Let the number of edges between the vertices a_1 and a_2 be $p_{a_1 a_2}$. Since, $\mathscr{G}$ is complete FMG, $(a_1, a_2)_{\mu^j} = \min\{\sigma(a_1), \sigma(a_2)\}$ for each $j = 1, 2, \ldots, p_{a_1 a_2}$ and for all $a_1, a_2 \in \mathscr{V}$.

Let $P_1, P_2, \ldots, P_N$ be the point of intersections in $\mathscr{G}$.

For any edge (a_1, a_2) in a complete FMG, $I_{(a_1,a_2)} = \frac{(a_1,a_2)_{\mu^j}}{\min\{\sigma(a_1),\sigma(a_2)\}} = 1$.

Let P_1 be a point of intersection, then $\mathscr{I}_{P_1} = \frac{1+1}{2} = 1$.

Hence, $\mathscr{I}_{P_i} = 1$ for $i = 1, 2, \ldots, N$.

Now,

$$f(\mathcal{G}) = \frac{1}{1 + \mathcal{I}_{P_1} + \mathcal{I}_{P_2} + \cdots + \mathcal{I}_{P_N}}$$

$$= \frac{1}{1 + (1 + 1 + \cdots + 1)}$$

$$= \frac{1}{1 + N}.$$

2.2 Effective Edges and Considerable Edges

Effective edges and considerable edges play an important role to determine the number of points of intersections in a FPG.

Here two types of edges are mentioned in a FMG, namely effective and considerable edges.

The edge (a_1, a_2) is called effective, if $I_{(a_1, a_2)} \geq 0.5$.

We define another type of edge called considerable edge which is useful in modeling certain problems.

Definition 2.2 Let $\mathcal{G} = (\mathcal{V}, \sigma, \mu)$ be a FG and κ $(0 < \kappa < 0.5)$ be a real number. An edge (a_1, a_2) is said to be a **considerable edge** if $\frac{\mu(a_1, a_2)}{\min\{\sigma(a_1), \sigma(a_2)\}} \geq \kappa$. Otherwise, the edge is called non-considerable.

For FMG, a multiedge (a_1, a_2) is said to be considerable if $\frac{(a_1, a_2)_{\mu^j}}{\min\{\sigma(a_1), \sigma(a_2)\}} \geq \kappa$ for all $j = 1, 2, \ldots, p_{a_1 a_2}$.

The number κ is called **considerable number** of the FG $\mathcal{G} = (\mathcal{V}, \sigma, \mu)$ if $\frac{\mu(a_1, a_2)}{\min\{\sigma(a_1), \sigma(a_2)\}} \geq \kappa$ for all edges (a_1, a_2). The value of κ is provided by the decision-maker.

Therefore, for a given value of κ, there is a set of considerable edges.

There could be an intersection between considerable edges and effective edges. The number of points of intersection between them is determined from the following.

Theorem 2.2 *Let $\mathcal{G}$ be a FPG'with DOP greater than 0.5 and κ be a given considerable number. Then number of points of intersection between considerable edges in $\mathcal{G}$ is at most $[\frac{1}{\kappa}]$ or $\frac{1}{\kappa} - 1$ according as $\frac{1}{\kappa}$ is not an integer or an integer respectively.*

Proof For a considerable edge $((a_1, a_2), (a_1, a_2)_{\mu^j})$, $(a_1, a_2)_{\mu^j} \geq \kappa \min\{\sigma(a_1), \sigma(a_2)\}$.

Therefore, $I_{(a_1, a_2)} \geq \kappa$.

Let the number of points of intersection between the considerable edges in $\mathcal{G}$ be m.

Let the ith point of intersect be P_i and let it be the intersection between the considerable edges $((a_1, a_2), (a_1, a_2)_{\mu^j})$ and $((a_3, a_4), (a_3, a_4)_{\mu^i})$.

Thus, $\mathcal{I}_{P_i} = \frac{1}{2}[I_{(a_1, a_2)} + I_{(a_3, a_4)}] \geq \kappa$.

Hence, $\sum_{i=1}^{m} \mathscr{I}_{P_i} \geq mc$ and so $f(\mathscr{G}) \leq \frac{1}{1+m\kappa}$.

Since, $f(\mathscr{G}) > 0.5, 0.5 < f(\mathscr{G}) \leq \frac{1}{1+m\kappa}$, i.e. $0.5 < \frac{1}{1+m\kappa}$.

It follows that $m < \frac{1}{\kappa}$.

This inequality is true for some integer values of m which is given in the following expression:

$$m = \begin{cases} [\frac{1}{\kappa}], & \text{if } \frac{1}{\kappa} \text{ is not an integer} \\ \frac{1}{\kappa} - 1, & \text{if } \frac{1}{\kappa} \text{ is an integer,} \end{cases}$$

$\square$

An important result is established below.

Theorem 2.3 *For any FG $\mathscr{G}$, if $f(\mathscr{G}) > 0.5$, then the number of points of intersection between effective edges in $\mathscr{G}$ is at most one.*

Proof Let, if possible, P_1 and P_2 be two points of intersection between effective edges in $\mathscr{G}$.

Then $I_{(a_1,a_2)} \geq 0.5$ for any effective edge (a_1, a_2).

If P_1 be the point of intersection between the effective edges $((a_1, a_2), (a_1, a_2)_{\mu^j})$ and $((a_3, a_4), (a_3, a_4)_{\mu^i})$, then $\frac{I_{(a_1,a_2)}+I_{(a_3,a_4)}}{2} \geq 0.5$.

Therefore, $\mathscr{I}_{P_1} \geq 0.5$ and similarly, $\mathscr{I}_{P_2} \geq 0.5$.

Thus, $1 + \mathscr{I}_{P_1} + \mathscr{I}_{P_2} \geq 2$.

Hence, $f(\mathscr{G}) = \frac{1}{1+\mathscr{I}_{P_1}+\mathscr{I}_{P_2}} \leq 0.5$. It contradicts the assumption $f(\mathscr{G}) > 0.5$.

Thus, the maximum number of points of intersection between the effective edges in $\mathscr{G}$ is one. $\square$

The following example justifies the above result.

Example 2.2 Let us consider two FPGs as shown in Fig. 2.5. In Fig. 2.5a, there is only one point of intersection between the effective edges (a_1, a_3) and (a_2, a_4). Let $\sigma(a_1) = \sigma(a_2) = \sigma(a_3) = \sigma(a_4) = 1$ and $\mu(a_1, a_3) = 0.99$, $\mu(a_2, a_4) = 0.99$, other values are not required. The DOP for this graph is 0.50251, i.e. $f(\mathscr{G}) > 0.5$ and the number of points of intersection is 1.

In Fig. 2.5b, there are two points of intersection between the effective multi-edges $(a_1, a_3), (a_2, a_4)$. For this FG, let $\sigma(a_1) = \sigma(a_2) = \sigma(a_3) = \sigma(a_4) = 1$ and $(a_1, a_3)_{\mu^1} = 0.5$, $(a_2, a_4)_{\mu^1} = 0.5$, $(a_1, a_3)_{\mu^2} = 0.5$, $(a_2, a_4)_{\mu^2} = 0.5$. The DOP of this graph is 0.5. Again, if there is no crossing, then $f(\mathscr{G})$ must be greater than 0.5.

These examples verify the statement of Theorem 2.3.

Theorem 2.4 *Let $f(\mathscr{G}) > 0.67$ for a FG G. Then no two effective edges intersect in $\mathscr{G}$.*

Proof If possible, let two effective fuzzy edges intersect at any point P.

Now, for any effective edge (a, b), $I_{(a,b)} \geq 0.5$.

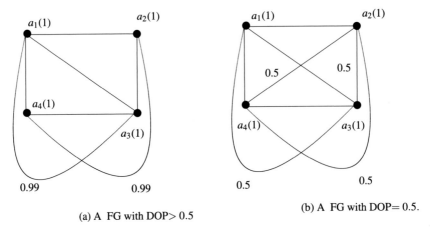

(a) A FG with DOP> 0.5

(b) A FG with DOP= 0.5.

Fig. 2.5 Fuzzy planar graphs with different DOP

We must have $\mathscr{I}_P = 0.5$ for least values of $I_{(a,b)}$ and $I_{(c,d)}$.
Hence, $f(\mathscr{G}) = \frac{1}{1+0.5} < 0.67$, a contradiction.
Thus, $\mathscr{G}$ has no point of intersection between the effective edges.   □

Theorem 2.4 motivates us to introduce a new type of FPG, called **strong FPG** whose DOP is more than or equal to 0.67. This FPG is also called **0.67-FPG**.

If DOP of a FG $\mathscr{G}$ is 1, its underlying graph is crisp planar graph and if it tends to 1, then the geometrical structure of $\mathscr{G}$ violates the (crisp) planarity. Therefore, strong FPGs are more significant and need further investigation.

2.3 Fuzzy Faces in Fuzzy Graphs

In a crisp planar graph, every bounded region is a face along with an unbounded region. There exists no crossing between edges in planar graphs. But, in a FPG there may be one or more crossings between the edges. So, it is very difficult to identify any bounded region in FPG. But, a FPG with DOP at lest 0.67 has no crossing among the edges. For this type of FPG ,we define fuzzy faces. To define fuzzy faces for other types of FPG ,further investigation is required.

The crisp complete graph K_5 and complete bipartite graph $K_{3,3}$ are the smallest non-planar graphs. These graphs cannot be drawn in a plain paper without crossings of edges. So, any graph containing K_5 or $K_{3,3}$ as a subgraph is non-planar. But, in our new concept, the complete FG is a FPG with some DOP.

Lemma 2.1 *The complete FG $\mathscr{K}_5$ and complete fuzzy bipartite graph $\mathscr{K}_{3,3}$ are FPGs with DOP 0.5.*

Fig. 2.6 An example of a FPG with DOP more than 0.5

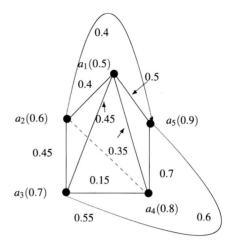

Proof Let $\{a_1, a_2, a_3, a_4, a_5\}$ be the set of vertices of $\mathcal{K}_5$. Since $\mathcal{K}_5$ is a CFG, so for all $a_i, a_j, i \neq j, \mu(a_i, a_j) = \min\{\sigma(a_i), \sigma(a_j)\}$.

For CFG $\mathcal{G}$ DOP is $f(\mathcal{G}) = \frac{1}{1+N}$, where N is the number of points of intersection of the edges in $\mathcal{G}$.

The CFG $\mathcal{K}_5$ cannot be drawn without crossing of a pair of edges. So, for this graph DOP is $\frac{1}{1+1} = 0.5$. Hence, $\mathcal{K}_5$ is a FPG with DOP 0.5.

Similarly, it can be proved that the DOP of the complete fuzzy bipartite graph $K_{3,3}$ is 0.5. $\qquad \square$

Let us consider a FG $\mathcal{G}_5$ with five vertices whose underlying graph is a complete graph, shown in Fig. 2.6. In this graph also, there is only one point of intersection between two fuzzy edges $((a_2, a_4), 0.15), ((a_1, a_3), 0.45)$. The DOP of this FG $\mathcal{G}_5$ is 0.63492. Note that this graph has five vertices and each pair of vertices is connected by a fuzzy edge. But, $\mathcal{G}_5$ is not a CFG.

From Lemma 2.1 and Fig. 2.6, it is observed that the DOP of a FG whose underlying graph is K_5 may or may not be greater than 0.5.

The fuzzy face of a FPG is defined below.

Like crisp graph, fuzzy face in a FG is a region bounded by fuzzy edges except for the outer region. If the membership values of all edges in the boundary of a fuzzy face are 1, it becomes a crisp face. If one such edge is removed or membership value becomes 0, then no such fuzzy face exists. Thus, the existence of a fuzzy face depends on the minimum value of the strengths of the fuzzy edges in its boundary. A fuzzy face along with its membership value is defined below.

Definition 2.3 Let $\mathcal{G} = (\mathcal{V}, \sigma, \mu)$ be a FPG. A **fuzzy face** of $\mathcal{G}$ is a region bounded by the set of fuzzy edges $E' \subseteq \mathcal{V} \times \mathcal{V}$ of $\mathcal{G}$. The **membership value of the fuzzy face** is

$$\min \left\{ \frac{(a, b)}{\min\{\sigma(a), \sigma(b)\}}, \text{ for all } (a, b) \in E' \right\}.$$

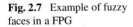

Fig. 2.7 Example of fuzzy
faces in a FPG

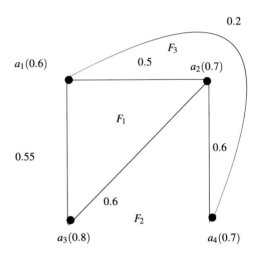

If the membership value of a fuzzy face is more than 0.5, then it is called a **strong fuzzy face**,, otherwise a weak face. The faces bounded by fuzzy edges are called **inner fuzzy faces**. Every FPG has an infinite region which is called **outer fuzzy face**.

Example 2.3 Let us consider the FG of Fig. 2.7. In this FG, there are three faces F_1, F_2, and F_3. The face F_1 is bounded by the edges $((a_1, a_2), 0.5)$, $((a_2, a_3), 0.6)$, $((a_1, a_3), 0.55)$ and its membership value is 0.833. Similarly, the edges of the face F_3 are $((a_1, a_2), 0.5)$, $((a_2, a_4), 0.6)$, $((a_1, a_4), 0.2)$ with membership value 0.33. We observed that F_2 is the outer fuzzy face and its membership value is 0.33. So F_1 is a strong fuzzy face and F_2, F_3 are weak fuzzy faces.

Since the membership value of every strong fuzzy face is more than 0.5, every edge of a strong fuzzy face must be strong.

2.4 Fuzzy Dual Graph

Every crisp planar graph has its dual, but, this is not true for FPG in general. Only for strong FPG ,the dual has been defined. The fuzzy dual graph $\mathcal{G}'$ for a FPG $\mathcal{G}$ is constructed by the following rule:
For each face of a strong FPG $\mathcal{G}$, there is a vertex in $\mathcal{G}'$ and there is an edge between two vertices in $\mathcal{G}'$ if the regions in $\mathcal{G}$ have a common edge.
The formal definition is given below.

Definition 2.4 Let $\mathcal{G} = (\mathcal{V}, \sigma, \mu)$ be a strong FPG and $F_1, F_2, \ldots, F_k$ be the strong fuzzy faces of $\mathcal{G}$. The **fuzzy dual graph** of $\mathcal{G}$ is a FPG $\mathcal{G}' = (\mathcal{V}', \sigma', \mu')$, where set of vertices is $\mathcal{V}' = \{x_i, i = 1, 2, \ldots, k\}$. For each face F_i in $\mathcal{G}$, there is a vertex x_i

Fig. 2.8 Illustration of a
FPG and its dual

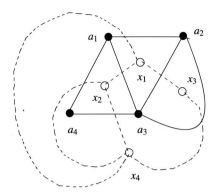

in $\mathscr{G}'$ for $i = 1, 2, \ldots, k$. If (u, v) is a common edge between two faces F_i and F_j then the corresponding vertices x_i and x_j are connected by an edge.

The membership values of vertices are given by the mapping $\sigma' : \mathscr{V}' \to [0, 1]$ such that $\sigma'(x_i) = \max\{\mu(u, v) : (u, v)$ is an edge of the boundary of the strong fuzzy face $F_i\}$.

The membership value of the fuzzy edge (x_i, x_j) in $\mathscr{G}'$ is the membership value of the common edge between the strong fuzzy faces F_i and F_j.

In case of a strong pendant edge in $\mathscr{G}$, a self-loop in $\mathscr{G}'$ is considered and its membership value is the membership value of the pendant edge.

It may be noted that the fuzzy dual graph of a strong FPG is also a strong FPG with DOP 1.

Thus, the fuzzy face of a fuzzy dual graph can be similarly described as was in the case of strong fuzzy planar graphs.

Example 2.4 A strong FPG $\mathscr{G} = (\mathscr{V}, \sigma, \mu)$ is considered in Fig. 2.8, where $\mathscr{V} = \{a_1, a_2, a_3, a_1\}$ is given. For this graph, let $\sigma(a_1) = 0.6, \sigma(a_2) = 0.7, \sigma(a_3) = 0.8, \sigma(a_4) = 0.9$ and the edges and their membership values be $\{((a_1, a_2), 0.5), ((a_1, a_3), 0.4), ((a_1, a_4), 0.55), ((a_2, a_3), 0.45), ((a_2, a_3), 0.6), ((a_3, a_4), 0.7)\}$.

The faces of $\mathscr{G}$ are

F_1 bounded by $((a_1, a_2), 0.5), ((a_1, a_3), 0.4), ((a_2, a_3), 0.45))$,

F_2 bounded by $((a_1, a_4), 0.55), ((a_3, a_4), 0.7), ((a_1, a_3), 0.4))$,

F_3 bounded by $((a_2, a_3), 0.45), ((a_2, a_3), 0.6))$, and

outer fuzzy face F_4 surrounded by $((a_1, a_2), 0.5), ((a_2, a_3), 0.6), ((a_3, a_4), 0.7), ((a_1, a_4), 0.55))$.

The fuzzy dual graph is constructed as follows. Note that for this graph all the fuzzy faces are strong. For each strong fuzzy face, we consider a vertex for the fuzzy dual graph. Thus, the vertex set of $\mathscr{G}'$ is $\mathscr{V}' = \{x_1, x_2, x_3, x_4\}$. Now, $\sigma'(x_1) = \max\{0.5, 0.4, 0.45\} = 0.5, \sigma'(x_2) = \max\{0.55, 0.7, 0.4\} = 0.7, \sigma'(x_3) = \max\{0.45, 0.6\} = 0.6$ and $\sigma'(x_4) = \max\{0.5, 0.6, 0.7, 0.55\} = 0.7$.

There are two common edges (a_1, a_4) and (a_3, a_4) between the faces F_2 and F_4 in $\mathscr{G}$. Hence, there are two edges in $\mathscr{G}'$ between the vertices x_2 and x_4. The membership values of these edges are given by $(x_2, x_4)_1 = (a_3, a_4) = 0.7$, $(x_2, x_4)_2 = (a_1, a_4) = 0.55$.

The membership values of other edges in $\mathscr{G}'$ are determined as $(x_1, x_3) = (a_2, a_3) = 0.45$, $(x_1, x_2) = (a_1, a_3) = 0.4$, $(x_1, x_4) = (a_1, a_2) = 0.5$, $(x_3, x_4) = (a_2, a_3) = 0.6$.

Thus, the edge set of the fuzzy dual graph is $E' = \{((x_1, x_3), 0.45), ((x_1, x_2), 0.4), ((x_1, x_4), 0.5), ((x_3, x_4), 0.6), ((x_2, x_4), 0.7), ((x_2, x_4), 0.55)\}$.

The fuzzy dual graph $\mathscr{G}'$ is shown in Fig. 2.8. The dotted lines represent the edges of the fuzzy dual graph.

If n, e, f denote the numbers of vertices, edges, and regions of a connected crisp planar graph G, and if n^*, e^*, f^* are the corresponding numbers in the dual graph G^*, then $n^* = f$, $e^* = e$ and $f^* = n$.

Similar result also holds in FPG which are stated below.

Theorem 2.5 *Let $\mathscr{G}$ be a strong FPG and let the number of vertices, edges, and strong faces be n, p, m respectively. Also, let n', p', m' be the number of vertices, edges, and strong faces in the dual graph $\mathscr{G}'$. Then*
(i) $n' = m$,
(ii) $p' = p$,
(iii) $m' = n$.

The number of strong fuzzy faces in the dual FG of a strong FPG $\mathscr{G}$ is less than or equal to the number of vertices of $\mathscr{G}$ as all fuzzy faces of a fuzzy dual graph may not be a strong fuzzy face.

An example is considered for supporting the statement. Let $\mathscr{G} = (\mathscr{V}, \sigma, \mu)$ be a strong FPG, where $\mathscr{V} = \{a_1, a_2, a_3, a_4\}$, $\sigma(a_1) = 0.8$, $\sigma(a_2) = 0.7$, $\sigma(a_3) = 0.9$, $\sigma(a_4) = 0.3$ and the set of edges be $\{((a_1, a_2), 0.7), ((a_2, a_3), 0.7), ((a_3, a_4), 0.2), ((a_2, a_4), 0.2), ((a_1, a_4), 0.2)\}$. The fuzzy dual graph is $\mathscr{G}' = (\mathscr{V}', \sigma', \mu')$, where $\mathscr{V}' = \{x_1, x_2, x_3\}$, $\sigma'(x_1) = 0.7$, $\sigma'(x_2) = 0.7$, $\sigma'(x_3) = 0.7$ and set of edges $\{((x_1, x_2), 0.2), ((x_1, x_3), 0.2), ((x_1, x_3), 0.7), ((x_2, x_3), 0.2), ((x_2, x_3), 0.7)\}$.

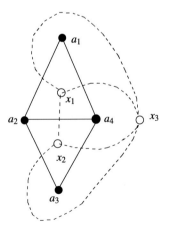

Fig. 2.9 An example of a fuzzy dual graph with strong face

Here, the number of strong fuzzy face is one while the number of fuzzy faces is three (see Fig. 2.9).

2.5 Isomorphism on Fuzzy Planar Graphs

In this section, the isomorphism between FGs is investigated. If there is an isomorphism between two FGs and one of them be a FPG, then the other FG is also a FPG. This is proved in the following theorem. Most of this work is taken from [16, 17].

Theorem 2.6 *Let* $\mathscr{G}$ *be a FPG and* $\mathscr{G}_1$ *be any FG. If there exists an* **isomorphism** $h : \mathscr{G} \to \mathscr{G}_1$, *then* $\mathscr{G}_1$ *is also a FPG with same DOP as of* $\mathscr{G}$.

Proof By definition of isomorphism, the membership values of edges and vertices of $\mathscr{G}$ and $\mathscr{G}_1$ are same. Also, the order and size of $\mathscr{G}$ and $\mathscr{G}_1$ are same [11].

Then, $\mathscr{G}_1$ can be drawn as $\mathscr{G}$. Hence, the number of intersection between the edges is same and hence the DOP of $\mathscr{G}_1$ is also same as $\mathscr{G}$. □

For a crisp graph, the dual of dual of a planar graph is the original planar graph. But, in FG theory this result is not valid, i.e. the fuzzy dual of a fuzzy dual is not isomorphic to the original FPG. The membership values of vertices of a fuzzy dual graph are the maximum membership values of its bounding edges of the corresponding fuzzy faces in the FPG. Thus, the membership values of the vertices are not preserved in the fuzzy dual graph. But, edge membership values are preserved in a fuzzy dual graph. This result is established in the following theorem.

Theorem 2.7 *Let* $\mathscr{G}_2$ *be the fuzzy dual of a strong FPG* $\mathscr{G}$ *without weak edges. Then a co-weak isomorphism exists between* $\mathscr{G}$ *and* $\mathscr{G}_2$.

Proof Let $\mathscr{G}_1$ be the fuzzy dual of $\mathscr{G}$ and $\mathscr{G}_2$ be the same for $\mathscr{G}_1$. Now, we show that there is a co-weak isomorphism between $\mathscr{G}_2$ and $\mathscr{G}$.

By Theorem 2.5, the number of vertices of $\mathscr{G}_2$ and $\mathscr{G}$ are same. By definition of fuzzy dual graph, the membership value of an edge in a fuzzy dual graph is equal to the membership value of an edge in FPG. Thus, one can construct a co-weak isomorphism from $\mathscr{G}_2$ to $\mathscr{G}$. □

Theorem 2.7 is illustrated by considering the following example.

Let us consider three FGs $\mathscr{G} = (\mathscr{V}, \sigma, \mu)$, $\mathscr{G}_1 = (\mathscr{V}_1, \sigma_1, \mu_1)$, and $\mathscr{G}_2 = (\mathscr{V}_2, \sigma_2, \mu_2)$ and they are shown in Fig. 2.10a, b, and c respectively. Here, $\mathscr{G}$ is a strong FG, G_1 is the dual of $\mathscr{G}$ and $\mathscr{G}_2$ is that of $\mathscr{G}_1$. Now, a bijective mapping can be defined from $\mathscr{V}_2$ to $\mathscr{V}$. Similarly, we can extend the mapping from the edge set of $\mathscr{G}_2$ to the edge set of $\mathscr{G}$. It is observed that $\mu_2(a) \le \mu(b)$, for all $a \in \mathscr{V}_2$ and $b \in \mathscr{V}$ under the mapping and edge membership values are equal under the mapping. Thus, the mapping is said to satisfy the co-weak isomorphism property.

Two FPGs with same number of vertices may be isomorphic. There is a very interesting result between DOP of two FPGs.

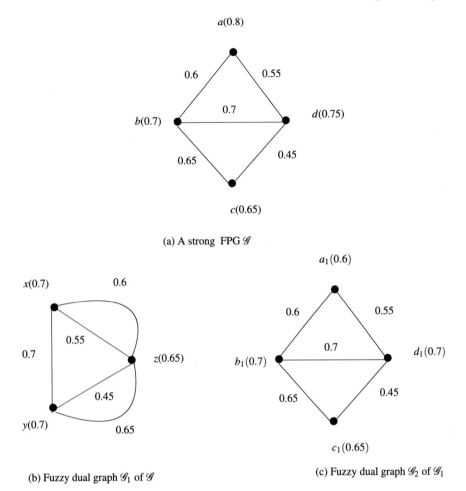

(a) A strong FPG $\mathscr{G}$

(b) Fuzzy dual graph $\mathscr{G}_1$ of $\mathscr{G}$

(c) Fuzzy dual graph $\mathscr{G}_2$ of $\mathscr{G}_1$

Fig. 2.10 Dual of dual is co-weak isomorphic for a FPG

Theorem 2.8 *Let $\mathscr{G}_1$ and $\mathscr{G}_2$ be two weak isomorphic FPGs with degrees of planarity $f(\mathscr{G}_1)$ and $f(\mathscr{G}_2)$ respectively. If the membership values of the intersecting edges are same, then $f(\mathscr{G}_1) = f(\mathscr{G}_2)$.*

Proof Let $\mathscr{G}_1 = (\mathscr{V}, \sigma_1, \mu_1)$ and $\mathscr{G}_2 = (\mathscr{V}, \sigma_2, \mu_2)$ be weak isomorphic FGs.

Since $\mathscr{G}_1$ and $\mathscr{G}_2$ are weak isomorphic, $\sigma_1(a) = \sigma_2(b)$ for some a in $\mathscr{V}_1$ and some b in $\mathscr{V}_2$.

For simplicity, it is assumed that there is only one point of intersection in both the graphs. Let the edges (a_1, b_1) and (c_1, d_1) be intersected in $\mathscr{G}_1$, and the corresponding edges in $\mathscr{G}_2$ be (a_2, b_2) and (c_2, d_2).

Then the intersecting value at the point of intersection in $\mathscr{G}_1$ is
$\frac{1}{2}\left[\frac{\mu(a_1,b_1)}{\sigma(a_1)\wedge\sigma(b_1)} + \frac{\mu(c_1,d_1)}{\sigma(c_1)\wedge\sigma(d_1)}\right]$. Similarly, that in $\mathscr{G}_2$ is $\frac{1}{2}\left[\frac{\mu(a_2,b_2)}{\sigma(a_2)\wedge\sigma(b_2)} + \frac{\mu(c_2,d_2)}{\sigma(c_2)\wedge\sigma(d_2)}\right]$.
Now, $f(\mathscr{G}_1) = f(\mathscr{G}_2)$, if $\mu(a_1, b_1) = \mu(a_2, b_2)$.

Now, if the sum of the intersecting values of $\mathscr{G}_1$ and $\mathscr{G}_2$ are same, then the DOPs of $\mathscr{G}_1$ and $\mathscr{G}_2$ are equal.

Thus, for equality of DOPs, the membership values of intersecting edges in both the graphs must be equal.

$\square$

Corollary 2.1 *Let $\mathscr{G}_1$ and $\mathscr{G}_2$ be two co-weak isomorphic FGs with DOPs $f(\mathscr{G}_1)$ and $f(\mathscr{G}_2)$ respectively. If the minimum of membership values of the end vertices of corresponding intersecting edges are same, then $f(\mathscr{G}_1) = f(\mathscr{G}_2)$.*

References

1. N. Abdul-jabbar, J.H. Naoom, E.H. Ouda, Fuzzy dual graph. J. Al-Nahrain Univ. **12**(4), 168–171 (2009)
2. M. Akram, Single-valued neutrosophic planar graphs. Int. J. Algebra Stat. **5**(2), 157–167 (2016). https://doi.org/10.20454/ijas.2016.1207
3. M. Akram, S. Samanta, M. Pal, Application of bipolar fuzzy sets in planar graphs. Int. J. Appl. Comput. Math. **3**(2), 773–785 (2017)
4. M. Akram, J.M. Dar, A. Farooq, Planar graphs under pythagorean fuzzy environment. Mathematics **6**, 278 (2018). https://doi.org/10.3390/math6120278
5. M. Akram, A. Bashir, S. Samanta, Complex pythagorean fuzzy planar graphs. Int. J. Appl. Comput. Math. **6**, 58 (2020). https://doi.org/10.1007/s40819-020-00817-2
6. N. Alshehri, M. Akram, Intuitionistic fuzzy planar graphs, in *Discrete Dynamics in Nature and Society*, Vol. 2014, Art. ID 397823 (2014). https://doi.org/10.1155/2014/397823.
7. G. Ghorai, M. Pal, Faces and dual of *m*-polar fuzzy planar graphs. J. Intell. Fuzzy Syst. **31**, 2041–2049 (2016)
8. G. Ghorai, M. Pal, A study on m-polar fuzzy planar graphs. Int. J. Comput. Sci. Math. **7**(3), 283–292 (2016)
9. A. Muneera, T. Nageswara Rao, S.M. Shaw, Inverse domination in fuzzy planar graphs. Int. J. Pure Appl. Math. **120**(6), 4293–4303 (2018)
10. N.A. Jabbar, J.H. Naoom, E.H. Ouda, Fuzzy dual graph. J. Al-Nahrain Univ. **12**(4), 168–171 (2009)
11. A. Nagoorgani, K. Radha, On regular fuzzy graphs. J. Phys. Sci. **12**, 33–40 (2008)
12. G. Nirmala, K. Dhanabal, Special planar fuzzy graph configurations. Int. J. Sci. Res. Publ. **2**(7), 1–4 (2012)
13. M. Pal, S. Samanta, A. Pal (eds.), *Handbook of Research on Advanced Applications of Graph Theory in Modern Society*, vol. 591. (IGI Global, USA, 2020). https://doi.org/10.4018/978-1-5225-9380-5
14. T. Pramanik, S. Samanta, M. Pal, Interval-valued fuzzy planar graphs. Int. J. Mach. Learn. Cyber. **7**(4), 653–664 (2016). https://doi.org/10.1007/s13042-014-0284-7
15. T. Pramanik, S. Mondal, S. Samanta, M. Pal, A study on bipolar fuzzy planar graph and its application in image shrinking. J. Intell. Fuzzy Syst. **34**(3), 1863–1874 (2018). https://doi.org/10.3233/JIFS-171209
16. S. Samanta, A. Pal, M. Pal, Concept of fuzzy planar graphs, in *Proceedings of Science and Information Conference, 7–9, 2013* (London, 2013), pp. 557–563

17. S. Samanta, A. Pal, M. Pal, New concepts of fuzzy planar graphs. Int. J. Adv. Res. Artif. Intell. **3**(1), 52–59 (2014)
18. S. Samanta, M. Pal, Some more results on bipolar fuzzy sets and bipolar fuzzy intersection graphs. J. Fuzzy Math. **22**(2), 253–262 (2014)
19. S. Samanta, M. Pal, Fuzzy planar graphs. IEEE Trans. Fuzzy Syst. **23**(6), 1936–1942 (2016)

Chapter 3
Fuzzy Cut Vertices and Fuzzy Trees

In 1847, Kirchhoff developed the theory of trees, to solve a system of simultaneous linear equations. Also, in 1857, Cayley worked on this theory and enumerated the isomers of saturated hydrocarbons $C_n H_{2n+2}$ for different values of n. Nowadays, trees are frequently used to construct the directory of a computer, searching of files and folders, preparation of dictionary, sorting of mails according to PIN code or email addresses in and many other areas. The fuzzy tree characterized by Sunitha et al. [17] using the concept of strong edges. But, finding strong edges is comparatively harder as it requires a knowledge of all edges of the graph. In 2003, Bhutani et al. [6] further investigated fuzzy tree properties with strongness of edges.

This chapter focuses on a detailed discussion on fuzzy trees including forests, bridges, cut vertices, and end vertices of a FG .

The contributions of this chapter are mainly from [1–8, 10–12, 16–18].

For more works on fuzzy tree, see [9, 13–15, 19].

3.1 Some Basic Definitions

Before going to the definition of fuzzy tree, some related terms are recalled first:

Let $\mathscr{G} = (\mathscr{V}, \sigma, \mu)$ be a FG of a crisp graph $G^* = (V, E)$. A **fuzzy path** in $\mathscr{G}$ is a sequence of distinct vertices $a_0, a_1, \ldots, a_n$ such that $\mu(a_{i-1}, a_i) > 0, 1 \leq i \leq n$. Note that the underlying graph of fuzzy path is a crisp path. Like crisp cycle, the fuzzy cycle is defined. A fuzzy path becomes a **fuzzy cycle** if its end vertices a_0 and a_n coincide. The **strength of a fuzzy path** is the minimum of the membership values of all the edges in the path and the corresponding edge (which attains the minimum) is called the **weakest edge**.

There lies a big question in FG about the connectivity between two edges because the membership values (we can say the degree of existence) of the edges being different, maybe very less. So, a new term called **strength of connectedness** between

M. Pal et al., *Modern Trends in Fuzzy Graph Theory*,
https://doi.org/10.1007/978-981-15-8803-7_3

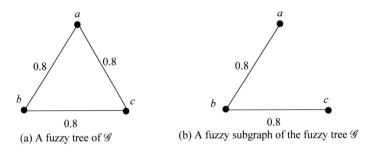

Fig. 3.1 Fuzzy tree and fuzzy subgraph

two vertices a and b in $\mathscr{G}$ is defined and it is the maximum of the strengths of all paths between a and b and it is denoted by $CONN_{\mathscr{G}}(a, b)$.

A **fuzzy subgraph** $H = (\mathscr{V}, \sigma', \mu')$ is called a **partial fuzzy subgraph** of FG $\mathscr{G}$ if $\sigma'(a) \leq \sigma(a)$ for all $a \in \mathscr{V}$ and $\mu'(a, b) \leq \mu(a, b)$ for all edge (a, b) of H.

A fuzzy subgraph H of a FG $\mathscr{G} = (\mathscr{V}, \sigma, \mu)$ is called a **full fuzzy subgraph** of $\mathscr{G}$ if $\sigma(a) > 0$ for all $a \in \mathscr{V}$ and $\mu(a, b) > 0$ for all edge (a, b) of $\mathscr{G}$.

Two vertices a and b in $\mathscr{G}$ are called **neighbors** if $\mu(a, b) > 0$.

Let $\mathscr{G}_1 = (\mathscr{V}, \sigma_1, \mu_1)$ and $\mathscr{G}_2 = (\mathscr{V}, \sigma_2, \mu_2)$ be two fuzzy graphs. The FG $\mathscr{G}_2$ is said to be **spanning subgraph** of $\mathscr{G}_1$ if $\sigma_1(a) = \sigma_2(a)$ for all $a \in \mathscr{V}$ and $\mu_2(a, b) < \mu_1^*(a, b)$ for all a, b. Note that the sets of vertices of two FGs are same only the membership values of edges are strictly less than the other.

If $\mathscr{G}$ is a cycle or a tree, a full fuzzy subgraph of $\mathscr{G}$ is called an *F*-**cycle** or an *F*-**tree**, respectively. In a cycle, there must be at least three vertices and a nontrivial tree has at least one vertex.

The FG $\mathscr{G}_2$ is said to be a **fuzzy tree** if it has a spanning fuzzy subgraph $H = (\mathscr{V}, \sigma', \mu')$ which is an *F*-tree and if $(a, b) \in \mathscr{G}_2$ but $(a, b) \notin H$, then $\mu_2(a, b) < CONN_H(a, b)$. Notice the sign "<". If it is replaced by $\leq$, then the graph of Fig. 3.1a is a fuzzy tree because then it has a fuzzy subgraph as shown in Fig. 3.1b.

From the definition, it follows that fuzzy tree and crisp tree are structure-wise different. That is, the underlying graph of a fuzzy tree is not necessarily a tree. As per definition, the fuzzy tree may have a cycle in crisp sense.

Definition 3.1 (*Fuzzy forest*) The fuzzy graph $\mathscr{G}_2 = (\mathscr{V}, \sigma_2, \mu_2)$ is called a **fuzzy forest** if it has a partial fuzzy spanning subgraph $H = (\mathscr{V}, \sigma', \mu')$ which is a forest such that for all edges (a, b) not in H, $\mu_2(a, b) < CONN_H(a, b)$. In other words, if $(a, b) \in \mathscr{G}_2$ but not in H, there is a path in H between the vertices a and b whose strength is greater than $\mu(a, b)$.

It is obvious that a forest is a fuzzy forest. For example, the FG of Fig. 3.2 is a fuzzy forest, while the FG of Fig. 3.3 is not a fuzzy forest.

Note that, if "<" is replaced by "$\leq$" in the definition of fuzzy forest, then the FG of Fig. 3.4 would be a fuzzy forest since it has spanning fuzzy subgraph shown in Fig. 3.5.

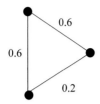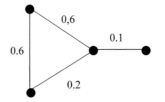

Fig. 3.2 An example of fuzzy forest

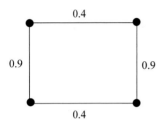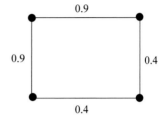

Fig. 3.3 Not a fuzzy forest

Fig. 3.4 A fuzzy forest

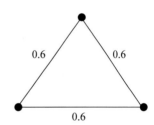

Fig. 3.5 Spanning subgraph
of the graph of Fig. 3.4

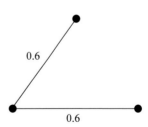

Followings are the results of fuzzy forests.

Theorem 3.1 *The FG $\mathscr{G}$ is a fuzzy forest if and only if in any cycle of $\mathscr{G}$, there is an edge (a, b) such that $\mu(a, b) < CONN_{\mathscr{G}-(a,b)}(a, b)$.*

Theorem 3.2 *Let $\mathscr{G}$ be a FG such that there is at most one strongest path between every pair of vertices. Then $\mathscr{G}$ is a fuzzy forest.*

3.2 Fuzzy Cut Vertex and Fuzzy Bridge

In a connected crisp graph, a vertex a is called a cut vertex if its removal disconnects the graph. Similarly, an edge (a, b) is called a bridge if its removal disconnects the graph. But, in FG the concepts of fuzzy cut vertex and bridge are different. In FG, the removal of fuzzy cut vertex or bridge reduces the connectedness of the FG.

The fuzzy bridge and fuzzy cut vertex are defined below and are illustrated by examples.

Definition 3.2 (*Fuzzy bridge*) An edge (a, b) is said to be a **fuzzy bridge** of a FG $\mathscr{G} = (\mathscr{V}, \sigma, \mu)$ if the strength of connectedness between a pair of vertices is reduced after the removal of (a, b), i.e. there is at least a pair of vertices $u, v \in \mathscr{V}$ such that $CONN_{\mathscr{G}}(u, v) > CONN_{\mathscr{G}-(a,b)}(u, v)$.

Thus, the edge (a, b) is a bridge if and only if there exists vertices u, v such that (a, b) is an edge of every strongest path from u to v.

Definition 3.3 (*Fuzzy cut vertex*) A vertex c is said to be a **fuzzy cut vertex** of a FG $\mathscr{G} = (\mathscr{V}, \sigma, \mu)$ if removal of it reduces the strength of connectedness between some other pair of vertices, i.e. $CONN_{\mathscr{G}}(a, b) > CONN_{\mathscr{G}-c}(a, b)$ for $a, b \in \mathscr{V}$, where a, b, c all are distinct.

Thus, c is a cut vertex if and only if there exist two vertices $a, b \in \mathscr{V}$ other than c such that c is on every strongest path from a to b.

Example 3.1 Let us consider the FG $\mathscr{G} = (\mathscr{V}, \sigma, \mu)$ containing three vertices a, b, c and three edges $(a, b), (b, c)$, and (c, a) with membership values $1, 0.5, 0.5$ respectively (see Fig. 3.6). In this graph, the edge (a, b) is a fuzzy bridge, since $CONN_{\mathscr{G}}(a, b) = 1 > CONN_{\mathscr{G}-(a,b)}(a, b) = 0.5$.

But, there is no cut vertex in this FG.

Example 3.2 Let us consider a FG $\mathscr{G} = (\mathscr{V}, \sigma, \mu)$, where $\mathscr{V} = \{a_1, a_2, a_3, a_4\}$. The set of edges is $\{(a_1, a_2), (a_2, a_3), (a_3, a_4), (a_4, a_1), (a_1, a_3)(a_2, a_4)\}$ and the membership values of the edges are $0.5, 1, 0.7, 1, 0.5, 0.7$ (see Fig. 3.7). In this FG the vertex a_4 is a cut vertex, since $CONN_{\mathscr{G}}(a_1, a_3) = 0.7 > CONN_{\mathscr{G}-a_4}(a_1, a_3) = 0.5$ (see Fig. 3.7).

Fuzzy bridges are related to the weakest edge of the graph. The following theorem connects fuzzy bridge and fuzzy cycle.

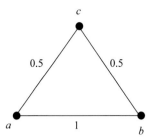

Fig. 3.6 A FG $\mathscr{G}$ where (a, b) is a fuzzy bridge

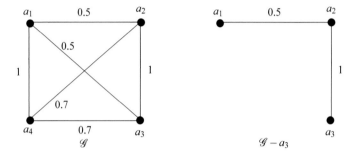

Fig. 3.7 A FG $\mathscr{G}$ where a_4 is a fuzzy cut vertex

Theorem 3.3 *If (a, b) is a fuzzy bridge in a FG $\mathscr{G}$, then it is not the weakest edge of a cycle.*

Proof Let (a, b) be a fuzzy bridge in $\mathscr{G}$. If possible let, (a, b) be the weakest edge of the graph. Now strength of connectedness between a and b is greater than the membership value of (a, b). According to the definition of fuzzy bridge, removal of the bridge reduces the strength of connectedness. If the edge is weakest, then no further reduction in the strength of connectedness is possible. Hence, our assumption is not correct. Thus, (a, b) is not the weakest edge. $\square$

Theorem 3.4 *Let (a, b) be a fuzzy bridge of a FG $\mathscr{G}$, then $CONN_{\mathscr{G}}(a, b) = \mu(a, b)$.*

Proof Let (a, b) be a fuzzy bridge of the FG $\mathscr{G} = (\mathscr{V}, \sigma, \mu)$. If, possible let, $CONN_{\mathscr{G}}(a, b) \geq \mu(a, b)$. Then there is a strongest path from a to b whose strength is greater than $\mu(a, b)$. Also, the strength of all edges of this strongest path is greater than $\mu(a, b)$. Thus, the path along with the edge (a, b) forms a cycle for which the (a, b) is weakest edge. This contradicts that (a, b) is a fuzzy bridge. $\square$

A common vertex of two bridges always has a lot of importance. It can be found out from the following theorem.

Theorem 3.5 *A vertex in a FG $\mathscr{G}$ is a fuzzy cut vertex if and only if it is a common vertex of two fuzzy bridges.*

Proof Let $\mathcal{G}$ be a FG and u be a fuzzy cut vertex with (a, u) and (u, b) be adjacent edges. As u is fuzzy cut vertex, its removal reduces the strength of connectedness. If possible, one of the edges, say (a, u), is not a fuzzy bridge. Then that edge may be the weakest edge of $\mathcal{G}$. Hence, if the vertex u is removed, the strength of connectedness does not reduce. Thus, the edges (a, u) and (u, b) must be fuzzy bridges.

Conversely, let u be the common vertex of two fuzzy bridges (a, u) and (u, b). We have to prove that u is fuzzy cut vertex. If two incident edges are fuzzy bridges, removal of both the edges reduces the strength of connectedness. Thus, the converse part is obvious.   □

Theorem 3.6 *Let* $\mathcal{G} = (\mathcal{V}, \sigma, \mu)$ *be a complete fuzzy graph. Then* $CONN_{\mathcal{G}}(a, b) = \mu(a, b)$ *for all* $a, b \in \mathcal{V}$.

Proof Let a, b be any two vertices of $\mathcal{G}$. The by definition μ^2
$$\mu^2(a, b) = \max_{c \in \mathcal{V}}[\min\{\mu(a, c), \mu(c, b)\}]$$
$$= \max[\min\{\sigma(a), \sigma(b), \sigma(c)\}] = \min\{\sigma(a), \sigma(b)\} = \mu(a, b).$$
Similarly, $\mu^3(a, b) = \mu(a, b)$, and in general, it can be prove that, $\mu^p(a, b) = \mu(a, b)$ for any positive integer p.
Hence, $CONN_{\mathcal{G}}(a, b) = \sup\{\mu^k(a, b) : \text{for all } k \geq 1\} = \mu(a, b)$.   □

The following result is proved in [2].

Lemma 3.1 *There are no fuzzy cut vertices in a complete fuzzy graph.*

The following theorem is proved in [17].

Theorem 3.7 *If* $\mathcal{G} = (\mathcal{V}, \sigma, \mu)$ *is a fuzzy tree, then* $\mathcal{G}$ *is not complete.*

Theorem 3.8 *If* c *is a cut vertex, then it has at least two strong neighbors.*

Proof Let $\mathcal{G}$ be a FG and c be a cut vertex. So, the removal of cut vertex c reduces the connectedness between two vertices say a and b. Thus, there exists a strongest path $P(a, \ldots, x, c, y, \ldots, b)$ from a to b which passes through c. Now, if the edge (x, c) is not strong, then obviously $\mu(x, c) < CONN_{\mathcal{G}}(x, c)$ assuming the edge (x, c) is removed. Thus, there exists a path Q from x to c, not involving (x, c), that is stronger than (x, c). Let p be the vertex just preceding c on Q. Since the strength of Q is at most $\mu(p, c)$, it must be $\mu(p, c) > \mu(x, c)$. If (p, c) is not strong, then this argument can be repeated. Since the graph is finite, the process terminates after finite number of repetitions, and we find a vertex p such that (p, c) is strong. Similarly, we find a q such that (c, q) is strong. If $p = q$, then there is a path from a to $p = q$ to b that is stronger than P. So, $CONN_{\mathcal{G}}(a, b)$ is not reduced after removal of c, a contradiction. Hence, the vertex c has at least two strong neighbors.   □

A special type of vertex called fuzzy end vertex which is associated with the cut vertex of a FG is defined here.

Definition 3.4 (*Fuzzy end vertices*) A vertex $a \in \mathcal{V}$ is called a **fuzzy end vertex** or end vertex of $\mathcal{G} = (\mathcal{V}, \sigma, \mu)$ if it has at most one strong neighbor in $\mathcal{G}$.

Fig. 3.8 A FG $\mathcal{G}$ where
vertex a is a fuzzy end vertex

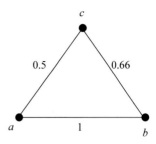

Obviously, if a is a fuzzy end vertex in the support of $\mathcal{G}$, then a is a fuzzy end vertex of $\mathcal{G}$ because a has only one neighbor in $\mathcal{G}$.

For example, in the fuzzy graph $\mathcal{G}$ shown in Fig. 3.8, the vertex a is a fuzzy end vertex since it has only strong neighbor b in $\mathcal{G}$. In this case, $\mu(a, b) = 1 > CONN_{\mathcal{G}-(a,b)}(a, b) = 0.5$. But $\mu(a, c) = 0.5 < CONN_{\mathcal{G}-(a,c)}(a, c) = 0.66$.

3.3 Multimin and Locamin F-cycles

The concepts of multimin and locamin have been defined for a FG . An F-cycle C is called multimin if C has more than one weakest edge. Again, C will be called locamin if every vertex of C lies on the weakest edge. Since a cycle has at least three vertices, locamin implies multimin.

The subgraph of a fuzzy cycle, that is F-cycle may be multimin and locamin if the following conditions hold.

Theorem 3.9 *An F-cycle is multimin if and only if it has no fuzzy end vertices.*

Proof There is a strong edge in a multimin F-cycle. So, for any vertex c, both of the edges incident to c are strong, and hence c must not be a fuzzy end vertex.

Conversely, let there be only one weakest edge (x, y) on the F-cycle. In this case, b is not a strong neighbor of a or vice versa (for some vertices b, a in F-cycle). So, only the other neighbors of a and b are strong. Hence, the vertices a and b are fuzzy end vertices.   □

Theorem 3.10 *A multimin F-cycle is locamin if and only if it has no cut vertices.*

Proof If (a, b) is the weakest edge of an F-cycle C, then a and b cannot be cut vertices of C. Thus, if C is locamin so that every vertex of C lies on the weakest edge, so no vertex of C can be a cut vertex.

Conversely, if C is not locamin, let a, b, c be three consecutive vertices of C such that neither (a, b) nor (b, c) is the weakest edge. Hence, it is obvious that removing the vertex b reduces $CONN_{\mathcal{G}}(b, c)$, and hence b is a cut vertex.   □

Theorem 3.11 *An F-cycle C is multimin if and only if it has at least one vertex which is neither a cut vertex nor a fuzzy end vertex.*

Proof It is true that if C is multimin it has no fuzzy end vertex, and it is obvious that a vertex on the weakest edge of C cannot be a cut vertex.

Conversely, if C is not a multimin F-cycle, it has the weakest edge. Thus, the vertices that lie on this weakest edge must be fuzzy end vertices. Again, all the other vertices of C must be cut vertices. Therefore, every vertex of C is either a fuzzy end vertex or a cut vertex.   □

3.4   Fuzzy Trees

The crisp tree is a connected graph without any cycle. Two or more trees form a forest. So, if we remove any edge or vertex form a crisp tree, it becomes disconnected. That is, the number of components will increase. This means every edge (or vertex) of a tree is a bridge (or cut vertex). But structurally, a fuzzy tree is completely different from a crisp tree. Although, if the underlying graph of a fuzzy graph is a tree (in crisp sense), then the fuzzy graph is also a fuzzy tree. A path in a fuzzy graph is a fuzzy tree. In crisp sense, no cycle is tree, but, a cycle may be a fuzzy tree. Like different types of fuzzy graphs, different types of fuzzy trees occur in literature. But, most of the works on fuzzy trees are based on membership values of vertices and edges.

Definition 3.5 Let $\mathscr{G} = (\mathscr{V}, \sigma, \mu)$ be a fuzzy tree. Then there exists a fuzzy spanning subgraph $T = (\mathscr{V}, \sigma, \nu)$ which is a tree called a **fuzzy spanning tree** of $\mathscr{G}$, where for all edges (a, b) not in T, $\mu(a, b) < CONN_T(a, b)$.

Definition 3.6 A **maximum spanning tree** of a connected FG $\mathscr{G} = (\mathscr{V}, \sigma, \mu)$ is a spanning subgraph $T = (\mathscr{V}, \sigma, \nu)$ of $\mathscr{G}$, which is a tree such that $CONN_{\mathscr{G}}(a, b)$ is the strength of the unique strongest $a - b$ path in T for all $a, b\mathscr{G}$.

Some important theorems related to the fuzzy tree are introduced here.

In a crisp tree, every internal vertex (other than leaves) is a cut vertex. The similar result is available in the fuzzy tree also.

Theorem 3.12 *Every internal vertex of a fuzzy tree is a fuzzy cut vertex.*

Proof If any of the internal vertices in a fuzzy three is deleted, the incident edges are automatically removed. We know that fuzzy trees are fuzzy graphs with a spanning tree such that all the edges which are not in the spanning tree have less connectivity than the strength of fuzzy graphs. Thus, the removal of such internal vertices, definitely removes some fuzzy bridges. Hence, the internal vertices of a fuzzy tree are fuzzy cut vertices.   □

Note 3.1 End vertices of a maximum spanning tree are not cut vertices.

Theorem 3.13 *If (a, b) is a bridge in a fuzzy tree $T = (\mathcal{V}, \sigma, \mu)$, then $CONN_T(a, b)$ is equal to $\mu(a, b)$.*

Proof Let T be a fuzzy tree and (a, b) be a bridge in T. So, there is a spanning tree such that all the edges which are not in spanning tree have less connectivity than the strength of fuzzy graph. Also let $CONN_{\mathcal{G}}(a, b)$ be not equal to $\mu(u, v)$. In that case, $\mu(a, b)$ is less than $CONN_{\mathcal{G}}(a, b)$. Therefore, (a, b) is not a bridge, a contradiction. Hence, the result follows. $\square$

In crisp graph, there is a unique path between any pair of vertices. The similar result for FG is presented below.

Theorem 3.14 *A FG $\mathcal{G}$ is a fuzzy tree if and only if there exists a unique strong path in $\mathcal{G}$ between any two vertices of $\mathcal{G}$.*

Proof Suppose there is a strong path P between any two vertices a and b in $\mathcal{G}$. Let $\mathcal{G}'$ be a spanning F-tree in $\mathcal{G}$. This path P lies wholly in $\mathcal{G}'$. Since $\mathcal{G}'$ is a tree, there is a unique path in $\mathcal{G}'$ from a to b. Thus, the path P is unique.

Conversely, let $\mathcal{G}$ be not a fuzzy tree. There is a cycle C in $\mathcal{G}$ such that $\mu(a, b) \geq CONN_{\mathcal{G}-(a,b)}(a, b)$ for any edge of C. That is, each edge on C is strong. Therefore, for any two vertices a and b on C, there are two strong paths between a and b. This is a contradiction. Hence, the result. $\square$

Theorem 3.15 *In a fuzzy tree, a strong a-b path is a path of maximum strength between the vertices a and b.*

Proof Let $\mathcal{G}'$ be an F-tree. Let P be the unique strong a-b path. Since every edge of P is strong, it is in $\mathcal{G}'$. Suppose P is not a path of maximum strength. Then let P' be such a path. Hence, the paths P' and P are distinct, so P and the reversal of P' form a cycle. In $\mathcal{G}'$, there is no cycle as $\mathcal{G}'$ is an F-tree. So, some edges (a', b') of Q' must not satisfy to be in $\mathcal{G}'$. By definition of $\mathcal{G}'$, it is true that $\mu(a', b') < CONN_{\mathcal{G}}(a', b')$.

Therefore, there exists a u'-v' path from in $\mathcal{G}'$. So every edge (a', b') of P' that fails to be in $\mathcal{G}'$ can be replaced by a path in $\mathcal{G}'$. This suggests an a-b path R in $\mathcal{G}'$. As a consequence, it was constructed by replacing edges (a', b') of P' by paths stronger than these edges, R is at least as strong as P'. Thus, R is not the same as P, so P and the reversal of R form a cycle, and this cycle is in $\mathcal{G}'$, which is not possible. $\square$

Fuzzy bridges are related to the weakest edges of the graph. The following theorem relates fuzzy bridge and fuzzy cycle.

Theorem 3.16 *If (a, b) is a fuzzy bridge in a FG $\mathcal{G}$, then it is not the weakest edge of a cycle.*

Proof Let (a, b) be a fuzzy bridge in $\mathcal{G}$. If possible let, (a, b) be the weakest edge of the graph. Therefore, $CONN_{\mathcal{G}}(a, b) > \mu(a, b)$. According to the definition of fuzzy bridge, removal of the bridge (a, b) reduces the strength of connectedness. If the edge is weakest, then it is not possible to further reduce the strength of connectedness. Hence, our assumption is not correct. Thus, (a, b) is not the weakest edge of a cycle. $\square$

References

1. P. Bhattacharyya, Some remarks on fuzzy graphs. Pattern Recognit. Lett. **6**, 297–302 (1987)
2. K.R. Bhutani, On automorphisms of fuzzy graphs. Pattern Recognit. Lett. **9**, 159–162 (1989)
3. K.R. Bhutani, A. Rosenfeld, Fuzzy end vertices in fuzzy graphs. Inf. Sci. **152**, 323–326 (2003)
4. K.R. Bhutani, A. Rosenfeld, Geodesics in fuzzy graphs. Electron. Notes Discret. Math. **15**, 51–54 (2003)
5. K.R. Bhutani, A. Rosenfeld, Strong arcs in fuzzy graphs. Inf. Sci. **152**, 319–322 (2003)
6. K.R. Bhutani, A. Battou, On M-strong fuzzy graphs. Inf. Sci. **155**(1–2), 103–109 (2003)
7. K.R. Bhutani, J. Moderson, A. Rosenfeld, On degrees of end nodes and cut nodes in fuzzy graphs. Iranian J. Fuzzy Syst. **1**(1), 53–60 (2004)
8. R.A. Borzooei, H. Rashmanlou, S. Samanta, M. Pal, A study on fuzzy labeling graphs. J. Intell. Fuzzy Syst. **30**, 3349–3355 (2016). https://doi.org/10.3233/IFS-152082
9. J. Gao, M. Lu, Fuzzy quadratic minimum spanning tree problem. Appl. Math. Comput. **164**(3), 773–788 (2005)
10. J.N. Mordeson, P.S. Nair, Cycles and cocycles of fuzzy graphs. Inf. Sci. **90**, 39–49 (1996)
11. J.N. Mordeson, S. Mathew, Advanced Topics in Fuzzy Graph Theory. Springer, Cham (2019). https://doi.org/10.1007/978-3-030-04215-8
12. J.N. Mordeson, P.S. Nair, *Fuzzy Graphs and Fuzzy Hypergraphs* (Physica-Verlag, Heidelberg, 2000)
13. A. Nagorgani, D.R. Subahashini, Fuzzy labeling trees. Int. J. Pure Appl. Math. **90**(2), 131–141 (2014)
14. S.M.A. Nayeem, M. Pal, Shortest path problem on a network with imprecise edge weight. Fuzzy Optim. Decis. Making **4**(4), 293–312 (2005)
15. S.M.A. Nayeem, M. Pal, Diameter constrained fuzzy minimum spanning tree problem. Int. J. Comput. Intell. Syst. **6**(6), 1040–1051 (2013)
16. S. Sahoo, M. Pal, Intuitionistic fuzzy labeling graphs. TWMS J. Appl. Eng. Math. **8**(2), 466–476 (2018)
17. M.S. Sunitha, A. Vijayakumar, A characterization of fuzzy trees. Inf. Sci. **113**, 293–300 (1999)
18. M.S. Sunitha, S. Mathew, Fuzzy graph theory: a survey. Ann. Pure Appl. Math. **4**(1), 92–110 (2013)
19. J. Zhou, L. Chen, K. Wang, F. Yang, Fuzzy α-minimum spanning tree problem: definition and solutions. Int. J. Gen. Syst. **45**(3), 311–335 (2016)

Chapter 4
Fuzzy Competition Graphs

In 1968, Cohen [4] first introduced competition graph while working on a problem in ecology. Competition graph (CompG) is one kind of intersection graph and it is originally used for modeling ecological problems. Though this graph can be used to model various types of problems in the real world, Cohen used the crisp competition graph to model the ecological problem and hence the vertices and edges of such graphs were precisely defined. Since the real world is full of uncertainty, so it is better to model such competitions by using the concept of fuzzy graphs. For example, in ecology, species have different characteristics such as non-vegetarian, vegetarian, weak, strong, small, large, etc. Also, the preys may or may not be digestive, tasty, harmful, etc. There are no specific measures for the terms tasty, digestible, harmful, etc. these are linguistic terms and these terms may be formulated mathematically with the help of fuzzy sets. Thus, preys and interrelationship between the species and preys can be modeled using a FG . In ecology, the food web is a very important ecological system for investigation of the ecological balance. So, motivated from the concept of food web and uncertainty involved in the ecological system, fuzzy competition graphs were defined. These graphs were also generalized to model other kinds of problems involved in the real world. Such generalizations are fuzzy k-competition graphs, p-competition fuzzy graphs, and m-step fuzzy competition graphs. The concept of fuzzy neighborhood graphs has also been incorporated.

The contribution of this chapter is from the papers [6, 8, 8–10, 15]. For more results on competition graphs, readers may consult with [1–3, 5, 7, 11, 17, 18].

4.1 Types of Fuzzy Competition Graphs

The concept of competition graphs (CompG) is extended to fuzzy competition graphs (FCompG) by incorporating the uncertainty in the vertices and edges. The contents of this section are from [12]. First of all, the fuzzy out-neighborhood (out-nbd) and fuzzy in-neighborhood (in-nbd) of a vertex are defined.

M. Pal et al., *Modern Trends in Fuzzy Graph Theory*, https://doi.org/10.1007/978-981-15-8803-7_4

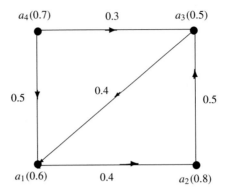

Fig. 4.1   Fuzzy out- and in-nbd of the vertices a_2 and a_1 of a FG

Definition 4.1 For a directed FG $\overrightarrow{\mathscr{G}} = (\mathscr{V}, \sigma, \mu)$, the **fuzzy out-nbd** of the vertex $a \in \mathscr{V}$ is the fuzzy set $\mathscr{N}^+(a) = (X_a^+, m_a^+)$, where $X_a^+ = \{b|\mu\overrightarrow{(a, b)} > 0\}$ and $m_a^+ : X_a^+ \to [0, 1]$ and is defined by $m_a^+(b) = \mu\overrightarrow{(a, b)}$.

Similarly, the **fuzzy in-nbd** of the vertex a of a directed FG $\overrightarrow{\mathscr{G}} = (\mathscr{V}, \sigma, \mu)$ is the fuzzy set $\mathscr{N}^-(a) = (X_a^-, m_a^-)$, where $X_a^- = \{b|\mu\overrightarrow{(b, a)}) > 0\}$ and $m_a^- : X_a^- \to [0, 1]$ and is defined by $m_a^-(b) = \mu\overrightarrow{(b, a)}$.

Example 4.1 Let $\mathscr{V} = \{a_1, a_2, a_3, a_4\}$ be the set of vertices of the directed FG $\overrightarrow{\mathscr{G}}$. The membership values of the vertices are taken as $\sigma(a_1) = 0.6$, $\sigma(a_2) = 0.8$, $\sigma(a_3) = 0.5$, $\sigma(a_4) = 0.7$ and that edges are $\mu\overrightarrow{(a_1, a_2)} = 0.4$, $\mu\overrightarrow{(a_4, a_1)} = 0.5$, $\mu\overrightarrow{(a_2, a_3)} = 0.5$, $\mu\overrightarrow{(a_4, a_3)} = 0.3$, and $\mu\overrightarrow{(a_3, a_1)} = 0.4$.

So, the out-nbd of the vertex a_2 and in-nbd of the vertex a_1 are respectively, $\mathscr{N}^+(a_2) = \{(a_3, 0.5)\}$. $\mathscr{N}^-(a_1) = \{(a_3, 0.4), (a_4, 0.5)\}$ (see Fig. 4.1).

The CompG $C(\overrightarrow{D})$ of a digraph $\overrightarrow{D} = (V, \overrightarrow{E})$ is an undirected graph $G = (V, E)$ with same set of vertices and there is an edge between two vertices $a, b \in V$ if there is a vertex $c \in V$ and directed edges $\overrightarrow{(a, c)}, \overrightarrow{(b, c)} \in \overrightarrow{E}(\overrightarrow{D})$. The definition of FCompG is given below.

Definition 4.2 *(Fuzzy competition graph)* The FCompG for a fuzzy digraph $\overrightarrow{\mathscr{G}} = (\mathscr{V}, \sigma, \mu)$ is denoted by $\mathscr{C}(\overrightarrow{\mathscr{G}})$ and it is an undirected FG $\mathscr{G} = (\mathscr{V}, \sigma, \mu)$ with the same set of vertices such that there exists an edge between two vertices a and b $(a, b \in \mathscr{V})$ in $\mathscr{C}(\overrightarrow{\mathscr{G}})$ if and only if $\mathscr{N}^+(a) \cap \mathscr{N}^+(b) \neq \emptyset$. The membership value of the edge (a, b) in $\mathscr{C}(\overrightarrow{\mathscr{G}})$ is determined by $\mu(a, b) = (\sigma(a) \wedge \sigma(b))h(E(\mathscr{N}_E^+(a) \cap \mathscr{N}_E^+(b)))$, where $E(\mathscr{N}_E^+(a) \cap \mathscr{N}_E^+(b))$ is the set of edges incident to the vertices of $\mathscr{N}_E^+(a) \cap \mathscr{N}_E^+(b)$.

Example 4.2 Let $\overrightarrow{\mathscr{G}}$ be a directed FG whose set of vertices with membership values be $\{(a_1, 0.3), (a_2, 0.6), (a_3, 0.4), (a_4, 0.5), (a_5, 0.4)\}$. The membership values of

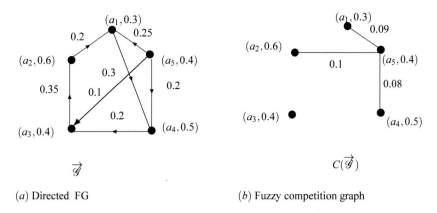

Fig. 4.2 Example of fuzzy competition graph

edges are taken as $\mu\overrightarrow{(a_2, a_1)} = 0.2$, $\mu\overrightarrow{(a_3, a_2)} = 0.35$, $\mu\overrightarrow{(a_4, a_3)} = 0.2$, $\mu\overrightarrow{(a_5, a_3)} = 0.1$, $\mu\overrightarrow{(a_5, a_4)} = 0.2$, $\mu\overrightarrow{(a_5, a_1)} = 0.25$, $\mu\overrightarrow{(a_1, a_4)} = 0.3$ (see Fig. 4.2(a)).

Notice that the common neighbor of the vertices a_1 and a_5 is a_4, so there is an undirected edge between the vertices a_1 and a_5 in the competition graph. Similarly, the vertices a_2 and a_5 are competitors, and a_4 and a_5 are competitors. So, (a_2, a_5) and (a_4, a_5) are other edges of the competition graph $C(\overrightarrow{\mathscr{G}})$.

The membership value of the edge (a_1, a_5) is $\sigma(a_1) \wedge \sigma(a_5) \times h(E(\mathscr{N}^+(a_1) \cap \mathscr{N}^+(a_5))) = 0.3 \times \max\{0.3, 0.2\} = 0.09$ since $\mathscr{N}^+(a_1) \cap \mathscr{N}^+(a_5) = \{(a_4, 0.5)\}$.

In a similar manner, other edge membership values are calculated. The FCompG for this FG is depicted in Fig. 4.2b.

If there is an edge between two vertices (species), this implies that they compete for at least one prey. So it is very crucial to know the strengths of the edges to describe the competition between vertices.

4.1.1 Fuzzy k-Competition Graphs

This is a generalization of FCompG. In this graph, a new condition is imposed for the existence of an edge between two vertices. For a given non-negative number k and a directed FG , a fuzzy k-competition graph is defined.

Definition 4.3 Let $\overrightarrow{\mathscr{G}} = (\mathscr{V}, \sigma, \mu)$ be a fuzzy digraph and k be a non-negative number. The **fuzzy k-competition graph** Fuzzy k-competition graph $\mathscr{C}_k(\overrightarrow{\mathscr{G}})$ for the directed FG $\overrightarrow{\mathscr{G}}$ is an undirected FG $\mathscr{G} = (\mathscr{V}, \sigma, \nu)$ which has an edge between two vertices $a, b \in \mathscr{V}$ in $\mathscr{C}_k(\overrightarrow{\mathscr{G}})$ if and only if $|\mathscr{N}^+(a) \cap \mathscr{N}^+(b)| > k$. The membership value of the edge (a, b) in $\mathscr{C}_k(\overrightarrow{\mathscr{G}})$ is determined by $\nu(a, b) = \frac{(k'-k)}{k'}[\sigma(a) \wedge \sigma(b)] h(E(\mathscr{N}^+(a) \cap \mathscr{N}^+(b)))$, where $k' = |E(\mathscr{N}^+(a) \cap \mathscr{N}^+(b))|$.

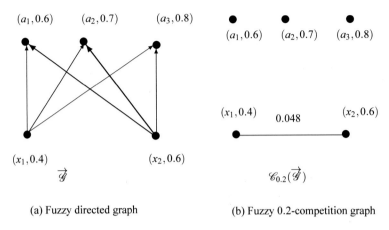

(a) Fuzzy directed graph   (b) Fuzzy 0.2-competition graph

Fig. 4.3 An example of a fuzzy 0.2-competition graph

A fuzzy k-competition graph becomes a FCompG for $k = 0$. A fuzzy 0.2-competition graph is illustrated in Fig. 4.3.

Example 4.3 Let $\overrightarrow{\mathcal{G}}$ be a directed FG. The vertices along with membership values of $\overrightarrow{\mathcal{G}}$ are taken as $(x_1, 0.4), (x_2, 0.6), (a_1, 0.6), (a_2, 0.7), (a_3, 0.8)$ and the membership values of the directed edges are considered as $\mu\overrightarrow{(x_1, a_1)} = 0.3, \mu\overrightarrow{(x_1, a_2)} = 0.35,$ $\mu\overrightarrow{(x_1, a_3)} = 0.36, \mu\overrightarrow{(x_2, a_1)} = 0.6, \mu\overrightarrow{(x_2, a_2)} = 0.5, \mu\overrightarrow{(x_2, a_3)} = 0.45,$ This graph is depicted in Fig. 4.3a. In this example, only the vertices x_1 and x_2 are competiting each other. So, there must be an edge between x_1 and x_2 in $\mathcal{C}(\mathcal{G})$. The membership value of the edge is calculated below.

Here, $k = 0.2$ and $k' = \min\{0.3, 0.35, 0.36\} = 0.3$. Hence, $\mu(x_1, x_2) = \frac{0.3 - 0.2}{0.3} \times 0.4 \times 0.36 = 0.048$. The fuzzy 0.2-competition graph of the graph of Fig. 4.3a is drawn in the Fig. 4.3b.

Theorem 4.1 Let $\overrightarrow{\mathcal{G}} = (\mathcal{V}, \sigma, \mu)$ be a fuzzy digraph and for two vertices $a, b \in \mathcal{V}$ the fuzzy set $\mathcal{N}^+(a) \cap \mathcal{N}^+(b)$ be a singleton set. Then the edge (a, b) in $C(\overrightarrow{\mathcal{G}})$ is strong if and only if $|\mathcal{N}^+(a) \cap \mathcal{N}^+(b)| > 0.5$.

Proof Here, $\overrightarrow{\mathcal{G}} = (\mathcal{V}, \sigma, \overrightarrow{\mu})$ is a fuzzy digraph. Let $\mathcal{N}^+(a) \cap \mathcal{N}^+(b) = \{(c, m)\}$, where m is the membership value of the vertex c. Here, $|\mathcal{N}^+(a) \cap \mathcal{N}^+(b)| = m = h(\mathcal{N}^+(a) \cap \mathcal{N}^+(b))$. So, $\mu(a, b) = m \times \{\sigma(a) \wedge \sigma(b)\}$. Hence, by definition, the edge (a, b) is strong in $C(\overrightarrow{\mathcal{G}})$ if and only if $m > 0.5$. □

Suppose all the edges of a fuzzy digraph are strong. Then all the edges of its corresponding FCompG may or may not be strong. This case is discussed below.

Let $\overrightarrow{\mathcal{G}}$ be a fuzzy digraph and a, b be its two vertices with $\sigma(a) = 0.3, \sigma(b) = 0.4$. Suppose, the common neighbor of a, b be c with $\sigma(c) = 0.2$. Let $\overrightarrow{\mu}(a, c) =$

0.2, $\overrightarrow{\mu}(b, c) = 0.15$. Obviously, the edges $\overrightarrow{(a, c)}$ and $\overrightarrow{(b, c)}$ are strong. Now, the membership value of the edge (a, b) in the corresponding competition graph is $0.3 \times 0.15 = 0.045 < 0.5$. Hence, (a, b) is not a strong edge.

But still, a result can be found if all the edges are strong of a fuzzy digraph.

Theorem 4.2 *If all edges of a fuzzy digraph* $\overrightarrow{\mathscr{G}} = (\mathscr{V}, \sigma, \overrightarrow{\mu})$ *are strong, then* $\frac{\mu(a,b)}{(\sigma(a) \wedge \sigma(b))^2} > 0.5$ *for all edges* (a, b) *in* $C(\overrightarrow{\mathscr{G}})$.

Proof Since each edge of $\overrightarrow{\mathscr{G}}$ is strong therefore $\frac{\overrightarrow{\mu}(a,b)}{\sigma(a) \wedge \sigma(b)} > 0.5$ for all edge (a, b) in $\overrightarrow{\mathscr{G}}$.

Let the FCompG for $\overrightarrow{\mathscr{G}}$ be $C(\overrightarrow{\mathscr{G}}) = (\mathscr{V}, \sigma, \mu)$.

Let us consider two cases.

Case 1. Let $\mathscr{N}^+(a) \cap \mathscr{N}^+(b) = \emptyset$ for all $a, b \in \mathscr{V}$. Therefore, there is no edge in $C(\overrightarrow{\mathscr{G}})$ between the vertices a and b.

Case 2. Let $\mathscr{N}^+(a) \cap \mathscr{N}^+(b) \neq \emptyset$.

Let $\mathscr{N}^+(a) \cap \mathscr{N}^+(b) = \{(a_1, m_1), (a_2, m_2), \ldots, (a_z, m_z)\}$, where m_i be the membership values of a_i, $i = 1, 2, \ldots, z$ respectively.

So $m_i = min\{\overrightarrow{\mu}(a, a_i), \overrightarrow{\mu}(b, a_i)\}$, $i = 1, 2, \ldots, z$.

Let $h(\mathscr{N}^+(a) \cap \mathscr{N}^+(b)) = max\{m_i, i = 1, 2, \ldots, z\} = m_{max}$.

$\mu(a, b) = (\sigma(a) \wedge \sigma(b)) h(\mathscr{N}^+(a) \cap \mathscr{N}^+(b)) = m_{max} \times \sigma(a) \wedge \sigma(b)$.

Hence, $\frac{\mu(a,b)}{(\sigma(a) \wedge \sigma(b))^2} = \frac{m_{max}}{\sigma(a) \wedge \sigma(b)} > 0.5$. □

Again, if all the edges of $\overrightarrow{\mathscr{G}}$ are strong, then for all edges (a, b) in $C(\overrightarrow{\mathscr{G}})$, $\frac{\mu(a,b)}{(\sigma(a) \wedge \sigma(b))^2} > 0.5$.

Thus, if the height of intersection between two out neighborhoods of two vertices of a fuzzy digraph is greater than 0.5, the edge in corresponding the FCompG is strong. But, this result is not true in thecorresponding fuzzy k-competition graph. This is justified below.

Theorem 4.3 *Let* $\overrightarrow{\mathscr{G}} = (\mathscr{V}, \sigma, \mu)$ *be a fuzzy digraph. For an edge* (a, b) *in* $\overrightarrow{\mathscr{G}}$, *if* $h(\mathscr{N}^+(a) \cap \mathscr{N}^+(b)) = 1$ *and* $|\mathscr{N}^+(a) \cap \mathscr{N}^+(b)| > 2k$, *then* (a, b) *is strong in* $C_k(\overrightarrow{\mathscr{G}})$.

Proof Here, $h(\mathscr{N}^+(a) \cap \mathscr{N}^+(b)) = 1$ and $|\mathscr{N}^+(a) \cap \mathscr{N}^+(b)| > 2k$.

Now, $\mu(a, b) = \frac{k'-k}{k'} \sigma(a) \wedge \sigma(b) h(\mathscr{N}^+(a) \cap \mathscr{N}^+(b))$, where $k' = |\mathscr{N}^+(a) \cap \mathscr{N}^+(b)|$.

So, $\mu(a, b) = \frac{k'-k}{k'} \sigma(a) \wedge \sigma(b)$.

Hence, $\frac{\mu(a,b)}{\sigma(a) \wedge \sigma(b)} = \frac{k'-k}{k'} > 0.5$ as $k' > 2k$. Hence, the edge (a, b) is strong. □

4.1.2 Fuzzy Neighborhood Graph

Let G be a crisp graph and b be a vertex. The neighborhood of b is defined as $N(b) = \{a : ab \in E\}$ If there is a loop at b, then $b \in N(b)$. The set $N(b)$ is called as the open neighborhood of b. The closed neighborhood of b is $N[b] = N(b) \cup \{b\}$. This concept can also be applied in FG to define fuzzy open and closed neighborhood of a vertex.

Definition 4.4 Fuzzy open-nbd of a vertex b of a FG $\mathcal{G} = (\mathcal{V}, \sigma, \nu)$ is the fuzzy set $\mathcal{N}(b) = (X_b, m_b)$, where $X_b = \{a | \nu(b, a) > 0\}$ and $m_b : X_b \to [0, 1]$ is defined by $m_b(a) = \nu(b, a)$. For each vertex $b \in \mathcal{V}$, let the fuzzy singleton set, $A_b = (\{b\}, \sigma')$ such that $\sigma' : \{b\} \to [0, 1]$ defined by $\sigma'(b) = \sigma(b)$.

Likewise the **fuzzy closed-nbd** of a vertex b is $\mathcal{N}[b] = \mathcal{N}(b) \cup A_b$.

Like in crisp graph, fuzzy open-nbd graphs and fuzzy closed-nbd graphs can be defined.

Definition 4.5 Fuzzy open-nbd graph of a FG $\mathcal{G} = (\mathcal{V}, \sigma, \mu)$ is another FG $\mathcal{N}(\mathcal{G}) = (\mathcal{V}, \sigma, \nu')$ with same set of vertices such that there exists an edge between the vertices a and b in $\mathcal{N}(\mathcal{G})$ if and only if $\mathcal{N}(a) \cap \mathcal{N}(b) \neq \emptyset$ in $\mathcal{G}$ and $\nu' : \mathcal{V} \times \mathcal{V} \to [0, 1]$ is such that $\nu'(a, b) = [\sigma(a) \wedge \sigma(b)] h(\mathcal{N}(a) \cap \mathcal{N}(b))$.

Definition 4.6 Fuzzy closed-nbd graph of $\mathcal{G} = (\mathcal{V}, \sigma, \mu)$ is a FG $\mathcal{N}[\mathcal{G}] = (\mathcal{V}, \sigma, \nu')$ with same set of vertices and two vertices a and b are connected by an edge in $\mathcal{N}[\mathcal{G}]$ if and only if $\mathcal{N}[a] \cap \mathcal{N}[b] \neq \emptyset$ in $\mathcal{G}$ and the edge membership value $\nu' : \mathcal{V} \times \mathcal{V} \to [0, 1]$ and it is defined as $\nu'(a, b) = [\sigma(a) \wedge \sigma(b)] h(\mathcal{N}[a] \cap \mathcal{N}[b])$.

Definition 4.7 Fuzzy (k)-nbd graph of $\mathcal{G} = (\mathcal{V}, \sigma, \mu)$ is a FG $\mathcal{N}_k(\mathcal{G}) = (\mathcal{V}, \sigma, \nu')$ with same set of vertices and there exists an edge between two vertices a and b in $\mathcal{N}_k(\mathcal{G})$ if and only if $|\mathcal{N}(a) \cap \mathcal{N}(b)| > k$ in $\mathcal{G}$ and $\nu' : \mathcal{V} \times \mathcal{V} \to [0, 1]$ such that $\nu'(a, b) = \frac{(k_1 - k)}{k_1} [\sigma(a) \wedge \sigma(b)] h(\mathcal{N}(a) \cap \mathcal{N}(b))$, where $k_1 = |\mathcal{N}(a) \cap \mathcal{N}(b)|$.

Definition 4.8 Fuzzy $[k]$-nbd graph of $\mathcal{G}$ is a FG $\mathcal{N}_k[\mathcal{G}] = (\mathcal{V}, \sigma, \nu')$ whose fuzzy vertex set is same as of $\mathcal{G}$ and there is an edge between vertices a and b in $\mathcal{N}_k[\mathcal{G}]$ if and only if $|\mathcal{N}[a] \cap \mathcal{N}[b]| > k$ in $\mathcal{G}$ and $\nu' : \mathcal{V} \times \mathcal{V} \to [0, 1]$ such that $\nu'(a, b) = \frac{(k_1 - k)}{k_1} [\sigma(a) \wedge \sigma(b)] h(\mathcal{N}[a] \cap \mathcal{N}[b])$, where $k_1 = |\mathcal{N}[a] \cap \mathcal{N}[b]|$.

Fuzzy (0.2)-nbd and **fuzzy $[0.2]$-nbd** graphs are described in the following example.

Example 4.4 Let $\mathcal{G} = (\mathcal{V}, \sigma, \mu)$ be a FG and the set vertices with membership values be $\{(a_1, 0.3), (a_2, 0.6), (a_3, 0.4), (a_4, 0.5), (a_5, 0.4)\}$. The edge membership values be taken as $\mu(a_1, a_2) = 0.3$, $\mu(a_2, a_3) = 0.4$, $\mu(a_3, a_4) = 0.4$, $\mu(a_4, a_5) = 0.4$, $\mu(a_5, a_1) = 0.3$, $\mu(a_1, a_4) = 0.3$ and for other edges these are 0. The fuzzy open-nbd graph $\mathcal{N}(\mathcal{G})$, the fuzzy closed-nbd graph $\mathcal{N}[\mathcal{G}]$, the fuzzy (0.2)-nbd graph $\mathcal{N}_{0.2}(\mathcal{G})$ and the fuzzy $[0.2]$-nbd graph $\mathcal{N}_{0.2}[\mathcal{G}]$ of the FG $\mathcal{G}$ are displayed in Fig. 4.4.

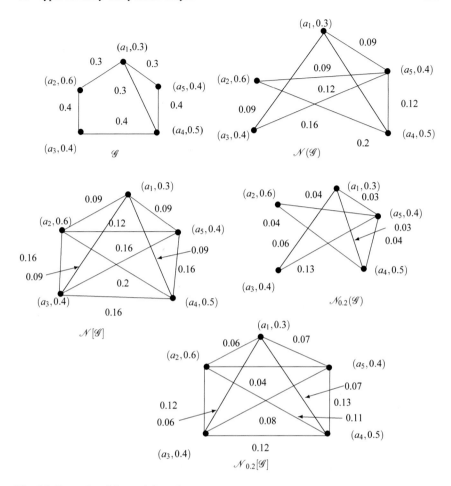

Fig. 4.4 Examples of fuzzy nbd graphs

Theorem 4.4 *For each edge of a FG $\mathscr{G}$, there is an edge in $\mathscr{N}[\mathscr{G}]$.*

Proof Let (a, b) be an edge of the FG $\mathscr{G}$. Let the closed nbd graph of $\mathscr{G}$ be $\mathscr{N}[\mathscr{G}] = (\mathscr{V}, \sigma, \nu)$.

Then, $a, b \in \mathscr{N}[a]$ and $a, b \in \mathscr{N}[b]$. So $a, b \in \mathscr{N}[a] \cap \mathscr{N}[b]$.

Since $\mathscr{N}[a] \cap \mathscr{N}[b]$ has at least two elements, so $h(\mathscr{N}[a] \cap \mathscr{N}[b]) \neq 0$ and, hence, $\nu(a, b) = \sigma(a) \wedge \sigma(b) h(\mathscr{N}[a] \cap \mathscr{N}[b]) > 0$.

Thus, for each edge (a, b) in $\mathscr{G}$, there exists an edge in $\mathscr{N}[\mathscr{G}]$. $\square$

Definition 4.9 The underlying FG of a directed FG $\overrightarrow{\mathscr{G}}$ is denoted by $\mathscr{U}(\mathscr{G})$ and is defined as $\mathscr{U}(\mathscr{G}) = (\mathscr{V}, \sigma, \nu)$, where the membership value of an edge (a, b) in $\mathscr{U}(\mathscr{G})$ is given by $\nu(a, b) = \min\{\overrightarrow{\mu(a, b)}, \overrightarrow{\mu(b, a)}\}$ for all $a, b \in \mathscr{V}$.

The relations between fuzzy (k)-nbd, fuzzy $[k]$-nbd and fuzzy k-competition graphs are established in the following theorems.

Theorem 4.5 *Let $\overrightarrow{\mathscr{G}}$ be a loop less symmetric fuzzy digraph. Then $\mathscr{C}_k(\overrightarrow{\mathscr{G}}) = \mathscr{N}_k(\mathscr{U}(\mathscr{G}))$.*

Proof Let $\overrightarrow{\mathscr{G}} = (\mathscr{V}, \sigma, \overrightarrow{\mu})$ be the directed FG and $\mathscr{U}(\mathscr{G}) = (\mathscr{V}, \sigma, \mu)$ be the underlying FG.

Since their sets of vertices are same, let $\mathscr{C}_k(\overrightarrow{\mathscr{G}}) = (\mathscr{V}, \sigma, \nu)$ and $\mathscr{N}_k(\mathscr{U}(\mathscr{G})) = (\mathscr{V}, \sigma, \nu')$.

Also, sets of vertices of $\overrightarrow{\mathscr{G}}$ and $\mathscr{C}_k(\overrightarrow{\mathscr{G}})$ are same. Same statement is also valid for an underlying FG of the directed FG.

Hence, $\mathscr{N}_k(\mathscr{U}(\mathscr{G}))$ and $\overrightarrow{\mathscr{G}}$ have same sets of vertices. Now, we have to prove that $\nu(a, b) = \nu'(a, b)$ for all $a, b \in \mathscr{V}$.

If $\nu(a, b) = 0$ in $\mathscr{C}_k(\overrightarrow{\mathscr{G}})$, then $|\mathscr{N}^+(a) \cap \mathscr{N}^+(b)| \leq k$.

Since, $\overrightarrow{\mathscr{G}}$ is a symmetric fuzzy graph, $|\mathscr{N}[a] \cap \mathscr{N}[b]| \leq k$ in $\mathscr{U}(\mathscr{G})$. So $\nu'(a, b) = 0$.

If $|\mathscr{N}^+(a) \cap \mathscr{N}^+(b)| > k$, then $\nu(a, b) > 0$ in $\mathscr{C}_k(\overrightarrow{\mathscr{G}})$.

So $\nu(a, b) = \frac{(k_1-k)}{k_1}[\sigma(a) \wedge \sigma(b)]h(\mathscr{N}^+(a) \cap \mathscr{N}^+(b))$, where $k_1 = |\mathscr{N}^+(a) \cap \mathscr{N}^+(b)|$.

Again, $\overrightarrow{\mathscr{G}}$ is a symmetric fuzzy digraph, $|\mathscr{N}[a] \cap \mathscr{N}[b]| > k$ in $\mathscr{U}(\mathscr{G})$.

Therefore, $\nu'(a, b) = \frac{(k_2-k)}{k_2}[\sigma(a) \wedge \sigma(b)]h(\mathscr{N}[a] \cap \mathscr{N}[b])$, where $k_2 = |\mathscr{N}[a] \cap \mathscr{N}[b]|$.

It is obvious that $h(\mathscr{N}^+(a) \cap \mathscr{N}^+(b))$ in $\overrightarrow{\mathscr{G}}$ and $h(\mathscr{N}[a] \cap \mathscr{N}[b])$ in $\mathscr{U}(\mathscr{G})$ are same, as $\overrightarrow{\mathscr{G}}$ is a symmetric digraph.

Thus, $k_1 = k_2$. Hence, $\nu(a, b) = \nu'(a, b)$ for all $a, b \in \mathscr{V}$. $\qquad\square$

Similar relation is established between fuzzy $[k]$-nbd graph and fuzzy k-competition graph.

Theorem 4.6 *Let $\overrightarrow{\mathscr{G}}$ be a symmetric fuzzy digraph and each vertex has a loop. Then $\mathscr{N}_k[\mathscr{U}(\mathscr{G})] = \mathscr{C}_k(\overrightarrow{\mathscr{G}})$, where $\mathscr{U}(\mathscr{G})$ is the loop less underlying FG $\overrightarrow{\mathscr{G}}$.*

Proof Let $\mathscr{C}_k(\overrightarrow{\mathscr{G}}) = (\mathscr{V}, \sigma, \nu)$ and $\mathscr{N}_k[\mathscr{U}(\mathscr{G})] = (\mathscr{V}, \sigma, \nu')$.

The vertex sets of $\overrightarrow{\mathscr{G}}$ and $\mathscr{C}_k(\overrightarrow{\mathscr{G}})$ are same. Also, the underlying FG and directed FG have same sets of vertices.

Hence, $\mathscr{N}_k[\mathscr{U}(\mathscr{G})]$ has the same fuzzy vertex set as $\overrightarrow{\mathscr{G}}$.

Now, we need to show that $\nu(a, b) = \nu'(a, b)$ for all $a, b \in \mathscr{V}$. Since each vertex of $\overrightarrow{\mathscr{G}}$ has a loop, so out-nbd of each vertex contains the vertex itself.

Thus, if $\nu(a, b) = 0$ in $\mathscr{C}_k(\overrightarrow{\mathscr{G}})$, then $|\mathscr{N}^+(a) \cap \mathscr{N}^+(b)| \leq k$.

Again, since $\overrightarrow{\mathscr{G}}$ is symmetric, $|\mathscr{N}(a) \cap \mathscr{N}(b)| \leq k$ in $\mathscr{U}(\mathscr{G})$. So $\nu'(a, b) = 0$.

If $|\mathscr{N}^+(a) \cap \mathscr{N}^+(b)| > k$, $\nu(a, b) > 0$ in $\mathscr{C}_k(\overrightarrow{\mathscr{G}})$.

Therefore, $v(a,b) = \frac{(k_1-k)}{k_1}[\sigma(a) \wedge \sigma(b)]h(\mathcal{N}^+(a) \cap \mathcal{N}^+(b))$, where $k_1 = |\mathcal{N}^+(a) \cap \mathcal{N}^+(b)|$.

As $\overrightarrow{\mathcal{G}}$ is a symmetric fuzzy digraph, $|\mathcal{N}(a) \cap \mathcal{N}(b)| > k$ in $\mathcal{U}(\mathcal{G})$. So $v' = \frac{(k_2-k)}{k_2}[\sigma(a) \wedge \sigma(b)]h(\mathcal{N}(a) \cap \mathcal{N}(b))$, where $k_2 = |\mathcal{N}(a) \cap \mathcal{N}(b)|$.

It is obvious that $h(\mathcal{N}^+(a) \cap \mathcal{N}^+(b))$ in $\overrightarrow{\mathcal{G}}$ is equal to $h(\mathcal{N}(a) \cap \mathcal{N}(b))$ in $\mathcal{U}(\mathcal{G})$ and $k_1 = k_2$ as $\overrightarrow{\mathcal{G}}$ is symmetric. Hence, $v(a,b) = v'(a,b)$ for all $a,b \in \mathcal{V}$.   □

4.1.3 p-Competition Fuzzy Graph

Another variation of FCompG which is similar to fuzzy k-competition graph is p-competition fuzzy graph which is discussed in this section. In k-competition FG, k is a real number, but here p is a positive integer. So, 'p' is a restrictive case of 'k'.

Definition 4.10 For a given positive integer p and a directed FG $\overrightarrow{\mathcal{G}} = (\mathcal{V}, \sigma, \overrightarrow{\mu})$, the p-**competition FG** is an undirected FG $\mathcal{C}^p(\overrightarrow{\mathcal{G}}) = (\mathcal{V}, \sigma, \mu)$ with same set of vertices. There is an edge between the vertices $a, b \in \mathcal{V}$ in $\mathcal{C}^p(\overrightarrow{\mathcal{G}})$ if and only if $|supp(\mathcal{N}^+(a) \cap \mathcal{N}^+(b))| \geq p$ and the membership value of the edge (a,b) in $\mathcal{C}^p(\overrightarrow{\mathcal{G}})$ is determined by $\mu(a,b) = \frac{(n-p)+1}{n}[\sigma(a) \wedge \sigma(b)] h(\mathcal{N}^+(a) \cap \mathcal{N}^+(b))$, where $n = |supp(\mathcal{N}^+(a) \cap \mathcal{N}^+(b))|$.

A **2-competition FG** is illustrated below.

Example 4.5 Let $\mathcal{G}$ be a FG with set of vertices $\{x_1, x_2, x_3, a_1, a_2, a_3\}$ (see Fig. 4.5a). Let the membership values of the vertices be $\sigma(x_1) = 0.7, \sigma(x_2) = 0.8, \sigma(x_3) = 0.9, \sigma(a_1) = 0.75, \sigma(a_2) = 0.85, \sigma(a_3) = 0.95$ and that for the edges be $\mu(\overrightarrow{x_1,a_1}) = 0.7, \overrightarrow{\mu}(x_1,a_2) = 0.7, \mu(\overrightarrow{x_2,a_1}) = 0.6, \mu(\overrightarrow{x_2,a_2}) = 0.8, \mu(\overrightarrow{x_2,a_3}) = 0.8, \mu(\overrightarrow{x_3,a_3}) = 0.9. \mu(\overrightarrow{a_2,a_3}) = 1. \mu(\overrightarrow{a_2,a_1}) = 1$. The 2-competition FG for the digraph of Fig. 4.5a is shown in Fig. 4.5b.

Fuzzy k-competition graph and p-competition FG are two generalizations of FCompG. The positive real number k for fuzzy k-competition graph is compared with the cardinality of a fuzzy set, whereas the parameter p of p-competition FG is related to the cardinality of a crisp set and it is a non-negative integer.

Theorem 4.7 Let $\overrightarrow{\mathcal{G}}$ be a fuzzy digraph and (a,b) be an edge. The edge (a,b) is strong if $h(\mathcal{N}^+(a) \cap \mathcal{N}^+(b)) = 1$ in $C^{[\frac{\kappa}{2}]}(\overrightarrow{\mathcal{G}})$, where $\kappa = |supp(\mathcal{N}^+(a) \cap \mathcal{N}^+(b))|$.

Proof Let $\mathcal{G} = (\mathcal{V}, \sigma, \overrightarrow{\mu})$ be the $[\frac{n}{2}]$-FCompG for the digraph $\overrightarrow{\mathcal{G}} = (\mathcal{V}, \sigma, \mu)$, where $\kappa = |supp(\mathcal{N}^+(a) \cap \mathcal{N}^+(b))|$.

Also, let $h(\mathcal{N}^+(a) \cap \mathcal{N}^+(b)) = 1$ for $a,b \in \mathcal{V}$.

Now, $\mu(a,b) = \frac{\kappa-[\frac{\kappa}{2}]+1}{\kappa}\sigma(a) \wedge \sigma(b)$. Thus, $\frac{\mu(a,b)}{\sigma(a) \wedge \sigma(b)} = \frac{\kappa-[\frac{\kappa}{2}]+1}{\kappa} > 0.5$. Hence, (a,b) is a strong edge.   □

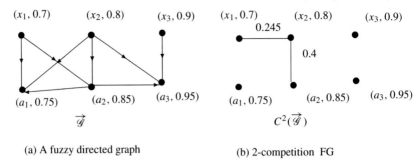

(a) A fuzzy directed graph

(b) 2-competition FG

Fig. 4.5 An example of a 2-competition FG

4.2 *m*-Step Fuzzy Competition Graph

There is another extension of fuzzy competition graph — *m*-step FCompG. In a competition graph, there is an edge between vertices a and b, if there is a common neighbor. But, in an *m*-step competition graph, there is an edge between two vertices a and b if there is a common neighbor at a distance m from both the vertices.

If there is a directed walk from a to b in a digraph $\overrightarrow{D}$ of length m, then b is called an *m*-step neighbor of a. Again, if a vertex c is an *m*-step neighbor of both vertices a and b, then c is called an *m*-step common neighbor of the vertices a and b.

Suppose a digraph $\overrightarrow{D}$ and a positive integer m is given. The *m*-step digraph of $\overrightarrow{D}$ is denoted by $\overrightarrow{D}^m$, where $\mathscr{V}(\overrightarrow{D}^m) = \mathscr{V}(\overrightarrow{D})$ and there is an edge (a, b) in $\overrightarrow{D}^m$ if and only if there exists a directed walk of length m from a to b in $\overrightarrow{D}$.

The *m*-step CompG of $\overrightarrow{D}$, denoted by $C^m(\overrightarrow{D})$, has the same set of vertices as $\overrightarrow{D}$ and there is an edge between vertices a and b in $C^m(\overrightarrow{D})$ if and only if a and b have an *m*-step common neighbor in $\overrightarrow{D}$. Note that $C^1(D)$ is the ordinary CompG of $\overrightarrow{D}$.

Now, the above concept is extended for fuzzy graphs. First of all, the *m*-step fuzzy digraph is defined.

Definition 4.11 For a fuzzy digraph $\overrightarrow{\mathscr{G}} = (\mathscr{V}, \sigma, \mu)$, the *m***-step fuzzy digraph** is denoted by $\overrightarrow{\mathscr{G}}_m = (\overrightarrow{\mathscr{G}}, \sigma, \nu)$, where the fuzzy vertex set of $\overrightarrow{\mathscr{G}}$ is same as the fuzzy vertex set of $\overrightarrow{\mathscr{G}}_m$ and the vertices a and b are connected by a directed edge in $\overrightarrow{\mathscr{G}}_m$ if there is a directed path $\overrightarrow{P}^m_{(a,b)}$ in $\overrightarrow{\mathscr{G}}$.

Definition 4.12 Fuzzy *m*-step out-nbd of a vertex b of a directed FG $\overrightarrow{\mathscr{G}} = (\mathscr{V}, \sigma, \mu)$ is the fuzzy set $\mathscr{N}^+_m(b) = (X^+_b, \rho^+_b)$, where $X^+_b = \{a|$ there is a directed fuzzy path from b to a of length m, $\overrightarrow{P}^m_{b,a}\}$ and $\rho^+_b : X^+_b \to [0, 1]$ defined by $\rho^+_b(a) = \min \{\mu(\overrightarrow{c, d}), \overrightarrow{(c, d)}$ is an edge of $\overrightarrow{P}^m_{b,a}\}$.

Fuzzy *m*-step in-nbd of a vertex b of a directed FG $\overrightarrow{\mathscr{G}} = (\mathscr{V}, \sigma, \mu)$ is the fuzzy set $\mathscr{N}^-_m(b) = (X^-_b, \rho^-_b)$, where $X^-_b = \{a|$ there is a directed fuzzy path from

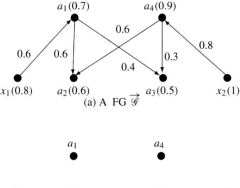

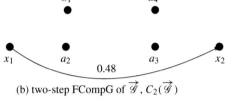

(b) two-step FCompG of $\overrightarrow{\mathcal{G}}$, $C_2(\overrightarrow{\mathcal{G}})$

Fig. 4.6 An example of two-step FCompG

a to b of length m, $\overrightarrow{P}^{\,m}_{a,b}\}$ and $\rho^-_b : X^-_b \rightarrow [0, 1]$ defined by $\rho^-_b(a) = \min \{\mu(\overrightarrow{c, d}), (\overrightarrow{c, d})$ is an edge of $\overrightarrow{P}^{\,m}_{a,b}\}$.

Example 4.6 In Fig. 4.6a, two-step out-nbd of the vertex x_1 is the fuzzy set $\{a_2(0.6), a_3(0.4)\}$ in $\overrightarrow{\mathcal{G}}$ and two-step in-nbd of a_2 is $\{x_1(0.6), x_2(0.6)\}$ in $\overrightarrow{\mathcal{G}}$.

Definition 4.13 Let $\overrightarrow{\mathcal{G}} = (\mathcal{V}, \sigma, \mu)$ be a fuzzy digraph. The *m*-**step FCompG** of the fuzzy digraph $\overrightarrow{\mathcal{G}} = (\mathcal{V}, \sigma, \mu)$ is an undirected graph, denoted by $C_m(\overrightarrow{\mathcal{G}})$ and defined by $C_m(\overrightarrow{\mathcal{G}}) = (\mathcal{V}, \sigma, \nu)$, where the membership value of an edge (a, b) is determined by $\nu(a, b) = \sigma(a) \wedge \sigma(b)\, h(\mathcal{N}^+_m(a) \cap \mathcal{N}^+_m(b))$ for all $a, b \in \mathcal{V}$.

Example 4.7 Let $\overrightarrow{\mathcal{G}} = (\mathcal{V}, \sigma, \mu)$ be a directed FG , where set of vertices be $\mathcal{V} = \{x_1, x_2, a_1, a_2, a_3, a_4\}$. The membership values of the vertices are taken as $\sigma(x_1) = 0.8, \sigma(x_2) = 1, \sigma(a_1) = 0.7, \sigma(a_2) = 0.6, \sigma(a_3) = 0.5, \sigma(a_4) = 0.9$. Also, the membership values of the edges are considered as $\mu(\overrightarrow{x_1, a_1}) = 0.6, \mu(\overrightarrow{x_2, a_4}) = 0.8, \mu(\overrightarrow{a_1, a_2}) = 0.6, \mu(\overrightarrow{a_1, a_3}) = 0.4, \mu(\overrightarrow{a_4, a_2}) = 0.6, \mu(\overrightarrow{a_4, a_3}) = 0.3$.

This directed FG is shown in Fig. 4.6a.

For this graph, the two-step out-nbd for the vertex x_1 is $\{a_2(0.6), a_3(0.4)\}$ and that of x_2 is $\{a_2(0.6), a_3(0.3)\}$. Therefore $\mathcal{N}^+_2(x_1) \cap \mathcal{N}^+_2(x_2) = \{a_2(0.6), a_3(0.3)\}$.

Hence, $h(\mathcal{N}^+_2(x_1) \cap \mathcal{N}^+_2(x_2)) = 0.6$. So, the membership value of (x_1, x_2) in two-step FCompG is $\sigma(x_1) \wedge \sigma(x_2)\, h(\mathcal{N}^+_2(x_1) \cap \mathcal{N}^+_2(x_2)) = 0.8 \times 0.6 = 0.48$. The corresponding two-step FCompG is shown in Fig. 4.6b.

In connection with problems in ecology, one can say that an edge $(\overrightarrow{a, b})$ in a fuzzy digraph represents that a predator a attacks the prey b, or a country a attacks a country b.

But, sometimes it happens that a vertex needs help of several intermediate vertices $a_1, a_2, \ldots, a_{m-1}$ to attack another vertex. In such situations, m-step FCompGs are suitable for modeling the problems. The connection among such vertices is shown by a directed path $\overrightarrow{P}{}^{m}_{a,b}$ in the fuzzy digraph. Thus, an m-step vertex in a fuzzy digraph is the m-step out-nbd of some vertices.

Now, the strength of a vertex (maybe a prey) in a fuzzy digraph (in ecology) is defined as follows.

Definition 4.14 Let $\overrightarrow{\mathscr{G}} = (\mathscr{V}, \sigma, \mu)$ be a fuzzy digraph. Let b be a common vertex of m-**step out-nbds** of vertices $a_1, a_2, \ldots, a_k$. Also, let (c_i, d_i) be the edge on the path $\overrightarrow{P}{}^{m}_{(a_i, b)}$ whose membership value is minimum among other edges on the same path, for all $i = 1, 2, \ldots, k$.

The m-step vertex $b \in \mathscr{V}$ is called strong if $\mu\overrightarrow{(c_i, d_i)} > 0.5$ for all $i = 1, 2, \ldots, k$.

The strength $s : \mathscr{V} \to [0, 1]$ of a vertex b is denoted by $s(b)$ and is measured from the expression:

$$s(b) = \frac{\sum_{i=1}^{k} \mu\overrightarrow{(c_i, d_i)}}{k}.$$

Example 4.8 In Fig. 4.6a, the strength of the vertex b is $\dfrac{0.6 + 0.6}{2} = 0.6 > 0.5$. Also, b is a strong two-step vertex.

Theorem 4.8 *For a strong vertex b of $\overrightarrow{\mathscr{G}}$, $s(b) > 0.5$.*

Proof For a strong vertex b, $\overrightarrow{\mu}(c_i, d_i) > 0.5$ for all edges (c_i, d_i) on the path $\overrightarrow{P}{}^{m}_{(a_i, b)}$. So, by definition $s(b) > 0.5$. $\square$

The converse of the above theorem is not true, *i.e.* if $s(b) > 0.5$, then all vertices may not be strong. This is explained by the following example.

Let $\mu\overrightarrow{(c_p, d_p)} = 0.3$ and $\mu\overrightarrow{(c_q, d_q)} = 0.9$ for some p and q, and other edges be $\mu\overrightarrow{(c_i, d_i)} > 0.5$.

For this case, $s(b) > 0.5$, but there is an edge whose membership value is less than 0.5.

Theorem 4.9 *If all vertices of $\overrightarrow{\mathscr{G}}$ are strong, then all the edges of $C_m(\overrightarrow{\mathscr{G}})$ are also strong.*

Proof Let $\overrightarrow{\mathscr{G}} = (\mathscr{V}, \sigma, \overrightarrow{\mu})$ be a directed FG and all vertices of $\overrightarrow{\mathscr{G}}$ be strong.

Let $C_m(\overrightarrow{\mathscr{G}}) = (\mathscr{V}, \sigma, v)$, where $v(a, b) = \sigma(a) \wedge \sigma(b) \, h(\mathscr{N}_m^+(a) \cap \mathscr{N}_m^+(b))$ for all edges (a, b) in $C_m(\overrightarrow{\mathscr{G}})$.

Let us consider two cases:

Case 1. Let $\mathscr{N}_m^+(a) \cap \mathscr{N}_m^+(b) = \emptyset$. In this case, there is no edge between the vertices a and b in $C_m(\overrightarrow{\mathscr{G}})$.

Case 2. Let $\mathscr{N}_m^+(a) \cap \mathscr{N}_m^+(b) \neq \emptyset$.

Clearly, $h(\mathcal{N}_m^+(a) \cap \mathcal{N}_m^+(b)) > 0.5$ in $\overrightarrow{\mathcal{G}}$ as all vertices are strong.

Therefore, the membership value of any edge (a, b) in $C_m(\overrightarrow{\mathcal{G}})$ is $\sigma(a) \wedge \sigma(b) h(\mathcal{N}_m^+(a) \cap \mathcal{N}_m^+(b))$ and, hence, all the edges are strong. $\square$

The following relationship has been established between m-step FCompG of a fuzzy digraph and FCompG of m-step fuzzy digraph.

Theorem 4.10 *Let $\overrightarrow{\mathcal{G}} = (\mathcal{V}, \sigma, \mu)$ be a directed FG and $\overrightarrow{\mathcal{G}}_m$ be the m-step fuzzy digraph of $\overrightarrow{\mathcal{G}}$. Then*

1. $C(\overrightarrow{\mathcal{G}}_m) = C_m(\overrightarrow{\mathcal{G}})$.
2. $C_m(\overrightarrow{\mathcal{G}})$ *has no edge if $m > |\mathcal{V}|$.*

Proof (1) Here, $\overrightarrow{\mathcal{G}}$ is a fuzzy digraph and $\overrightarrow{\mathcal{G}}_m$ is the m-step fuzzy digraph of $\overrightarrow{\mathcal{G}}$.

Let $\overrightarrow{\mathcal{G}} = (\mathcal{V}, \sigma, \overrightarrow{\mu})$ and $\overrightarrow{\mathcal{G}}_m = (\mathcal{V}, \sigma, \overrightarrow{v})$. Also, let $C(\overrightarrow{\mathcal{G}}_m) = (\mathcal{V}, \sigma, v)$ and $C_m(\overrightarrow{\mathcal{G}}) = (\mathcal{V}, \sigma, \mu)$.

By definition, the sets of vertices of these graphs are same.

Now, we have to prove that the sets of edges of both the graphs $C(\overrightarrow{\mathcal{G}}_m)$ and $C_m(\overrightarrow{\mathcal{G}})$ are same.

Let (a, b) be an edge in $C(\overrightarrow{\mathcal{G}}_m)$. So, there must exist fuzzy directed edges $\overrightarrow{(a, a_1)}, \overrightarrow{(b, a_1)}; \overrightarrow{(a, a_2)}, \overrightarrow{(b, a_2)}; \ldots \overrightarrow{(a, a_k)}, \overrightarrow{(b, a_k)}$ in $\overrightarrow{\mathcal{G}}_m$, for some positive integer k.

Now, in $\overrightarrow{\mathcal{G}}_m$, $N^+(a) \cap N^+(b) = \{(a_i, m_i)|i = 1, 2, \ldots, k\}$, where $m_i = \overrightarrow{v}(a, a_i) \wedge \overrightarrow{v}(b, a_i)$. Let $M = max\{m_i|i = 1, 2, \ldots, k\}$. Hence, $v(a, b) = \sigma(a) \wedge \sigma(b) h(N^+(a) \cap N^+(b)) = M \times \sigma(a) \wedge \sigma(b)$.

An edge $\overrightarrow{(a, a_i)}$ exists in $\overrightarrow{\mathcal{G}}_m$ implies that there exists a fuzzy directed path from a to a_i of length m, $\overrightarrow{P}_{(a,a_i)}^m$ in $\overrightarrow{\mathcal{G}}$ and $\overrightarrow{v}(a, a_i) = min\{\overrightarrow{\mu}(u, v) \mid (u, v) \text{ is an edge in } \overrightarrow{P}_{(a,a_i)}^m\}$.

Thus, the edge (a, b) is also available in $C_m(\overrightarrow{\mathcal{G}})$. Also, $h(N_m^+(a) \cap N_m^+(b)) = M$ in $\overrightarrow{\mathcal{G}}$.

Hence, finally, $\mu(a, b) = \sigma(a) \wedge \sigma(b) h(N_m^+(a) \cap N_m^+(b)) = M \times \sigma(a) \wedge \sigma(b)$. This shows that there is an edge in $C_m(\overrightarrow{\mathcal{G}})$.

Similarly, for each edge in $C_m(\overrightarrow{\mathcal{G}})$, there is an edge in $C(\overrightarrow{\mathcal{G}}_m)$. Thus, $C(\overrightarrow{\mathcal{G}}_m) = C_m(\overrightarrow{\mathcal{G}})$.

(2) Let $\overrightarrow{\mathcal{G}} = (\mathcal{V}, \sigma, \overrightarrow{\mu})$ be a directed FG and $C_m(\overrightarrow{\mathcal{G}}) = (\mathcal{V}, \sigma, v)$ be the corresponding m-step FCompG, where $v(a, b) = \sigma(a) \wedge \sigma(b) h(\mathcal{N}_m^+(a) \cap \mathcal{N}_m^+(b))$.

If $m > |\mathcal{V}|$, then there does not exist any directed fuzzy path of length m in $\overrightarrow{\mathcal{G}}$. So, $\mathcal{N}_m^+(a) \cap \mathcal{N}_m^+(b) = \emptyset$. Thus, there is no edge in $C_m(\overrightarrow{\mathcal{G}})$. This proves the result. $\square$

4.3 *m*-Step Fuzzy Neighborhood Graph

Here, we describe a particular case of *m*-step fuzzy graph called *m*-step fuzzy nbd graph.

Definition 4.15 Fuzzy *m*-step nbd of a vertex b in a FG $\mathscr{G} = (\mathscr{V}, \sigma, \mu)$ is a fuzzy set $\mathscr{N}_m(b) = (X_b, \rho_b)$, where $X_b = \{a|$ there exists a fuzzy path from b to a of length m, $P^m_{b,a}\}$ and $\rho_b : X_b \to [0, 1]$ is defined by $\rho_b(a) = \min \{\mu(c, d), (c, d)$ is an edge of $P^m_{b,a}\}$.

Definition 4.16 For the FG $\mathscr{G} = (\mathscr{V}, \sigma, \mu)$, the *m*-**step fuzzy nbd graph** is denoted by $N_m(\mathscr{G})$ and defined by $N_m(\mathscr{G}) = (\mathscr{V}, \sigma, \eta)$, where $\eta : \mathscr{V} \times \mathscr{V} \to [0, 1]$ is the membership value of the edge (a, b) and is given by $\eta(a, b) = \sigma(a) \wedge \sigma(b) \, h(\mathscr{N}_m(a) \cap \mathscr{N}_m(b))$ for all $a, b \in \mathscr{V}$.

Theorem 4.11 *Let $\overrightarrow{\mathscr{G}}$ be a fuzzy digraph without any parallel edge. Then $C_m(\overrightarrow{\mathscr{G}}) = N_m(\mathscr{G})$ for $m > 1$, where $\mathscr{G}$ is the underlying FG of $\overrightarrow{\mathscr{G}}$.*

Proof Let $\overrightarrow{\mathscr{G}} = (\mathscr{V}, \sigma, \overrightarrow{\mu})$ be a fuzzy digraph without parallel edges. Let $\mathscr{G} = (\mathscr{V}, \sigma, \mu)$ be the underlying FG of $\overrightarrow{\mathscr{G}}$. Also, let $C_m(\overrightarrow{\mathscr{G}}) = (\mathscr{V}, \sigma, \nu)$ be the *m*-step FCompG and $N_m(\mathscr{G}) = (\mathscr{V}, \sigma, \eta)$ be the *m*-step fuzzy nbd graph. Since $\overrightarrow{\mathscr{G}}$ has no parallel edges, $\overrightarrow{\mu}(a, b) = \mu(a, b)$ for all vertices $a, b \in \mathscr{V}$.

Now, we will show that $\nu(a, b) = \eta(a, b)$ for all edges (a, b).

Since $N^+_m(a) \cap N^+_m(b)$ in $\overrightarrow{\mathscr{G}}$ is equal to $N_m(a) \cap N_m(b)$ in $\mathscr{G}$ and $\nu(a, b) = \sigma(a) \wedge \sigma(b) \, h(N^+_m(a) \cap N^+_m(b))$, $\eta(a, b) = \sigma(a) \wedge \sigma(b) \, h(N_m(a) \cap N_m(b))$. Hence, $\nu(a, b) = \eta(a, b)$. $\qquad\square$

4.4 Fuzzy Economic Competition Graph

Nowadays, people are transferring money online from different sources (e.g. banks, ATMs, POSs, PayPal, Paytm, etc.) to different destinations within a very short time. To explain this graph, let us consider three funding agencies F_1, F_2, F_3 for running scientific projects. Also, it is assumed that five projects say $P_1, P_2, \cdots, P_5$ are running in five different institutions, but one same project may be funded by more than one funding agencies (see Fig. 4.7). The funding agencies and projects are considered as vertices of the graph and there is a directed edge from the funding agency to the project (e.g. the funding agencies for the project P_1 are F_1 and F_2, so there are directed edges (F_1, P_1) and (F_2, P_1), and so on).

This section is based on the work by Samanta et al. [14] in 2015. Here, the fuzzy economic competition graph and *m*-step fuzzy economic competition graph are introduced.

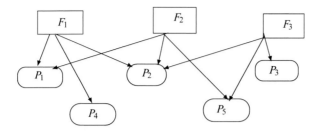

Fig. 4.7 An example of economic competition graph

Definition 4.17 Let $\overrightarrow{\mathscr{G}} = (\mathscr{V}, \sigma, \overrightarrow{\mu})$ be a fuzzy digraph and $E(\overrightarrow{\mathscr{G}})$ be its corresponding **fuzzy economic competition graph**. $E(\overrightarrow{\mathscr{G}})$ is an undirected FG $\mathscr{G} = (\mathscr{V}, \sigma, \phi)$ which has an edge between two vertices $a, b \in \mathscr{V}$ in $E(\overrightarrow{\mathscr{G}})$ if and only if $\mathscr{N}^-(a) \cap \mathscr{N}^-(b) \neq \emptyset$ in $\overrightarrow{\mathscr{G}}$ and the membership value $\phi(a, b)$ of the edge (a, b) in $\mathscr{C}(\overrightarrow{\mathscr{G}})$ is given by $\phi(a, b) = (\sigma(a) \wedge \sigma(b)) h(\mathscr{N}^-(a) \cap \mathscr{N}^-(b))$.

Definition 4.18 Let $\overrightarrow{\mathscr{G}} = (\mathscr{V}, \sigma, \overrightarrow{\mu})$ be a fuzzy digraph. The **m-step fuzzy economic competition graph** is denoted by $E_m(\overrightarrow{\mathscr{G}})$ and is defined by $E_m(\overrightarrow{\mathscr{G}}) = (\mathscr{V}, \sigma, \phi)$, where $\phi : \mathscr{V} \times \mathscr{V} \rightarrow [0, 1]$ represents the membership value of an edge and is given by $\phi(a, b) = \sigma(a) \wedge \sigma(b) h(\mathscr{N}_m^-(a) \cap \mathscr{N}_m^-(b))$ for all $a, b \in \mathscr{V}$.

Example 4.9 A fuzzy digraph $\overrightarrow{\mathscr{G}}$ is depicted in Fig. 4.8a. The fuzzy economic competition graph $(E(\overrightarrow{\mathscr{G}}))$ of $\overrightarrow{\mathscr{G}}$ and two-step fuzzy economic competition graph $(E_2(\overrightarrow{\mathscr{G}}))$ are shown in Fig. 4.8b and c respectively.

Theorem 4.12 *Let $\overrightarrow{\mathscr{G}_1}$ be a fuzzy sub-digraph of $\overrightarrow{\mathscr{G}}$. Then*
(i) $\mathscr{C}_m(\overrightarrow{\mathscr{G}_1}) \subset \mathscr{C}_m(\overrightarrow{\mathscr{G}})$.
(ii) $E_m(\overrightarrow{\mathscr{G}_1}) \subset E_m(\overrightarrow{\mathscr{G}})$.

Proof Let $\overrightarrow{\mathscr{G}} = (\mathscr{V}, \sigma, \overrightarrow{\mu})$ and $\overrightarrow{\mathscr{G}_1} = (\mathscr{V}_1, \sigma_1, \overrightarrow{\mu_1})$, where $\mathscr{V}_1 \subset \mathscr{V}$, $\sigma_1(a) \leq \sigma(a)$ for all $a \in \mathscr{V}_1$ and $\overrightarrow{\mu_1}(c, d) \leq \overrightarrow{\mu}(c, d)$ for all $c, d \in \mathscr{V}_1$.
   (i) The fuzzy vertex set of $\mathscr{C}_m(\overrightarrow{\mathscr{G}_1})$ is the subset of $\mathscr{C}_m(\overrightarrow{\mathscr{G}})$ as $\mathscr{V}_1 \subset \mathscr{V}$. Now for any fuzzy edge (c, d) in $\mathscr{C}_m(\overrightarrow{\mathscr{G}_1})$, $\mathscr{N}_m^+(c) \cap \mathscr{N}_m^+(d)$ is a fuzzy subset of the same in $\mathscr{C}_m(\overrightarrow{\mathscr{G}})$. So, $\overrightarrow{\mu_1}(c, d) \leq \overrightarrow{\mu}(c, d)$ for all $c, d \in \mathscr{V}_1$. Hence, the result is proved.
   (ii) The proof is similar to (i).   □

These results can also be extended to fuzzy m-step nbd graphs.

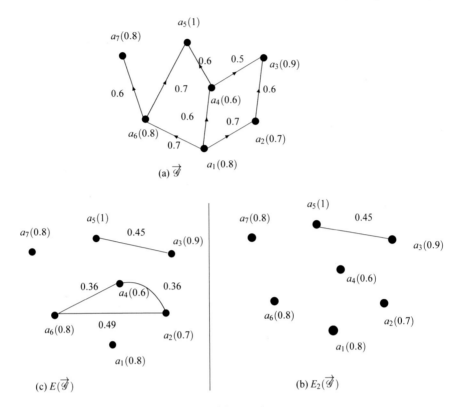

Fig. 4.8 Example of fuzzy economic competition graphs

4.5 Fuzzy Competition Number

The parameter fuzzy competition number of a FCompG is defined below.
    The work of this section is from [15].

Definition 4.19 Let $C(\overrightarrow{\mathscr{G}})$ be the FCompG for a fuzzy digraph $\overrightarrow{\mathscr{G}}$. The **fuzzy competition number** of a FG $\mathscr{G}^*$ is the minimum number of isolated fuzzy vertices such that $\mathscr{G}^*$ along with these isolated fuzzy vertices (say $\mathscr{G}_1$) becomes a FCompG of $\overrightarrow{\mathscr{G}}$, i.e. $C(\overrightarrow{\mathscr{G}}) = \mathscr{G}_1$.

Let us consider the following example to illustrate this concept.

Example 4.10 Let $\mathscr{G}^* = (\mathscr{V}, \sigma, \mu)$ be a FG with sets of vertices and edges $\mathscr{V} = \{(a_1, 0.4), (a_2, 0.6), (a_3, 0.1)\}$ and $E = \{((a_1, a_2), 0.2), ((a_2, a_3), 0.1), ((a_3, a_1), 0.1)\}$ respectively (see Fig. 4.9a). Now, we consider an isolated fuzzy vertex $a_4(0.4)$ and an edge $((a_3, a_4), 0.04)$. Let the updated FG be $\mathscr{G}_1 = \mathscr{G}^* \cup \{(a_4, 0.4), ((a_3, a_4), 0.04)\}$. This graph is shown in Fig 4.9b. It is easy to check that

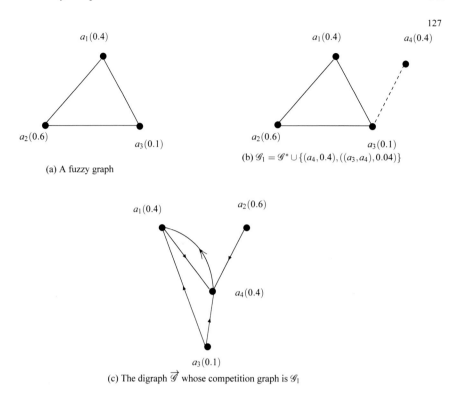

(a) A fuzzy graph

(b) $\mathscr{G}_1 = \mathscr{G}^* \cup \{(a_4, 0.4), ((a_3, a_4), 0.04)\}$

(c) The digraph $\overrightarrow{\mathscr{G}}$ whose competition graph is $\mathscr{G}_1$

Fig. 4.9 Illustration of fuzzy competition number

$C(\overrightarrow{\mathscr{G}}) = \mathscr{G}_1$, where the diagraph $\overrightarrow{\mathscr{G}}$ is shown in Fig. 4.9c. In this case, only one isolated vertex is sufficient to make $\mathscr{G}^*$ a FCompG. Thus, the fuzzy competition number of $\mathscr{G}^*$ is 1.

An important theorem related to strongly isolated vertices is given below.

Theorem 4.13 *Let P_n be a fuzzy path. The path P_n along with a strong isolated vertex may be a FCompG of a fuzzy directed tree of $n + 1$ vertices.*

Proof Let $P_n = (\mathscr{V}, \sigma, \mu)$ be a fuzzy path whose vertices be $\mathscr{V} = \{a_1, a_2, \ldots, a_n\}$. Let us consider a vertex x not contained in P_n. Now, construct the directed FG whose FCompG is $P_n \cup \{x\}$. We construct the directed fuzzy edges $\overrightarrow{(a_i, x)}, i = 1, 2, \ldots, n$. Obviously, this directed FG is a directed fuzzy tree T_{n+1} having $n + 1$ vertices. Also, $C(T_{n+1}) = P_n \cup \{x\}$. □

This result is extended as follows.

Some disjoint paths along with a strong isolated vertex may be a FCompG of a fuzzy directed tree. Let the disjoint paths be $P_{m_1}, P_{m_2}, \ldots, P_{m_n}$. The number of vertices of the corresponding directed FG is $m_1 + m_2 + \ldots + m_n + 1$.

From Theorem 4.13, it is easy to observe that the fuzzy competition number of a path is 1. Like crisp graph, the fuzzy competition numbers can be found as follows.

The competition number of a fuzzy chordal graph which has no strong isolated vertex is 1.

If a FG is triangle-free, then the fuzzy competition number is equal to two more than the difference between the number of edges and vertices.

The competition number of a fuzzy complete graph is one.

4.6 Isomorphism in Fuzzy Competition Graph

Isomorphism on fuzzy graphs is a well-studied topic. Here, the isomorphism in fuzzy digraphs is discussed. This work is also from [14].

A **homomorphism of fuzzy digraphs** $\overrightarrow{\mathcal{G}} = (\mathcal{V}, \sigma, \mu)$ and $\overrightarrow{\mathcal{G}'} = (\mathcal{V}', \sigma', \mu')$ is a mapping $f : \mathcal{V} \to \mathcal{V}'$ such that $\sigma(a) \leq \sigma'(f(a))$ for all $a \in \mathcal{V}$ and $\mu\overrightarrow{(a, b)} \leq \mu'\overrightarrow{(f(a), f(b))}$ for all $a, b \in \mathcal{V}$.

A **weak isomorphism** between fuzzy digraphs $\overrightarrow{\mathcal{G}} = (\mathcal{V}, \sigma, \mu)$ and $\overrightarrow{\mathcal{G}'} = (\mathcal{V}', \sigma', \mu')$ is a bijective homomorphism $f : \mathcal{V} \to \mathcal{V}'$ such that $\sigma(a) = \sigma'(f(a))$ for all $a \in \mathcal{V}$.

A **co-weak isomorphism** between fuzzy digraphs $\overrightarrow{\mathcal{G}} = (\mathcal{V}, \sigma, \mu)$ and $\overrightarrow{\mathcal{G}'} = (\mathcal{V}', \sigma', \mu')$ is a bijective homomorphism $f : \mathcal{V} \to \mathcal{V}'$ such that $\mu\overrightarrow{(a, b)} = \mu'\overrightarrow{(f(a), f(b))}$ for all $a, b \in \mathcal{V}$.

An isomorphism between fuzzy graphs of fuzzy digraphs is a bijective homomorphism $f : \mathcal{V} \to \mathcal{V}'$ such that $\sigma(a) = \sigma'(f(a))$ for all $a \in \mathcal{V}$ and $\mu\overrightarrow{(a, b)} = \mu'\overrightarrow{(f(a), f(b))}$ for all $a, b \in \mathcal{V}$.

An isomorphism between fuzzy graphs is an equivalence relation. But, if there is an isomorphism between two FGs and one is a FCompG, then the other will also be a FCompG. But, the corresponding digraphs may not be isomorphic in this case. This result can be seen from the following.

Remark 4.1 Let $\overrightarrow{\mathcal{G}_1}$ and $\overrightarrow{\mathcal{G}_2}$ be two fuzzy digraphs such that $C(\overrightarrow{\mathcal{G}_1})$ and $C(\overrightarrow{\mathcal{G}_2})$ are isomorphic. But, the fuzzy digraphs $\overrightarrow{\mathcal{G}_1}$ and $\overrightarrow{\mathcal{G}_2}$ may not be isomorphic.

This remark is justified in the following example.

Let $\overrightarrow{\mathcal{G}_1}$ and $\overrightarrow{\mathcal{G}_2}$ be two fuzzy digraphs shown in Fig. 4.10a and b. Obviously, these fuzzy digraphs are not isomorphic because the degree of vertices is distinct. But, the fuzzy competition graphs corresponding to these digraphs are isomorphic.

Another similar result is stated below:

If two fuzzy digraphs are isomorphic, their corresponding FCompGs must be isomorphic. This is established in the following theorem.

Theorem 4.14 *If two fuzzy digraphs $\overrightarrow{\mathcal{G}_1}$ and $\overrightarrow{\mathcal{G}_2}$ are isomorphic, then $C(\overrightarrow{\mathcal{G}_1})$ and $C(\overrightarrow{\mathcal{G}_2})$ are also isomorphic.*

Fig. 4.10 Fuzzy competition graphs are isomorphic, but their corresponding digraphs are not isomorphic

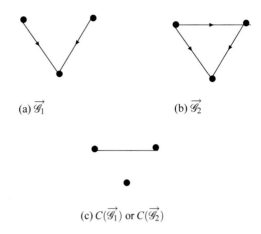

(a) $\overrightarrow{\mathscr{G}_1}$ (b) $\overrightarrow{\mathscr{G}_2}$

(c) $C(\overrightarrow{\mathscr{G}_1})$ or $C(\overrightarrow{\mathscr{G}_2})$

Proof Suppose, the fuzzy digraphs $\overrightarrow{\mathscr{G}_1}$ and $\overrightarrow{\mathscr{G}_2}$ are isomorphic. Therefore, there exists one to one correspondence between vertices and edges. Also, the membership values of the vertices and edges remain same. Also, the adjacency of vertices also remains unchanged. Hence, $C(\overrightarrow{\mathscr{G}_1})$ and $C(\overrightarrow{\mathscr{G}_2})$ are isomorphic.   □

In a similar way, one can prove that if two fuzzy digraphs $\overrightarrow{\mathscr{G}_1}$ and $\overrightarrow{\mathscr{G}_2}$ are isomorphic, then $C_k(\overrightarrow{\mathscr{G}_1})$ and $C_k(\overrightarrow{\mathscr{G}_2})$ are also isomorphic.

References

1. M. Akram, S. Siddique, Neutrosophic competition graphs with applications. J. Intell. Fuzzy Syst. **33**(2), 921–935 (2017)
2. M. Akram, M. Sarwar, New Applications of m-polar fuzzy competition graphs. New Math. Nat. Comput. **14**(2), 249–276 (2018)
3. R.A. Borzooei, H. Rashmanlou, S. Samanta, M. Pal, New concept of vague competition graphs. J. Intell. Fuzzy Syst. **31**, 69–75 (2016)
4. J.E. Cohen, *Interval Graphs and Food Webs: A Finding and a Problem, Document 17696-PR* (RAND Corporation, Santa Monica, 1968)
5. A. Habib, M. Akram, A. Farooq, q-Rung orthopair fuzzy competition graphs with application in the soil ecosystem. Mathematics **7**, 91 (2019). https://doi.org/10.3390/math7010091
6. M. Pal, S. Samanta, A. Pal, Fuzzy k-competition graphs, in *Science and Information Conference*, London (2013), pp. 572–576
7. M. Pal, An introduction to intersection graphs, chapter 2, in *An Handbook of Research on Advanced Applications of Graph Theory in Modern Society*, eds. by M. Pal, S. Samanta, A. Pal (IGI Global, USA, 2020), pp. 24–65
8. T. Pramanik, S. Samanta, M. Pal, S. Mondal, B. Sarkar, Interval-valued fuzzy ϕ-tolerance competition graphs. SpringerPlus **5**, 1981 (2016). https://doi.org/10.1186/s40064-016-3463-z
9. T. Pramanik, S. Samanta, B. Sarkar, M. Pal, Fuzzy ϕ-tolerance competition graphs. Soft Comput. **21**, 3723–3734 (2017). https://doi.org/10.1007/s00500-015-2026-5

10. T. Pramanik, G. Muhiuddin, A.M. Alanazi, M. Pal, An extension of fuzzy competition graph and its uses in manufacturing industries. Mathematics **8**, 1008 (2020)
11. S. Sahoo, M. Pal, Intuitionistic fuzzy competition graphs. J. Appl. Math. Comput. **52**, 37–57 (2016). https://doi.org/10.1007/s12190-015-0928-0
12. S. Samanta, M. Pal, Fuzzy k-competition graphs and p-competition fuzzy graphs. Fuzzy Eng. Inf. **5**(2), 191–204 (2013)
13. S. Samanta, M. Pal, Some more results on bipolar fuzzy sets and bipolar fuzzy intersection graphs. J. Fuzzy Math. **22**(2), 253–262 (2014)
14. S. Samanta, M. Akram, M. Pal, m-step fuzzy competition graphs. J. Appl. Math. Comput. **47**(1–2), 461–472 (2015)
15. S. Samanta, M. Pal, A. Pal, Some more results on fuzzy k-competition graphs. Int. J. Adv. Res. Artif. Intell. **3**(1), 60–67 (2014). https://doi.org/10.14569/IJARAI.2014.030109
16. S. Samanta, B. Sarkar, Representation of competitions by generalized fuzzy graphs. Int. J. Comput. Intell. Syst. **11**(1), 1005–1015 (2018)
17. M. Sarwar, M. Akram, Novel concepts of bipolar fuzzy competition graphs. J. Appl. Math. Comput. **54**(1–2), 511–547 (2016). https://doi.org/10.1007/s12190-016-1021-z
18. A.A. Talebi, H. Rashmanlou, S.H. Sadati, Interval-valued intuitionistic fuzzy competition graph. J. Multiple-Valued Logic Soft Comput. **34**(3/4), 335–364 (2020)

Chapter 5
Fuzzy Threshold Graph

In 1973, Chvatal and Hammer [2] introduced threshold graphs and applied them in set packing problems. This particular class of graphs has many applications in several applied areas such as computer science, scheduling theory, psychology, etc. These graphs are also used in control theory, particularly to control the flow of information between processors, traffic lights controlling system to manage the flow of the traffic, etc. Ordman [4] established the use of these graphs in resource allocation problems. In this chapter, fuzzy threshold graph (FTG) is defined and some of its important properties are investigated. Here, we use the notation $\mathscr{C}_n$, $\mathscr{P}_n$, $\mathscr{K}_n$ for fuzzy cycle, fuzzy path, complete fuzzy graph respectively with n vertices. This work is mainly from [8, 9]. Related works are available in [1, 3, 6, 10, 11].

5.1 Alternating 4-Cycle of a Graph

The crisp threshold graph is defined by Chvatal and Hammer [2] in 1973. The threshold graph can be defined in many ways.

Let M be a real number and $wt(a)$ be the weight of the vertex a. A graph G is said to be **threshold graph** if for any two vertices v, u, there is an edge uv if and only if $wt(a) + wt(b) > M$.

The threshold graph can also be defined as follows:

Let T be a real number and $wt(a)$ be the weight of the vertex a. A graph $G = (V, E)$ is called **threshold graph** if for any vertex set $U \subseteq V$, S is independent if and only if

$$\sum_{a \in U} wt(a) \leq T.$$

So $G = (V, E)$ is a threshold graph whenever one can assign weights to the vertices such that a set of vertices is stable if and only if its total weight is bounded by a certain threshold.

© The Editor(s) (if applicable) and The Author(s), under exclusive license to Springer Nature Singapore Pte Ltd. 2020
M. Pal et al., *Modern Trends in Fuzzy Graph Theory*,
https://doi.org/10.1007/978-981-15-8803-7_5

Note that these two definitions are different. In the first definition, M is the "threshold" for the property on an edge, whereas in the second definition T is the threshold for being independent.

The forbidden graph characterization for threshold graph is stated below:
A threshold graph is a graph none of whose induced subgraphs with four vertices are isomorphic to $2K_2$, C_4, P_4.

We already defined incidence matrix [5] of a graph in Chap. 1, but for ready references, the same is defined below.

The incidence matrix of a graph G with n vertices is a matrix of order $n \times n$ in which $a_{ij} = 1$ if the vertices i and j are adjacent and it is 0, otherwise. Therefore, the elements of an incidence matrix are either 0 or 1, i.e. 0–1 matrix.

In [2], Chvatal and Hammer showed that the m constraints

$$\sum_{j=1}^{n} a_{ij} x_j \le 1, \qquad i = 1, 2, \dots, m$$
$$x_j = 0 \text{ or } 1, j = 1, 2, \dots, n$$

are equivalent to only one constraint

$$\sum_{j=1}^{n} c_j x_j \le d,$$
$$x_j = 0 \text{ or } 1, j = 1, 2, \dots, n$$

if and only if the intersection graph corresponding to the matrix (a_{ij}) is a threshold graph.

A **split graph** is a graph in which the vertices can be partitioned into a clique and an independent set.

Now, we fuzzify the threshold graph.

Definition 5.1 A FG $\mathcal{G} = (\mathcal{V}, \sigma, \mu)$ is called a **fuzzy threshold graph** (FTG) if there is a non-negative real number T such that $\sum_{a \in U} \sigma(a) \le T$ if and only if $U \subset \mathcal{V}$ is an independent set in $\mathcal{G}$.

Example 5.1 Let $\mathcal{G} = (\mathcal{V}, \sigma, \mu)$ be a FG, where $\mathcal{V} = \{a_1, a_2, a_3, a_4, a_5\}$. Let the membership values of vertices be $\sigma(a_1) = 0.1, \sigma(a_2) = 0.5, \sigma(a_3) = 0.2, \sigma(a_4) = 0.4, \sigma(a_5) = 0.2$. Also, let threshold be $T = 0.5$. Here, the set $\{a_1, a_3, a_5\}$ is an independent set in crisp sense. And $\sigma(a_1) + \sigma(a_3) + \sigma(a_5) = 0.5 = T$. So, this graph is a FTG. (see Fig. 5.1).

Alternating 4-cycle of a graph. Let $G = (V, E)$ be a graph with set of vertices $\{a_1, a_2, a_3, a_4\}$. Also, let $(a_1, a_2), (a_3, a_4) \in E$ and $(a_1, a_3), (a_2, a_4) \notin E$. By including or excluding the edges $(a_1, a_4), (a_2, a_3)$, it is easy to see that the vertices of alternating 4-cycle induce a matching $2K_2$, or a square C_4, or a path P_4.

By extending this idea fuzzy alternating 4-cycle is defined.

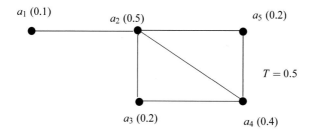

Fig. 5.1   An example of FTG

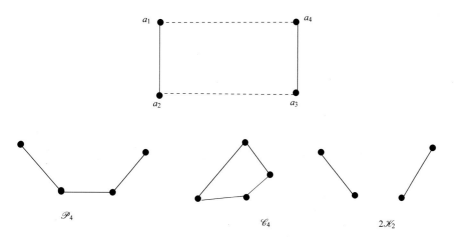

Fig. 5.2   Induced subgraphs of fuzzy alternating 4-cycle

Definition 5.2   Let $\mathscr{G} = (\mathscr{V}, \sigma, \mu)$ be a FG   and $\mathscr{V} = \{a_1, a_2, a_3, a_4\}$. Also, let $\sigma(a_i) > 0$ for all $i = 1, 2, 3, 4$ and $\mu(a_1, a_2) > 0, \mu(a_3, a_4) > 0$ and $\mu(a_1, a_3) = \mu(a_2, a_4) = 0$. The arrangement of this four vertices is called **fuzzy alternating 4-cycle**. This fuzzy alternating 4-cycle induces a matching $2\mathscr{K}_2$ (when $\mu(a_1, a_4) = 0, \mu(a_2, a_3) = 0$), a square $\mathscr{C}_4$ (when $\mu(a_1, a_4) > 0, \mu(a_2, a_3) > 0$) or a path $\mathscr{P}_4$ (when one of $\mu(a_1, a_4), \mu(a_2, a_3)$ is zero and other is non-zero).

Fuzzy alternating 4-cycle may also induce a fuzzy a square $\mathscr{C}_4$ if the membership values of edges do not have exactly one minimum value. The fuzzy alternating 4-cycle and its induced FGs are shown in Fig. 5.2.

Definition 5.3   An alternating 4-cycle is called a strong alternating 4-cycle if fuzzy $\mathscr{C}_4$ be induced from it.

Degree partition is a very important term in graph theory. For the graph $G = (V, E)$ having n vertices and distinct positive degrees of vertices $\delta_1 < \delta_2 < \cdots < \delta_m$, $\delta_0 = 0$ (assume that there are some vertices with degree 0, though it in reality does not happen) and $\delta_{m+1} = |V| - 1, \quad m + 1 \leq n$, *degree partition* is the sequence $D_i = \{a \in V : \deg(a) = \delta_i\}$ for $i = 0, 1, \ldots, m$. Here, we define degree partition in a FG.

Definition 5.4 Let $\mathscr{G} = (\mathscr{V}, \sigma, \mu)$ be a FG and let the distinct positive degrees of the vertices be $\lambda_1 < \lambda_2 < \cdots < \lambda_p$, (for some p) and let $\lambda_0 = 0$ (even if no isolated vertex exists), $\lambda_{p+1} = |\mathscr{V}| - \lambda_1$. Let $\mathscr{D}_i = \{a \in \mathscr{V} : i \leq \deg(a) < i + 1\}$ for non-negative integer $i \leq p$. The sequence of set of vertices $\mathscr{D}_0, \mathscr{D}_1, \ldots, \mathscr{D}_p$ is called the **degree partition of the FG** $\mathscr{G}$.

Theorem 5.1 *A FTG does not have any strong fuzzy alternating 4-cycle.*

Proof Let $\mathscr{G} = (\mathscr{V}, \sigma, \mu)$ be a FTG. Suppose $\mathscr{G}$ has a strong fuzzy alternating 4-cycle. Therefore, there exists four vertices a_1, a_2, a_3, a_4 with $\mu(a_1, a_2) > 0$, $\mu(a_3, a_4) > 0$ and $\mu(a_1, a_4) = 0 = \mu(a_2, a_3)$. Since T is the threshold of FTG, then $\sigma(a_1) + \sigma(a_2) > T, \sigma(a_3) + \sigma(a_4) > T$ (since the vertices a_1, a_2, a_3, a_4 constitute a strong fuzzy alternating 4-cycle).

$\quad \sigma(a_1) + \sigma(a_4) \leq T, \sigma(a_2) + \sigma(a_3) \leq T$ (since $\mu(a_1, a_4) = 0 = \mu(a_2, a_3)$).

The above inequalities are not consistent. Therefore, the vertices a_1, a_2, a_3, a_4 do not constitute a strong fuzzy alternating 4-cycle. Hence, FTG does not have a strong fuzzy alternating 4-cycle. □

Theorem 5.2 *A FTG is a fuzzy split graph.*

Proof Let $\mathscr{G} = (\mathscr{V}, \sigma, \mu)$ be a FTG. Also, let the largest clique in $\mathscr{G}$ be K. If the edge (a_1, a_2) is strong in $|\mathscr{V} - K|$, there exists distinct vertices a_3, a_4 in K such that $\mu(a_1, a_4) = 0 = \mu(a_2, a_3)$ (as K is maximal). These vertices form a strong fuzzy alternating 4-cycle, a contradiction. Thus, $|\mathscr{V} - K|$ must be a stable set. Hence, $\mathscr{G}$ is a fuzzy split graph. □

Theorem 5.3 *A FTG can be constructed from one vertex graph by the following ways:*
(i) repeatedly adding a fuzzy isolated vertex to the graph
(ii) repeatedly adding a fuzzy dominating vertex to the graph.

Proof A FTG is a fuzzy split graph (Theorem 5.2). So, it is sufficient to prove that $\mathscr{G}$ contains a fuzzy isolated vertex or a fuzzy dominating vertex if for such a vertex is removed, then the graph still remains a fuzzy split graph.

Let $\mathscr{S}$ and $\mathscr{K}$ be a stable set and a fuzzy clique of the FTG of $\mathscr{G}$ respectively. Then $\mathscr{G} = (\mathscr{K}, \mathscr{S})$. If $\mathscr{S}$ has only fuzzy isolated vertex, then the result holds. If $\mathscr{S}$ does not contain any fuzzy isolated vertex, then any vertex $a \in \mathscr{S}$ with the smallest neighborhood has some neighbor $b \in \mathscr{K}$. Since $\mathscr{K}$ is a fuzzy clique, therefore, b is a dominating vertex of $\mathscr{G}$. □

Two vertices a and b are said to be **incomparable** if one of the following two conditions are true.
(i) there is no path from a to b and no path from b to a,
(ii) they are not present in the same tree.
We now define a strong comparability of two vertices in a FG.

Definition 5.5 Let $\mathscr{G} = (\mathscr{V}, \sigma, \mu)$ be a FG. Two vertices a and b are called **strong comparable** if there is a path from a to b or b to a whose every edge is strong.

5.2 Threshold Dimension

Let us define a parameter for a threshold graph.

The minimum number k of threshold subgraphs $T_1, T_2, \ldots, T_k$ of G that cover the set of edges of G is called **threshold dimension**, and it is denoted by $t(G)$.

For a FG, the threshold dimension is defined below.

Let $\mathscr{T}_1, \mathscr{T}_2, \ldots, \mathscr{T}_k$ be a family of k fuzzy threshold subgraphs of a FG $\mathscr{G}$. If $\mathscr{T}_1 \cup \mathscr{T}_2 \cup \cdots \cup \mathscr{T}_k = \mathscr{G}$, then the subgraphs $\mathscr{T}_1, \mathscr{T}_2, \ldots, \mathscr{T}_k$ cover the set of edges of $\mathscr{G}$.

Definition 5.6 The minimum number k of fuzzy threshold subgraphs $\mathscr{T}_1, \mathscr{T}_2, \ldots, \mathscr{T}_k$ of $\mathscr{G}$ that cover the set of edges of $\mathscr{G}$, is called the **threshold dimension** $\tilde{t}(\mathscr{G})$ of a FG $\mathscr{G} = (\mathscr{V}, \sigma, \mu)$.

The threshold dimension of a FG is well defined because every fuzzy edge and also fuzzy isolated vertices form a fuzzy threshold subgraph. The threshold dimension is also bounded and it is bounded by the number of edges of the graph.

Let us denote the stability number of a FG $\mathscr{G}$ by $\alpha(\mathscr{G})$, which is the order of the largest stable set of $\mathscr{G}$.

Theorem 5.4 *For every FG $\mathscr{G} = (\mathscr{V}, \sigma, \mu)$ having n vertices, $\tilde{t}(\mathscr{G}) \leq (n - \alpha(\mathscr{G}))$. Again, if $\mathscr{G}$ is triangle-free, then $\tilde{t}(\mathscr{G}) = (n - |supp(\mathscr{S})|)$, where $\mathscr{S}$ is the stable set with largest number of vertices.*

Proof Let $\mathscr{G} = (\mathscr{V}, \sigma, \mu)$ be a FG with n vertices and $\mathscr{S} = (S, \sigma)$ be a maximum stable set (stable set containing maximum number of vertices). For each vertex $a \in \mathscr{V} - S$, we consider the star centered at a. Each such star is a FTG. If we add one or more weak fuzzy arc of stable set to the stars then they satisfy the condition of FTG. So all such stars together with weak arcs of stable sets cover the edge set of $\mathscr{G}$. Thus, $\tilde{t}(\mathscr{G}) \leq |\mathscr{V} - S|$. Again we know that $|\mathscr{V}| = n$ and $\alpha(\mathscr{G}) \leq |S|$, $\mathscr{V}$, S being crisp sets. Thus, $\tilde{t}(\mathscr{G}) \leq (n - \alpha(\mathscr{G}))$.

It is true that $|S| = |supp(\mathscr{S})|$. So $\tilde{t}(\mathscr{G}) \leq (n - |supp(\mathscr{S})|)$. Again, if $\mathscr{G}$ is fuzzy triangle-free, then every FTG is a star or star together with weak edges. So $\tilde{t}(\mathscr{G}) \geq (n - |supp(\mathscr{S})|)$. Hence, $\tilde{t}(\mathscr{G}) = (n - |supp(\mathscr{S})|)$. □

5.3 Threshold Partition Number

Now, we introduce another concept related to threshold subgraphs.

Let $G = (V, E)$ be a graph. The minimum number of edge disjoint threshold subgraphs which cover all the edges of G is called **threshold partition number** and it is denoted by $tp(G)$.

Fuzzy threshold partition number is as important as threshold dimension.

Definition 5.7 The **fuzzy threshold partition number** $\tilde{tp}(\mathscr{G})$ of the FG $\mathscr{G}$ is the minimum number of fuzzy threshold subgraphs, not containing common strong arcs, cover edge set of $\mathscr{G}$.

Example 5.2 Here, we give an example of a FG whose fuzzy threshold dimension number is 2 and fuzzy partition number is 3. Let $\mathscr{G} = (\mathscr{V}, \sigma, \mu)$ be a FG and the membership values of the vertices be $a(0.1)$, $a_1(0.1)$, $b(0.5)$, $b_1(0.5)$, $c(0.4)$, $c_1(0.4)$, $e(0.4)$, $d(0.2)$, $d_1(0.2)$, $f(0.3)$. The strong edges are shown in Fig. 5.3.

Theorem 5.5 *For every fuzzy triangle-free graph* $\mathscr{G}$, $\tilde{t}(\mathscr{G}) = \tilde{tp}(\mathscr{G})$.

Proof Let $\mathscr{G} = (\mathscr{V}, \sigma, \mu)$ be a FG. We know that the edge set can be covered by $\tilde{t}(\mathscr{G})$ number of stars. If a strong arc belongs to more than one star then, we delete it from all but one of the stars. This gives a fuzzy threshold partition of size $\tilde{t}(\mathscr{G})$. $\square$

The **Ferrers digraph** [7] is a digraph related to a threshold graph. A digraph $\overrightarrow{G} = (\mathscr{V}, \overrightarrow{E})$ is said to be a Ferrers digraph if it does not contain vertices a_1, a_2, a_3, a_4, not necessarily distinct, satisfying $\overrightarrow{(a_1, a_2)}, \overrightarrow{(a_3, a_4)} \in \overrightarrow{E}$ and $\overrightarrow{(a_1, a_4)}, \overrightarrow{(a_3, a_2)} \notin \overrightarrow{E}$. Fuzzy Ferrers digraph is an important graph related to FTG.

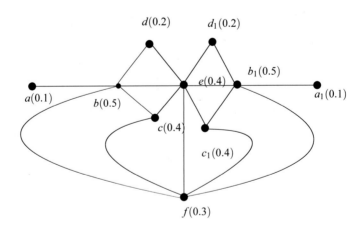

Fig. 5.3 A FG with $\tilde{t}(\mathscr{G}) = 2$ and $\tilde{tp}(\mathscr{G}) = 3$

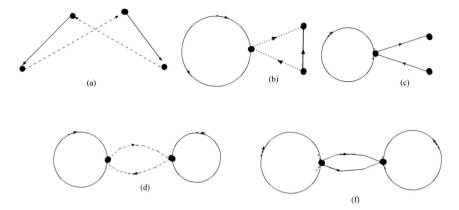

Fig. 5.4 Forbidden configurations in fuzzy Ferrers digraph

Definition 5.8 A fuzzy diagraph $\overrightarrow{\mathscr{G}} = (\mathscr{V}, \sigma, \overrightarrow{\mu})$ is called a **fuzzy Ferrers digraph** if it does not contain vertices a_1, a_2, a_3, a_4, not necessarily distinct, satisfying $\overrightarrow{\mu}(a_1, a_2)$, $\overrightarrow{\mu}(a_3, a_4)$ are non-zero and $\overrightarrow{\mu}(a_1, a_4)$, $\overrightarrow{\mu}(a_3, a_2)$ are zero.

Forbidden configurations are shown in Fig. 5.4. Here solid arrow represents that edge membership values are non-zero and dotted arrow represents that edge membership values are zero.

Definition 5.9 Let $\overrightarrow{\mathscr{G}} = (\mathscr{V}, \sigma, \overrightarrow{\mu})$ be a fuzzy diagraph. The underlying FG of $\overrightarrow{\mathscr{G}}$ is denoted by $\mathscr{U}(\mathscr{G})$ and is defined as $\mathscr{U}(\mathscr{G}) = (\mathscr{V}, \sigma, \mu)$ where $\mu(a_1, a_2) = \min\{\overrightarrow{\mu}(a_1, a_2), \overrightarrow{\mu}(a_2, a_1)\}$ for all $a_1, a_2 \in \mathscr{V}$.

Theorem 5.6 *If $\overrightarrow{\mathscr{G}}$ is a symmetric fuzzy Ferrers digraph, then its underlying fuzzy undirected loop less FG $\mathscr{U}(\mathscr{G})$ may or may not be FTG.*

Proof If $\overrightarrow{\mathscr{G}}$ is a symmetric fuzzy Ferrers digraph, then its underlying fuzzy undirected loop less FG $\mathscr{U}(\mathscr{G})$ has no fuzzy alternating 4-cycle. So it may contain a fuzzy cycle. Hence, $\mathscr{U}(\mathscr{G})$ maybe a FTG. If $\mathscr{U}(\mathscr{G})$ does not have a strong fuzzy alternating 4-cycle then $\mathscr{U}(\mathscr{G})$ must be FTG. □

References

1. M. Andelic, S.K. Simic, Some notes on the threshold graphs. Discrete Math. **310**, 2241–2248 (2010)
2. V. Chvatal, P. L. Hammer, Set-packing problems and threshold graphs, CORR, 73–21, University of Waterloo, Canada (1973)
3. R.M. Makwana, V.K. Thakar, N.C. Chauhan, Extraction of illumination invariant features using fuzzy threshold based approach. Int. J. Comput. Appl. Spl Issue, 25–31 (2011)

4. E.T. Ordman, Threshold coverings and resource allocation, in *16-th Southeastern Conference on Combinatorics, Graph Theory and Computing*, pp. 99–113 (1985)
5. M. Pal, Fuzzy matrices with fuzzy rows and fuzzy columns. J. Intell. Fuzzy Syst. **30**(1), 561–573 (2015). https://doi.org/10.3233/IFS-151780
6. M. Pal, An introduction to intersection graphs, Chapter 2, in *An Handbook of Research on Advanced Applications of Graph Theory in Modern Society*, eds. by M. Pal, S. Samanta, A. Pal, pp. 24–65. IGI Global, USA (2020)
7. U.N. Peled, N.V. Mahadev, *Threshold graphs and related topics* (North Holland, 1995)
8. T. Pramanik, M. Pal, S. Mondal, Interval-valued fuzzy threshold graph. Pacif. Sci. Rev. A: Nat. Sci. Eng. **18**, 66–71 (2016)
9. S. Samanta, M. Pal, Fuzzy threshold graphs, CiiT. Int. J. Fuzzy Syst. **3**(12), 360–364 (2011)
10. W. Tao, H. Jin, Y. Zhang, L. Liu, D. Wang, Image thresholding using graph cuts. Syst. Humans **38**(5), 1181–1195 (2008)
11. L. Yang, H. Mao, Intuitionistic fuzzy threshold graphs. J. Intell. Fuzzy Syst. **36**, 6641–6651 (2019). https://doi.org/10.3233/JIFS-18755

Chapter 6
Fuzzy Tolerance Graphs

List of abbreviation

FInv	fuzzy interval
FInvG	fuzzy interval graph
FTol	fuzzy tolerance
FTolG	fuzzy tolerance graph
TolComG	tolerance competition graph
FTolComG	fuzzy tolerance competition graph
FInvConG	fuzzy interval containment graph

Golumbic and Trenk first introduced tolerance graph [7] to generalize some well-known applications of interval graphs. In a tolerance graph, every vertex is represented by an interval and a tolerance such that an edge occurs if and only if the overlap of corresponding intervals is at least as large as the tolerance associated with one of the vertices. Resource allocation sharing tolerate among users and certain scheduling problems can be modeled by tolerance graphs. Tolerance graph is also applicable in constrained-based temporal reasoning, data transmission through networks to efficiently scheduling aircraft and crews as well as contributing to genetic analysis and studies of the brain. Due to the presence of uncertainty, intervals and tolerances are replaced by fuzzy intervals (FInvs) and fuzzy tolerances in fuzzy tolerance graphs. So, it is important to identify and study the nature of fuzzy tolerances in FInvs. In this chapter, a detailed analysis on fuzzy tolerance (FTol), fuzzy tolerance graph (FTolG), fuzzy bounded tolerance graph, FInv containment graph (FInvConG), and regular representation of FTolG, fuzzy unit tolerance graph, proper tolerance graph, and ϕ-tolerance competition graph (TolComG) are given.

The contributions of this chapter are from the articles [1, 10, 16, 19–21]. For more information interested readers may consult with [3, 6, 11, 15, 17].

6.1 Fuzzy Tolerance and Fuzzy Tolerance Graph

Let $\mathbb{I} = \{I_1, I_2, \ldots, I_n\}$ be a set of closed real intervals and $T = \{T_1, T_2, \ldots, T_n\}$ be another set of positive real numbers. For each interval, let us consider a vertex.

A graph $G = (V, E)$ along with two sets I and T is said to be a **tolerance graph** if, for each I_j, there is a vertex a_j in $\mathcal{V}$ and there is an edge in E between two vertices a_i and a_j if and only if

$$|I_{a_i} \cap I_{a_j}| \geq min\{T_{a_i}, T_{a_j}\},$$

where $|I_{a_i}|$ is the length of the interval I_{a_i}. The pair I and T is denoted by $\langle I, T \rangle$ and is called **tolerance representation** of the graph G. For a given tolerance representation, the corresponding graph is unique. But, the converse is not true, i.e., for a given tolerance graph there exists many different tolerance representations.

A tolerance representation $\langle I, T \rangle$ is said to be bounded if $T_a \leq |I_a|$ for all $a \in V$. A tolerance graph is called a **bounded tolerance graph** if it has a **bounded tolerance representation**.

By imposing the conditions on the length of the tolerances, we obtained two well-known intersection graphs. If all the tolerances T_x are equal and equal to a fixed positive constant c, then the tolerance graph becomes interval graph [12, 14]. If $T_a = |I_a|$ for all $a \in V$, then we obtain the permutation graph [12, 14] (or, equivalently, the interval containment graph). Therefore, one can conclude that interval graphs and permutation graphs are all bounded tolerance graphs.

So, there is a set of intervals for each tolerance graph. Therefore, before going to define FTolG, we introduce a fuzzy interval (FInv).

The following is the definition of FInv given in [5].

Definition 6.1 (*Fuzzy interval, Dubois and Prade* [5]) A **fuzzy intervals** $\mathscr{I}$ on a linearly ordered set X is defined as a normal, convex fuzzy subset of X. This means $\mathscr{I}(a) = 1$ for some $a \in X$ and $a \leq b \leq c$ implies $\mathscr{I}(b) \geq \mathscr{I}(a) \wedge \mathscr{I}(c)$.

Craine [4] thereafter defined FInv graph and examined some of its important properties as generalization of the crisp interval graph.

Definition 6.2 (*Fuzzy interval graph, Craine* [4]) A **FInv graph** is the fuzzy intersection of a finite family of FInvs.

The discussions in this section are from [21].

Definition 6.3 For a FInv, a **fuzzy tolerance** is an arbitrary FInv whose core length is a positive real number and it is denoted by $\mathscr{T}$. Let L be a real number and $|I_j - I_{j-1}| = L$, where $I_{j-1}, I_j \in \mathbb{R}$ then FTol is a fuzzy set of the interval $[I_{j-1}, I_j]$.

The support and core of a FTol are defined as tolerance support and tolerance core respectively and for a FTol $\mathscr{T}_i$, these are denoted by $s(\mathscr{T}_i)$ and $c(\mathscr{T}_i)$ respectively. It may be noted that a fuzzy tolerance may be a fuzzy number. The FTols are illustrated in Fig. 6.1.

The FTolG is defined as an extension of FInvG and it is defined below.

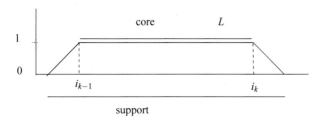

Fig. 6.1 Fuzzy tolerances

Definition 6.4 Suppose $\mathscr{I} = \{\mathscr{I}_1, \mathscr{I}_2, \ldots, \mathscr{I}_n\}$ is a finite set of FInvs defined on the real line and $\mathscr{T} = \{\mathscr{T}_1, \mathscr{T}_2, \ldots, \mathscr{T}_n\}$ be the corresponding set of **fuzzy tolerances**. For each FInv $\mathscr{I}_i$, $i = 1, 2, \ldots, n$, a crisp vertex v_i is defined. The set of such vertices is denoted by $\mathscr{V}$. A FTolG $\mathscr{G} = (\mathscr{V}, \sigma, \mu)$ is a FG , where $\sigma : \mathscr{V} \to [0, 1]$ and $\mu : \mathscr{V} \times \mathscr{V} \to [0, 1]$, the respective membership values of the vertex $v_i \in \mathscr{V}$ is given by $\sigma(v_i) = h(\mathscr{I}_i) = 1$ and the membership value of the edge (v_i, v_j) is determined by

$$\mu(v_i, v_j) = \begin{cases} 1, & \text{if } c(\mathscr{I}_i \cap \mathscr{I}_j) \geq \min\{c(\mathscr{T}_i), c(\mathscr{T}_j)\} \\ \frac{s_{ij} - \min\{s(\mathscr{T}_i), S(\mathscr{T}_j)\}}{s_{ij}} h_{ij}, & \text{otherwise if } s_{ij} \geq \min\{s(\mathscr{T}_i), s(\mathscr{T}_j)\} \\ 0, & \text{otherwise.} \end{cases}$$

$c(\mathscr{I}_i \cap \mathscr{I}_j)$ is the core of the intersection of the $\mathscr{I}_i$ and $\mathscr{I}_j$.
$\min\{c(\mathscr{T}_i), c(\mathscr{T}_j)\}$ is the minimum of the cores of the corresponding tolerances $\mathscr{T}_i$ and $\mathscr{T}_j$. Also $h_{ij} = h(\mathscr{I}_i \cap \mathscr{I}_j)$, $s_{ij} = s(\mathscr{I}_i \cap \mathscr{I}_j)$ and $s(\mathscr{I}_i \cap \mathscr{I}_j)$, $\min\{s(\mathscr{T}_i), s(\mathscr{T}_j)\}$ are the support of the intersection of intervals $\mathscr{I}_i$, $\mathscr{I}_j$ and minimum support of the tolerances $\mathscr{T}_i$, $\mathscr{T}_j$ respectively.

If the fuzzy tolerances are not incorporated with the FInvs, then the corresponding graph becomes FInvG. The FTolG has been shown in Figs. 6.2, 6.3, 6.4 for three different cases.

Let $\mathscr{I}_i$, $\mathscr{I}_j$ be two FInvs and their corresponding FTols be $\mathscr{T}_i$, $\mathscr{T}_j$. In the first case, minimum of two tolerance cores $\min\{c(\mathscr{T}_i), c(\mathscr{T}_j)\}$ is less than the intersection core $c(\mathscr{I}_i \cap \mathscr{I}_j)$ of the intervals $\mathscr{I}_i$ and $\mathscr{I}_j$. In this case, edge membership value is 1. If $\min\{c(\mathscr{T}_i), c(\mathscr{T}_j)\}$ is greater than $c(\mathscr{I}_i \cap \mathscr{I}_j)$ then second case arises where membership value of the edge is less than 1. Here, minimum of two tolerance supports $\min\{s(\mathscr{T}_i), s(\mathscr{T}_j)\}$ is less than the intersection support $s(\mathscr{I}_i \cap \mathscr{I}_j)$ of the intervals $\mathscr{I}_i$, $\mathscr{I}_j$. Otherwise, the third case arises where edge membership value is 0.

If $\langle I, T \rangle$ is a tolerance representation of a crisp graph G and a vertex is a point instead of an interval then the corresponding vertex is the isolated vertex of G. But, it may not be true in the fuzzy sense. If $\langle \mathscr{I}, \mathscr{T} \rangle$ is a FTol representation of $\mathscr{G}$ and f is a fuzzy number instead of FInv, then the corresponding vertex may not be isolated. The intersection of supports of a FInv and a fuzzy number may be greater than the

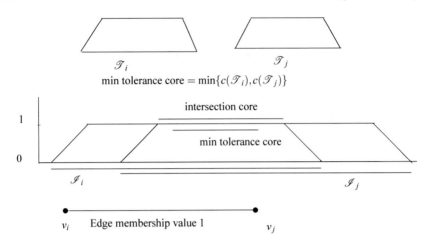

Fig. 6.2 Fuzzy tolerance graph with edge membership value 1

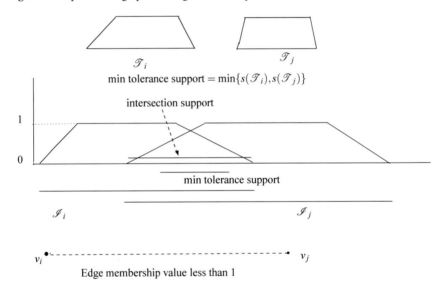

Fig. 6.3 FToIG with edge membership value less than 1

minimum support of their corresponding tolerances. So there is an edge of some positive strength.

Example 6.1 For illustration, let us consider three FInvs $\mathscr{I}_1$, $\mathscr{I}_2$, $\mathscr{I}_3$ and their corresponding FTols $\mathscr{T}_1$, $\mathscr{T}_2$, $\mathscr{T}_3$. Let the core and support of the interval $\mathscr{I}_1$ be $[2, 7]$ and $[1, 8]$; that of $\mathscr{I}_2$ be $[4, 8]$ and $[2.5, 9.5]$; that of $\mathscr{I}_3$ be $[9, 13]$ and $[6.5, 15.5]$. Let the length of core and support of $\mathscr{T}_1$, $\mathscr{T}_2$, $\mathscr{T}_3$ be $\{1, 1.2\}$, $\{0.8, 1\}$, $\{0.5, 2\}$ respectively.

It is easy to calculate that, $\mathscr{I}_1 \cap \mathscr{I}_2 = [4, 7]$, $\mathscr{I}_2 \cap \mathscr{I}_3 = \emptyset$, $\mathscr{I}_1 \cap \mathscr{I}_3 = \emptyset$.

Since $\mathscr{I}_1 \cap \mathscr{I}_2 \neq \emptyset$, so $h(\mathscr{I}_1 \cap \mathscr{I}_2) = 1$.

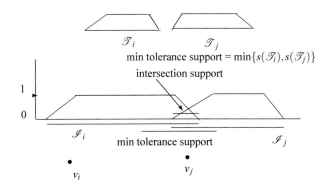

Fig. 6.4 Fuzzy tolerance graph with edge membership value 0

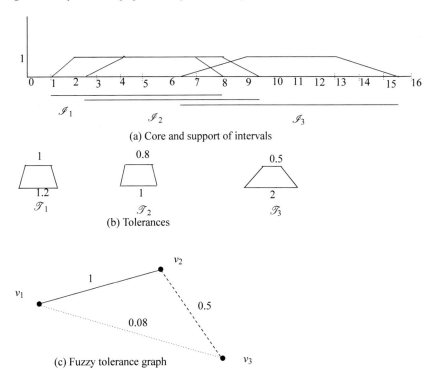

(a) Core and support of intervals

(b) Tolerances

(c) Fuzzy tolerance graph

Fig. 6.5 Example of FTolG

If the intersection of two fuzzy intervals, say $\mathscr{I}_i$ and $\mathscr{I}_j$ is empty, then draw the line segments from left support coordinates to left core coordinates (suppose they are L_i and L_j) and same for right coordinates (R_i and R_j) for each of two intervals. Now, find the coordinates of the line segments L_i and R_j or R_i and L_j which one is feasible. If there is no point of intersection between these line segments, then the $h(\mathscr{I}_i \cap \mathscr{I}_j) = 0$. Otherwise, the second coordinate of the point of intersection represents the height of $(\mathscr{I}_i \cap \mathscr{I}_j)$.

The end coordinates of the line segment L_3 are $(6.5, 0)$ and $(9, 1)$; and that of R_3 are $(13, 1)$ and $(15.5, 0)$. Similarly, the coordinates for R_2 are $(8, 1)$ and $(9.5, 0)$; for L_2 are $(2.5, 0)$ and $(4, 1)$ (see Fig. 6.5a). By routine calculation, we obtained the point of intersection between the line segments R_2 and L_3 is $(\frac{67}{8}, \frac{3}{4})$. Hence, $h(\mathscr{I}_2 \cap \mathscr{I}_3) = \frac{3}{4}$. Similarly, $h(\mathscr{I}_1 \cap \mathscr{I}_3) = 0.43$.

To calculate the membership value of the edge $\mu(v_1, v_2)$, we calculate intersection of the core of fuzzy intervals, which is $[2, 7] \cap [4, 8] = [4, 7]$. Here, the length of the interval $7 - 4 = 3$ and it is greater than the minimum of core of corresponding fuzzy tolerances. Thus, $\mu(v_1, v_2) = 1$.

The core of the fuzzy intervals $\mathscr{I}_2$ and $\mathscr{I}_3$ are disjoint and, in this case, we calculate the intersection of the support and it is $[2.5, 9.5] \cap [6.5, 15.5] = [6.5, 9.5]$, and length is $9.5 - 6.5 = 3$. Thus, $\mu(v_2, v_3) = \frac{3 - min\{1,2\}}{3} \times h(\mathscr{I}_2 \cap \mathscr{I}_3) = 0.5$.

Similarly, $\mu(v_1, v_3) = \frac{1.5 - min\{1.2,2\}}{1.5} \times h(\mathscr{I}_1 \cap \mathscr{I}_3) = 0.08$.

Thus, the membership values of the edges (v_1, v_2), (v_2, v_3) and (v_1, v_3) are $1, 0.5$, and 0.08 respectively and the graph is shown in Fig. 6.5c.

6.2   Fuzzy Bounded Tolerance Graph

Let G be a tolerance graph and its tolerance representation be $\langle I, T \rangle$. If $T_j \leq |I_j|$ for every $j = 1, 2, \ldots, n$, then G is called a [10] **bounded tolerance graph**.

Here, we extended the concept of (crisp) bounded tolerance graph into fuzzy bounded tolerance graph as follows.

Definition 6.5 Let $\mathscr{I} = \{\mathscr{I}_1, \mathscr{I}_2, \ldots, \mathscr{I}_n\}$ be a finite set of FInvs defined on a real line and let the corresponding fuzzy tolerances be $\mathscr{T} = \{\mathscr{T}_1, \mathscr{T}_2, \ldots, \mathscr{T}_n\}$. If FInv $\mathscr{I}_i$ with core $c(\mathscr{I}_i)$ and support $s(\mathscr{I}_i)$ has tolerance with core $c(\mathscr{T}_i)$ and support $s(\mathscr{T}_i)$ such that $c(\mathscr{I}_i) \geq c(\mathscr{T}_i)$ and $s(\mathscr{I}_i) \geq s(\mathscr{T}_i)$ of a FTolG, then the corresponding representation is said to be **fuzzy bounded tolerance representation** and the FG is called **fuzzy bounded tolerance graph**.

It may be noted that if $\langle \mathscr{I}, \mathscr{T} \rangle$ is a fuzzy bounded tolerance representation then no FInv $\mathscr{I}_v$, $v \in \mathscr{V}$ is a fuzzy number.

Theorem 6.1 *If $\mathscr{G}$ is a FInvG, then $\mathscr{G}$ is a FTolG with constant core and constant support of tolerance.*

Proof Let $\mathscr{G}$ be a FInvG and the FInv for the vertex a is $\mathscr{I}_a$. Let the cores of FInvs $\mathscr{I}_a$ and $\mathscr{I}_b$ be denoted by $c(\mathscr{I}_a)$ and $c(\mathscr{I}_b)$ and that of supports be $s(\mathscr{I}_a)$ and $s(\mathscr{I}_b)$. Also, $c(\mathscr{I}_a \cap \mathscr{I}_b) = c(\mathscr{I}_a) \wedge c(\mathscr{I}_b)$ and $s(\mathscr{I}_a \cap \mathscr{I}_b) = s(\mathscr{I}_a) \wedge s(\mathscr{I}_b)$. Let κ_1 and κ_2 be positive real numbers such that $\kappa_1 < |c(\mathscr{I}_a \cap \mathscr{I}_b)|$ and $\kappa_2 < |s(\mathscr{I}_a \cap \mathscr{I}_b)|$ for all $a, b \in \mathscr{V}$ with $\kappa_1 \leq \kappa_2$.

Thus, the intervals $\{\mathscr{I}_v \mid v \in \mathscr{V}\}$ together with tolerances with core k_1 and support k_2 give a FTol representation. $\qquad\square$

Theorem 6.2 *If $\mathscr{G}$ is a FTolG and both core and support of tolerances are constant, then $\mathscr{G}$ is fuzzy bounded tolerance graph.*

Proof Let $\mathscr{G}$ be a FTolG with $\mathscr{I}$ and $\mathscr{T}$ be the FInvs and FTols. Let k_1 and k_2 be two positive real numbers and $c(\mathscr{T}_i) = k_1$ and $s(\mathscr{T}_i) = k_2$ for all i.

If $c(\mathscr{I}_i) \geq k_1$ and $s(\mathscr{I}_i) \geq k_2$ for all i, then $\mathscr{G}$ is a fuzzy bounded tolerance graph. If $c(\mathscr{I}_i) < k_1$ for any i and $s(\mathscr{I}_j) < k_2$ for any j, then we take $c(\mathscr{I}_i) = k_1$ and $s(\mathscr{I}_j) = k_2$, to make $\mathscr{G}$ be bounded. In this case, adjacency relation remains unchanged. $\qquad\square$

Theorem 6.3 *Let $\mathscr{G}$ be a FTolG. Then the cut level graphs $\mathscr{G}^t$, for all $t \in [0, 1]$ are tolerance graphs.*

Proof Let $\mathscr{G}$ be a FTolG with fuzzy tolerance representation be $\langle \mathscr{I}, \mathscr{T} \rangle$. For each $t \in [0, 1]$, $\mathscr{I}_i^t \in \mathscr{I}^t$ for all $i = 1, 2, ..., n$. So it is a crisp interval. Also, for $t \in [0, 1]$, $\mathscr{T}_i^t \in \mathscr{T}^t$ is crisp length, i.e., a crisp tolerance.

So crisp intervals together with crisp tolerances produced a crisp tolerance graph. Hence, $\mathscr{G}^t$ for $t \in [0, 1]$ is a tolerance graph. $\qquad\square$

Now, we define some related graphs. A graph is said to be **weakly chordal graph** [7] if it does not contain any induced subgraph isomorphic to C_n for any n greater than 4. A graph G is said to be **perfect** if the chromatic number of H is equal to the number of vertices in the largest clique of all induced subgraphs H of G. A graph G is called **alternately orientable** [7] if this is an orientation of G which is transitive on every chordless cycle of length greater than 3, i.e., the directions of the oriented edges must alternate.

Now we prove the following result.

Theorem 6.4 *If $\mathscr{G}$ is a FTolG then each cut level graph is*
(i) weakly chordal,
(ii) alternately orientable,
(iii) perfect graph.

Proof It is known that each cut level graph of FTolG is a tolerance graph (Theorem 6.3). Also, the tolerance graphs are weakly chordal, alternately orientable, perfect graph [7]. So each cut level graph of a FTolGs is weakly chordal, alternately orientable, and perfect graph. $\qquad\square$

6.3 Fuzzy Interval Containment Graph

A graph $G = (V, E)$ is called an **interval containment graph** [7] if there exists a set of real intervals $I = \{I_a : a \in V\}$ such that for $(a, b) \in E$ one of the intervals I_a, I_b contains the other. This representation is known as **interval containment representation**.

Let us define FInvConG.

Definition 6.6 Let $\mathscr{I} = \{\mathscr{I}_1, \mathscr{I}_2, \ldots \mathscr{I}_n\}$ be a finite set of FInvs on real line and there be a vertex v_i for the FInv $\mathscr{I}_i$ for all i. The FInvConG is the FG $\mathscr{G} = (\mathscr{V}, \sigma, \mu)$, where $\sigma : \mathscr{V} \to [0, 1]$ and $\mu : \mathscr{V} \times \mathscr{V} \to [0, 1]$. The vertex membership value σ and edge membership value μ are defined as $\sigma(v_i) = h(\mathscr{I}_i) = 1$ for all $i = 1, 2, \ldots, n$ and $\mu : \mathscr{V} \times \mathscr{V} \to [0, 1]$ and

$$\mu(v_i, v_j) = \begin{cases} 1, & \text{if core and support of one of } \mathscr{I}_i, \mathscr{I}_j \text{ contain the other} \\ \frac{1}{2}\left(\frac{c(\mathscr{I}_i \cap \mathscr{I}_j)}{\min\{c(\mathscr{I}_i), c(\mathscr{I}_j)\}} + \frac{s(\mathscr{I}_i \cap \mathscr{I}_j)}{\min\{s(\mathscr{I}_i), s(\mathscr{I}_j)\}}\right) h(\mathscr{I}_i \cap \mathscr{I}_j), & \text{otherwise,} \end{cases}$$

where $c(\mathscr{I}_i \cap \mathscr{I}_j)$, $s(\mathscr{I}_i \cap \mathscr{I}_j)$ are respectively the core and support of the intersection of the intervals $\mathscr{I}_i$, $\mathscr{I}_j$ and $\min\{c(\mathscr{I}_i), c(\mathscr{I}_j)\}$, $\min\{s(\mathscr{I}_i), s(\mathscr{I}_j)\}$ are the minimum of core and support of the intervals $\mathscr{I}_i$ and $\mathscr{I}_j$.

Example 6.2 Let us consider an example of four FInvs $\mathscr{I}_1, \mathscr{I}_2, \mathscr{I}_3, \mathscr{I}_4$. Let the core and the support of $\mathscr{I}_1$ be $[2, 7]$ and $[1, 8]$; that of $\mathscr{I}_2$ be $[4, 6]$ and $[3, 7]$; that of $\mathscr{I}_3$ be $[9, 13]$ and $[7.5, 14.5]$; that of $\mathscr{I}_4$ be $[12, 16]$ and $[11, 17]$. So $h(I_1 \cap I_2) = 1$, $h(I_3 \cap I_4) = 1$, and $h(I_1 \cap I_3) = 0.2$. Clearly, the edge membership values of the edges (v_1, v_2), (v_1, v_3), and (v_3, v_4) are respectively 1, 0.007, and 0.42. The diagrammatic representation of this example is depicted in Fig. 6.6.

Theorem 6.5 *If $\mathscr{G}$ is a FInvConG with edge membership value 1 or 0, then $\mathscr{G}$ has a FTol representation with core (support) of an interval equals to the core (support) of corresponding tolerance.*

Proof Let $V = \{v_1, v_2, \ldots, v_n\}$ be a set of crisp vertices corresponding to the set of FInvs $\mathscr{I} = \{\mathscr{I}_1, \mathscr{I}_2, \ldots, \mathscr{I}_n\}$ on real line. Let $\mathscr{G} = (\mathscr{V}, \sigma, \mu)$ be a FInvConG of the FInvs.

Let $\mu(v_i, v_j) = 1$ in $\mathscr{G}$. Then core and support of one interval contain the other. So intersection of the two cores equals to the minimum of two cores. Let $\mathscr{G}' = (\mathscr{V}, \sigma, \mu')$ be a FTolG of $\mathscr{I}$ together with fuzzy tolerances $\mathscr{T} = \{\mathscr{T}_1, \mathscr{T}_2, \ldots, \mathscr{T}_n\}$ such that the cores of the fuzzy tolerances equal to the cores of the intervals and supports of the fuzzy tolerances equal to the supports of the intervals. So $c(\mathscr{I}_i \cap \mathscr{I}_j) = \min\{\mathscr{I}_i, \mathscr{I}_j\} = \min\{\mathscr{T}_i, \mathscr{T}_j\}$. Then $\mu'(v_i, v_j) = 1$ in $\mathscr{G}'$.

Let $\mu(v_k, v_l) = 0$ in $\mathscr{G}$. Then there is no common part of cores and supports. Now in $\mathscr{G}'$, $c(\mathscr{I}_k \cap \mathscr{I}_l) = 0$ and $s(\mathscr{I}_k \cap \mathscr{I}_l) = 0$. So $\mu'(v_k, v_l) = 0$ in $\mathscr{G}'$.

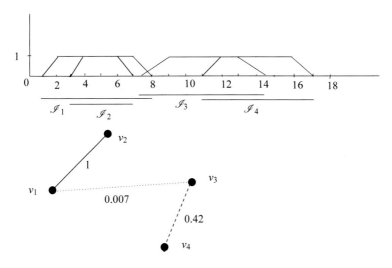

Fig. 6.6 FInvConG

Hence, if $\mathscr{G}$ is a FInvConG with edge membership value 1 or 0, then $\mathscr{G}$ has a FTol representation with cores of FInvs equal to the cores of corresponding tolerances and supports of the intervals equal to the supports of the tolerances.   □

If $\mathscr{G}$ is a FTolG with a FTol representation with core of an interval equals to the core of the corresponding tolerance and support of the interval equals to the support of the corresponding tolerance, then it is clear that $\mathscr{G}$ has a FInvConG representation using same FInvs.

6.4 Regular Representation of Tolerance Competition Graph

A FTolG is said to have **regular representation** if the following three properties are satisfied:

(1) Any FTol core and support length larger than the core and support lengths of the corresponding FInv can be set to infinity without changing the adjacency of vertices in the FTolG.

(2) The lengths of all core and support of FTols are distinct.

(3) The cores and supports of different FInvs never share the end points.

Theorem 6.6 *Every FTolG has a regular representation.*

Proof Let $\mathscr{I}$ and $\mathscr{T}$ be the finite family of FInvs on real line and FTols. Let $\mathscr{V} = \{v_1, v_2, \ldots, v_n\}$ be the crisp vertex set for the set of intervals $\mathscr{I}$. Let $\mathscr{G} = (\mathscr{V}, \sigma, \mu)$

be a FTolG and $\langle \mathscr{I}, \mathscr{T} \rangle$ be the tolerance representation of it. Let $L(c(\mathscr{I}_v))$ and $R(c(\mathscr{I}_v))$ be the left and right end points of the core $c(\mathscr{I}_v)$ and that of support $s(\mathscr{I}_v)$ be $L(s(\mathscr{I}_v))$ and $R(s(\mathscr{I}_v))$ respectively for each $v \in \mathscr{V}$.

For each vertex v_i, let $c(\mathscr{T}_i) > c(\mathscr{I}_i)$ and $s(\mathscr{T}_i) > s(\mathscr{I}_i)$. There exists an edge (v_i, v_j) in $\mathscr{G}$ if $c(\mathscr{I}_i \cap \mathscr{I}_j) \geq \min\{c(\mathscr{T}_i), c(\mathscr{T}_j)\} = c(\mathscr{T}_j)$, otherwise $s(\mathscr{I}_i \cap \mathscr{I}_j) > \min\{s(\mathscr{T}_i), s(\mathscr{T}_j)\} = s(\mathscr{T}_j)$. So tolerance core and support for v_i can be set to infinity without changing the adjacency of vertices in the FTolG. Thus, any FTol core and support length larger than the core and support length of the corresponding FInv can be set to infinity without changing the adjacency of vertices in the FTolG.

Let ε be the smallest positive number appearing in the union of the sets (i) $\{|L(c(\mathscr{I}_x)) - L(c(\mathscr{I}_y))|\}$, (ii) $\{|R(c(\mathscr{I}_x)) - R(c(\mathscr{I}_y))|\}$, (iii) $\{|L(c(\mathscr{I}_x)) - R(c(\mathscr{I}_y))|\}$, (iv) $\{c(\mathscr{T}_x)\}$, (v) $\{|c(\mathscr{T}_x) - c(\mathscr{T}_y)|\}$, (vi) $\{c(\mathscr{T}_x) - |c(\mathscr{I}_x) \cap c(\mathscr{I}_y)|\}$.

If x and y be two distinct vertices with $c(\mathscr{T}_x) = c(\mathscr{T}_y)$, choosing one of them, say x, replacing $c(\mathscr{T}_x)$ by $c(\mathscr{T}'_x) = c(\mathscr{T}_x) - \frac{\varepsilon}{2}$, we leave $c(\mathscr{T}_y)$ unchanged. We show that this gives a representation of $\mathscr{G}$ with one fewer repeated tolerance core. If $(x, z) \in E$, then $|c(\mathscr{I}_x) \cap c(\mathscr{I}_z)| \geq \min\{c(\mathscr{T}_x), c(\mathscr{T}_z)\} \geq \min\{c(\mathscr{T}'_x), c(\mathscr{T}_z)\}$ is one possible case. If $(x, z) \notin E$, then $|c(\mathscr{I}_x) \cap c(\mathscr{I}_z)| < \min\{c(\mathscr{T}_x), c(\mathscr{T}_z)\}$ is a possible case and by our choice of ε, $\min\{c(\mathscr{T}_x), c(\mathscr{T}_z)\} - |c(\mathscr{I}_x) \cap c(\mathscr{I}_z)| \geq \varepsilon$. Thus, $|c(\mathscr{I}_x) \cap c(\mathscr{I}_z)| \leq \min\{c(\mathscr{T}_x), c(\mathscr{T}_z)\} - \varepsilon < \min\{c(\mathscr{T}'_x), c(\mathscr{T}_z)\}$ as desired. If necessary, we recompute ε and repeat the process until all tolerance cores are distinct. The similar process can be done for the case of tolerance support.

Now, suppose cores of two different FInvs share an end point. Let $S = \{L(c(\mathscr{I}_v)), R(c(\mathscr{I}_v)) \; \forall v \in \mathscr{V}\}$ be the set of core end points. Let x and y be the distinct elements of $\mathscr{V}$ for which p is an end point of $c(\mathscr{I}_x)$ and $c(\mathscr{I}_y)$. If there exist $x \in \mathscr{V}$ with $R(c(\mathscr{I}_x)) = p$, taking one whose length $c(\mathscr{I}_x)$ is the longest and replacing $c(\mathscr{I}_x)$ by $c(\mathscr{I}'_x) = [L(c_x) + \frac{\varepsilon}{2}, R(c_x) + \frac{\varepsilon}{2}]$ otherwise, we take $x \in \mathscr{V}$ with $L(c(\mathscr{I}_x)) = p$ and $|c(\mathscr{I}_x)|$ as large as possible and replace $c(\mathscr{I}_x)$ by $c(\mathscr{I}'_x) = [L(c(\mathscr{I}_x)) + \frac{\varepsilon}{2}, R(c(\mathscr{I}_x)) + \frac{\varepsilon}{2}]$. So, it is easy to see that this gives a representation of $\mathscr{G}$ with one fewer pair of elements sharing an end point. All tolerances are still distinct. If necessary, we recompute ε and repeat the process until all end points of the core are distinct. Similar argument can be applied to the support also. $\square$

6.5 Fuzzy Unit Tolerance Graph and Fuzzy Proper Tolerance Graph

If in a tolerance graph the lengths of all intervals are same (unit), then the tolerance graph is called **unit tolerance graph** [1]. Similarly, a tolerance graph is called a **proper tolerance graph** [1] if no interval is properly contains in another.

Now, we define fuzzy unit interval graph and fuzzy proper interval graph then that in tolerance representation.

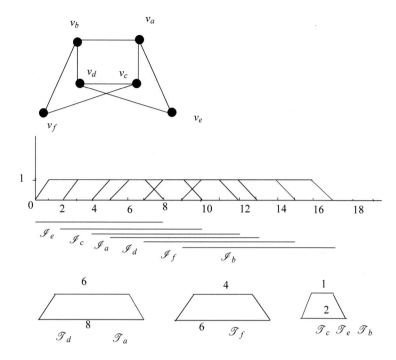

Fig. 6.7 Fuzzy unit tolerance graph

Definition 6.7 A FInvG $\mathscr{G}$ with a fuzzy interval representation is called **fuzzy unit interval graph** if the lengths of cores and supports are same for all FInvs. Similarly, $\mathscr{G}$ is called **fuzzy proper interval graph** if cores and supports of FInvs do not properly contain the cores and supports of other FInvs.

Definition 6.8 A FTolG with a tolerance representation is called a **fuzzy unit tolerance graph** if lengths of cores and length of supports of all FInvs are same.

A FTolG with a tolerance representation is called a **fuzzy proper tolerance graph** if the core and support of a FInv do not contain the core and support of another FInv.

Clearly, the class of fuzzy unit tolerance graph is a subset of the class of fuzzy proper tolerance graph. In the following, we consider a FG and its fuzzy unit tolerance representation.

Example 6.3 Let us consider six FInvs with corresponding tolerances. The cores and supports of FInvs are given respectively as $\mathscr{I}_a$ [5, 11] [4, 12]; $\mathscr{I}_b$ [10, 16] [9, 17]; $\mathscr{I}_c$ [3, 9] [2, 10]; $\mathscr{I}_d$ [6, 12] [5, 13]; $\mathscr{I}_e$ [1, 7] [0, 8]; $\mathscr{I}_f$ [8, 14] [7, 15]. The lengths of cores and supports of the corresponding tolerances are given as $\mathscr{T}_a$ {6, 8}, $\mathscr{T}_b$ {1, 2}, $\mathscr{T}_c$ {1, 2}, $\mathscr{T}_d$ {6, 8}, $\mathscr{T}_e$ {1, 2}, $\mathscr{T}_f$ {4, 6}. Here, core lengths of the FInvs are same (= 6) and support lengths are same (= 8) for all intervals. So the FInvs together with fuzzy tolerances represent fuzzy unit tolerance graph. The corresponding graph is shown in Fig. 6.7.

Theorem 6.7 *Any fuzzy proper or unit tolerance representation may have fuzzy bounded tolerances.*

Proof We fix a fuzzy unit or proper tolerance representation $\langle \mathscr{I}, \mathscr{T} \rangle$ of a fuzzy proper or unit tolerance graph. Assumed that all end points of all cores and supports in this representation are distinct. Now, we replaced $c(\mathscr{T}_a)$ by $|c(\mathscr{I}_a)|$ for each $a \in \mathscr{V}$ when $c(\mathscr{T}_a) \geq |c(\mathscr{I}_a)|$ and $s(\mathscr{T}_a)$ replace by $|s(\mathscr{I}_a)|$ for each $a \in \mathscr{V}$ when $s(\mathscr{T}_a) \geq |s(\mathscr{I}_a)|$. Since there are no containments of core and support of FInvs, so the adjacency does not change.   □

The fuzzy out-neighborhood and in-neighborhood are defined as follows:

Definition 6.9 Fuzzy out-neighborhood of a vertex v of a directed FG $\overrightarrow{\mathscr{G}} = (\mathscr{V}, \sigma, \overrightarrow{\mu})$ is the fuzzy set $\mathscr{N}^+(v) = (X_v^+, m_v^+)$, where $X_v^+ = \{u | \overrightarrow{\mu}(v, u) > 0\}$ and $m_v^+ : X_v^+ \to [0, 1]$ defined by $m_v^+(u) = \overrightarrow{\mu}(v, u)$.

Similarly, the fuzzy in-neighborhood is defined.

Let $\overrightarrow{\mathscr{G}} = (\mathscr{V}, \sigma, \overrightarrow{\mu})$ be a fuzzy directed graph. The in-neighborhood of a vertex a is a fuzzy set $\mathscr{N}^-(a) = (X_a^-, m_a^-)$, where $X_a^- = \{b | \overrightarrow{\mu}(b, a) > 0\}$ and $m_a^- : X_a^- \to [0, 1]$ and m_a^- is defined as $m_a^-(b) = \overrightarrow{\mu}(b, a)$.

6.6 Fuzzy ϕ-Tolerance Competition Graph

Fuzzy ϕ-**tolerance competition graph** is a combination of FTolG and competition graph. First of all, fuzzy ϕ-TolComG is defined below. The following results are from [17].

Definition 6.10 (*Fuzzy ϕ-TolComG*) Let ϕ be a given function defined by $\phi : \mathbb{N} \times \mathbb{N} \to \mathbb{N}$. The fuzzy ϕ-TolComG $TC_\phi(\overrightarrow{\mathscr{D}})$ defined for a fuzzy digraph $\overrightarrow{\mathscr{D}} = (\mathscr{V}, \sigma, \overrightarrow{\mu})$ is an undirected FG $TC_\phi(\overrightarrow{\mathscr{D}}) = (\mathscr{V}, \sigma, \mu')$ which has the same fuzzy vertex set as in $\overrightarrow{\mathscr{D}}$ and

$$
\mu'(u, v) = \begin{cases} h\left(\mathscr{N}^+(u) \cap \mathscr{N}^+(v)\right), & \text{if } c\left(\mathscr{N}^+(u) \cap \mathscr{N}^+(v)\right) \geq \\ & \qquad \phi\{c(\mathscr{T}_u), c(\mathscr{T}_v)\} \\ \frac{s(\mathscr{N}^+(u) \cap \mathscr{N}^+(v)) - \phi\{s(\mathscr{T}_u), s(\mathscr{T}_v)\} + 1}{s(\mathscr{N}^+(u) \cap \mathscr{N}^+(v))} & \text{if } s\left(\mathscr{N}^+(u) \cap \mathscr{N}^+(v)\right) \geq \\ \qquad \times h\left(\mathscr{N}^+(u) \cap \mathscr{N}^+(v)\right), & \qquad \phi\{s(\mathscr{T}_u), s(\mathscr{T}_v)\} \\ 0, & \text{otherwise.} \end{cases}
$$

The notion of edge clique covering [18] is used to characterize the competition graph. A collection of cliques of a graph G is called an **edge clique cover** (ECC) if every edge of G is in at least one of these cliques. The smallest number of cliques that cover all edges of a graph G is called the **ECC number** of G and is denoted by $\theta_e(G)$.

These concepts have been generalized in [8, 9] and lead to another type of ECC called p-**edge clique cover**. A p-**edge clique cover** (p-ECC) of a graph G is the family of sets $\{S_1, S_2, \ldots, S_k\}$ such that $S_{i_1} \cap S_{i_2} \cap \cdots \cap S_{i_p}$ (the subscripts must be distinct) is either empty or induces a clique of G and these p sets form an ECC of G. The p-**edge clique cover number** is the smallest p for which there is a p-ECC, and is denoted by $\theta_e^p(G)$.

Again, Brigham et al. [2] extended this definition. Let ϕ be a symmetric function such that $\phi : \mathbb{N} \times \mathbb{N} \to \mathbb{N}$, and $T = (t_1, t_2, \ldots, t_n)$ be an n-tuple of non-negative integers (not necessarily distinct). For a graph $G = (V, E)$, $V = \{a_1, a_2, \ldots, a_n\}$ a ϕ-T-edge clique cover (ϕ-T-ECC) is a collection $S_1, S_2, \ldots, S_k$ of subsets of V such that $(a_i, a_j) \in E$ if and only if at least $\phi(t_i, t_j)$ of the sets S_i, $(i = 1, 2, \ldots, k)$ contain both a_i and a_j. The size of the smallest ϕ-T-ECC of G taken over all vectors T is called the ϕ-T-**edge clique cover number** of G and it is denoted $\theta_\phi(G)$.

The above concepts can be extended for FGs also. Let $\phi : \mathbb{N} \times \mathbb{N} \to \mathbb{N}$ be a symmetric function, and $\mathscr{T} = \{\mathscr{T}_1, \mathscr{T}_2, \ldots, \mathscr{T}_n\}$ be the FTols. A ϕ-$\mathscr{T}$-edge clique cover (ϕ-$\mathscr{T}$-ECC) of a graph FG $\mathscr{G} = (\mathscr{V}, \sigma, \mu)$, $\mathscr{V} = \{a_1, a_2, \ldots, a_n\}$, is a collection $S_1, S_2, \ldots, S_k$ of subsets of $\mathscr{V}$ such that $\mu(a_i, a_j) > 0$ if and only if at least $\phi(c(\mathscr{T}_i), c(\mathscr{T}_j))$ (here the core lengths of fuzzy tolerances $\mathscr{T}_i$ and $\mathscr{T}_j$ are considered as integers) of the sets S_i, $(i = 1, 2, \ldots, k)$ contain both a_i and a_j. The size of the smallest ϕ-$\mathscr{T}$-ECC of $\mathscr{G}$ taken over all tolerances $\mathscr{T}$ is called ϕ-$\mathscr{T}$-**edge clique cover number** and this number is denoted $\theta_\phi(\mathscr{G})$.

Now, we investigate the above graphs as they have many interesting properties.

Theorem 6.8 *Let* $\phi : \mathbb{N} \times \mathbb{N} \to \mathbb{N}$ *be a symmetric function. Let* $\overrightarrow{\mathscr{G}} = (\mathscr{V}, \sigma, \overrightarrow{\mu})$ *be a fuzzy digraph with n vertices $\alpha_1, \alpha_2, \ldots, \alpha_n$ of $\mathscr{V}$. Then $TC_\phi(\overrightarrow{\mathscr{G}})$ is a fuzzy ϕ-TolComG if and only if $\theta_\phi(\overrightarrow{\mathscr{G}}) \leq |\mathscr{V}|$, i.e., $\theta_\phi(\overrightarrow{\mathscr{G}}) \leq n$.*

Proof Let $TC_\phi(\overrightarrow{\mathscr{G}}) = (\mathscr{V}, \sigma, \overrightarrow{\mu})$ be a fuzzy ϕ-TolComG of the fuzzy digraph $\overrightarrow{\mathscr{G}} = (\mathscr{V}, \sigma, \overrightarrow{\mu})$. Let $\mathscr{V} = \{\alpha_1, \alpha_2, \ldots, \alpha_n\}$ and $S_i = (X_i, m_i)$ where $S_i = \{\alpha_j : \mu(\alpha_j, \alpha_i) > 0\}$ and $m_i : X_i \to [0, 1]$ such that $m_i(\alpha_j) = \sigma(\alpha_j) = h(\alpha_j), \alpha_j \in X_i$. It is obvious that since the number of vertices of $\mathscr{G}$ is n, there are at most n number of S_i's. Let $\mathscr{T}_1, \mathscr{T}_2, \ldots, \mathscr{T}_n$ be the FTols corresponding to the vertices of $\mathscr{G}$.

Now, $\mu(\alpha_r, \alpha_s) > 0$ if and only if either

$$c\left(\mathscr{N}^+(\alpha_r) \cap \mathscr{N}^+(\alpha_s)\right) \geq \phi\left\{c(\mathscr{T}_r), c(\mathscr{T}_s)\right\}$$

$$\text{or } s\left(\mathscr{N}^+(\alpha_r) \cap \mathscr{N}^+(\alpha_s)\right) > \phi\left\{s(\mathscr{T}_r), s(\mathscr{T}_s)\right\}.$$

So, at most n sets $S_1, S_2, \ldots, S_n$ constitute a family of ϕ-$\mathscr{T}$-ECC of maximum size n, i.e., $\theta_\phi(\overrightarrow{\mathscr{G}}) \leq |\mathscr{V}|$.

Conversely, let $\{S_1, S_2, \ldots, S_k\}$ be a ϕ-$\mathscr{T}$-ECC of a FG $\mathscr{G}$. Here, the size of ϕ-$\mathscr{T}$-ECC can be at most n, i.e., $k \leq n$, where each S_i is defined by $S_i = \{\alpha_j : \mu(\alpha_i, \alpha_j) > 0\}$. Define a fuzzy digraph $\overrightarrow{\mathscr{G}} = (\mathscr{V}, \sigma, \overrightarrow{\mu})$ by $\overrightarrow{\mu}(\alpha_i, \alpha_j) = \mu(\alpha_i, \alpha_j)$, if $\alpha_j \in S_i$.

Then in $\mathscr{G}$ either

$$c\left(\mathscr{N}^+(\alpha_i) \cap \mathscr{N}^+(\alpha_j)\right) \geq \phi\left\{c(\mathscr{T}_i), c(\mathscr{T}_j)\right\}$$
$$\text{or } s\left(\mathscr{N}^+(\alpha_i) \cap \mathscr{N}^+(\alpha_j)\right) \geq \phi\left\{s(\mathscr{T}_i), s(\mathscr{T}_j)\right\}.$$

Thus, $\mathscr{G}$ is a fuzzy ϕ-TolComG.   □

For different types of the function ϕ, different types of FTolComGs can be constructed.

Some particular cases of ϕ are discussed in the subsequent sections.

6.6.1 Fuzzy Min-Tolerance Competition Graph

Let the function $\phi : \mathbb{N} \times \mathbb{N} \to \mathbb{N}$ be defined by $\phi(m, n) = \min\{m, n\}$. Then the fuzzy ϕ-TolComG is called **fuzzy min-TolComG**.

Example 6.4 For illustration, consider the fuzzy digraph depicted in the Fig. 6.8a and the fuzzy min-TolComG given in Fig. 6.8b of the digraph shown in Fig. 6.8a. Let the core $c(\mathscr{T}_{v_i})$ and the support $s(\mathscr{T}_{v_i})$ of FTols $\mathscr{T}_{v_i}$, $i = 1, 2, 3, \ldots, 9$ associated to the vertices v_i, $i = 1, 2, 3, \ldots, 9$ be 0, 2, 1, 2, 1, 1, 2, 3, 1 and 2, 3, 4, 3, 3, 4, 2, 5, 1 respectively. Now,

$$\mathscr{N}^+(v_9) = \{v_1(1), v_4(1), v_5(0.76), v_6(0.33), v_7(1), v_8(0.45)\}$$
$$\mathscr{N}^+(v_2) = \{v_1(0.08), v_9(0.52)\}$$
$$\mathscr{N}^+(v_3) = \{v_4(1), v_9(0.67)\}$$
$$\mathscr{N}^+(v_6) = \{v_5(1)\}$$
$$\mathscr{N}^+(v_7) = \{v_8(0.13)\}.$$

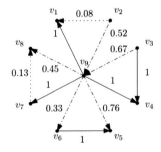

Fig. (a): Fuzzy digraph

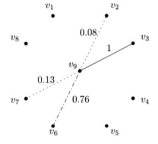

Fig. (b): Fuzzy min-tolerance competition graph

Fig. 6.8 a Fuzzy digraph; **b** Fuzzy min-TolComG

By routine calculation, we obtain the following results:

$$\mathscr{N}^+(v_2) \cap \mathscr{N}^+(v_9) = \{v_1(0.08)\}; \ c\left(\mathscr{N}^+(v_2) \cap \mathscr{N}^+(v_9)\right) = 0; \ s\left(\mathscr{N}^+(v_2) \cap \mathscr{N}^+(v_9)\right) = 1,$$
$$\mathscr{N}^+(v_3) \cap \mathscr{N}^+(v_9) = \{v_4(1)\}; \ c\left(\mathscr{N}^+(v_3) \cap \mathscr{N}^+(v_9)\right) = 1; \ s\left(\mathscr{N}^+(v_3) \cap \mathscr{N}^+(v_9)\right) = 1,$$
$$\mathscr{N}^+(v_6) \cap \mathscr{N}^+(v_9) = \{v_5(0.76)\}; \ c\left(\mathscr{N}^+(v_5) \cap \mathscr{N}^+(v_6)\right) = 0; \ s\left(\mathscr{N}^+(v_5) \cap \mathscr{N}^+(v_6)\right) = 1,$$
$$\mathscr{N}^+(v_7) \cap \mathscr{N}^+(v_9) = \{v_8(0.13)\}; \ c\left(\mathscr{N}^+(v_5) \cap \mathscr{N}^+(v_6)\right) = 0; \ s\left(\mathscr{N}^+(v_5) \cap \mathscr{N}^+(v_6)\right) = 1,$$
$$\mathscr{N}^+(v_2) \cap \mathscr{N}^+(v_3) = \{v_9(0.52)\}; \ c\left(\mathscr{N}^+(v_5) \cap \mathscr{N}^+(v_6)\right) = 0; \ s\left(\mathscr{N}^+(v_5) \cap \mathscr{N}^+(v_6)\right) = 1.$$

$$\text{Therefore, } \mu(v_2, v_9) = \frac{1-1+1}{1} \times 0.08 = 0.08,$$
$$\mu(v_3, v_9) = 1,$$
$$\mu(v_6, v_9) = \frac{1-1+1}{1} \times 0.76 = 0.76,$$
$$\mu(v_7, v_9) = \frac{1-1+1}{1} \times 0.13 = 0.13,$$
$$\mu(v_2, v_3) = 0.$$

Theorem 6.9 *If $\mathscr{G}$ is a fuzzy bipartite graph, then it is a fuzzy min-TolComG.*

Proof Let $\mathscr{G} = (B, \sigma, \mu)$ be a fuzzy bipartite graph such that $B = B_1 \cup B_2$ and $B_1 \cap B_2 = \emptyset$, in crisp sense. Let $B_1 = \{\alpha_i : i = 1, 2, \ldots, \kappa\}$ and $B_2 = \{\beta_j : j = \kappa + 1, \kappa + 2, \ldots, n\}$. Theorem 6.8 is used to prove this result. That is, we show that there exists θ_ϕ (a min-$\mathscr{T}$-ECC) of $\mathscr{G}$ such that $\theta_\phi \leq |B|$. Then $\mathscr{G}$ is a fuzzy min-TolComG.

Let us consider the tolerance of the vertex α_i ($i = 1, 2, \ldots, \kappa$) of B_1 as $\mathscr{T}_{\alpha_i}$, where $c(\mathscr{T}_{\alpha_i}) = 1$ and $s(\mathscr{T}_{\alpha_i}) \geq 1$ and same for each vertex β_j of B_2 be $c(\mathscr{T}_{\alpha_i}) = \kappa + 1$ and $s(\mathscr{T}_{\alpha_i}) \geq \kappa + 1$, $\alpha_i \in B_1$ and $\beta_j \in B_2$, $i, j = 1, 2, \ldots, n$.

Now consider the fuzzy sets $S_i = (X_i, m_i) \cup A_i$, where $X_i = \{\beta_j \in B_2| \mu(\alpha_i, \beta_j) > 0\}$ and $A_i = (\alpha_i, \sigma(\alpha_i))$ with $m_i : X_i \to [0, 1]$ defined by $m_i(\beta_j) = \sigma(\beta_j)$, $i = 1, 2, \ldots, \kappa$. Then it is easy to see that there are at least $\min\{s(\mathscr{T}_{\alpha_i}), s(\mathscr{T}_{\beta_j})\}$ of sets S_i ($i = 1, 2, \ldots, \kappa$) that contain both α_i, β_j. Therefore, $\theta_\phi \leq \kappa < n$. So, there is a ϕ-$\mathscr{T}$-ECC and hence the given fuzzy bipartite graph is a min-TolComG. $\square$

We discussed min-TolComG for fuzzy diagraph. Also, min-TolComG for FInv digraph is studied and it has many applications in real life situations. For this study, the function ϕ is defined by $\phi : \mathbb{R} \times \mathbb{R} \to \mathbb{R}$ (set of real numbers) such that $\phi(x, y) = \min\{x, y\}$. One application of such graph is considered in Example 6.5.

Example 6.5 Let us consider a science museum whose opening time is from 12 P.M. to 4 P.M., i.e., in terms of interval the working time of the museum $I \equiv [12 - 16]$. Assumed that the museum authority is not strict about opening and closing of

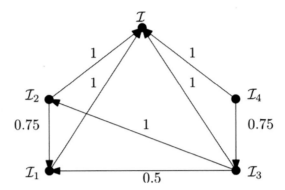

Fig. 6.9 Fuzzy interval digraph model for the museum show problem stated in Example 6.5

the museum, i.e., the opening time may be 11:30 A.M. to any time close to 12 P.M. and similarly, the closing time may extend to 4:45 P.M. So, the timing of the museum is considered as a fuzzy time interval $\mathscr{I} \equiv [11:30 \sim 12 - 16 \sim 16:45]$ instead of the exact time interval $I \equiv [12 - 16]$. It is also assumed that in the museum there is only one theatre for conducting a science show. And the science show may start with a minimum number of audience with some flexibility of time. The degree of confidence of occurring of an event on a certain time is assigned as membership value of the time. The membership values of each interval mentioned here are shown in Fig. 6.10. Now, consider there are four groups S_1, S_2, S_3, S_4 of school students who are allowed to visit the museum in the time intervals $\mathscr{I}_1 = [11:15 \sim 11:45 - 13:30 \sim 13:45]$, $\mathscr{I}_2 = [13:00 \sim 13: 30 - 14:15 \sim 14:30]$, $\mathscr{I}_3 = [13:15 \sim 14:00 - 15:00 \sim 15:30]$, $\mathscr{I}_4 = [14:45 \sim 15:15 - 16:15 \sim 16:30]$ respectively. The membership values of all these fuzzy time intervals are shown in Fig. 6.10. Also, the tolerance time of four groups S_1, S_2, S_3, S_4 are taken as 10–15 min, 5–10 min, 10–20 min, and 10–15 min of their scheduled times. Therefore, the core and support lengths of tolerances of the four groups are $\langle 10, 15\rangle$, $\langle 5, 10\rangle$, $\langle 10, 20\rangle$, and $\langle 10, 15\rangle$ respectively. The problem of the authority is to decide how strongly the group of students compete to view the show and depending on these strengths the authority would decide how they can be managed. This problem can be modeled as a fuzzy min-TolComG for the FInv.

The diagrammatic representation of this problem as a fuzzy min-TolComG is depicted in Fig. 6.11. The strengths of two groups are shown as the fuzzy membership value of the corresponding edges between two groups.

From the graph of Fig. 6.11, it is shows that there is no competitions between S_1 and S_4, and also between S_2 and S_4. On the other hand, the groups S_2 and S_3 have maximum competition (as membership value is 1), i.e., no one can tolerate the situation. The degree of competitions among the groups (S_1, S_2), (S_1, S_3), (S_3, S_4) are 60%, 27%, 52% respectively.

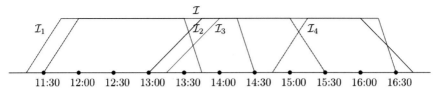

Fig. 6.10 Fuzzy interval representation of the FInv digraph of Fig. 6.9

Fig. 6.11 Fuzzy min-TolComG of the FInv digraph of Fig. 6.9

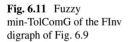

Theorem 6.10 *For $\phi : \mathbb{R} \times \mathbb{R} \to \mathbb{R}$, defined by $\phi(a, b) = \min\{a, b\}$, every fuzzy min-TolComG of a FInv digraph has a regular representation.*

Proof 1. Let $\mathscr{F}_{\mathscr{I}} = \{\mathscr{I}_1, \mathscr{I}_2, \ldots, \mathscr{I}_n\}$ be a finite family of real FInvs and $\mathscr{T} = \{\mathscr{T}_1, \mathscr{T}_2, \ldots, \mathscr{T}_n\}$ be the corresponding fuzzy tolerances. Let $\mathscr{G} = (\mathscr{V}, \sigma, \mu)$ be a fuzzy min-TolComG. Then $\sigma(\mathscr{I}_i) = h(\mathscr{I}_i) = 1$ and

$$
\mu(\mathscr{I}_i, \mathscr{I}_j) = \begin{cases}
h\left(\mathscr{N}^+(\mathscr{I}_i) \cap \mathscr{N}^+(\mathscr{I}_j)\right), & \text{if } c\left(\mathscr{N}^+(\mathscr{I}_i) \cap \mathscr{N}^+(\mathscr{I}_j)\right) \geq \\
& \quad \min\{c(\mathscr{T}_i), c(\mathscr{T}_j)\} \\
\frac{s(\mathscr{N}^+(\mathscr{I}_i) \cap \mathscr{N}^+(\mathscr{I}_j)) - \min\{s(\mathscr{T}_i), s(\mathscr{T}_j)\} + 1}{s(\mathscr{N}^+(\mathscr{I}_i) \cap \mathscr{N}^+(\mathscr{I}_j))} & \text{if } s\left(\mathscr{N}^+(\mathscr{I}_i) \cap \mathscr{N}^+(\mathscr{I}_j)\right) \geq \\
\quad \times h\left(\mathscr{N}^+(\mathscr{I}_i) \cap \mathscr{N}^+(\mathscr{I}_j)\right), & \quad \min\{s(\mathscr{T}_i), s(\mathscr{T}_j)\} \\
0, & \text{otherwise.}
\end{cases}
$$

of FInv digraph $\overrightarrow{\mathscr{D}} = (\mathscr{V}, \sigma, \overrightarrow{\mu})$. Now, for the ith vertex, let $c(\mathscr{T}_i) > c(\mathscr{I}_i)$ and $s(\mathscr{T}_i) > s(\mathscr{I}_i)$. Then $\mu(\mathscr{I}_i, \mathscr{I}_j) > 0$ if either

$$
c(\mathscr{N}^+_{v_i} \cap \mathscr{N}^+_{v_j}) \geq \min\{c(\mathscr{T}_i), c(\mathscr{T}_j)\} = c(\mathscr{T}_j)
$$
$$
\text{or, } s(\mathscr{N}^+_{v_i} \cap \mathscr{N}^+_{v_j}) \geq \min\{s(\mathscr{T}_i), s(\mathscr{T}_j)\} = s(\mathscr{T}_j).
$$

This implies that one can choose the tolerance core and support for v_i as much large as possible without changing adjacency between the vertices v_i and v_j.

2. Now, we prove lengths of the cores and support of all FTols are distinct. If possible let, $c(\mathscr{T}_i) = c(\mathscr{T}_j)$ for two distinct FTols $\mathscr{T}_i$ and $\mathscr{T}_j$.

Now, there exists a number $\varepsilon > 0$ (however small) we can write $c(\mathscr{T}_i') = c(\mathscr{T}_i) + \varepsilon$.

Then $\min\{c(\mathscr{T}_i), c(\mathscr{T}_j)\} = c(\mathscr{T}_i) = c(\mathscr{T}_j)$ remains same if we replace T_i by $\mathscr{T}_i'$ because $\min\{c(\mathscr{T}_i'), c(\mathscr{T}_j)\} = c(\mathscr{T}_i)$.

So, the adjacency relation of the vertices remains unchanged. Similarly, the result can be verified for the lengths of support also.

3. If possible let, two FInvs $\mathscr{I}_i$ and $\mathscr{I}_j$ share a common end point. Then for a positive number (however small it may be) we replace $\mathscr{I}_i$ by $\mathscr{I}_i'$ such that left end point $L(\mathscr{I}_i') = L(\mathscr{I}_i) - \frac{\varepsilon}{2}$ and right end point $R(\mathscr{I}_i') = R(\mathscr{I}_i) + \frac{\varepsilon}{2}$. Therefore, the length of support for the interval $s(\mathscr{I}_i') = s(\mathscr{I}_i) + \varepsilon$ and hence $s\left(\mathscr{N}^+(\mathscr{I}_i') \cap \mathscr{N}^+(\mathscr{I}_j)\right) = s\left(\mathscr{N}^+(\mathscr{I}_i) \cap \mathscr{N}^+(\mathscr{I}_j)\right) + f(\varepsilon)$, where $f(\varepsilon)$ is a function of ε.

So, $s\left(\mathscr{N}^+(\mathscr{I}_i') \cap \mathscr{N}^+(\mathscr{I}_j)\right) > s\left(\mathscr{N}^+(\mathscr{I}_i) \cap \mathscr{N}^+(\mathscr{I}_j)\right) > \min\{s(\mathscr{T}_i), s(\mathscr{T}_j)\}$.

From this relation, one can conclude that such changes in end points of a FInv doe not make any change in the adjacency relation. $\square$

6.6.2 Fuzzy Unit Min-Tolerance Competition Graph

This is a special case of **unit min-tolerance competition graphs**. Fuzzy unit min-TolComG is a particular case of fuzzy min-TolComG of a FInv digraph where lengths of all interval cores and support are same.

Theorem 6.11 *For every fuzzy unit min-TolComG, there is a fuzzy bounded tolerance representation.*

Proof In a fuzzy unit min-TolComG, the lengths of interval cores and supports are equal. Therefore, no interval contains another interval and hence $\mathscr{N}^+(\mathscr{I}_i) \cap \mathscr{N}^+(\mathscr{I}_j) \subseteq \mathscr{I}_i \cap \mathscr{I}_j$.

Then

$$c(\mathscr{N}^+(\mathscr{I}_i) \cap \mathscr{N}^+(\mathscr{I}_j)) \le c(\mathscr{I}_i \cap \mathscr{I}_j)$$
$$\text{and } s(\mathscr{N}^+(\mathscr{I}_i) \cap \mathscr{N}^+(\mathscr{I}_j)) \le s(\mathscr{I}_i \cap \mathscr{I}_j).$$

So, $\quad |\min\{c(\mathscr{I}_i), c(\mathscr{I}_j)\}| \ge |c(\mathscr{I}_i \cap \mathscr{I}_j)| \ge |c(\mathscr{N}^+(\mathscr{I}_i) \cap \mathscr{N}^+(\mathscr{I}_j))| \ge \min\{c(\mathscr{T}_i), c(\mathscr{T}_j)\}$. $\square$

Theorem 6.12 *Let $\overrightarrow{\mathscr{D}}$ be a directed FInvG. Then the fuzzy min-TolComG $TC_{\min}(\overrightarrow{\mathscr{D}})$ of $\overrightarrow{\mathscr{D}}$ is a FTolG after deleting all its isolated vertices.*

Proof Since $TC_{\min}(\overrightarrow{\mathscr{D}})$ is a fuzzy min-TolComG, then it has a fuzzy edge, if either

$$c(\mathscr{N}^+(\mathscr{I}_i) \cap \mathscr{N}^+(\mathscr{I}_j)) \ge \min\{c(\mathscr{T}_i), c(\mathscr{T}_j)\}$$
$$\text{or, } s(\mathscr{N}^+(\mathscr{I}_i) \cap \mathscr{N}^+(\mathscr{I}_j)) \ge \min\{s(\mathscr{T}_i), s(\mathscr{T}_j)\}.$$

Since, $\mathcal{N}^+(\mathcal{I}_i) \cap \mathcal{N}^+(\mathcal{I}_j) \subseteq \mathcal{I}_i \cap \mathcal{I}_j$, then

$$c(\mathcal{N}^+(\mathcal{I}_i) \cap \mathcal{N}^+(\mathcal{I}_j)) \neq 0 \Rightarrow c(\mathcal{I}_i \cap \mathcal{I}_j) \neq 0$$
$$\text{and } s(\mathcal{N}^+(\mathcal{I}_i) \cap \mathcal{N}^+(\mathcal{I}_j)) \neq 0 \Rightarrow s(\mathcal{I}_i \cap \mathcal{I}_j) \neq 0.$$

Also, $$|c(\mathcal{N}^+(\mathcal{I}_i) \cap \mathcal{N}^+(\mathcal{I}_j))| \leq |c(\mathcal{I}_i \cap \mathcal{I}_j)|$$
$$\text{and } |s(\mathcal{N}^+(\mathcal{I}_i) \cap \mathcal{N}^+(\mathcal{I}_j))| \leq |s(\mathcal{I}_i \cap \mathcal{I}_j)|.$$

Therefore, a FG is obtained after removing all isolated vertices from FTolCG. The membership values of the edges will be non-zero if and only if

$$c(\mathcal{I}_i \cap \mathcal{I}_j) \geq \min\{c(\mathcal{T}_i), c(\mathcal{T}_j)\}$$
$$\text{or, } s(\mathcal{I}_i \cap \mathcal{I}_j) \geq \min\{s(\mathcal{T}_i), s(\mathcal{T}_j)\}$$

So, this graph is a FTolG. $\qquad\square$

Theorem 6.13 *For each $t \in [0, 1]$, the cut level graph of a fuzzy min-TolComG is a fuzzy min-TolComG.*

Proof Let $TC_{\min}$ be a fuzzy min-TolComG. Let the tolerance representation of $TC_{\min}$ be $\langle \mathcal{I}, \mathcal{T} \rangle$, where $\mathcal{I} = \{\mathcal{I}_1, \mathcal{I}_2, \ldots, \mathcal{I}_n\}$ on real line and $\mathcal{T} = \{\mathcal{T}_1, \mathcal{T}_2, \ldots, \mathcal{T}_n\}$ be the corresponding tolerances. Then, for each $t \in [0, 1]$, t-cut level family $\mathcal{I}^t$ of $\mathcal{I}$ is $\mathcal{I}^t = \{\mathcal{I}_i^t | \mathcal{I}_i \in \mathcal{I}, i = 1, 2, \ldots, n\}$. So, the sets of vertices and edges of $TC_{\min}$ are respectively $\mathcal{V} = \{\mathcal{I}_i^t | \mathcal{I}_i \in \mathcal{I}, i = 1, 2, \ldots, n\}$ and $E = \{\mu^t\}$. Hence, $TC_{\min}$ is a min-TolComG. $\qquad\square$

Theorem 6.14 *If there exists, in crisp sense, an embedding of a FG $\mathcal{G} = (\mathcal{V}, \sigma, \mu)$ on a surface which has no more faces than vertices and in which two vertices u, v be such that $\mu(u, v) > 0$ if and only if they have more than one face in common, then $\mathcal{G}$ is a fuzzy min-TolComG.*

Proof The core and support length of tolerances to each vertex $v \in \mathcal{V}$ are assigned as $c(\mathcal{T}_v) = 1$ and $s(\mathcal{T}_v) = 2$. To avoid ambiguity, each vertex is assigned to a particular face, then the FG $\mathcal{G}$ is a fuzzy min-TolComG. $\qquad\square$

6.6.3 Fuzzy Max-Tolerance Competition Graph

The function ϕ can also be taken as $\phi(m, n) = \max\{m, n\}$, where $\phi : \mathbb{N} \times \mathbb{N} \to \mathbb{N}$. In this case, the fuzzy ϕ-TolComG is called the fuzzy max-TolComG.

Theorem 6.15 *Let $\mathcal{G}$ be a fuzzy bipartite graph. Then $\mathcal{G}$ may or may not be a fuzzy max-TolComG.*

Proof Specifically speaking that the fuzzy bipartite graph $\mathcal{K}_{3,3}$ is not a fuzzy max-TolComG. To prove this, we are trying to construct a max-$\mathcal{T}$-ECC of the fuzzy bipartite graph $\mathcal{K}_{3,3} = (B_1, B_2, \sigma, \mu)$, where $B_1 = \{\alpha_1, \alpha_2, \alpha_3\}$ and $B_2 = \{\beta_1, \beta_2, \beta_3\}$.

Let the maximum support length of all the tolerances of the vertices be 3. Without loss of generality, let the tolerance support length of α_1 is 3. Then α_1 is in exactly four sets of a max-$\mathcal{T}$-ECC. Then to be a FTolComG each of $\beta_1, \beta_2, \beta_3$ must be in three of four sets. It could happen if any two of them must be in at most two sets of max-$\mathcal{T}$-ECC. This implies that at least two of $\beta_1, \beta_2, \beta_3$ have tolerance support length 3. Therefore, α_2, α_3 must be in at least three sets with each of β_1, β_2. Hence, it is not possible to create max-$\mathcal{T}$-ECC for the FG $\mathcal{K}_{3,3}$. Hence, $\mathcal{K}_{3,3}$ is not a max-TolComG.   □

6.6.4 Fuzzy Sum-Tolerance Competition Graph

By considering the function ϕ as $\phi(a, b) = a + b$, where $\phi : \mathbb{R} \times \mathbb{R} \to \mathbb{R}$, one can define another fuzzy ϕ-TolComG known as fuzzy sum-TolComG.

We saw that the FG $\mathcal{K}_{3,3}$, which is not a fuzzy max-tolerance graph, could be a **fuzzy sum-tolerance graph**.

Theorem 6.16 *The fuzzy bipartite graph $\mathcal{K}_{3,3}$ is a fuzzy sum-TolComG.*

Proof Let the partitions of the vertices of the FG $\mathcal{K}_{3,3}$ be B_1 and B_2, where $B_1 = \{\alpha_1, \alpha_2, \alpha_3\}$ and $B_2 = \{\beta_1, \beta_2, \beta_3\}$.

The tolerances $\mathcal{T}_{\alpha_1}, \mathcal{T}_{\alpha_2}, \mathcal{T}_{\alpha_3}, \mathcal{T}_{\beta_1}, \mathcal{T}_{\beta_2}, \mathcal{T}_{\beta_3}$ are assigned to the vertices $\alpha_1, \alpha_2, \alpha_3, \beta_1, \beta_2, \beta_3$ respectively, such that $c(\mathcal{T}_{\alpha_1}) = c(\mathcal{T}_{\alpha_2}) = \frac{1}{5}$, $c(\mathcal{T}_{\alpha_3}) = \frac{2}{7}$, $c(\mathcal{T}_{\beta_1}) = c(\mathcal{T}_{\beta_2}) = \frac{1}{6}$, $c(\mathcal{T}_{\beta_3}) = 1$. Then obviously, $\{S_1, S_2, S_3, S_4, S_5, S_6\}$ is a sum-$\mathcal{T}$-ECC, where $S_1 = \{\alpha_1, \alpha_3, \beta_1, \beta_3\}$, $S_2 = \{\alpha_2, \alpha_3, \beta_2, \beta_3\}$, $S_3 = \{\alpha_1, \beta_2\}$, $S_4 = \{\alpha_1, \beta_3\}$, $S_5 = \{\alpha_2, \beta_3\}$, $S_6 = \{\alpha_2, \beta_1\}$. Since, $\theta_\phi(\mathcal{K}_{3,3}) \leq 6$, therefore $\mathcal{K}_{3,3}$ is a fuzzy sum-TolComG.   □

References

1. K.P. Bogart, P.C. Fishburn, G. Isaak, L. Langley, Proper and unit tolerance graphs. Discret. Appl. Math. **60**, 99–117 (1995)
2. R.C. Brigham, F.R. McMorris, R.P. Vitray, Tolerance competition graphs. Linear Algebra Appl. **217**, 41–52 (1995)
3. D. Catanzaro, S. Chaplick, S. Felsner, B.V. Halldórsson, M.M. Halldórsson, T. Hixon, J. Stacho, Max point-tolerance graphs. Discret. Appl. Math. **216**(1), 84–97 (2017)
4. W.L. Craine, Characterizations of fuzzy interval graphs. Fuzzy Sets Syst. **68**, 181–193 (1994)
5. D. Dubois, H. Prade, *Fuzzy Sets and Systems. Theory and Applications* (Academic, New York, 1980)

6. S. Felsner, Tolerance graphs and orders, in *Graph-Theoretic Concepts in Computer Science*. WG 1992. Lecture Notes in Computer Science, ed. by E.W. Mayr, vol. 657 (Springer, Berlin, 1993)

7. M.C. Gulumbic, A. Trenk, *Tolerance Graphs* (Cambridge University Press, Cambridge, 2004)

8. S. Kim, T.A. McKee, F.R. McMorris, F.S. Roberts, p-Competition Graphs, DIMACS Technical Report 89-19, Center for Discrete Mathematics and Theoretical Computer Science, Rutgers University, Piscataway, N.J. 08855,1989

9. S. Kim, T.A. McKee, F.R. McMorris, F.S. Roberts, p-competition numbers. Discret. Appl. Math. **46**, 87–92 (1993)

10. G.B. Mertzios, I.S.S. Zaks, The recognition of tolerance and bounded tolerance graphs, in *Symposiam on Theoritial Aspects of Computer Science* (2000), pp. 585–596

11. van H. Nguyen, D. Eberard, W.M. Favre, L. Krhenbhl, Tolerance synthesis using bond graph inversion and fuzzy logic, in *ICM International Conference on Mechatronics* (Vicenza, Italy, 2013), pp. 442–447. https://doi.org/10.1109/ICMECH.2013.6518577ff

12. M. Pal, Intersection graphs: an introduction. Ann. Pure Appl. Math. **4**(1), 41–93 (2013)

13. M. Pal, Interval-valued fuzzy matrices with interval-valued fuzzy rows and columns. Fuzzy Inf. Eng. **7**(3), 335–368 (2015). https://doi.org/10.1016/j.fiae.2015.09.006

14. M. Pal, An introduction to intersection graphs, Chapter 2, in *An Handbook of Research on Advanced Applications of Graph Theory in Modern Society*, ed. M. Pal, S. Samanta, A. Pal (IGI Global, USA, 2020), pp. 24–65

15. S. Paul, On central max-point-tolerance graphs. AKCE Int. J. Graphs Comb. (2020). https://doi.org/10.1016/j.akcej.2020.01.003

16. T. Pramanik, S. Samanta, M. Pal, S. Mondal, B. Sarkar, Interval-valued fuzzy ϕ-tolerance competition graphs. SpringerPlus **5**, 1981 (2016). https://doi.org/10.1186/s40064-016-3463-z

17. T. Pramanik, S. Samanta, B. Sarkar, M. Pal, Fuzzy ϕ-tolerance competition graphs. Soft Comput. **21**, 3723–3734 (2017). https://doi.org/10.1007/s00500-015-2026-5

18. F.S. Roberts, Applications of edge coverings by cliques. Discret. Appl. Math. **10**, 93–109 (1985)

19. S. Sahoo, M. Pal, Intuitionistic fuzzy competition graphs. J. Appl. Math. Comput. **52**(1–2), 37–57 (2016)

20. S. Sahoo, M. Pal, Intuitionistic fuzzy tolerance graphs with application. J. Appl. Math. Comput. **55**, 495–511 (2017)

21. S. Samanta, M. Pal, Fuzzy tolerance graphs. Int. J. Latest Trends Math. **1**(2), 57–67 (2011)

Chapter 7
Coloring of Fuzzy Graph

List of abbreviation

FG	Fuzzy Graph
CI	Chromatic Index
FCP	Fuzzy Chromatic Polynomial
FCN	Fuzzy Chromatic Number
FFCN	Fuzzy Fractional Chromatic Number
RFG	Radio Fuzzy Graoh
FFC	Fuzzy Fractional Coloring
BC	Basic Color

Graph coloring is a very important problem in graph theory. It has many applications to solve real life problem, viz. scheduling, registrar allocation, traffic light signaling, city planning, frequency assignment, and many more. This chapter gives new concepts on coloring of fuzzy graph (FG). Mainly, there are two types of colorings, vertex/node coloring and edge coloring. Both the methods of coloring are described along with examples. A new method of coloring, namely fuzzy fractional coloring of a FG is introduced here along with fuzzy fractional chromatic number. The main objective is to understand how vertices and edges of a FG can be colored using this new method of fuzzy fractional coloring. Two applications of coloring of graph is presented in Chap. 10. A very good application on coloring of graph is described in [20]. The work presented here is mainly from [2, 14, 16, 21].

Results and applications not mentioned here can be found in [1, 3–13, 15, 17–19, 22, 23].

7.1   Vertex Coloring of Fuzzy Graphs

In graph coloring problem, a graph is colored using a minimum number of colors such that two adjacent vertices have different colors. This is called the **optimal coloring problem** and this only initiated work in the said field. There are many variations of coloring a graph designed by the researchers from time to time by imposing different conditions and/or constraints. The method of coloring a FG that has been introduced here involves consideration of fuzzy colors instead of crisp colors.

7.1.1   Fuzzy Color

It is well known that if two different colors are mixed, a new color is produced. Thus, if a color is mixed with white color, then the density of the color will be changed. Suppose $z \leq 1$ units of a color c_k is mixed with $1 - z$ units of white color, then the mixture is termed a standard mixture of the color c_k. The resultant color is termed as **fuzzy color** of the color c_k with membership value z, whereas c_k is termed as the **basic color**.

Definition 7.1 Let $C = \{c_1, c_2, \ldots, c_n\}$ be a set of basic colors. The fuzzy set (C, f) is called the set of fuzzy colors, where $f : C \rightarrow [0, 1]$, is a membership function of the color c_i. The color $\widetilde{c_i} = (c_i, f(c_i))$ is the fuzzy color corresponding to the basic color c_i and f_i its membership value. In contrast to fuzzy color, the membership value of all basic colors are taken as 1. Therefore, basic color is a fuzzy color whose membership value is 1.

As per the definition of fuzzy color, it is obvious that many different fuzzy colors can be prepared from a basic color. For example, a basic color red is considered. A "fuzzy red" color can be prepared from this red color by mixing 0.9 units of red color with 0.1 units of white color. This 'fuzzy red' color is denoted by $(red, 0.9)$. Correspondingly, many other fuzzy red colors, say, $(red, 0.75)$ can be prepared by mixing 0.75 units of red color with 0.25 units of white color, and so on. Note that $(red, 1)$ denotes the original red color, and it is nothing but the basic color red.

Now, coloring of FG is describes in the following section.

7.1.2   Procedure to Color Fuzzy Graph

Suppose $\mathcal{G} = (\mathcal{V}, \sigma, \mu)$ be a simple connected FG and $\{c_1, c_2, \ldots, c_k\}$ be the set of basic colors. In FGs, there are two types edges, viz. strong and weak edges. In strong edge, the relationship between the end vertices is strong. A FG is classified according to the type of edges as (i) all edges are strong, (ii) some edges are strong, (iii) all edges are weak.

Fig. 7.1 A FG (Weak
(strong) edges are shown by
dotted (solid) lines)

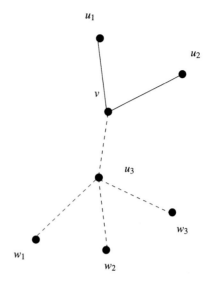

Case 1. All edges are strong.
If all edges a FG are strong, then coloring of this FG is alike to the coloring a
crisp graph. That is, two vertices are colored by two separate colors if a strong edge
between the end vertices exists.

Case 2. Some edges are strong.
Let v be a vertex and $N(v) = \{u_i, i = 1, 2, \ldots, n\}$. For simplicity, it is assumed that
u_1, u_2 be two vertices such that there exists exactly two strong edges as $(v, u_1), (v, u_2)$
connected to v and every other edges $(v, u_i), i = 3, 4, \ldots, n$ are weak edges. In this
case, let v be colored by a color say, $(c, 1)$, then u_1 will get colored by a different
color other than $(c, 1)$, say, $(c_2, 1)$. Likewise, u_2 will get colored by a different color
other than that of v.

We are to color vertex u_3 now keeping in mind that (v, u_3) is a weak edge.

Case 2.1 Adjacent vertices of u_3 are not colored.
As (v, u_3) is weak edge, u_3 gets a fuzzy color of the color of v. If the color of v is
$(c, 1)$, then u_3 gets $(c, f(c))$, where $f(c)$ can be computed by $f(c) = 1 - I_{(v,u_3)}$.

Case 2.2 Every adjacent vertices of u_3 are colored.
If an edge (u_3, w) incident to u_3 is strong, then u_3 can not be colored by the color
of w. That is, if color of w is $(c_w, f(c_w))$, then u_3 can not be colored by any fuzzy
color of c_w. Suppose, u_3 has some weak incident edges, say $(u_3, w_i), i = 1, 2, \ldots, q$
(see Fig. 7.1). Without loss of generality, it is assumed that the color of w_i is
$(x_i, f(x_i)), i = 1, 2, \ldots, q$, where $f(x_i), i = 1, 2, \ldots, q$ are the membership values
of the color $x_i, i = 1, 2, \ldots, q$ and $x_i, i = 1, 2, \ldots, q$ may or may not be identical.
Now, the color of u_3 is found out.

Let $M = \max\{1 - I_{(u_3, w_i)}, 1 - I_{(v, u_3)}$ where $i = 1, 2, \ldots, q\}$. Also let M be attained for the edge (u_3, w_p), i.e. $M = 1 - I_{(u_3, w_p)}$. If the color of w_p is $(x_p, f(x_p))$, then u_3 gets the color (x_p, M).

Case 2.3 Some of the adjacent vertices of u_3 are colored.
The adjacent vertices which are not colored are not involved in the coloring of u_3. So, the adjacent vertices which are colored, are involved in the coloring of u_3. The method of coloring of u_3 is alike to Case 2.2. After coloring of u_3, every other vertex are colored in a similar manner.

Case 3. All edges are weak.
Here pick one vertex y and color the vertex by any basic color, say $(c_y, 1)$. Then every other vertex gets some fuzzy color of c_y. The membership values of the fuzzy colors are computed as the method described in Case 2.2. The adjacent vertices of y are colored first and then every vertex is colored accordingly.

7.1.3 Fuzzy Chromatic Number

Unlike coloring of crisp graph, in fuzzy coloring, two types of colors, viz. basic colors and mixed colors are used to color a FG. The minimum number of basic colors required to color a FG is called the **fuzzy chromatic number** (FCN). The FCN of a FG $\mathscr{G}$ is denoted by $\gamma(\mathscr{G})$. The FCN of some well-known FGs is discussed below.

Let us consider two fuzzy stars in which first fuzzy star (see Fig. 7.2a) contains only weak edges. As all edges are weak, therefore, one basic color is sufficient to color all vertices. In this case, the central vertex is colored by $(R, 1)$ and other vertices are colored by different fuzzy colors, say, $(R, 0.a)$, $(R, 0.b)$, $(R, 0.c)$, $(R, 0.d)$ of R, where a, b, c, d are non-zero integers. Therefore, FCN of this FG is 1. In second FG (see Fig. 7.2b), all edges are strong. Consequently, FCN is 2 which is equal to the FCN of any crisp star. In the figures, fuzzy colors are mentioned within brackets.

Let us consider four FGs each of whose underlying crisp graph is K_4. In Fig. 7.3a, all edges are strong. Therefore, this FG is four chromatic like K_4. In Fig. 7.3b, edges adjacent to only one vertex (say u) are weak and all other edges are strong. The other vertices are colored by $(G, 1)$, $(B, 1)$, $(Y, 1)$. The coloring of such weak vertex depends on the weak incident edges whose other end vertices are colored. That weak vertex is colored by the fuzzy color of one of the three basic colors. The membership value of the fuzzy color is determined by calculating the strengths of such weak edges. Let m be the minimum among the strengths of the weak edges (here it is 0.4) and it attained in the edge (u, v), where the vertex v is colored by $(B, 1)$. Then the vertex u is colored by the fuzzy color $(B, 1 - m)$, i.e. $(B, 0.6)$. Thus, it is three chromatic. Other two FGs are colored in a similar way (see Fig. 7.3c and d).

The relationship between FCN ($\gamma(\mathscr{G})$) of a FG and the chromatic number of corresponding underlying graph ($\chi(G^*)$) is stated in the following theorem.

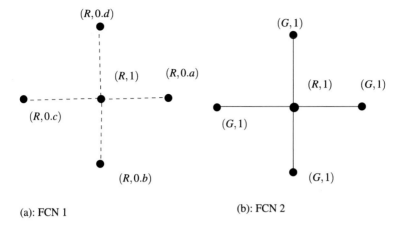

(a): FCN 1

(b): FCN 2

Fig. 7.2 FCN of a fuzzy star (dotted (solid) lines represent weak (strong) edges and a, b, c, d are non-zero integers)

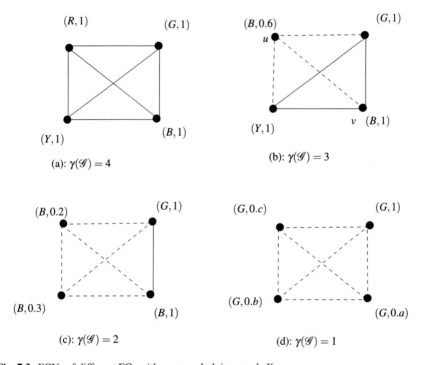

(a): $\gamma(\mathscr{G}) = 4$

(b): $\gamma(\mathscr{G}) = 3$

(c): $\gamma(\mathscr{G}) = 2$

(d): $\gamma(\mathscr{G}) = 1$

Fig. 7.3 FCNs of different FGs with same underlying graph K_4

Theorem 7.1 *If $\mathscr{G}$ be a FG and G^* be its underlying crisp graph, then $\gamma(\mathscr{G}) \leq \chi(G^*)$.*

The least number of weak edges of a FG can be determined if the FCN of the FG and chromatic number of the underlying graph are known.

Theorem 7.2 *If $\mathscr{G}$ is a FG and G^* be its underlying graph, then $\chi(G^*) - \gamma(\mathscr{G})$ represents the least number of weak edges in $\mathscr{G}$.*

Proof Let $\omega = \chi(G^*) - \gamma(\mathscr{G})$. Now, there are two possible cases.
Case 1. $\omega = 0$.
If $\omega = 0$, $\chi(G^*) = \gamma(\mathscr{G})$. This signifies that every edge in $\mathscr{G}$ is strong. Therefore, the graph has no weak edge.

Case 2 $\omega > 0$.
Let (u, v_1) be an edge in $\mathscr{G}$ and also (u', v_1') be the corresponding edge in $\mathscr{G}$. In $\mathscr{G}$, two distinct colors are required to color the vertices u, v_1. In $\mathscr{G}$, if (u', v_1') is a strong edge, then u', v_1' also need to be colored by two different basic colors. If the edge is not strong, then u' and v_1' may be colored by fuzzy colors identical to one basic color unless any other strong edge is connected with u'. Therefore, u' and v_1' are colored by fuzzy colors generated from one basic color, if every edges incident to u' or v_1' are not strong edges. Consequently, two end vertices in a FG are colored by fuzzy colors formed by identical basic color if, at least, the corresponding edge of the end vertices is weak. This signifies that $\mathscr{G}$ has at least ω weak edges. Hence, $\chi(G) - \gamma(\mathscr{G})$ is the least number of weak edges.   □

7.1.3.1 Strength Cut Graph of a Fuzzy Graph

Strength of an edge (x, y) in a FG $\mathscr{G} = (\mathscr{V}, \sigma, \mu)$ is defined as $I_{(x,y)} = \dfrac{\mu(x, y)}{\sigma(x) \wedge \sigma(y)}$.
Again, the **strength of a vertex** is the greatest value among the strengths of its incident edges and its membership value. Mathematically, the strength of a vertex $u \in \mathscr{V}$ is denoted by I_u and defined by $I_u = \max\{\theta_u, \sigma(u)\}$, where $\theta_u = \max\{I_{(u,v)} \mid (u, v)$ is an edge in $\mathscr{G}, v \in \mathscr{V}\}$.
Now, the **strength cut graph** of a FG is a crisp graph which is introduced below.

Definition 7.2 Let $\mathscr{G} = (\mathscr{V}, \sigma, \mu)$ be a FG. For $0 \leq \gamma \leq 1$, the γ-strength cut graph of a FG $\mathscr{G}$ is a crisp graph $\mathscr{G}^\gamma = (V^\gamma, E^\gamma)$, where sets of vertices and edges are defined by $V^\gamma = \{v \in V \mid I_v \geq \gamma\}$ and $E^\gamma = \{(u, v), u, v \in V \mid I_{(u,v)} \geq \gamma\}$.

Example 7.1 Let $\mathscr{G} = (\mathscr{V}, \sigma, \mu)$ be a FG and $\mathscr{V} = \{a, b, c, d, e\}$ (see Fig. 7.4a). Let the vertex and edge membership values be $\sigma(a) = 0.6$, $\sigma(b) = 0.4$, $\sigma(c) = 0.7$, $\sigma(d) = 0.8$, $\sigma(e) = 1$ and $\mu(a, b) = 0.35$, $\mu(a, c) = 0.5$, $\mu(a, d) = 0.5$, $\mu(a, e) = 0.55$, $\mu(b, c) = 0.3$, $\mu(b, d) = 0.3$, $\mu(b, e) = 0.2$, $\mu(c, d) = 0.7$, $\mu(c, e) = 0.6$, $\mu(d, e) = 0.75$. Now, the corresponding 0.5-cut graph and 0.5-strength cut graph of $\mathscr{G}$ are depicted in Fig. 7.4b and c respectively.

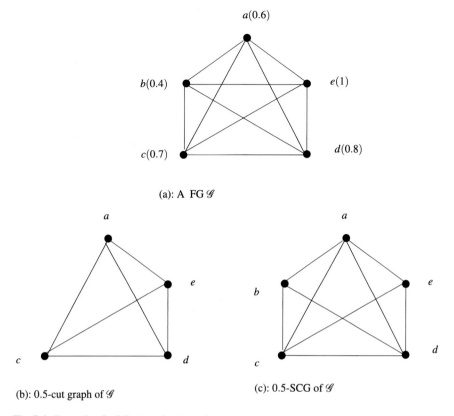

(a): A FG $\mathscr{G}$

(b): 0.5-cut graph of $\mathscr{G}$ (c): 0.5-SCG of $\mathscr{G}$

Fig. 7.4 Example of a 0.5-strength cut graph

The relation among distinct strength cut graphs of a FG $\mathscr{G}$ is established in the following theorem.

Theorem 7.3 *Let $\mathscr{G}$ be a FG. If $0 \leq \gamma \leq \delta \leq 1$, then $\mathscr{G}^{\delta} \subseteq \mathscr{G}^{\gamma}$.*

The following result follows from Theorem 7.3.

Theorem 7.4 *Let $\mathscr{G}$ be a FG. If $0 \leq \gamma \leq \delta \leq 1$, then $\chi(\mathscr{G}^{\delta}) \leq \chi(\mathscr{G}^{\gamma})$, where $\mathscr{G}^{\gamma}$ is a strength cut graph.*

The following result is obvious and left as an exercise.

Lemma 7.1 *If $\mathscr{G}$ be a FG, then $\chi(\mathscr{G}^{0.5}) = \gamma(\mathscr{G})$.*

7.2 Fuzzy Chromatic Polynomial

Like chromatic polynomial of a crisp graph, the chromatic polynomial of FG has also been defined and it is called **fuzzy chromatic polynomial** (FCP). Such polynomial is defined for γ-cut graph of a FG.

Note that $\mathcal{G}^0$ is a crisp complete graph whose vertex set is $\mathcal{V}$ and obviously, all edges are connected with all other edges.

Also, FCP is defined as FGs with fuzzy vertices and fuzzy edges [2]. The definition of chromatic polynomial for γ-cut graph $\mathcal{G}^\gamma$ is defined below:

The λ-coloring of a graph means coloring of the maximum number of vertices using λ number of colors, $\lambda \leq n$, the number of vertices of the graph.

Definition 7.3 ([2]) The number of distinct λ-coloring on the vertices of $\mathcal{G}^\gamma$ is called the **chromatic polynomial** of $\mathcal{G}^\gamma$ and it is denoted by $\psi(\mathcal{G}^\gamma, \lambda)$.

Let L be the set of positive membership values of all vertices and edges of a FG $\mathcal{G}$. Assume that $L = \{\gamma_1, \gamma_2, \ldots, \gamma_k\}$, where $\gamma_1 \leq \gamma_2 \leq \cdots \leq \gamma_k$. This set L is called level set or fundamental set of $\mathcal{G}$. Also, let $I = L \cup \{0\}$.

Definition 7.4 The FCP of a FG $\mathcal{G}$ is defined as the **chromatic polynomial** of its crisp graphs $\mathcal{G}^\gamma$, for $\gamma \in I$. This polynomial is denoted by $\tilde{\psi}_\gamma(\mathcal{G}, \lambda)$. That is,

$$\tilde{\psi}_\gamma(\mathcal{G}, \lambda) = \psi(\mathcal{G}^\gamma, \lambda), \text{ for all } \gamma \in I. \tag{7.1}$$

Note that $\mathcal{G}^\gamma$ is a crisp graph for any $\gamma \in I$ and the degree of the chromatic polynomial of a crisp graph is equal to the number of vertices. Thus, the degree of FCP $\tilde{\psi}_\gamma(\mathcal{G}, \lambda)$ of $\mathcal{G}$ is the number of vertices of $\mathcal{G}$,

Example 7.2 Let us consider a FG $\mathcal{G}$ of Fig. 7.5. For this graph, the set of vertices and edges are $\mathcal{V} = \{a_1(0.8), a_2(0.4), a_3(0.6), a_4(0.6), a_5(0.4)\}$ and $E^* = \{a_1a_2(0.2), a_1a_3(0.3), a_1a_5(0.2), a_2a_3(0.2), a_3a_4(0.3), a_3a_5(0.3), a_4a_5(0.2)\}$, the number within parentheses represents membership value.

When $\gamma = 0.6$, the reduced graph is $\mathcal{G}^{0.6}$, whose vertex set is $\{a_1, a_3, v_4\}$ and there is no edges (see Fig. 7.5b). And this graph can be colored in λ^3 different ways and hence its chromatic polynomial is λ^3. Thus, for $\gamma = 0.6$, $\tilde{\psi}_\gamma(\mathcal{G}, \lambda) = \lambda^3$.

Let us consider another $\gamma(=0.3)$. In this case, the reduced graph is $\mathcal{G}^{0.3}$, and its vertex set is $\{a_1, a_2, a_3, a_4, a_5\}$ and edge set is $\{a_1a_3, a_3a_4, a_3a_5\}$ (see Fig. 7.5c). First, we color the vertex a_3 in λ different ways. So, the adjacent vertices a_1, a_4, a_5 can be colored in $\lambda - 1$ different ways. The isolated vertex a_2 can be colored in λ different ways. Thus, the chromatic polynomial for $\mathcal{G}^{0.3}$ is $\lambda^2(\lambda - 1)^3$.

By similar method, one can determine the chromatic polynomials for other values of γ. Finally, the FCP for the graph of Fig. 7.5 is written below.

$$\tilde{\psi}_\gamma(\mathscr{G}, \lambda) = \begin{cases} \lambda(\lambda - 1)(\lambda - 2)(\lambda - 3)(\lambda - 4) & \gamma = 0 \\ \lambda(\lambda - 1)(\lambda - 2)^3 & \gamma = 0.2 \\ \lambda^2(\lambda - 1)^3 & \gamma = 0.3 \\ \lambda^5 & \gamma = 0.4 \\ \lambda^3 & \gamma = 0.6 \\ \lambda & \gamma = 0.8 \end{cases}$$

Theorem 7.5 *Let $\mathscr{G}$ be a FG defined on G^*. If $\gamma = \min L$, where L is the level set of $\mathscr{G}$, then $\tilde{\psi}_\gamma(\mathscr{G}, \lambda) = \psi(G^*, \lambda)$.*

Proof Since $\gamma = \min L$, i.e. γ is the minimum value of L, so $\mathscr{G}^\gamma$ contains all vertices and edges of $\mathscr{G}$. Thus, $\mathscr{G}^\gamma = G^*$. Hence, the result. □

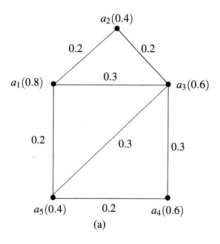

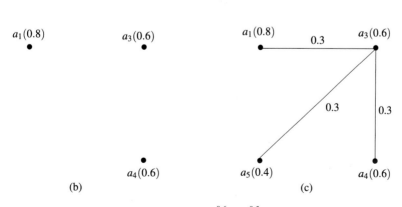

Fig. 7.5 **a** A FG with fuzzy vertices and edges; **b** $\mathscr{G}^{0.6}$; **c** $\mathscr{G}^{0.3}$

Theorem 7.6 *Let $\mathscr{G}$ be a FG having n vertices and G^* be its underlying crisp graph. If $\gamma \in I$ and $\beta = \min L$, then*

$$\tilde{\psi}_\gamma(\mathscr{G}, \lambda) = \begin{cases} \psi(K_n, \lambda), & \gamma = 0 \\ \psi(G^*, \lambda), & \gamma = \beta \\ \psi(\mathscr{G}^\gamma, \lambda), & \beta < \gamma \leq 1 \end{cases}$$

Remark 7.1 The FCP of a FG is computed based on γ-cut graph. That is, FCP is nothing but the chromatic polynomial of an appropriate crisp graph. So, the FCP of a FG is a function of multiple polynomials and number of such polynomials depends on the size of level set L. So, more research is required to find FCP for a given FG directly.

7.3 Edge Coloring of Fuzzy Graph

Like vertex coloring, edge coloring is also an important problem in graph theory and it has many applications. To color edges of a FG, we also use fuzzy colors.

Let $\mathscr{G} = (\mathscr{V}, \sigma, \mu)$ be a connected FG and $C = (c_1, c_2, \ldots, c_k)$ be a set of basic colors. Now, if two edges are adjacent, then the (fuzzy) colors of the edges are different with distinct basic colors; otherwise, the edges have fuzzy colors whose basic colors may be identical.

The **color of an edge** is denoted by $(c_i, f_{e_j}(c_i))$, where c_i is the basic color assigned to the edge $e_j = (u, v)$ and $f_{e_j}(c_i)$ is the membership value of the color c_i whose value is determined by

$$f_{e_j}(c_i) = \frac{\mu(u, v)}{\sigma(u) \wedge \sigma(v)}, \tag{7.2}$$

where $\sigma(u)$ and $\sigma(v)$ are the membership values of vertices u and v respectively. Also, $\mu(u, v)$ is the membership value of the edge e_j.

We use depth-first search technique is used to colored the edges of a FG.

First of all, the vertex "1" is considered for coloring. Every edge incident to this vertex is colored in such a way that two incident edges have different colors. The intensity of the colors depend on the f-values defined in (7.2). Next, the neighbors of "1" are considered for coloring except the vertices considered earlier. Then consider another vertex from the neighborhood of these vertices and repeat this process until all the edges are colored.

7.4 Chromatic Index of Fuzzy Graph

The least number of basic colors used to color a FG is called the **chromatic index** (CI) of a FG. Suppose, such least number of basic colors be N. This CI is not sufficient to mention the strengths of the edges. So, we redefined the CI as a number with two components, say (N, W), where W is the weight and we call it fuzzy CI. The weight is defined as

$$W = \sum_{i=1}^{N} \{\max_{j} f_{e_j}(c_i)\},$$

where the basic color c_i is assigned to the edge e_j for some j, and the intensity (or membership value) of the color c_i is $f_{e_j}(c_i)$. This weight is meaningful only when it is very high or very low value, i.e. every edge is strong, or every edge is weak. Therefore, the weights need further restrictions.

Example 7.3 Let us consider the graph of Fig. 7.6. In this FG, three basic colors red, black, and green are assigned to the vertices, i.e. for this case $N = 3$. The intensity of colors is the strength of the corresponding edge. The red color is assigned to the edges CA and BD with membership values 0.5 and 0.83 respectively. The black color is assigned to the edges BC and AD with membership value 1. The green color is given to the edges CD and AB with membership values 1 and 0.71 respectively. Therefore, the weight $W = 0.83 + 1 + 1 = 2.83$. Hence, CI of this FG is $(3, 2.83)$.

Lemma 7.2 *The CI of a complete fuzzy graph is (N, N).*

The upper value of the weight is stated below.

Lemma 7.3 *Let (N, W) be the CI of a FG, then $0 < W \le N$.*

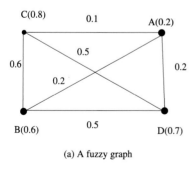

(a) A fuzzy graph

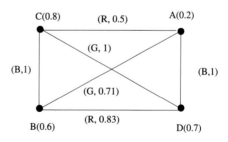

(b) Edge colouring of the fuzzy graph shown in part (a)

Fig. 7.6 Edge coloring of FGs

7.4.1   Strong Chromatic Index of Edge Coloring of Fuzzy Graph

The weight of the fuzzy CI is significant only if the weight is very large or very small. Suppose $(3, 2.9)$ and $(3, 0.3)$ are fuzzy CIs of two FGs. From these CIs, one can conclude that the edges of the second graph are not strong. But, if the weight is near about half of its upper bound (i.e. $N/2$), then it does not make any clear conclusion. So, we need a modification of CI and hence strong CI is defined to explain such case. Let $\mathcal{G} = (\mathcal{V}, \sigma, \mu)$ be a connected FG. To color the edges of a FG sometimes we use fuzzy colors whose membership values are more than 0.5. Such colors are called strong colors and this leads us to define strong CI. The **strong CI** is denoted by $\gamma_s(\mathcal{G}) = (M_s, W_s)$, where M_s is the number of basic colors required to color $\mathcal{G}$ and W_s is the sum of membership values of the basic colors.

Example 7.4 Let us consider a FG whose set of vertices and edges be $\{a(0.7), b(0.5), c(0.4), d(0.6), e(0.8)\}$ and $\{ab(0.4), ac(0.1), ad(0.2), ae(0.7)\}$. In this graph, four basic colors, viz. red, yellow, green, and blue are used to color the edges, and the membership values of the colors are 0.8, 0.25, 0.33, 1 respectively (calculated by $\frac{\mu(x,y)}{\sigma(x) \wedge \sigma(y)}$). Thus, $M_s = 2$ and $W_s = (0.8 + 1) = 1.8$. Hence, the strong CI of this FG is $(2, 1.8)$.

Theorem 7.7 Let $\mathcal{G}$ be a FG, and the CI and strong CI of $\mathcal{G}$ be (N, W) and (M_s, W_s), then
(i) $N \geq M_s$ and $W \geq W_s$,
(ii) $2W_s - M$ is either zero or positive.

Theorem 7.8 Let $\mathcal{G}$ be a FG and its strong CI be (M_s, W_s). Then $\dfrac{M_s}{2} \leq W_s \leq M_s$ is true.

Proof Let $\mathcal{G} = (\mathcal{V}, \sigma, \mu)$ be a FG and its strong CI be (M_s, W_s). Therefore, the FG $\mathcal{G}$ is colored by M_s number of strong basic color and the membership value of each such strong basic colors is at least 0.5. Thus, $W_s = \{0.5 + 0.5 + \cdots M_s \text{ times}\} = \dfrac{M_s}{2}$.

So, the least value of strong weight is $\dfrac{M_s}{2}$. Also, the maximum intensity of a color is 1. So, $W_s \leq \{1 + 1 + 1 + \cdots M \text{ times}\} = M_s$. Hence, the result.   □

Lemma 7.4 Let $\mathcal{G}$ be a FG which is not complete and its CI, strong CI be (N, W) and (M_s, W_s) respectively. Then $\dfrac{W - W_s}{N - M_s} \leq 0.5$.

Lemma 7.5 Let $\mathcal{G}$ be a fuzzy cycle with CI (N, W). If the fuzzy cycle has even number of edges, then $N = 2$ and $0 < W \leq 2$. If the fuzzy cycle has an odd number of edges then $N = 3, 0 < W \leq 3$.

Note 7.1 Let $\mathcal{G}$ be a bipartite FG and Δ be the maximum among the degrees of the vertices, then its CI is (Δ, W).

Theorem 7.9 *Let $\mathscr{G}$ be a FG with each vertex membership value be 1 and (M_s, W_s) be its strong CI. Then, M_s lies between $[\frac{1}{2}\Delta'(\mathscr{G})]$ and $[\Delta'(\mathscr{G}) + 1]$. ($\Delta'(x) = \max \sum_{y \in V, \mu(x,y) > 0.5} \mu(x, y)$ and $[x]$ represents the greatest integer function)*

Proof Suppose $\mathscr{G}$ be a FG with vertex membership values 1 and (M_s, W_s) be the strong CI. Let $\mathscr{G}$ has a vertex u such that $d(u) = \sum_{y \in V, \mu(u,y) > 0.5} \mu(u, y)$. Also, the membership values of vertices are 1. Therefore, strong edges refer to the edges with membership values more than 0.5. As the CI of the graph is (M_s, W_s), $d \leq W_s$ and the number of strong edges incident to u must be less than or equal to M_s. Again, $\Delta'(x) = \max_{x \in V} d(x)$. Therefore, $N \geq [\frac{1}{2}\Delta'(\mathscr{G})]$.

For upper bound, the least number of colors required to color a fuzzy cycle of three vertices is 2+1, while the maximum degree is 2. This particular example can be generally assumed for an odd cycle of degree n. The CI for such cycle is $(n + 1, W)$. Therefore, the maximum value for CI of such cycle is (N, W) where $N = [\Delta'(x) + 1]$. If edge membership values are maximum, then the CI must have its greatest value $([\Delta'(x) + 1], [\Delta'(x) + 1])$. For other graphs, according to Vizing's theorem, it is well known that the crisp graphs have its maximum CI as $\Delta'(x) + 1$. Consequently, the result for FGs that N lies between $[\frac{1}{2}\Delta'(\mathscr{G})]$ and $[\Delta'(\mathscr{G}) + 1]$ is true. $\qquad\square$

7.5 Fuzzy Fractional Coloring

The n-coloring of a FG $\mathscr{G}$ (designated by $\gamma(G) = n$) can be thought of as a way of allowing one color for each vertex, from a set of n colors in such a way that two strong adjacent vertices have different colors. It can be defined in another way by using the idea of homomorphism on FGs.

A proper n-coloring of a FG can be specified as a FG homomorphism from $\mathscr{G}$ to K_n in such a way that the vertices which get the identical color in $\mathscr{G}$ map to a single vertex in K_n. Since no vertex in K_n is adjacent to itself, no adjacent vertices in $\mathscr{G}$ get the identical color.

In a proper coloring, if it is taken the inverse image of a single vertex in K_n, this gets the set of every vertex in $\mathscr{G}$ assigned with a certain color. It is always an independent set. Therefore, the FCN for simply coloring a FG is the smallest number of an independent set which is required to cover the vertex-set of the FG.

Theorem 7.10 *If $\mathscr{G}$ and $\mathscr{H}$ are two FGs such that $\gamma(\mathscr{H}) = n$, then $\mathscr{G} \circ K_n$ is a fuzzy subgraph of $\mathscr{G} \circ \mathscr{H}$.*

The concept of coloring has been generalized to fuzzy fractional coloring (FFC) (or a set of coloring) here. Using this, fuzzy fractional chromatic number (FFCN) is defined, which may take non-integer values as well.

Let $\mathscr{G}$ be a given FG and consider two integers p and q such that $0 < q \leq p$. A **proper p/q coloring** is a mapping that allows vertices to map to a set of q

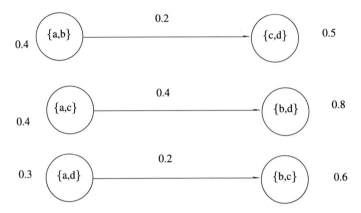

Fig. 7.7 Kneser FG with six vertices

distinct colors from a set of p colors in such a way that adjacent vertices get disjoint arrangements sets of color. Consequently, simply coloring or n-coloring is identical as $n/1$ coloring.

It may be noted that $p/q \neq x/y$, where $py = qx$. p/q means from p objects we have to select q number of objects which is identical to $\binom{p}{q}$. This may not be identical to the number of y objects chosen from x objects which is identical to $\binom{x}{y}$.

Let us consider one interesting graph class known as Kneser FG.

Definition 7.5 The **Kneser FG** is a FG whose vertices correspond to the r-element subsets of a set of n elements, and two vertices are adjacent if and only if their corresponding sets are distinct. It is denoted by $KG_{n,r}$.

Example 7.5 Here, an illustration of Kneser FG $KG_{4,2}$ (as displayed in Fig. 7.7) on the set of vertices $\{a, b, c, d\}$ is given. Here, $n = 4$ and $r = 2$. So, there are $\binom{4}{2}$ number of vertices and it has three edges since there are $\frac{\binom{4}{2} \times \binom{2}{2}}{2} = 3$ separate pair of sets.

As a proper n-coloring of a FG is seen as a homomorphism between a FG and $KG_{p,q}$, correspondingly a proper p/q coloring of a FG can be seen as a homomorphism from the FG $\mathscr{G} = (\mathscr{V}, \sigma, \mu)$ and Kneser graph $KG_{p,q}$.

The **FFCN** is the least of every rational number p/q such that there occurs a proper p/q coloring of $\mathscr{G}$. It may not be clear that the FFCN may be a rational number for an arbitrary FG. To show this, we give a different definition of FFC.

Definition 7.6 Let $\ell(G)$ be the set of all independent sets of a FG $\mathscr{G} = (\mathscr{V}, \sigma, \mu)$. A mapping $f : \ell(G) \to [0, 1]$ is called FFC of $\mathscr{G}$ if $\sum\limits_{v \in S, S \in \ell(G)} f(S) \geq 1$, for each $v \in \mathscr{V}$. The sum of the functional values over all independent sets is said to be the weight of the FFC. The least possible weight of a FFC is denoted by $\gamma_F(G)$ and is called the **FFCN** of $\mathscr{G}$.

For a FG $\mathcal{G}$, FFCN is less than or equal to FCN, i.e. $\gamma_F(G) \leq \gamma(G)$.

Definition 7.7 A FFC in a FG $\mathcal{G} = (\mathcal{V}, \sigma, \mu)$ is said to be **regular** if $\sum\limits_{v \in S, S \in \ell(G)} f(S)$ $= 1$, for each $v \in \mathcal{V}$.

Theorem 7.11 *The FFCN of a fuzzy subgraph H is less than or equal to the FFCN of the FG $\mathcal{G}$, i.e. $\gamma_F(H) \leq \gamma_F(\mathcal{G})$.*

Proof Let $H = (B, \sigma_1, \mu_1)$ be a fuzzy subgraph of the FG $\mathcal{G} = (\mathcal{V}, \sigma, \mu)$. Let us consider two cases.

Cases 7.1 $B = \mathcal{V}$.
In this case, the vertices of H and $\mathcal{G}$ are same, so the number of strongly adjacent vertices are also same. Therefore, for this case, $\gamma_F(H) = \gamma_F(\mathcal{G})$.

Cases 7.2 $B \subset \mathcal{V}$.
In this case, the number of vertices in H is less than the number of vertices in $\mathcal{G}$. Thus, the number of strongly adjacent vertices in H is also less than the number of strongly adjacent vertices in $\mathcal{G}$. Therefore, any proper FFC of $\mathcal{G}$, restricted to B, is a proper FFC of H. Hence, $\gamma_F(H) < \gamma_F(G)$. $\qquad\square$

7.5.1 FFC of a Fuzzy Graph

The method of FFC of a FG is given stepwise.
Input: A FG $\mathcal{G}$.
Output: Colors are assigned to the vertex set of $\mathcal{G}$.
Step 1: First, determine all strong adjacent vertices in $\mathcal{G}$.
Step 2: Remove all non-strong adjacent vertices.
Step 3: Next, find out all independent sets in between strongly adjacent vertices.
Step 4: Assign a color to an independent set and do it for all the independent sets.
Step 5: Allocate a fraction to each independent set in such a way that $\sum\limits_{v \in S, S \in \ell(G)} f(S)$ ≥ 1.
Step 6: For non-strongly adjacent vertices, assign the same color which is assigned to the vertex adjacent to it.

Step 5 can be modeled as a linear programming problem:

$$\text{minimize} \quad z = \mathbf{cx}$$
$$\text{subject to} \quad A\mathbf{x} \geq \mathbf{1}$$
$$\text{and} \quad \mathbf{x} > \mathbf{0},$$

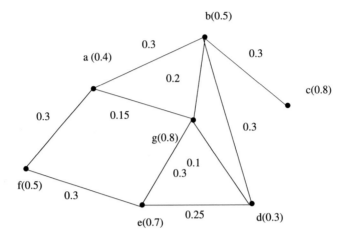

Fig. 7.8 A FG $\mathcal{G}_1$

where
$\mathbf{x} = [x_1, x_2, \ldots, x_n]^T$, each x_j, $j = 1, 2, \ldots n$ represents the weight of the colors
$\mathbf{1} = [1, 1, \ldots, 1]^T = \mathbf{c}^T$,
$\mathbf{0} = [0, 0, \ldots, 0]$
and the matrix A can be constructed in the following way

1. The rows are indexed by the vertices and columns are indexed by all independent sets.
2. Each row is essentially the characteristic function of the corresponding independent sets, with entries equal to 1 on columns corresponding to the vertices in the independent set, and 0 otherwise.

The number z is nothing but the FFCN of $\mathcal{G}$.

Note 7.2 The FFCN of FG and crisp fractional chromatic number of the underlying crisp graph are equal only when every edge in the FG are strong.

Example 7.6 To describe the method of FFC, a FG $\mathcal{G}_1$ is considered which is shown in Fig. 7.8.

In this graph, ag, bg, gd, ge all are non-strongly adjacent edges of $\mathcal{G}$. So, all these edges are removed from G_1 and the reduced graph is shown in Fig. 7.9.

Now, the independent sets are $S_1 = \{a, c, d\}$, $S_2 = \{a, c, e\}$, $S_3 = \{c, d, f\}$, $S_4 = \{b, e\}$, $S_5 = \{b, f\}$. Assign a color to each independent set and the weights can be found out by solving the following linear programming program (LPP).

Minimize $\quad z = x_1 + x_2 + x_3 + x_4 + x_5$

subject to

$$\begin{bmatrix} 1 & 1 & 0 & 0 & 0 \\ 0 & 0 & 1 & 1 & 0 \\ 1 & 1 & 0 & 0 & 1 \\ 1 & 0 & 0 & 0 & 1 \\ 0 & 1 & 1 & 0 & 0 \\ 0 & 0 & 0 & 1 & 1 \end{bmatrix} \begin{bmatrix} x_1 \\ x_2 \\ x_3 \\ x_4 \\ x_5 \end{bmatrix} \geq \begin{bmatrix} 1 \\ 1 \\ 1 \\ 1 \\ 1 \\ 1 \end{bmatrix}$$

and $\quad x_1, x_2, x_3, x_4, x_5 > 0$

The solution of the above problem is $x_1 = x_2 = x_3 = x_4 = x_5 = \frac{1}{2}$. Note that all values are same and hence the weight of each color is $\frac{1}{2}$. By definition, the FFCN is $\sum x_j = \frac{5}{2}$.

The graph $\mathscr{G}_2$ colored by FFC is displayed in Fig. 7.10.

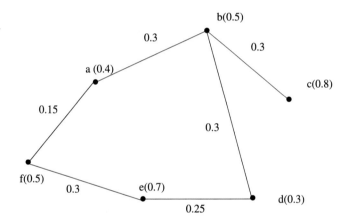

Fig. 7.9 FG $\mathscr{G}_2$ corresponding to the FG $\mathscr{G}_1$ of Fig. 7.8

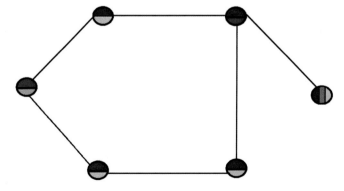

Fig. 7.10 FG with fuzzy fractional colors

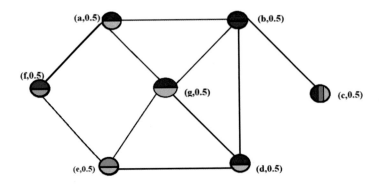

Fig. 7.11 Fractional coloring of the FG $\mathscr{G}_1$

Now, to color the FG $\mathscr{G}_1$ of Fig. 7.8, the edges $(a, g), (e, g), (d, g), (b, g)$ are considered. The strengths of these edges are computed as $0.375, 0.428, 0.333, 0.4$. The least strength occurs at the edge (d, g). Therefore, the color of the vertex g is same as the color of d. Hence, the FFC of $\mathscr{G}_1$ of Fig. 7.8 is depicted in Fig. 7.11, where the membership values of the colored vertices are calculated by

$$\frac{\text{Total \ weight}}{\text{Number \ of \ colors}}.$$

According to the rule, each of the vertices a, b, d, e, f, g gets two colors with equal weight $\frac{1}{2}$ and the membership value of each of these vertex is $\frac{\frac{1}{2}+\frac{1}{2}}{1+1} = \frac{1}{2}$.

The vertex c gets three colors and the weight of each color is $\frac{1}{2}$. The membership value of the vertex c is $\frac{\frac{1}{2}+\frac{1}{2}+\frac{1}{2}}{1+1+1} = \frac{1}{2}$.

Theorem 7.12 *For a FG $\mathscr{G}$, $\gamma_F(\mathscr{G}^{0.5}) = \chi_F(G^*)$.*

Theorem 7.13 *Let $\mathscr{G} = (\mathscr{V}_1, \sigma_1, \mu_1)$ and $H = (\mathscr{V}_2, \sigma_2, \mu_2)$ be two FGs and P be their disjoint union. Then the following results hold goods.*
(i) $\omega(P) = \max\{\omega(G), \omega(H)\}$,
(ii) $\gamma(P) = \max\{\gamma(G), \gamma(H)\}$,
(iii) $\gamma_F(P) = \max\{\gamma_F(G), \gamma_F(H)\}$.

References

1. F.S. Al-Anzia, Y.N. Sotskov, A. Allahverdi, G.V. Andreev, Using mixed graph coloring to minimize total completion time in job shop scheduling. Appl. Math. Comput. **182**(2), 1137–1148 (2006)
2. M.A. Ashebo, V.N.S.R. Repalle, Fuzzy chromatic polynomial of fuzzy graphs with crisp and fuzzy vertices using α-cuts. Adv. Fuzzy Syst. **2019**, 11 (2019), Article ID 5213020. https://doi.org/10.1155/2019/5213020

3. L.S. Bershtein, A.V. Bozhenuk, Fuzzy coloring for fuzzy graphs, in *10th IEEE International Conference on Fuzzy Systems* (2001), pp. 1101–1103

4. K.R. Bhutani, A. Battou, On M-strong fuzzy graphs. Inf. Sci. **155**(1–2), 103–109 (2003)

5. K.R. Bhutani, A. Rosenfeld, Fuzzy end nodes in fuzzy graphs. Inf. Sci. **152**, 323–326 (2003)

6. K.R. Bhutani, A. Rosenfeld, Geodesies in fuzzy graphs. Electron. Notes Discrete Math. **15**, 49–52 (2003)

7. K.R. Bhutani, A. Rosenfeld, Strong arcs in fuzzy graphs. Inf. Sci. **152**, 319–322 (2003)

8. K.R. Bhutani, J. Moderson, A. Rosenfeld, On degrees of end nodes and cut nodes in fuzzy graphs. Iranian J. Fuzzy Syst. **1**(1), 53–60 (2004)

9. J. Culberson, I. Gent, Frozen development in graph coloring. Theor. Comput. Sci. **265**(1–2), 227–264 (2001)

10. C. Eslahchi, B.N. Onaghe, Vertex Strength, of Fuzzy Graphs. Int. J. Math. Math. Sci. **2006**, 1–9 (2006). Article ID 43614. https://doi.org/10.1155/IJMMS/2006/43614

11. M. Gamache, A. Hertz, J.O. Ouellet, A graph coloring model for a feasibility problem in monthly crew scheduling with preferential bidding. Comput. Oper. Res. **34**(8), 2384–2395 (2007)

12. R.H. Goetschel, Fuzzy colorings of fuzzy hypergraphs. Fuzzy Sets Syst. **94**(2), 185–204 (1998)

13. R. Lewis, J. Thompson, On the application of graph colouring techniques in round-robin sports scheduling. Comput. Oper. Res. **38**(1), 190–204 (2011)

14. R. Mahapatra, S. Samanta, T. Allahviranloo, M. Pal, Radio fuzzy graphs and assignment of frequency in radio stations. Comput. Appl. Math. **38**(3), 1–20 (2019)

15. T. Mahapatra, M. Pal, Fuzzy colouring of *m*-polar fuzzy graph and its application. J. Intell. Fuzzy Syst. **35**(6), 6379–6391 (2018)

16. T. Mahapatra, G. Ghorai, M. Pal, Fuzzy fractional coloring of fuzzy graph with its application. J. Ambient Intell. Humaniz. Comput. (2020). https://doi.org/10.1007/s12652-020-01953-9

17. P.S. Nair, S.C. Cheng, Cliques and fuzzy cliques in fuzzy graphs, in *IFSA World Congress and 20th NAFIPS International Conference*, vol. 4 (2001), pp. 2277–2280

18. P.S. Nair, Perfect and precisely perfect fuzzy graphs, in *NAFIPS 2008 - 2008 Annual Meeting of the North American Fuzzy Information Processing Society*, New York City, NY, USA (2008). https://doi.org/10.1109/NAFIPS.2008.4531246

19. S.K. Pal, S.S. Sarma, Graph coloring approach for hiding of information. Procedia Technol. **4**, 272–277 (2012)

20. A. Saha, M. Pal, T.K. Pal, Selection of programme slots of television channels for giving advertisement: a graph theoretic approach. Inf. Sci. **177**(12), 2480–2492 (2007)

21. S. Samanta, T. Pramanik, M. Pal, Fuzzy colouring of fuzzy graphs. Afrika Matematica **27**(1–2), 37–50 (2016)

22. L.A. Zadeh, Fuzzy sets. Inf. Control **8**(3), 338–353 (1965)

23. L.A. Zadeh, Similarity relations and fuzzy orderings. Inf. Sci. **3**(2), 177–200 (1971)

Chapter 8
m-Polar Fuzzy Graphs

Fuzzy graph is a generalization of crisp graph which investigates uncertainty in various types of networks. To deal with multiple information precisely, graphical models can be applied. To discuss the multipolar nature of objects, the theory of fuzzy graphs have been extended to *m*-polar fuzzy graphs.

An *m*-polar fuzzy graph (mPFG) is a generalization of the notion of bipolar fuzzy graph which in turn generalizes the concept of fuzzy graph. Construction of it completely depends on the concept of *m*-polar fuzzy set. So it is both a convenient (because it is computationally tractable) and intuitive notion. In this chapter, the concept of *m*-polar fuzzy sets is reviewed. Then mPFGs are considered and defined isomorphism of mPFGs. We also describe several isomorphic properties of *m*-polar fuzzy graphs. Various methods of their constructions are described. The notion of strong mPFGs are introduced and examined the notions of self-complementary and weak self-complementary strong mPFGs. We define the density of mPFG and balanced of mPFGs. Most of the results in the next few sections are due to Ghorai and Pal [12–14].

Other results and applications not mentioned here can be found in [1–10, 15–35].

8.1 *m*-Polar Fuzzy Sets and *m*-Polar Fuzzy Graphs

Since "multipolar information" exists in the data of real world problems, the notion of mPFGs is introduced with the help of *m*-polar fuzzy sets in this section.

First, we recall the definition of *m*-polar fuzzy set given by Chen et al. [11] in 2014.

Let m be an natural number. We consider $[0, 1]^m$, the mth power of $[0, 1]$ as a **poset** with point-wise order relation.

© The Editor(s) (if applicable) and The Author(s), under exclusive license
to Springer Nature Singapore Pte Ltd. 2020
M. Pal et al., *Modern Trends in Fuzzy Graph Theory*,
https://doi.org/10.1007/978-981-15-8803-7_8

$\leq$ is defined by $a \leq b \Leftrightarrow p_i(a) \leq p_i(b)$ for $a, b \in [0, 1]^m$, $(i = 1, 2, \ldots, m)$ and $p_i : [0, 1]^m \to [0, 1]$ is the ith **projection mapping**.

Definition 8.1 ([11]) An *mPF* set (or a $[0, 1]^m$-set) on X is a mapping $\mathscr{A} : X \to [0, 1]^m$. $m(X)$ denotes the set of all *mPF* sets on X.

Here, $\mathbf{0} = (0, 0, \ldots, 0)$ and $\mathbf{1} = (1, 1, \ldots, 1)$ are the smallest and largest elements in $[0, 1]^m$ respectively.

Definition 8.2 Let $\mathscr{A}$ and $\mathscr{B}$ are two *mPF* sets in X. Then $\mathscr{A} \cup \mathscr{B}$ and $\mathscr{A} \cap \mathscr{B}$ are defined as below:

$p_i \circ (\mathscr{A} \cup \mathscr{B})(a) = p_i \circ \mathscr{A}(a) \vee p_i \circ \mathscr{B}(a)$ and
$p_i \circ (\mathscr{A} \cap \mathscr{B})(a) = p_i \circ \mathscr{A}(a) \wedge p_i \circ \mathscr{B}(a)$ for $a \in X$, $(i = 1, 2, \ldots, m)$.
$\mathscr{A} \subseteq \mathscr{B} \Leftrightarrow p_i \circ \mathscr{A}(a) \leq p_i \circ \mathscr{B}(a)$ for $a \in X$, $(i = 1, 2, \ldots, m)$.
$\mathscr{A} = \mathscr{B} \Leftrightarrow p_i \circ \mathscr{A}(a) = p_i \circ \mathscr{B}(a)$ for $a \in X$, $(i = 1, 2, \ldots, m)$.

Definition 8.3 Let $\mathscr{A}$ be an *mPF* set on a set X. An *mPF* **relation** on $\mathscr{A}$ is an *mPF* set $\mathscr{B}$ of $X \times X$ such that $\mathscr{B}(a, b) \leq \mathscr{A}(a) \wedge \mathscr{A}(b)$ for all $a, b \in X$ i.e. for each $i = 1, 2, \ldots, m$, for all $a, b \in X$, $p_i \circ \mathscr{B}(a, b) \leq p_i \circ \mathscr{A}(a) \wedge p_i \circ \mathscr{A}(b)$.
$\mathscr{B}$ on X is called **symmetric** if $\mathscr{B}(a, b) = \mathscr{B}(b, a)$ for all $a, b \in X$.

Here, we assume the following:

On a given non-empty set $\mathscr{V}$, we define a relation $\sim$ on $\mathscr{V} \times \mathscr{V}$ by for all $(a, b), (c, d) \in \mathscr{V} \times \mathscr{V}$, $(a, b) \sim (c, d) \Leftrightarrow a = c, b = d$ or $a = d, b = c$. One can check that $\sim$ is **reflexive**, symmetric, and **transitive**. Hence, $\sim$ is an **equivalence relation** on $\mathscr{V} \times \mathscr{V}$. Let $[(a, b)]$ denote the equivalence classes of (a, b) w.r.t. $\sim$. Then $[(a, b)] = \{(a, b), (b, a)\}$. Let $\Omega_{\mathscr{V}} = \{[a, b] | a, b \in \mathscr{V}, a \neq b\}$. When $\mathscr{V}$ is understood, we often use Ω for $\Omega_{\mathscr{V}}$. Let $W \subseteq \Omega$ and a graph is a pair consisting of $\mathscr{V}$ and E. For $a, b \in \mathscr{V}$, let ab denote $[(a, b)]$. Then clearly $ab = ba$. We assume that a graph has no loops or parallel edges.

Next, we define the notion of mPFG.

Definition 8.4 An *m*-**polar fuzzy graph** (mPFG) of a graph $\mathscr{G}^* = (\mathscr{V}, E)$ is a triplet $\mathscr{G} = (\mathscr{V}, \mathscr{A}, \mathscr{B})$, where $\mathscr{A} : \mathscr{V} \to [0, 1]^m$ is an *mPF* set in $\mathscr{V}$ and $\mathscr{B} : \Omega \to [0, 1]^m$ is an *mPF* set in Ω such that $\mathscr{B}(ab) \leq \mathscr{A}(a) \wedge \mathscr{A}(b)$ for all $a, b \in \Omega$ and $\mathscr{B}(ab) = \mathbf{0}$ for all $ab \in (\Omega - E)$.

Example 8.1 Let us consider the graph $\mathscr{G}^* = (\mathscr{V}, E)$. where $\mathscr{V} = \{a_1, a_2, a_3\}$ and $E = \{a_1 a_2, a_2 a_3, a_3 a_1\}$. Let $\mathscr{A}$ be a 3-polar fuzzy subset of $\mathscr{V}$ and $\mathscr{B}$ be a 3-polar fuzzy subset of Ω, where $\mathscr{A}(a_1) = (0.5, 0.8, 0.9)$, $\mathscr{A}(a_2) = (0.8, 0.7, 0.5)$, $\mathscr{A}(a_3) = (0.4, 0.7, 0.5)$ and $\mathscr{B}(a_1 a_2) = (0.4, 0.7, 0.4)$, $\mathscr{B}(a_2 a_3) = (0.4, 0.6, 0.5)$, $\mathscr{B}(a_3 a_1) = (0.4, 0.6, 0.5)$. Then it can be easily verified that $\mathscr{G} = (\mathscr{V}, \mathscr{A}, \mathscr{B})$ is a 3-polar FG.

Definition 8.5 The mPFG $\mathscr{H} = (\mathscr{P}, \mathscr{C}, \mathscr{D})$ is called an *mPF* **subgraph** of $\mathscr{G} = (\mathscr{V}, \mathscr{A}, \mathscr{B})$ induced by $\mathscr{P}$ if $\mathscr{P} \subseteq \mathscr{V}$, $\mathscr{C}(a) = \mathscr{A}(a)$ for all $a \in \mathscr{P}$ and $\mathscr{D}(ab) = \mathscr{B}(ab)$ for all $a, b \in \mathscr{P}$.

Example 8.2 Let $\mathcal{H} = (\mathcal{P}, \mathcal{C}, \mathcal{D})$ where $\mathcal{P} = \{a, b\}$, $\mathcal{C}(a) = (0.4, 0.7, 0.8)$, $\mathcal{C}(b) = (0.7, 0.6, 0.4)$ and $\mathcal{D}(ab) = (0.3, 0.6, 0.3)$. Then H is a 3-polar fuzzy subgraph of the graph $\mathcal{G}$ of Example 8.1.

Next, we define certain operations on mPFGs. Here, we assume that $\mathcal{G}_1^* = (\mathcal{V}_1, E_1)$ and $\mathcal{G}_2^* = (\mathcal{V}_2, E_2)$ are the (crisp) graphs.

Definition 8.6 Let $\mathcal{A}_1$ and $\mathcal{A}_2$ be *mPF* subsets of $\mathcal{V}_1$ and $\mathcal{V}_2$ and $\mathcal{B}_1$ and $\mathcal{B}_2$ be *mPF* subsets of E_1 and E_2 respectively. Then the **Cartesian product** of two mPFGs $\mathcal{G}_1 = (\mathcal{V}_1, \mathcal{A}_1, \mathcal{B}_1)$ and $\mathcal{G}_2 = (\mathcal{V}_2, \mathcal{A}_2, \mathcal{B}_2)$ of the graphs $\mathcal{G}_1^* = (\mathcal{V}_1, E_1)$ and $\mathcal{G}_2^* = (\mathcal{V}_2, E_2)$ is denoted by $\mathcal{G}_1 \times \mathcal{G}_2 = (\mathcal{V}_1 \times \mathcal{V}_2, \mathcal{A}_1 \times \mathcal{A}_2, \mathcal{B}_1 \times \mathcal{B}_2)$ and defined by for $i = 1, 2, \ldots, m$

1. $p_i \circ (\mathcal{A}_1 \times \mathcal{A}_2)(a_1, a_2) = p_i \circ \mathcal{A}_1(a_1) \wedge p_i \circ \mathcal{A}_2(a_2)$ for all $(a_1, a_2) \in \mathcal{V}_1 \times \mathcal{V}_2$.
2. $p_i \circ (\mathcal{B}_1 \times \mathcal{B}_2)((a, a_2)(a, b_2)) = p_i \circ \mathcal{A}_1(a) \wedge p_i \circ \mathcal{B}_2(a_2 b_2)$ for all $a \in \mathcal{V}_1$ and $a_2 b_2 \in E_2$.
3. $p_i \circ (\mathcal{B}_1 \times \mathcal{B}_2)((a_1, b)(b_1, b)) = p_i \circ \mathcal{B}_1(a_1 b_1) \wedge p_i \circ \mathcal{A}_2(b)$ for all $b \in \mathcal{V}_2$ and $a_1 b_1 \in E_1$.

Here, $E = \{(a, a_2)(a, b_2) | a \in \mathcal{V}_1 \text{ and } a_2 b_2 \in E_2\} \cup \{(a_1, b)(b_1, b) | b \in \mathcal{V}_2 \text{ and } a_1 b_1 \in E_1\}$ is the edge set of $\mathcal{G}_1 \times \mathcal{G}_2$.

Proposition 8.1 *The Cartesian product of two mPFGs $\mathcal{G}_1$ and $\mathcal{G}_2$ is an mPFG.*

Proof Let $a \in \mathcal{V}_1$ and $a_2 b_2 \in E_2$. Then for $i = 1, 2, \ldots, m$

$$
\begin{aligned}
p_i \circ (\mathcal{B}_1 \times \mathcal{B}_2)((a, a_2)(a, b_2)) &= p_i \circ \mathcal{A}_1(a) \wedge p_i \circ \mathcal{B}_2(a_2 b_2) \\
&\leq \wedge \{p_i \circ \mathcal{A}_1(a), p_i \circ \mathcal{A}_2(a_2) \wedge p_i \circ \mathcal{A}_2(b_2)\} \\
&= \wedge \{p_i \circ \mathcal{A}_1(a) \wedge p_i \circ \mathcal{A}_2(a_2), p_i \circ \mathcal{A}_1(a) \wedge p_i \circ \mathcal{A}_2(b_2)\} \\
&= p_i \circ (\mathcal{A}_1 \times \mathcal{A}_2)(a, a_2) \wedge p_i \circ (\mathcal{A}_1 \times \mathcal{A}_2)(a, b_2).
\end{aligned}
$$

Let $b \in \mathcal{V}_2$ and $a_1 b_1 \in E_1$. Then for $i = 1, 2, \ldots, m$

$$
\begin{aligned}
p_i \circ (\mathcal{B}_1 \times \mathcal{B}_2)((a_1, b)(b_1, b)) &= p_i \circ \mathcal{B}_1(a_1 b_1) \wedge p_i \circ \mathcal{A}_2(b) \\
&\leq \wedge \{p_i \circ \mathcal{A}_1(a_1) \wedge p_i \circ \mathcal{A}_1(b_1), p_i \circ \mathcal{A}_2(b)\} \\
&= \wedge \{p_i \circ \mathcal{A}_1(a_1) \wedge p_i \circ \mathcal{A}_2(b), p_i \circ \mathcal{A}_1(b_1) \wedge p_i \circ \mathcal{A}_2(b)\} \\
&= p_i \circ (\mathcal{A}_1 \times \mathcal{A}_2)(a_1, b) \wedge p_i \circ (\mathcal{A}_1 \times \mathcal{A}_2)(b_1, b).
\end{aligned}
$$

$\square$

Definition 8.7 Let $\mathcal{A}_1$ and $\mathcal{A}_2$ be *mPF* subsets of $\mathcal{V}_1$ and $\mathcal{V}_2$ and $\mathcal{B}_1$ and $\mathcal{B}_2$ be *mPF* subsets of E_1 and E_2 respectively. Then the **composition** two mPFGs $\mathcal{G}_1 = (\mathcal{V}_1, \mathcal{A}_1, \mathcal{B}_1)$ and $\mathcal{G}_2 = (\mathcal{V}_2, \mathcal{A}_2, \mathcal{B}_2)$ of the graphs $\mathcal{G}_1^* = (\mathcal{V}_1, E_1)$ and $\mathcal{G}_2^* = (\mathcal{V}_2, E_2)$ is denoted by $\mathcal{G}_1[\mathcal{G}_2] = (\mathcal{V}_1 \times \mathcal{V}_2, \mathcal{A}_1 \circ \mathcal{A}_2, \mathcal{B}_1 \circ \mathcal{B}_2)$ and defined by for $i = 1, 2, \ldots, m$,

1. $p_i \circ (\mathscr{A}_1 \circ \mathscr{A}_2)(a_1, a_2) = p_i \circ \mathscr{A}_1(a_1) \wedge p_i \circ \mathscr{A}_2(a_2)$ for all $(a_1, a_2) \in \mathscr{V}_1 \times \mathscr{V}_2$.
2. $p_i \circ (\mathscr{B}_1 \circ \mathscr{B}_2)((a, a_2)(a, b_2)) = p_i \circ \mathscr{A}_1(a) \wedge p_i \circ \mathscr{B}_2(a_2 b_2)$ for all $a \in \mathscr{V}_1$, for all $a_2 b_2 \in E_2$.
3. $p_i \circ (\mathscr{B}_1 \circ \mathscr{B}_2)((a_1, b)(b_1, b)) = p_i \circ \mathscr{B}_1(a_1 b_1) \wedge p_i \circ \mathscr{A}_2(b)$ for all $b \in \mathscr{V}_2$, for all $a_1 b_1 \in E_1$.
4. $p_i \circ (\mathscr{B}_1 \circ \mathscr{B}_2)((a_1, a_2)(b_1, b_2)) = p_i \circ \mathscr{A}_2(a_2) \wedge p_i \circ \mathscr{A}_2(b_2) \wedge p_i \circ \mathscr{B}_1(a_1 b_1)$ for all $(a_1, a_2)(b_1, b_2) \in E^0 \setminus E$.

Here, $E^0 = E \cup \{(a_1, a_2)(b_1, b_2) : a_1 b_1 \in E_1, a_2 \neq b_2\}$ where E is same as defined in Definition 8.6.

Proposition 8.2 *The composition $\mathscr{G}_1[\mathscr{G}_2]$ of two mPFGs $\mathscr{G}_1$ and $\mathscr{G}_2$ is an mPFG.*

Proof Let $a \in \mathscr{V}_1$, $a_2 b_2 \in E_2$ and $i = 1, 2, \ldots, m$. So

$$
\begin{aligned}
p_i \circ (\mathscr{B}_1 \times \mathscr{B}_2)((a, a_2)(a, b_2)) &= p_i \circ \mathscr{A}_1(a) \wedge p_i \circ \mathscr{B}_2(a_2 b_2) \\
&\leq \wedge \{p_i \circ \mathscr{A}_1(a), p_i \circ \mathscr{A}_2(a_2) \wedge p_i \circ \mathscr{A}_2(b_2)\} \\
&= \wedge \{p_i \circ \mathscr{A}_1(a) \wedge p_i \circ \mathscr{A}_2(a_2), p_i \circ \mathscr{A}_1(a) \wedge p_i \circ \mathscr{A}_2(b_2)\} \\
&= p_i \circ (\mathscr{A}_1 \times \mathscr{A}_2)(a, a_2) \wedge p_i \circ (\mathscr{A}_1 \times \mathscr{A}_2)(a, b_2).
\end{aligned}
$$

Let $b \in \mathscr{V}_2$ and $a_1 b_1 \in E_1$. Then

$$
\begin{aligned}
p_i \circ (\mathscr{B}_1 \times \mathscr{B}_2)((a_1, b)(b_1, b)) &= p_i \circ \mathscr{B}_1(a_1 b_1) \wedge p_i \circ \mathscr{A}_2(b) \\
&\leq \wedge \{p_i \circ \mathscr{A}_1(a_1) \wedge p_i \circ \mathscr{A}_1(b_1), p_i \circ \mathscr{A}_2(b)\} \\
&= \wedge \{p_i \circ \mathscr{A}_1(a_1) \wedge p_i \circ \mathscr{A}_2(b), p_i \circ \mathscr{A}_1(b_1) \wedge p_i \circ \mathscr{A}_2(b)\} \\
&= p_i \circ (\mathscr{A}_1 \times \mathscr{A}_2)(a_1, b) \wedge p_i \circ (\mathscr{A}_1 \times \mathscr{A}_2)(b_1, b).
\end{aligned}
$$

Let $(a_1, a_2)(b_1, b_2) \in E^0 \setminus E$. So $a_1 b_1 \in E_1$ and $a_2 \neq b_2$. So for $i = 1, 2, \ldots, m$,

$$
\begin{aligned}
p_i \circ (\mathscr{B}_1 \circ \mathscr{B}_2)((a_1, a_2)(b_1, b_2)) &= p_i \circ \mathscr{A}_2(a_2) \wedge p_i \circ \mathscr{A}_2(b_2) \wedge p_i \circ \mathscr{B}_1(a_1 b_1) \\
&\leq \wedge \{p_i \circ \mathscr{A}_2(a_2), p_i \circ \mathscr{A}_2(b_2), p_i \circ \mathscr{A}_1(a_1) \wedge p_i \circ \mathscr{A}_1(b_1)\} \\
&= \wedge \{p_i \circ \mathscr{A}_1(a_1) \wedge p_i \circ \mathscr{A}_2(a_2), p_i \circ \mathscr{A}_1(b_1) \wedge p_i \circ \mathscr{A}_2(b_2)\} \\
&= \wedge p_i \circ (\mathscr{A}_1 \times \mathscr{A}_2)(a_1, a_2) \wedge p_i \circ (\mathscr{A}_1 \times \mathscr{A}_2)(b_1, b_2).
\end{aligned}
$$

Hence, $\mathscr{G}_1[\mathscr{G}_2]$ is an mPFG. $\square$

Definition 8.8 Let $\mathscr{A}_1$ and $\mathscr{A}_2$ be *mPF* subsets of $\mathscr{V}_1$ and $\mathscr{V}_2$ and $\mathscr{B}_1$ and $\mathscr{B}_2$ be *mPF* subsets of E_1 and E_2 respectively. Then the **union of two mPFGs** $\mathscr{G}_1 = (\mathscr{V}_1, \mathscr{A}_1, \mathscr{B}_1)$ and $\mathscr{G}_2 = (\mathscr{V}_2, \mathscr{A}_2, \mathscr{B}_2)$ of the graphs $\mathscr{G}_1^* = (\mathscr{V}_1, E_1)$ and $\mathscr{G}_2^* = (\mathscr{V}_2, E_2)$ is denoted by $\mathscr{G}_1 \cup \mathscr{G}_2 = (\mathscr{V}_1 \cup \mathscr{V}_2, \mathscr{A}_1 \cup \mathscr{A}_2, \mathscr{B}_1 \cup \mathscr{B}_2)$ and defined by for $i = 1, 2, \ldots, m$,

1. $p_i \circ (\mathscr{A}_1 \cup \mathscr{A}_2)(a) = \begin{cases} p_i \circ \mathscr{A}_1(a) & \text{if } a \in \mathscr{V}_1 - \mathscr{V}_2 \\ p_i \circ \mathscr{A}_2(a) & \text{if } a \in \mathscr{V}_2 - \mathscr{V}_1 \\ p_i \circ \mathscr{A}_1(a) \vee p_i \circ \mathscr{A}_2(a) & \text{if } a \in \mathscr{V}_1 \cap \mathscr{V}_2. \end{cases}$

$$2. \ p_i \circ (\mathscr{B}_1 \cup \mathscr{B}_2)(ab) = \begin{cases} p_i \circ \mathscr{B}_1(ab) & if \ ab \in E_1 - E_2 \\ p_i \circ \mathscr{B}_2(ab) & if \ ab \in E_2 - E_1 \\ p_i \circ \mathscr{B}_1(ab) \vee p_i \circ \mathscr{B}_2(ab) & if \ ab \in E_1 \cap E_2. \end{cases}$$

Example 8.3 Consider the 3-polar FGs $\mathscr{G}_1$ and $\mathscr{G}_2$, where $\mathscr{V}_1 = \{a_1, a_2, a_3\}$, $E_1 = \{a_1a_2, a_1a_3, a_2a_3\}$, $\mathscr{V}_2 = \{a_1, a_2, a_4\}$ and $E_2 = \{a_1a_2, a_1a_4, a_2a_4\}$ such that
$\mathscr{A}_1(a_1) = (0.3, 0.4, 0.2)$, $\mathscr{A}_1(a_2) = (0.5, 0.4, 0.6)$, $\mathscr{A}_1(a_3) = (0.2, 0.6, 0.3)$,
$\mathscr{B}_1(a_1a_2) = (0.2, 0.4, 0.2)$,
$\quad \mathscr{B}_1(a_1a_3) = (0.1, 0.4, 0.2)$, $\mathscr{B}_1(a_2a_3) = (0.2, 0.4, 0.1)$, $\mathscr{A}_2(a_1)=(0.3, 0.2, 0.5)$,
$\mathscr{A}_2(a_2) = (0.6, 0.2, 0.5)$,
$\quad \mathscr{A}_2(a_4) = (0.5, 0.3, 0.4)$, $\mathscr{B}_2(a_1a_2) = (0.3, 0.2, 0.4)$, $\mathscr{B}_2(a_2a_4)=(0.5, 0.1, 0.2)$,
$\mathscr{B}_2(a_1a_4) = (0.2, 0.2, 0.4)$.
$\quad$ Then $\mathscr{G}_1 \cup \mathscr{G}_2 = (\mathscr{V}, \mathscr{A}_1 \cup \mathscr{A}_2, \mathscr{B}_1 \cup \mathscr{B}_2)$ is a 3-polar FG where $\mathscr{V} = \{a_1, a_2, a_3, a_4\}$,
$(\mathscr{A}_1 \cup \mathscr{A}_2)(a_1) = (0.3, 0.4, 0.5)$, $(\mathscr{A}_1 \cup \mathscr{A}_2)(a_2)=(0.6, 0.4, 0.6)$, $(\mathscr{A}_1 \cup \mathscr{A}_2)(a_3)$
$= (0.2, 0.6, 0.3)$,
$(\mathscr{A}_1 \cup \mathscr{A}_2)(a_4) = (0.5, 0.3, 0.4)$ and $(\mathscr{B}_1 \cup \mathscr{B}_2)(a_1a_2) = (0.3, 0.4, 0.4)$, $(\mathscr{B}_1 \cup \mathscr{B}_2)(a_1a_3) = (0.2, 0.4, 0.1)$,
$(\mathscr{B}_1 \cup \mathscr{B}_2)(a_1a_4) = (0.2, 0.2, 0.4)$, $(\mathscr{B}_1 \cup \mathscr{B}_2)(a_2a_3) = (0.1, 0.4, 0.2)$,
$(\mathscr{B}_1 \cup \mathscr{B}_2)(a_2a_4) = (0.5, 0.1, 0.2)$.

Proposition 8.3 *The union $\mathscr{G}_1 \cup \mathscr{G}_2$ of two mPFGs $\mathscr{G}_1$ and $\mathscr{G}_2$ is an mPFG.*

Proof Let $ab \in E_1 \cap E_2$. So

$$p_i \circ (\mathscr{B}_1 \cup \mathscr{B}_2)(ab) = p_i \circ \mathscr{B}_1(ab) \vee p_i \circ \mathscr{B}_2(ab)$$
$$\leq \vee\{p_i \circ \mathscr{A}_1(a) \wedge p_i \circ \mathscr{A}_1(b), \ p_i \circ \mathscr{A}_2(a) \wedge p_i \circ \mathscr{A}_2(b)\}$$
$$= p_i \circ (\mathscr{A}_1 \cup \mathscr{A}_2)(a) \wedge p_i \circ (\mathscr{A}_1 \cup \mathscr{A}_2)(b).$$

Similarly, for $ab \in E_1 - E_2$,
$p_i \circ (\mathscr{B}_1 \cup \mathscr{B}_2)(ab) \leq p_i \circ (\mathscr{A}_1 \cup \mathscr{A}_2)(a) \wedge p_i \circ (\mathscr{A}_1 \cup \mathscr{A}_2)(b)$.
Also if $ab \in E_2 - E_1$, then
$p_i \circ (\mathscr{B}_1 \cup \mathscr{B}_2)(ab) \leq p_i \circ (\mathscr{A}_1 \cup \mathscr{A}_2)(a) \wedge p_i \circ (\mathscr{A}_1 \cup \mathscr{A}_2)(b)$.
These are true for $i = 1, 2, \ldots, m$. Hence, $\mathscr{G}_1 \cup \mathscr{G}_2$ is an mPFG. $\qquad \square$

Proposition 8.4 *Let $\{\mathscr{G}_\alpha : \alpha \in I\}$ be an arbitrary family of mPFGs with the underlying set $\mathscr{V}$. Then $\bigcap\limits_{\alpha \in I} \mathscr{G}_\alpha$ is an mPFG.*

Proof We have $\bigcap\limits_{\alpha \in I} \mathscr{G}_\alpha = \left(\mathscr{V}, \bigcap\limits_{\alpha \in I} \mathscr{A}_\alpha, \bigcap\limits_{\alpha \in I} \mathscr{B}_\alpha \right)$ such that for $a, b \in \mathscr{V}$,

$$\left(\bigcap_{\alpha \in I} \mathscr{B}_\alpha\right)(ab) = \bigwedge_{\alpha \in I} \mathscr{B}_\alpha(ab)$$

$$\leq \bigwedge_{\alpha \in I}\{\mathscr{A}_\alpha(a) \wedge \mathscr{A}_\alpha(b)\}$$

$$\leq \left(\bigwedge_{\alpha \in I}\{\mathscr{A}_\alpha(a)\} \wedge \left(\bigwedge_{\alpha \in I} \mathscr{A}_\alpha(b)\}\right)\right)$$

$$= \left(\bigcap_{\alpha \in I} \mathscr{A}_\alpha\right)(a) \wedge \left(\bigcap_{\alpha \in I} \mathscr{A}_\alpha\right)(b).$$

Hence, $\bigcap_{\alpha \in I} \mathscr{G}_\alpha$ is an mPFG. $\qquad\qquad\qquad\qquad\qquad\qquad\qquad\qquad\qquad\qquad\qquad\square$

Definition 8.9 Let $\mathscr{A}_1$ and $\mathscr{A}_2$ be *mPF* subsets of $\mathscr{V}_1$ and $\mathscr{V}_2$ and $\mathscr{B}_1$ and $\mathscr{B}_2$ be *mPF* subsets of E_1 and E_2 respectively. Here we assume that $\mathscr{V}_1 \cap \mathscr{V}_2 = \emptyset$. Then the **join of two mPFGs** $\mathscr{G}_1 = (\mathscr{V}_1, \mathscr{A}_1, \mathscr{B}_1)$ and $\mathscr{G}_2 = (\mathscr{V}_2, \mathscr{A}_2, \mathscr{B}_2)$ of the graphs $\mathscr{G}_1^* = (\mathscr{V}_1, E_1)$ and $\mathscr{G}_2^* = (\mathscr{V}_2, E_2)$ is denoted by $\mathscr{G}_1 + \mathscr{G}_2 = (\mathscr{V}_1 \cup \mathscr{V}_2, \mathscr{A}_1 \cup \mathscr{A}_2, \mathscr{B}_1 \cup \mathscr{B}_2)$ and defined by for $i = 1, 2, \ldots, m$

1. $p_i \circ (\mathscr{A}_1 + \mathscr{A}_2)(a) = p_i \circ (\mathscr{A}_1 \cup \mathscr{A}_2)(a)$ if $a \in \mathscr{V}_1 \cup \mathscr{V}_2$,
2. $p_i \circ (\mathscr{B}_1 + \mathscr{B}_2)(ab) = p_i \circ (\mathscr{B}_1 \cup \mathscr{B}_2)(ab)$ if $ab \in E_1 \cup E_2$,
3. $p_i \circ (\mathscr{B}_1 + \mathscr{B}_2)(ab) = p_i \circ \mathscr{A}_1(a) \wedge p_i \circ \mathscr{A}_2(b)$ if $ab \in E$, where E is the set of all edges joining the vertices of $\mathscr{V}_1$ and $\mathscr{V}_2$.

Proposition 8.5 *The join* $\mathscr{G}_1 + \mathscr{G}_2$ *of two mPFGs* $\mathscr{G}_1$ *and* $\mathscr{G}_2$ *is an mPFG.*

Proof If $ab \in E_1 \cup E_2$, then the result follows from Proposition 8.3.
If $ab \in E$, then

$$p_i \circ (\mathscr{B}_1 + \mathscr{B}_2)(ab) = p_i \circ \mathscr{A}_1(a) \wedge p_i \circ \mathscr{A}_2(b)$$
$$\leq p_i \circ (\mathscr{A}_1 \cup \mathscr{A}_2)(a) \wedge p_i \circ (\mathscr{A}_1 \cup \mathscr{A}_2)(b)$$
$$= p_i \circ (\mathscr{A}_1 + \mathscr{A}_2)(a) \wedge p_i \circ (\mathscr{A}_1 + \mathscr{A}_2)(b).$$

Therefore, $\mathscr{G}_1 + \mathscr{G}_2$ is an mPFG. $\qquad\qquad\qquad\qquad\qquad\qquad\qquad\qquad\qquad\qquad\square$

Proposition 8.6 *Let* $\mathscr{G}_1^* = (\mathscr{V}_1, E_1)$ *and* $\mathscr{G}_2^* = (\mathscr{V}_2, E_2)$ *be crisp graphs and let* $\mathscr{V}_1 \cap \mathscr{V}_2 = \emptyset$. *Let* $\mathscr{A}_1, \mathscr{A}_2, \mathscr{B}_1$ *and* $\mathscr{B}_2$ *be mPF subsets of* $\mathscr{V}_1, \mathscr{V}_2, E_1$ *and* E_2 *respectively. Then* $\mathscr{G}_1 \cup \mathscr{G}_2 = (\mathscr{V}_1 \cup \mathscr{V}_2, \mathscr{A}_1 \cup \mathscr{A}_2, \mathscr{B}_1 \cup \mathscr{B}_2)$ *is an mPFG of* $\mathscr{G}_1^* \cup \mathscr{G}_2^*$ *if and only if* $\mathscr{G}_1 = (\mathscr{V}_1, \mathscr{A}_1, \mathscr{B}_1)$ *and* $\mathscr{G}_2 = (\mathscr{V}_2, \mathscr{A}_2, \mathscr{B}_2)$ *are mPFGs of* $\mathscr{G}_1^*$ *and* $\mathscr{G}_2^*$ *respectively.*

Proof Suppose $\mathscr{G}_1 \cup \mathscr{G}_2 = (\mathscr{V}_1 \cup \mathscr{V}_2, \mathscr{A}_1 \cup \mathscr{A}_2, \mathscr{B}_1 \cup \mathscr{B}_2)$ is an mPFG.
Let $xy \in E_1$. Then $xy \notin E_2$ and $x, y \in \mathscr{V}_1 \setminus \mathscr{V}_2$. Thus, for $i = 1, 2, \ldots, m$

$$p_i \circ \mathscr{B}_1(ab) = p_i \circ (\mathscr{B}_1 \cup \mathscr{B}_2)(ab)$$
$$\leq p_i \circ (\mathscr{A}_1 \cup \mathscr{A}_2)(a) \wedge p_i \circ (\mathscr{A}_1 \cup \mathscr{A}_2)(b)$$
$$\leq p_i \circ \mathscr{A}_1(a) \wedge p_i \circ \mathscr{A}_1(b).$$

Therefore, $\mathscr{G}_1 = (\mathscr{V}_1, \mathscr{A}_1, \mathscr{B}_1)$ is an mPFG. Similarly, we can show that $\mathscr{G}_2 = (\mathscr{V}_2, \mathscr{A}_2, \mathscr{B}_2)$ is an mPFG.

The converse follows by Proposition 8.3. □

As a consequence of Propositions 8.5 and 8.6, we have the following result.

Proposition 8.7 *Let* $\mathscr{G}_1^* = (\mathscr{V}_1, E_1)$ *and* $\mathscr{G}_2^* = (\mathscr{V}_2, E_2)$ *be crisp graphs and let* $\mathscr{V}_1 \cap \mathscr{V}_2 = \emptyset$. *Let* $\mathscr{A}_1, \mathscr{A}_2, \mathscr{B}_1$ *and* $\mathscr{B}_2$ *be mPF subsets of* $\mathscr{V}_1, \mathscr{V}_2, E_1$ *and* E_2 *respectively.* *Then* $\mathscr{G}_1 + \mathscr{G}_2 = (\mathscr{V}_1 \cup \mathscr{V}_2, \mathscr{A}_1 + \mathscr{A}_2, \mathscr{B}_1 + \mathscr{B}_2)$ *is an mPFG of* $\mathscr{G}_1^* + \mathscr{G}_2^*$ *if and only if* $\mathscr{G}_1 = (\mathscr{V}_1, \mathscr{A}_1, \mathscr{B}_1)$ *and* $\mathscr{G}_2 = (\mathscr{V}_2, \mathscr{A}_2, \mathscr{B}_2)$ *are mPFGs of* $\mathscr{G}_1^*$ *and* $\mathscr{G}_2^*$ *respectively.*

8.2 Isomorphisms of *m*-Polar Fuzzy Graphs

The concept of different types of isomorphisms of mPFGs are defined in this section.

Definition 8.10 Let $\mathscr{G}_1 = (\mathscr{V}_1, \mathscr{A}_1, \mathscr{B}_1)$ and $\mathscr{G}_2 = (\mathscr{V}_2, \mathscr{A}_2, \mathscr{B}_2)$ be two mPFGs. A **homomorphism** between $\mathscr{G}_1$ and $\mathscr{G}_2$ is a mapping $\psi : \mathscr{V}_1 \rightarrow \mathscr{V}_2$ satisfying the following $(i = 1, 2, \ldots, m)$

1. $p_i \circ \mathscr{A}_1(a_1) \leq p_i \circ \mathscr{A}_2(\psi(a_1))$ for all $a_1 \in \mathscr{V}_1$,
2. $p_i \circ \mathscr{B}_1(a_1 b_1) \leq p_i \circ \mathscr{B}_2(\psi(a_1)\psi(b_1))$ for all $a_1, b_1 \in \mathscr{V}_1$.

Definition 8.11 Let $\mathscr{G}_1 = (\mathscr{V}_1, \mathscr{A}_1, \mathscr{B}_1)$ and $\mathscr{G}_2 = (\mathscr{V}_2, \mathscr{A}_2, \mathscr{B}_2)$ be two mPFGs. An **Isomorphism of *m*-polar fuzzy graph** between $\mathscr{G}_1$ and $\mathscr{G}_2$ is a bijective mapping $\psi : \mathscr{V}_1 \rightarrow \mathscr{V}_2$ satisfying the following $(i = 1, 2, \ldots, m)$

1. $p_i \circ \mathscr{A}_1(a_1) = p_i \circ \mathscr{A}_2(\psi(a_1))$ for all $a_1 \in \mathscr{V}_1$,
2. $p_i \circ \mathscr{B}_1(a_1 b_1) = p_i \circ \mathscr{B}_2(\psi(a_1)\psi(b_1))$ for all $a_1 b_1 \in \mathscr{V}_1$.
   In this case, we write $\mathscr{G}_1 \cong \mathscr{G}_2$.

Definition 8.12 Let $\mathscr{G}_1 = (\mathscr{V}_1, \mathscr{A}_1, \mathscr{B}_1)$ and $\mathscr{G}_2 = (\mathscr{V}_2, \mathscr{A}_2, \mathscr{B}_2)$ be two mPFGs of the graphs $\mathscr{G}_1^* = (\mathscr{V}_1, E_1)$ and $\mathscr{G}_2^* = (\mathscr{V}_2, E_2)$ respectively. A **weak isomorphism** between $\mathscr{G}_1$ and $\mathscr{G}_2$ is a bijective mapping $\psi : \mathscr{V}_1 \rightarrow \mathscr{V}_2$ satisfying the following $(i = 1, 2, \ldots, m)$

1. ψ is a homomorphism, and
2. $p_i \circ \mathscr{A}_1(a_1) = p_i \circ \mathscr{A}_2(\psi(a_1))$ for all $a_1 \in \mathscr{V}_1$.
   From this definition, it follows that weak isomorphism preserves the weight of the vertices but not necessarily the weights of the edges.

Example 8.4 Let $\mathscr{V}_1 = \{a_1, b_1\}$, $\mathscr{V}_2 = \{c_1, d_1\}$, $E_1 = \{a_1 b_1\}$ and $E_2 = \{c_1 d_1\}$. Consider the 3-polar FGs $\mathscr{G}_1$ and $\mathscr{G}_2$ of the graphs $\mathscr{G}_1^* = (\mathscr{V}_1, E_1)$ and $\mathscr{G}_2^* = (\mathscr{V}_2, E_2)$ respectively defined as below:

$\mathscr{A}_1(a_1) = (0.3, 0.5, 0.6)$, $\mathscr{A}_1(b_1) = (0.4, 0.6, 0.8)$, $\mathscr{A}_2(c_1) = (0.4, 0.6, 0.8)$, $\mathscr{A}_2(d_1) = (0.3, 0.5, 0.6)$, $\mathscr{B}_1(a_1 b_1) = (0.2, 0.5, 0.4)$, $\mathscr{B}_2(c_1 d_1) = (0.3, 0.5, 0.5)$.

Define a mapping $\psi : \mathscr{V}_1 \rightarrow \mathscr{V}_2$ by $\psi(a_1) = d_1$, $\psi(b_1) = c_1$. Then we have

$p_1 \circ \mathscr{A}_1(a_1) = p_1 \circ \mathscr{A}_2(d_1), \quad p_2 \circ \mathscr{A}_1(a_1) = p_2 \circ \mathscr{A}_2(d_1), \quad p_3 \circ \mathscr{A}_1(a_1) = p_3 \circ \mathscr{A}_2(d_1),$

$p_1 \circ \mathscr{A}_1(b_1) = p_1 \circ \mathscr{A}_2(c_1), \quad p_2 \circ \mathscr{A}_1(b_1) = p_2 \circ \mathscr{A}_2(c_1), \quad p_3 \circ \mathscr{A}_1(b_1) = p_3 \circ \mathscr{A}_2(c_1).$

$p_1 \circ \mathscr{B}_1(a_1b_1) = 0.2 < 0.3 = p_1 \circ \mathscr{B}_2(c_1d_1), \qquad p_2 \circ \mathscr{B}_1(a_1b_1) = 0.5 = p_2 \circ \mathscr{B}_2(c_1d_1),$

$p_3 \circ \mathscr{B}_1(a_1b_1) = 0.4 < 0.5 = p_3 \circ \mathscr{B}_2(c_1d_1).$

Hence, $\mathscr{B}_1(a_1b_1) \neq \mathscr{B}_2(\psi(a_1)\psi(b_1)).$

Therefore, ψ is a weak isomorphism but not an isomorphism.

Definition 8.13 Let $\mathscr{G}_1 = (\mathscr{V}_1, \mathscr{A}_1, \mathscr{B}_1)$ and $\mathscr{G}_2 = (\mathscr{V}_2, \mathscr{A}_2, \mathscr{B}_2)$ be two mPFGs. A **co-weak isomorphism** between $\mathscr{G}_1$ and $\mathscr{G}_2$ is a bijective mapping $\psi : \mathscr{V}_1 \to \mathscr{V}_2$ satisfying the following $(i = 1, 2, \ldots, m)$:

1. ψ is a homomorphism,
2. $p_i \circ \mathscr{B}_1(a_1b_1) = p_i \circ \mathscr{B}_2(\psi(a_1b_1))$ for all $a_1, b_1 \in \mathscr{V}_1$.

   From this definition, it follows that a co-weak isomorphism preserves the weight of the edges but not necessarily the weights of the vertices.

Example 8.5 Let $\mathscr{V}_1 = \{a_1, b_1\}$, $\mathscr{V}_2 = \{c_1, d_1\}$, $E_1 = \{a_1b_1\}$ and $E_2 = \{c_1d_1\}$. Consider the 3-polar FGs $\mathscr{G}_1$ and $\mathscr{G}_2$ of the graphs $\mathscr{G}_1^* = (\mathscr{V}_1, E_1)$ and $\mathscr{G}_2^* = (\mathscr{V}_2, E_2)$ respectively defined as below:

$\mathscr{A}_1(a_1) = (0.3, 0.5, 0.6), \mathscr{A}_1(b_1) = (0.4, 0.6, 0.8), \mathscr{A}_2(c_1) = (0.5, 0.6, 0.7),$

$\mathscr{A}_2(d_1) = (0.4, 0.7, 0.6), \mathscr{B}_1(a_1b_1) = (0.2, 0.5, 0.3), \mathscr{B}_2(c_1d_1) = (0.2, 0.5, 0.3).$

Let us define a mapping $\psi : \mathscr{V}_1 \to \mathscr{V}_2$ by $\psi(a_1) = d_1, \psi(b_1) = c_1$. Then we have

$p_1 \circ \mathscr{A}_1(a_1) = 0.3 < 0.4 = p_1 \circ \mathscr{A}_2(d_1), \quad p_2 \circ \mathscr{A}_1(a_1) = 0.5 < 0.7 = p_2 \circ \mathscr{A}_2(d_1),$

$p_3 \circ \mathscr{A}_1(a_1) = 0.6 = p_3 \circ \mathscr{A}_2(d_1)$. Therefore, $\mathscr{A}_1(a_1) \neq \mathscr{A}_2(\psi(a_1))$.

Similarly, we can show that $\mathscr{A}_1(b_1) \neq \mathscr{A}_2(\psi(b_1))$.

But, $p_1 \circ \mathscr{B}_1(a_1b_1) = p_1 \circ \mathscr{B}_2(c_1d_1), \quad p_2 \circ \mathscr{B}_1(a_1b_1) = p_2 \circ \mathscr{B}_2(c_1d_1), \quad p_3 \circ \mathscr{B}_1(a_1b_1) = p_3 \circ \mathscr{B}_2(c_1d_1)$. Therefore, $\mathscr{B}_1(a_1b_1) = \mathscr{B}_2(\psi(a_1b_1))$. Therefore, the map ψ is a co-weak isomorphism but not an isomorphism.

We now make some definitions and observations.

1. If $\mathscr{G}_1 = \mathscr{G}_2 = \mathscr{G}$, then the **homomorphism** ψ of $\mathscr{G}$ into $\mathscr{G}$ is called an **endomorphism**. An isomorphism ψ over $\mathscr{G}$ is called an **automorphism**.
2. If $\mathscr{G}_1 = \mathscr{G}_2$, then the weak and co-weak isomorphisms actually become isomorphic.
3. Let $\mathscr{G}$ be an mPFG and $Aut(\mathscr{G})$ be the set of all automorphisms of $\mathscr{G}$. Let $\psi : \mathscr{V} \to \mathscr{V}$ be the identity mapping $\psi(x) = x$ for all $x \in \mathscr{V}$. Then clearly, $\psi \in Aut(\mathscr{G})$.

Definition 8.14 An *mPF* set $\mathscr{A}$ in a semigroup H is called an *m*-**polar fuzzy subsemigroup** of H if it satisfies

$$p_i \circ \mathscr{A}(ab) \geq p_i \circ \mathscr{A}(a) \wedge p_i \circ \mathscr{A}(b), \text{ for all } a, b \in H \ (i = 1, 2, \ldots, m).$$

An *mPF* set $\mathscr{A}$ in a group G is called a *m*-**polar subgroup** of G if it is a *m*-polar fuzzy sub-semigroup of G and satisfies

$$p_i \circ \mathscr{A}(a^{-1}) = p_i \circ \mathscr{A}(a), \text{ for all } a \in G \ (i = 1, 2, \ldots, m).$$

We state some propositions below without their proofs.

Proposition 8.8 *Let* $Aut(\mathscr{G})$ *be the set of all automorphisms of the mPFG* $\mathscr{G}$. *Then* $(Aut(\mathscr{G}), \circ)$ *forms a group.*

Proposition 8.9 *Let* ψ_1 *is a weak isomorphism of* $\mathscr{G}_1$ *onto* $\mathscr{G}_2$ *and* ψ_2 *is a weak isomorphism of* $\mathscr{G}_2$ *onto* $\mathscr{G}_3$ *where* $\mathscr{G}_j$, *(j = 1, 2, 3) are mPFGs. Then* $\psi_1 \circ \psi_2$ *is a weak isomorphism of* $\mathscr{G}_1$ *onto* $\mathscr{G}_3$.

8.3   Strong *m*-Polar Fuzzy Graphs

Here, we introduce strong mPFGs, **complement of mPFGs** and self-complementary mPFGs.

Definition 8.15 An *mPF* $\mathscr{G} = (\mathscr{V}, \mathscr{A}, \mathscr{B})$ is said to be strong if

$$p_i \circ \mathscr{B}(ab) = p_i \circ \mathscr{A}(a) \wedge p_i \circ \mathscr{A}(b) \text{ for all } ab \in E, (i = 1, 2, \ldots, m).$$

Example 8.6 Consider a graph $G^* = (\mathscr{V}, E)$ such that $\mathscr{V} = \{a, b, c\}$, $E = \{ab, bc, ca\}$. Let $\mathscr{A}$ be a 3-polar fuzzy set of $\mathscr{V}$ and $\mathscr{B}$ be a 3-polar fuzzy set of E defined by the following:
$\mathscr{A}(a) = (0.3, 0.5, 0.6)$, $\mathscr{A}(b) = (0.4, 0.6, 0.7)$, $\mathscr{A}(c) = (0.5, 0.4, 0.2)$, $\mathscr{B}(ab) = (0.3, 0.5, 0.6)$, $\mathscr{B}(bc) = (0.4, 0.4, 0.2)$, $\mathscr{B}(ca) = (0.3, 0.4, 0.2)$. Then we can easily check that $\mathscr{G} = (\mathscr{V}, \mathscr{A}, \mathscr{B})$ is a strong 3-polar fuzzy graph.

Proposition 8.10 *Let* $\mathscr{G}_1$ *and* $\mathscr{G}_2$ *be strong mPFGs. Then* $\mathscr{G}_1 \times \mathscr{G}_2$, $\mathscr{G}_1[\mathscr{G}_2]$ *and* $\mathscr{G}_1 + \mathscr{G}_2$ *are strong mPFGs.*

Proof The proof of this proposition follows from the Propositions 8.1, 8.2 and 8.5. ☐

We now give an example to show that the union of two strong mPFGs is not necessarily a strong mPFG.

Example 8.7 Let $\mathscr{V}_1 = \{a_1, b_1, c_1\} = \mathscr{V}_2$, $E_1 = \{a_1 b_1, b_1 c_1\}$, and $E_2 = \{b_1 c_1, c_1 a_1\}$. Let $\mathscr{G}_1$ and $\mathscr{G}_2$ be 3-polar fuzzy graphs with underlying vertices $\mathscr{V}_1$ and $\mathscr{V}_2$ respectively and they are defined as follows:
$\mathscr{A}_1(a_1) = (0.4, 0.6, 0.8)$, $\mathscr{A}_1(b_1) = (0.5, 0.8, 0.9)$, $\mathscr{A}_1(c_1) = (0.3, 0.5, 0.6)$, $\mathscr{B}_1(a_1 b_1) = (0.4, 0.6, 0.8)$,

$\mathscr{B}_1(b_1c_1) = (0.3, 0.5, 0.6), \quad \mathscr{A}_2(a_1) = (0.3, 0.5, 0.3), \mathscr{A}_2(b_1) = (0.7, 0.4, 0.5),$
$\mathscr{A}_2(c_1) = (0.8, 0.6, 0.1), \mathscr{B}_2(b_1c_1) = (0.7, 0.4, 0.5), \mathscr{B}_1(c_1a_1) = (0.3, 0.5, 0.3).$
Now

$$(\mathscr{A}_1 \cup \mathscr{A}_2)(c_1) \wedge (\mathscr{A}_1 \cup \mathscr{A}_2)(a_1) = (0.4, 0.6, 0.8) \neq (0.3, 0.5, 0.3) = (\mathscr{B}_1 \cup \mathscr{B}_2)(b_1c_1).$$

Hence, $\mathscr{G}_1 \cup \mathscr{G}_2$ is not strong.

Proposition 8.11 Let $\mathscr{G}_1 \times \mathscr{G}_2$ be strong mPFG. Then at least $\mathscr{G}_1$ or $\mathscr{G}_2$ must be strong.

Proof Suppose that both $\mathscr{G}_1$ and $\mathscr{G}_2$ are not strong mPFGs. Then there exists at least one $a_1b_1 \in E_1$ and at least one $a_2b_2 \in E_2$ such that

1. $\mathscr{B}_1(a_1b_1) < \mathscr{A}_1(a_1) \wedge \mathscr{A}_1(b_1)$, and $\mathscr{B}_2(a_2b_2) < \mathscr{A}_2(a_2) \wedge \mathscr{A}_2(b_2)$.
   Without loss of generality, we assume that
2. $\mathscr{B}_2(a_2b_2) \leq \mathscr{B}_1(a_1b_1) < \mathscr{A}_1(a_1) \wedge \mathscr{A}_1(b_1) \leq \mathscr{A}_1(a_1)$.

Let $E = \{(a, a_2)(a, b_2) : a \in \mathscr{V}_1, a_2b_2 \in E_2\} \cup \{(a_1, z)(b_1, z) : z \in \mathscr{V}_2, a_1b_1 \in E_1\}$.
Consider $(a, a_2)(a, b_2) \in E$. Then, by definition of $\mathscr{G}_1 \times \mathscr{G}_2$ and inequality (i), we have,
$(\mathscr{B}_1 \times \mathscr{B}_2)((a, a_2)(a, b_2)) = \mathscr{A}_1(a) \wedge \mathscr{B}_2(a_2b_2) < \mathscr{A}_1(a) \wedge \mathscr{A}_2(a_2) \wedge \mathscr{A}_2(b_2)$
and $(\mathscr{A}_1 \times \mathscr{A}_2)(a, a_2) = \mathscr{A}_1(a) \wedge \mathscr{A}_2(a_2), (\mathscr{A}_1 \times \mathscr{A}_2)(a, b_2) = \mathscr{A}_1(a) \wedge \mathscr{A}_2(b_2)$.
Thus, $(\mathscr{A}_1 \times \mathscr{A}_2)(a, a_2) \wedge (\mathscr{A}_1 \times \mathscr{A}_2)(a, b_2) = \mathscr{A}_1(a) \wedge \mathscr{A}_2(a_2) \wedge \mathscr{A}_2(b_2)$.
Hence, $(\mathscr{B}_1 \times \mathscr{B}_2)((a, a_2)(a, b_2)) = \mathscr{A}_1(a) \wedge \mathscr{B}_2(a_2b_2) < (\mathscr{A}_1 \times \mathscr{A}_2)(a, a_2) \wedge (\mathscr{A}_1 \times \mathscr{A}_2)(a, b_2)$
That is, $\mathscr{G}_1 \times \mathscr{G}_2$ is not strong mPFG, which is a contradiction.
Therefore, at least $\mathscr{G}_1$ or $\mathscr{G}_2$ must be strong mPFG if $\mathscr{G}_1 \times \mathscr{G}_2$ is strong.   □

Proposition 8.12 Let $\mathscr{G}_1[\mathscr{G}_2]$ is strong mPFG. Then at least $\mathscr{G}_1$ or $\mathscr{G}_2$ must be strong.

Proof The proof is same as above.   □

Proposition 8.13 Let $\mathscr{G} = (\mathscr{V}, \mathscr{A}, \mathscr{B})$ be a strong mPFG of a graph $G^* = (\mathscr{V}, E)$. If $\overline{\mathscr{G}} = (\mathscr{V}, \overline{\mathscr{A}}, \overline{\mathscr{B}})$ satisfies $\overline{\mathscr{A}} = \mathscr{A}$ and $\overline{\mathscr{B}}$ defined by for all $ab \in \Omega$, $(i=1, 2, \ldots, m)$
$$p_i \circ \overline{\mathscr{B}}(ab) = \begin{cases} 0 & if \ 0 < p_i \circ \mathscr{B}(ab) \leq 1 \\ p_i \circ \mathscr{A}(a) \wedge p_i \circ \mathscr{A}(b) \ if & p_i \circ \mathscr{B}(ab) = 0. \end{cases}$$
Then $\overline{\mathscr{G}}$ is a strong mPFG of $\overline{\mathscr{G}^*} = (\mathscr{V}, \Omega \setminus E)$.

Proof The mPF sets $\overline{\mathscr{A}}$ and $\overline{\mathscr{B}}$ satisfy $p_i \circ \overline{\mathscr{B}}(ab) \leq p_i \circ \mathscr{A}(a) \wedge p_i \circ \mathscr{A}(b)$ for all $ab \in \Omega, (i = 1, 2, \ldots, m)$.
Now, let $ab \in \Omega - (\Omega - E) = E$.
As $\mathscr{G}$ is strong mPFG, therefore $p_i \circ \mathscr{B}(ab) = p_i \circ \mathscr{A}(a) \wedge p_i \circ \mathscr{A}(b), (i = 1, 2, \ldots, m)$.
If $\mathscr{B}(ab) = 0$, then $p_i \circ \mathscr{B}(ab) = 0, (i = 1, 2, \ldots, m)$.
Therefore, $p_i \circ \overline{\mathscr{B}}(ab) = p_i \circ \mathscr{A}(a) \wedge p_i \circ \mathscr{A}(b) = p_i \circ \mathscr{B}(ab) = 0,$
$(i = 1, 2, \ldots, m)$.
Hence, $\overline{\mathscr{B}}(ab) = 0$.

If $0 < p_i \circ \mathscr{B}(ab) \leq 1$, $(i = 1, 2, \ldots, m)$ then $p_i \circ \overline{\mathscr{B}}(ab) = 0$, i.e. $\overline{\mathscr{B}}(ab) = 0$.
Hence, for all $ab \in \Omega - (\Omega - E) = E$, $\overline{\mathscr{B}}(ab) = 0$.
Therefore, $\overline{\mathscr{G}} = (\mathscr{V}, \overline{\mathscr{A}}, \overline{\mathscr{B}})$ is an mPFG of $G^* = (\mathscr{V}, \Omega \setminus E)$.
On the other hand, for all $ab \in \Omega - E$, we have by Definition 8.4, $\mathscr{B}(ab) = 0$,
i.e. $p_i \circ \mathscr{B}(ab) = 0$, $(i = 1, 2, \ldots, m)$.
Then we have $p_i \circ \overline{\mathscr{B}}(ab) = p_i \circ \mathscr{A}(a) \wedge p_i \circ \mathscr{A}(b)$, $(i = 1, 2, \ldots, m)$.
So, $\overline{\mathscr{G}}$ is a strong mPFG of $G^* = (\mathscr{V}, \Omega \setminus E)$.   □

Definition 8.16 The **strong mPFG** $\overline{\mathscr{G}} = (\mathscr{V}, \overline{\mathscr{A}}, \overline{\mathscr{B}})$ defined above, is called the complement of the strong mPFG $\mathscr{G} = (\mathscr{V}, \mathscr{A}, \mathscr{B})$.

Definition 8.17 A strong mPFG $\mathscr{G}$ is called **self-complementary** if $\mathscr{G} \cong \overline{\mathscr{G}}$.

In the following, we give an example of an mPFG which is self-complementary strong.

Example 8.8 Let $G^* = (\mathscr{V}, E)$ be a graph, where $\mathscr{V} = \{a, b, c, d\}$ and $E = \{ab, ac, cd\}$ and $\mathscr{G} = (\mathscr{V}, \mathscr{A}, \mathscr{B})$ be a strong 3-polar fuzzy graph of G^*, where $\mathscr{A}(a) = (0.1, 0.2, 0.3)$, $\mathscr{A}(b) = (0.1, 0.2, 0.3)$, $\mathscr{A}(c) = (0.1, 0.2, 0.3)$,
$\mathscr{A}(d) = (0.1, 0.2, 0.3)$, $\mathscr{B}(ab) = (0.1, 0.2, 0.3)$, $\mathscr{B}(ac) = (0.1, 0.2, 0.3)$,
$\mathscr{B}(cd) = (0.1, 0.2, 0.3)$.
Let $\overline{\mathscr{G}} = (\mathscr{V}, \overline{\mathscr{A}}, \overline{\mathscr{B}})$ be the complement of $\mathscr{G}$, where $\overline{\mathscr{A}} = \mathscr{A}$, $\overline{\mathscr{B}}(bd) = (0.1, 0.2, 0.3)$, $\overline{\mathscr{B}}(ad) = (0.1, 0.2, 0.3)$, $\overline{\mathscr{B}}(bc) = (0.1, 0.2, 0.3)$.
Let us now define a mapping $\psi : \mathscr{V} \to \mathscr{V}$ by $\psi(a) = b$, $\psi(b) = c$, $\psi(c) = d$, $\psi(d) = a$.
Then clearly, ψ is a bijective mapping and
$\mathscr{A}(a) = \overline{\mathscr{A}}(\psi(a))$, $\mathscr{A}(b) = \overline{\mathscr{A}}(\psi(b))$, $\mathscr{A}(c) = \overline{\mathscr{A}}(\psi(c))$, $\mathscr{A}(d) = \overline{\mathscr{A}}(\psi(d))$,
$\mathscr{B}(ab) = (0.1, 0.2, 0.3) = \overline{\mathscr{B}}(\psi(a)\psi(b))$, $\mathscr{B}(ac) = (0.1, 0.2, 0.3) = \overline{\mathscr{B}}(\psi(a)\psi(c))$,
$\mathscr{B}(cd) = (0.1, 0.2, 0.3) = \overline{\mathscr{B}}(\psi(c)\psi(d))$.
Hence, ψ is an isomorphism from $\mathscr{G}$ onto $\overline{\mathscr{G}}$, i.e. $\mathscr{G} \cong \overline{\mathscr{G}}$. So $\mathscr{G}$ is self-complementary.

Proposition 8.14 Let $\mathscr{G} = (\mathscr{V}, \mathscr{A}, \mathscr{B})$ be a strong mPFG of the graph $G^* = (\mathscr{V}, E)$ and $\overline{\mathscr{G}} = (\mathscr{V}, \overline{\mathscr{A}}, \overline{\mathscr{B}})$ be the complement of $\mathscr{G}$. Then $p_i \circ \overline{\mathscr{B}}(ab) = p_i \circ \mathscr{A}(a) \wedge p_i \circ \mathscr{A}(b) - p_i \circ \mathscr{B}(ab)$ for all $ab \in \Omega$, $(i = 1, 2, \ldots, m)$.

Proof Let $ab \in \Omega$. If $0 < p_i \circ \mathscr{B}(ab) \leq 1$, $(i = 1, 2, \ldots, m)$, then $ab \in E$ by Definition 8.4.
As $\mathscr{G}$ is strong, $p_i \circ \mathscr{A}(a) \wedge p_i \circ \mathscr{A}(b) - p_i \circ \mathscr{B}(ab) = 0 = p_i \circ \overline{\mathscr{B}}(ab)$, $(i = 1, 2, \ldots, m)$.
If $p_i \circ \mathscr{B}(ab) = 0$, $(i = 1, 2, \ldots, m)$ then

$$p_i \circ \mathscr{A}(a) \wedge p_i \circ \mathscr{A}(b) - p_i \circ \mathscr{B}(ab) = p_i \circ \mathscr{A}(a) \wedge p_i \circ \mathscr{A}(b)$$
$$= p_i \circ \overline{\mathscr{B}}(ab).$$

Hence the result.   □

Proposition 8.15 *Let $\mathscr{G}$ be a self-complementary strong mPFG. Then for all $ab \in \Omega$,* $(i = 1, 2, \ldots, m)$

$$\sum_{a \neq b} p_i \circ \mathscr{B}(ab) = \frac{1}{2} \sum_{a \neq b} p_i \circ \mathscr{A}(a) \wedge p_i \circ \mathscr{A}(b).$$

Proof Let $\mathscr{G} = (\mathscr{V}, \mathscr{A}, \mathscr{B})$ be a self-complementary strong mPFG. Then for all $ab \in E$, $(i = 1, 2, \ldots, m)$, $p_i \circ \mathscr{B}(ab) = p_i \circ \mathscr{A}(a) \wedge p_i \circ \mathscr{A}(b)$ and there exists an isomorphism $\psi : \mathscr{G} \to \overline{\mathscr{G}}$ such that
$p_i \circ \mathscr{A}(a) = p_i \circ \overline{\mathscr{A}}(a)$ for all $a \in \mathscr{V}$ and $p_i \circ \mathscr{B}(ab) = p_i \circ \overline{\mathscr{B}}(\psi(a)\psi(b))$ for all $ab \in \Omega$.

Let $ab \in \Omega$. Then by Proposition 8.14, we have for $i = 1, 2, \ldots, m$,
$p_i \circ \overline{\mathscr{B}}(\psi(a)\psi(b)) = p_i \circ \mathscr{A}(\psi(a)) \wedge p_i \circ \mathscr{A}(\psi(b)) - p_i \circ \mathscr{B}(\psi(a)\psi(b))$
i.e. $p_i \circ \mathscr{B}(ab) = p_i \circ \mathscr{A}(\psi(a)) \wedge p_i \circ \mathscr{A}(\psi(b)) - p_i \circ \mathscr{B}(\psi(a)\psi(b))$.
Therefore,

$$\sum_{a \neq b} p_i \circ \mathscr{B}(ab) + \sum_{a \neq b} p_i \circ \mathscr{B}(\psi(a)\psi(b)) = \sum_{a \neq b} p_i \circ \mathscr{A}(\psi(a)) \wedge p_i \circ \mathscr{A}(\psi(b)) = \sum_{a \neq b} p_i \circ \mathscr{A}(a) \wedge p_i \circ \mathscr{A}(b),$$

i.e.

$$2 \sum_{a \neq b} p_i \circ \mathscr{B}(ab) = \sum_{a \neq b} p_i \circ \mathscr{A}(a) \wedge p_i \circ \mathscr{A}(b),$$

This gives,

$$\sum_{a \neq b} p_i \circ \mathscr{B}(ab) = \frac{1}{2} \sum_{a \neq b} p_i \circ \mathscr{A}(a) \wedge p_i \circ \mathscr{A}(b).$$

$\square$

Proposition 8.16 *Let $\mathscr{G} = (\mathscr{V}, \mathscr{A}, \mathscr{B})$ be a strong mPFG of $G^* = (\mathscr{V}, E)$ and $p_i \circ \mathscr{B}(ab) = \frac{1}{2} p_i \circ \mathscr{A}(a) \wedge p_i \circ \mathscr{A}(b)$ for all $ab \in \Omega$, $(i = 1, 2, \ldots, m)$. Then $\mathscr{G}$ is self-complementary.*

Proof If $\mathscr{G} = (\mathscr{V}, \mathscr{A}, \mathscr{B})$ is a strong mPFG satisfies $p_i \circ \mathscr{B}(ab) = \frac{1}{2} p_i \circ \mathscr{A}(a) \wedge p_i \circ \mathscr{A}(b)$ for all $ab \in \Omega$, $(i = 1, 2, \ldots, m)$, then the identity mapping $I : \mathscr{V} \to \mathscr{V}$ is an isomorphism from $\mathscr{G}$ to $\overline{\mathscr{G}}$.

Clearly, I satisfies the first condition for isomorphism, i.e. $\mathscr{A}(a) = \overline{\mathscr{A}}(I(a))$ for all $a \in \mathscr{V}$ and by Proposition 8.14, we have for all $ab \in \Omega$, $i = 1, 2, \ldots, m$,
$p_i \circ \overline{\mathscr{B}}(I(a)(b)) = p_i \circ \overline{\mathscr{B}}(ab) = p_i \circ \mathscr{A}(a) \wedge p_i \circ \mathscr{A}(b) - p_i \circ \mathscr{B}(ab) = p_i \circ \mathscr{A}(a) \wedge p_i \circ \mathscr{A}(b) - \frac{1}{2} p_i \circ \mathscr{A}(a) \wedge p_i \circ \mathscr{A}(b) = \frac{1}{2} p_i \circ \mathscr{A}(a) \wedge p_i \circ \mathscr{A}(b) = p_i \circ \mathscr{B}(ab)$.
That is, $p_i \circ \overline{\mathscr{B}}(ab) = p_i \circ \mathscr{B}(ab)$ for all $ab \in \Omega$, $i = 1, 2, \ldots, m$.
Thus, I also satisfies the second condition for isomorphism.
Therefore, $\mathscr{G} \cong \overline{\mathscr{G}}$, i.e. $\mathscr{G}$ is self-complementary.

$\square$

From Propositions 8.15 and 8.16, we have the following.

Corollary 8.1 *Let $\mathscr{G} = (\mathscr{V}, \mathscr{A}, \mathscr{B})$ be a strong mPFG of $G^* = (\mathscr{V}, E)$. Then $\mathscr{G}$ is self-complementary if and only if $p_i \circ \mathscr{B}(ab) = \frac{1}{2} p_i \circ \mathscr{A}(a) \wedge p_i \circ \mathscr{A}(b)$ for all $ab \in \Omega$, $i = 1, 2, \ldots, m$.*

Proposition 8.17 *Let $\mathscr{G}_1$ and $\mathscr{G}_2$ be two strong mPFGs. Then $\mathscr{G}_1 \cong \mathscr{G}_2$ if and only if $\overline{\mathscr{G}_1} \cong \overline{\mathscr{G}_2}$.*

Proof Assume that, $\mathscr{G}_1 \cong \mathscr{G}_2$. Then there exists a bijective mapping $\psi : \mathscr{V}_1 \to \mathscr{V}_2$ satisfying
$\mathscr{A}_1(a) = \mathscr{A}_2(\psi(a))$ for all $a \in \mathscr{V}_1$ and $p_i \circ \mathscr{B}_1(ab) = p_i \circ \mathscr{B}_2(\psi(a)\psi(b))$ for all $ab \in \Omega_1$, $i = 1, 2, \ldots, m$.
Let $ab \in \Omega_1$. If $p_i \circ \mathscr{B}_1(ab) = 0$, $(i = 1, 2, \ldots, m)$ then

$$
\begin{aligned}
p_i \circ \overline{\mathscr{B}_1}(ab) &= p_i \circ \mathscr{A}_1(a) \wedge p_i \circ \mathscr{A}_1(b) \\
&= p_i \circ \mathscr{A}_2(\psi(a) \wedge p_i \circ \mathscr{A}_2(\psi(b)) \\
&= p_i \circ \overline{\mathscr{B}_2}(\psi(a)\psi(b)).
\end{aligned}
$$

If $0 < p_i \circ \mathscr{B}_1(ab) \leq 1$, $(i = 1, 2, \ldots, m)$ then $0 < p_i \circ \mathscr{B}_2(\psi(a)\psi(b)) \leq 1$.
Therefore, $p_i \circ \overline{\mathscr{B}_1}(ab) = 0 = p_i \circ \overline{\mathscr{B}_2}(\psi(a)\psi(b))$.
So, $\overline{\mathscr{G}_1} \cong \overline{\mathscr{G}_2}$.
Conversely, let $\overline{\mathscr{G}_1} \cong \overline{\mathscr{G}_2}$. Then there exists a bijective mapping $\psi : \mathscr{V}_1 \to \mathscr{V}_2$ satisfying
$\overline{\mathscr{A}_1}(a) = \overline{\mathscr{A}_2}(\psi(a))$ for all $a \in \mathscr{V}_1$ and $p_i \circ \overline{\mathscr{B}_1}(ab) = p_i \circ \overline{\mathscr{B}_2}(\psi(a)\psi(b))$ for all $ab \in \Omega_1$.
Let $ab \in \Omega_1$. If $p_i \circ \mathscr{B}_1(ab) = 0$, $(i = 1, 2, \ldots, m)$ then

$$
\begin{aligned}
p_i \circ \overline{\mathscr{B}_2}(\psi(a)\psi(b)) &= p_i \circ \overline{\mathscr{B}_1}(ab) \\
&= min\{p_i \circ \mathscr{A}_1(a), p_i \circ \mathscr{A}_1(b)\} \\
&= p_i \circ \overline{\mathscr{A}_1}(a) \wedge p_i \circ \overline{\mathscr{A}_1}(b) \\
&= p_i \circ \overline{\mathscr{A}_2}(\psi(a)) \wedge p_i \circ \overline{\mathscr{A}_2}(\psi(b)) \\
&= p_i \circ \mathscr{A}_2(\psi(a)) \wedge p_i \circ \mathscr{A}_2(\psi(b)).
\end{aligned}
$$

Again, $p_i \circ \overline{\mathscr{B}_2}(\psi(a)\psi(b)) = p_i \circ \mathscr{A}_2(\psi(a)) \wedge p_i \circ \mathscr{A}_2(\psi(b)) - p_i \circ \mathscr{B}_2(\psi(a)\psi(b))$
So, $p_i \circ \mathscr{B}_2(\psi(a)\psi(b)) = 0 = p_i \circ \mathscr{B}_1(ab))$, $(i = 1, 2, \ldots, m)$.
If $0 < p_i \circ \mathscr{B}_1(ab) \leq 1$, $(i = 1, 2, \ldots, m)$ then
$p_i \circ \overline{\mathscr{B}_2}(\psi(a)\psi(b)) = p_i \circ \overline{\mathscr{B}_1}(\psi(a)\psi(b)) = 0$.
Thus, we have,

$$p_i \circ \mathcal{B}_2(\psi(a)\psi(b)) = p_i \circ \mathcal{A}_2(\psi(a)) \wedge p_i \circ \mathcal{A}_2(\psi(b)) - 0$$
$$= p_i \circ \overline{\mathcal{A}_2}(\psi(a)) \wedge p_i \circ \overline{\mathcal{A}_2}(\psi(b))$$
$$= p_i \circ \overline{\mathcal{A}_1}(\psi(a)) \wedge p_i \circ \overline{\mathcal{A}_1}(\psi(b))$$
$$= p_i \circ \mathcal{B}_1(ab).$$

Hence, $\mathcal{G}_1 \cong \mathcal{G}_2$.   □

Theorem 8.1 *Let* $\mathcal{G}_1 = (\mathcal{V}_1, \mathcal{A}_1, \mathcal{B}_1)$ *and* $\mathcal{G}_2 = (\mathcal{V}_2, \mathcal{A}_2, \mathcal{B}_2)$ *be two mPFGs such that* $\mathcal{V}_1 \cap \mathcal{V}_2 = \emptyset$. *Then* $\overline{\mathcal{G}_1 + \mathcal{G}_2} \cong \overline{\mathcal{G}_1} \cup \overline{\mathcal{G}_2}$.

Proof To show that $\overline{\mathcal{G}_1 + \mathcal{G}_2} \cong \overline{\mathcal{G}_1} \cup \overline{\mathcal{G}_2}$, we need to show that there exists an isomorphism between $\overline{\mathcal{G}_1 + \mathcal{G}_2}$ and $\overline{\mathcal{G}_1} \cup \overline{\mathcal{G}_2}$.

We will show that the identity map $I : \mathcal{V}_1 \cup \mathcal{V}_2 \to \mathcal{V}_1 \cup \mathcal{V}_2$ is the required isomorphism between them. For this, we will show the following:

$\overline{(\mathcal{A}_1 + \mathcal{A}_2)}(a) = (\overline{\mathcal{A}_1} \cup \overline{\mathcal{A}_2})(a)$ for all $a \in \mathcal{V}_1 \cup \mathcal{V}_2$, and
$p_i \circ \overline{(\mathcal{B}_1 + \mathcal{B}_2)}(ab) = p_i \circ (\overline{\mathcal{B}_1} \cup \overline{\mathcal{B}_2})(ab),\ (i = 1, 2, \dots, m).$

Let $a \in \mathcal{V}_1 \cup \mathcal{V}_2$.

Then $\overline{(\mathcal{A}_1 + \mathcal{A}_2)}(a)$

$= \overline{(\mathcal{A}_1 + \mathcal{A}_2)}(a) = \overline{(\mathcal{A}_1 \cup \mathcal{A}_2)}(a)$

$= \begin{cases} \mathcal{A}_1(a) \ if \ a \in \mathcal{V}_1 - \mathcal{V}_2 \\ \mathcal{A}_2(a) \ if \ a \in \mathcal{V}_2 - \mathcal{V}_1 \end{cases}$

$= \begin{cases} \overline{\mathcal{A}_1}(a) \ if \ a \in \mathcal{V}_1 - \mathcal{V}_2 \\ \overline{\mathcal{A}_2}(a) \ if \ a \in \mathcal{V}_2 - \mathcal{V}_1 \end{cases}$

$= (\overline{\mathcal{A}_1} \cup \overline{\mathcal{A}_2})(a).$

Now, we have for $i = 1, 2, \dots, m$,

$p_i \circ \overline{(\mathcal{B}_1 + \mathcal{B}_2)}(ab)$

$= p_i \circ (\mathcal{A}_1 + \mathcal{A}_2)(a) \wedge p_i \circ (\mathcal{A}_1 + \mathcal{A}_2)(b) - p_i \circ (\mathcal{B}_1 + \mathcal{B}_2)(ab)$

$= \begin{cases} p_i \circ (\mathcal{A}_1 \cup \mathcal{A}_2)(a) \wedge p_i \circ (\mathcal{A}_1 \cup \mathcal{A}_2)(b) - p_i \circ (\mathcal{B}_1 \cup \mathcal{B}_2)(ab), & if \ ab \in E_1 \cup E_2 \\ p_i \circ (\mathcal{A}_1 \cup \mathcal{A}_2)(a) \wedge p_i \circ (\mathcal{A}_1 \cup \mathcal{A}_2)(b) - p_i \circ (\mathcal{A}_1)(a) \wedge p_i \circ (\mathcal{A}_2)(b), & if \quad ab \in E' \end{cases}$

$= \begin{cases} p_i \circ \mathcal{A}_1(a) \wedge p_i \circ \mathcal{A}_1(b) - p_i \circ \mathcal{B}_1(ab), & if \ ab \in E_1 - E_2 \\ p_i \circ \mathcal{A}_2(a) \wedge p_i \circ \mathcal{A}_2(b) - p_i \circ \mathcal{B}_2(ab), & if \ ab \in E_2 - E_1 \\ p_i \circ \mathcal{A}_1(a) \wedge p_i \circ \mathcal{A}_2(b) - p_i \circ (\mathcal{A}_1)(a) \wedge p_i \circ (\mathcal{A}_2)(b), & if \quad ab \in E' \end{cases}$

$= \begin{cases} p_i \circ \overline{\mathcal{B}_1}(ab), \ if \ ab \in E_1 - E_2 \\ p_i \circ \overline{\mathcal{B}_2}(ab), \ if \ ab \in E_2 - E_1 \\ \quad 0, \quad if \quad ab \in E' \end{cases}$

$= p_i \circ (\overline{\mathcal{B}_1} \cup \overline{\mathcal{B}_2})(ab).$ □

Theorem 8.2 *Let* $\mathcal{G}_1 = (\mathcal{V}_1, \mathcal{A}_1, \mathcal{B}_1)$ *and* $\mathcal{G}_2 = (\mathcal{V}_2, \mathcal{A}_2, \mathcal{B}_2)$ *be two mPFGs such that* $\mathcal{V}_1 \cap \mathcal{V}_2 = \emptyset$. *Then* $\overline{\mathcal{G}_1 \cup \mathcal{G}_2} \cong \overline{\mathcal{G}_1 + \mathcal{G}_2}$.

Proof Consider the identity map $I : \mathcal{V}_1 \cup \mathcal{V}_2 \to \mathcal{V}_1 \cup \mathcal{V}_2$. We will show that I is the required isomorphism between $\overline{\mathcal{G}_1 \cup \mathcal{G}_2}$ and $\overline{\mathcal{G}_1 + \mathcal{G}_2}$.

For this, we will show the following:

$\overline{(\mathcal{A}_1 \cup \mathcal{A}_2)}(a) = \overline{(\mathcal{A}_1 + \mathcal{A}_2)}(a)$ for all $a \in \mathcal{V}_1 \cup \mathcal{V}_2$, and
$p_i \circ \overline{(\mathcal{B}_1 \cup \mathcal{B}_2)}(ab) = p_i \circ \overline{(\mathcal{B}_1 + \mathcal{B}_2)}(ab),\ (i = 1, 2, \dots, m).$

Let $a \in \mathcal{V}_1 \cup \mathcal{V}_2$.
Then $\overline{\mathcal{A}_1 \cup \mathcal{A}_2}(a) = (\mathcal{A}_1 \cup \mathcal{A}_2)(a)$

$$= \begin{cases} \mathcal{A}_1(a), \text{ if } a \in \mathcal{V}_1 - \mathcal{V}_2 \\ \mathcal{A}_2(a), \text{ if } a \in \mathcal{V}_2 - \mathcal{V}_1 \end{cases}$$

$$= \begin{cases} \overline{\mathcal{A}_1}(a), \text{ if } a \in \mathcal{V}_1 - \mathcal{V}_2 \\ \overline{\mathcal{A}_2}(a), \text{ if } a \in \mathcal{V}_2 - \mathcal{V}_1 \end{cases}$$

$$= (\overline{\mathcal{A}_1} \cup \overline{\mathcal{A}_2})(a)$$

Now, we have for $i = 1, 2, \ldots, m$,

$p_i \circ \overline{(\mathcal{B}_1 \cup \mathcal{B}_2)}(ab)$
$= p_i \circ (\mathcal{A}_1 \cup \mathcal{A}_2)(a) \wedge p_i \circ (\mathcal{A}_1 \cup \mathcal{A}_2)(b) - p_i \circ (\mathcal{B}_1 \cup \mathcal{B}_2)(ab)$

$$= \begin{cases} p_i \circ \mathcal{A}_1(a) \wedge p_i \circ \mathcal{A}_1(b) - p_i \circ \mathcal{B}_1(ab), \text{ if } ab \in E_1 \cup E_2 \\ p_i \circ \mathcal{A}_2(a) \wedge p_i \circ \mathcal{A}_2(b) - p_i \circ \mathcal{B}_2(ab), \text{ if } ab \in E_2 - E_1 \\ \quad p_i \circ \mathcal{A}_1(a) \wedge p_i \circ \mathcal{A}_2(b) - 0, \qquad \text{if } a \in \mathcal{V}_1, b \in \mathcal{V}_2 \end{cases}$$

$$= \begin{cases} \quad\quad p_i \circ \overline{\mathcal{B}_1}(ab), \qquad\quad \text{if } ab \in E_1 - E_2 \\ \quad\quad p_i \circ \overline{\mathcal{B}_2}(ab), \qquad\quad \text{if } ab \in E_2 - E_1 \\ p_i \circ \mathcal{A}_1(a) \wedge p_i \circ \mathcal{A}_2(b) - 0, \text{ if } a \in \mathcal{V}_1, b \in \mathcal{V}_2 \end{cases}$$

$$= \begin{cases} \quad\quad p_i \circ \overline{\mathcal{B}_1}(ab), \qquad\quad \text{if } ab \in E_1 - E_2 \\ \quad\quad p_i \circ \overline{\mathcal{B}_2}(ab), \qquad\quad \text{if } ab \in E_2 - E_1 \\ p_i \circ \mathcal{A}_1(a) \wedge p_i \circ \mathcal{A}_2(b) - 0, \text{ if } \quad ab \in E' \end{cases}$$

$= p_i \circ (\overline{\mathcal{B}_1} + \overline{\mathcal{B}_2})(ab)$.
This completes the proof. $\square$

Theorem 8.3 *Let* $\mathcal{G}_1 = (\mathcal{V}_1, \mathcal{A}_1, \mathcal{B}_1)$ *and* $\mathcal{G}_2 = (\mathcal{V}_2, \mathcal{A}_2, \mathcal{B}_2)$ *be two strong mPFGs. Then* $\overline{\mathcal{G}_1 \circ \mathcal{G}_2} \cong \overline{\mathcal{G}_1} \circ \overline{\mathcal{G}_2}$.

Proof Let $\mathcal{G}_1 \circ \mathcal{G}_2 = (\mathcal{V}_1 \times \mathcal{V}_2, \mathcal{A}_1 \circ \mathcal{A}_2, \mathcal{B}_1 \circ \mathcal{B}_2)$ be an mPFG of the graph $G^* = (\mathcal{V}, E)$ where $\mathcal{V} = \mathcal{V}_1 \times \mathcal{V}_2$ and $E = \{(a, a_2)(a, b_2) : a \in \mathcal{V}_1, a_2 b_2 \in E_2\} \cup \{(a_1, c)(b_1, c) : c \in \mathcal{V}_2, a_1 b_1 \in E_1\} \cup \{(a_1, a_2)(b_1, b_2) : a_1 b_1 \in E_1, a_2 \neq b_2\}$.

We show that the identity map I is the required isomorphism between the graphs $\overline{\mathcal{G}_1 \circ \mathcal{G}_2}$ and $\overline{\mathcal{G}_1} \circ \overline{\mathcal{G}_2}$.

Let us consider the identity map $I : \mathcal{V}_1 \times \mathcal{V}_2 \to \mathcal{V}_1 \times \mathcal{V}_2$.

In order to show that I is the required isomorphism, we show that
$p_i \circ \overline{(\mathcal{B}_1 \circ \mathcal{B}_2)}(ab) = p_i \circ (\overline{\mathcal{B}_1} \circ \overline{\mathcal{B}_2})(ab), (i = 1, 2, \ldots, m)$.

Several cases may arise.

Case(i): Let $e = (a, a_2)(a, b_2)$ where $a \in \mathcal{V}_1, a_2 b_2 \in E_2$. Then $e \in E$.
Since $\mathcal{G}_1 \circ \mathcal{G}_2$ is strong mPFG, we have for $i = 1, 2, \ldots, m$
$p_i \circ \overline{(\mathcal{B}_1 \circ \mathcal{B}_2)}(e) = 0$ and
$p_i \circ (\overline{\mathcal{B}_1} \circ \overline{\mathcal{B}_2})(e) = p_i \circ \mathcal{A}_1(a) \wedge p_i \circ \overline{\mathcal{B}_2}(a_2 b_2) = 0$
(since $\mathcal{G}_2$ is strong and $a_2 b_2 \in E_2$, $p_i \circ \overline{\mathcal{B}_2}(a_2 b_2) = 0, i = 1, 2, \ldots, m$).

Case(ii): Let $e = (a, a_2)(a, b_2)$ where $a_2 \neq b_2$, $a_2 b_2 \notin E_2$. Then $e \notin E$.
So for $i = 1, 2, \ldots, m$, $p_i \circ \overline{(\mathcal{B}_1 \circ \mathcal{B}_2)}(e) = 0$ and
$p_i \circ (\overline{\mathcal{B}_1} \circ \overline{\mathcal{B}_2})(e)$
$= p_i \circ (\mathcal{A}_1 \circ \mathcal{A}_2)(a, a_2) \wedge p_i \circ (\mathcal{A}_1 \circ \mathcal{A}_2)(a, b_2)$
$= p_i \circ \mathcal{A}_1(a) \wedge p_i \circ \mathcal{A}_2(a_2) \wedge p_i \circ \mathcal{A}_2(b_2)$.

Again, since $a_2b_2 \in \overline{E_2}$, therefore for $i = 1, 2, \ldots, m$,

$p_i \circ \overline{(\mathscr{B}_1 \circ \mathscr{B}_2)}(e)$

$= p_i \circ \mathscr{A}_1(a) \wedge p_i \circ \overline{\mathscr{B}_2}(a_2b_2)$

$= p_i \circ \mathscr{A}_1(a) \wedge p_i \circ \mathscr{A}_2(a_2) \wedge p_i \circ \mathscr{A}_2(b_2)$.

Case(iii): Let $e = (a_1, c)(b_1, c)$ where $a_1b_1 \in E_1, c \in \mathscr{V}_2$.

Then $e \in E$. So $p_i \circ (\mathscr{B}_1 \circ \mathscr{B}_2)(e) = 0$ $(i = 1, 2, \ldots, m)$ as in Case(i).

Also, since $a_1b_1 \notin E_1$, therefore $p_i \circ \overline{(\mathscr{B}_1 \circ \mathscr{B}_2)}(e) = 0$ $(i = 1, 2, \ldots, m)$.

Case(iv): Let $e = (a_1, c)(b_1, c)$ where $a_1b_1 \notin E_1, c \in \mathscr{V}_2$. Then $e \notin E$.

Hence, $p_i \circ (\mathscr{B}_1 \circ \mathscr{B}_2)(e) = 0$ $(i = 1, 2, \ldots, m)$,

$p_i \circ \overline{(\mathscr{B}_1 \circ \mathscr{B}_2)}(e)$

$= p_i \circ (\mathscr{A}_1 \circ \mathscr{A}_2)(a_1, c) \wedge p_i \circ (\mathscr{A}_1 \circ \mathscr{A}_2)(b_1, c)$

$= p_i \circ \mathscr{A}_1(a_1) \wedge p_i \circ \mathscr{A}_1(b_1) \wedge p_i \circ \mathscr{A}_2(c)$ and

$p_i \circ \overline{(\mathscr{B}_1 \circ \mathscr{B}_2)}(e)$

$= p_i \circ \mathscr{A}_2(c) \wedge p_i \circ \overline{\mathscr{B}_1}(a_1b_1)$

$= p_i \circ \mathscr{A}_1(a_1) \wedge p_i \circ \mathscr{A}_1(b_1) \wedge p_i \circ \mathscr{A}_2(c)$ ($\mathscr{G}_1$ being strong).

Case(v): Let $e = (a_1, a_2)(b_1, b_2)$ where $a_1b_1 \in E_1, a_2 \neq b_2$. Then $e \in E$.

So we have $p_i \circ \overline{(\mathscr{B}_1 \circ \mathscr{B}_2)}(e) = 0$, $(i = 1, 2, \ldots, m)$ as in Case(i).

Also, since $a_1b_1 \in E_1$, we have $p_i \circ \overline{(\mathscr{B}_1 \circ \mathscr{B}_2)}(e) = 0$, $(i = 1, 2, \ldots, m)$.

Case(vi): Let $e = (a_1, a_2)(b_1, b_2)$ where $a_1b_1 \notin E_1, a_2 \neq b_2$. Then $e \notin E$ and hence $p_i \circ (\mathscr{B}_1 \circ \mathscr{B}_2)(e) = 0$ $(i = 1, 2, \ldots, m)$,

$p_i \circ \overline{(\mathscr{B}_1 \circ \mathscr{B}_2)}(e)$

$= p_i \circ (\mathscr{A}_1 \circ \mathscr{A}_2)(a_1, a_2) \wedge p_i \circ (\mathscr{A}_1 \circ \mathscr{A}_2)(b_1, b_2)$

$= p_i \circ \mathscr{A}_1(a_1) \wedge p_i \circ \mathscr{A}_1(b_1) \wedge p_i \circ \mathscr{A}_2(a_2) \wedge p_i \circ \mathscr{A}_2(b_2)\}$

and since $a_1b_1 \in \overline{E_1}$,

$p_i \circ \overline{(\mathscr{B}_1 \circ \mathscr{B}_2)}(e)$

$= p_i \circ \mathscr{A}_2(a_2) \wedge p_i \circ \mathscr{A}_2(b_2) \wedge p_i \circ \overline{\mathscr{B}_1}(a_1b_1)$

$= p_i \circ \mathscr{A}_1(a_1) \wedge p_i \circ \mathscr{A}_1(b_1) \wedge p_i \circ \mathscr{A}_2(a_2) \wedge p_i \circ \mathscr{A}_2(b_2)$ ($\mathscr{G}_1$ being strong by [13]).

Case(vii): Finally, let $e = (a_1, a_2)(b_1, b_2)$ where $a_1b_1 \notin E_1, a_2b_2 \notin E_2$.

Then $e \notin E$ and hence $p_i \circ (\mathscr{B}_1 \circ \mathscr{B}_2)(e) = 0$ $(i = 1, 2, \ldots, m)$,

$p_i \circ \overline{(\mathscr{B}_1 \circ \mathscr{B}_2)}(e)$

$= p_i \circ (\mathscr{A}_1 \circ \mathscr{A}_2)(a_1, a_2) \wedge p_i \circ (\mathscr{A}_1 \circ \mathscr{A}_2)(b_1, b_2)$.

Now, $a_1b_1 \in \overline{E_1}$ and if $a_2 = b_2 = c$, then we have the Case(iv).

Again, if $a_1b_1 \in \overline{E_1}$ and if $a_2 \neq b_2$, then we have Case(vi).

Thus, combining all the cases we have, for $a, b \in \mathscr{V}_1 \times \mathscr{V}_2$, $(i = 1, 2, \ldots, m)$

$p_i \circ \overline{(\mathscr{B}_1 \circ \mathscr{B}_2)}(ab) = p_i \circ (\overline{\mathscr{B}_1 \circ \mathscr{B}_2})(ab)$.   □

8.4   Weak Self-complement *m*-Polar Fuzzy Graphs

Self-complement mPFG is significant in the theory of mPFGs. If an mPFG is not self-complement then we still can say that it is self-complement in some weaker sense. In this section, we define weak self-complement mPFGs.

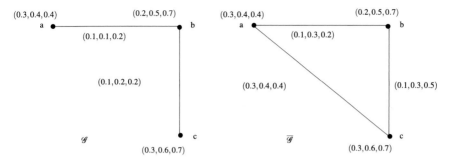

Fig. 8.1 Weak self-complement 3-polar fuzzy graphs

Definition 8.18 The mPFG $\mathcal{G} = (\mathcal{V}, \mathcal{A}, \mathcal{B})$ is said to be **weak self-complement** if there is a weak isomorphism between from $\mathcal{G}$ onto $\overline{\mathcal{G}}$. In other words, there exist a bijective homomorphism $\psi : \mathcal{G} \to \overline{\mathcal{G}}$ such that for $i = 1, 2, \ldots, m$

    (i) $p_i \circ \mathcal{A}(a) = p_i \circ \overline{\mathcal{A}}(\psi(a))$ for all $a \in \mathcal{V}$,

    (ii) $p_i \circ \mathcal{B}(ab) \leq p_i \circ \overline{\mathcal{B}}(\psi(a)\psi(b))$ for all $ab \in \Omega$.

Example 8.9 Let $\mathcal{G} = (\mathcal{V}, \mathcal{A}, \mathcal{B})$ be a 3-polar fuzzy graph of the graph $\mathcal{G}^* = (\mathcal{V}, E)$ where $\mathcal{V} = \{a, b, c\}$, $E = \{ab, bc\}$, $\mathcal{A}(a) = (0.3, 0.4, 0.4)$, $\mathcal{A}(b) = (0.2, 0.5, 0.7)$, $\mathcal{A}(c) = (0.3, 0.6, 0.7)$, $\mathcal{B}(ab) = (0.1, 0.1, 0.2)$, $\mathcal{B}(bc) = (0.1, 0.2, 0.2)$.

Then $\overline{\mathcal{G}} = (\mathcal{V}, \overline{\mathcal{A}}, \overline{\mathcal{B}})$, where $\overline{\mathcal{A}} = \mathcal{A}$ and $\overline{\mathcal{B}}(ab) = (0.1, 0.3, 0.2), \overline{\mathcal{B}}(bc) = (0.1, 0.3, 0.5), \overline{\mathcal{B}}(ca) = (0.3, 0.4, 0.4)$. We can check that the identity map is a weak isomorphism from $\mathcal{G}$ onto $\overline{\mathcal{G}}$ (see Fig. 8.1). Hence, $\mathcal{G}$ is a weak self-complement.

For a weak self-complement mPFG $\mathcal{G}$, we have the following inequality.

Theorem 8.4 *Let $\mathcal{G} = (\mathcal{V}, \mathcal{A}, \mathcal{B})$ be a weak self-complement mPFG. Then*

$$\sum_{a \neq b} p_i \circ \mathcal{B}(ab) \leq \frac{1}{2} \sum_{a \neq b} \min\{p_i \circ \mathcal{A}(a), p_i \circ \mathcal{A}(b)\}, \ (i = 1, 2, \ldots, m).$$

Proof Since $\mathcal{G}$ is a weak self-complement, therefore there exists a weak isomorphism $\psi : \mathcal{V} \to \mathcal{V}$ such that

    $p_i \circ \mathcal{A}(a) = p_i \circ \overline{\mathcal{A}}(\psi(a))$ for all $a \in \mathcal{V}$ and

$p_i \circ \mathcal{B}(ab) \leq p_i \circ \overline{\mathcal{B}}(\psi(a)\psi(b))$ for all $ab \in \Omega$, $(i = 1, 2, \ldots, m)$.

    Using the above we have,

    $p_i \circ \mathcal{B}(ab) \leq p_i \circ \overline{\mathcal{B}}(\psi(a)\psi(b)) = p_i \circ \mathcal{A}(a) \wedge p_i \circ \mathcal{A}(b) - p_i \circ \mathcal{B}(\psi(a)\psi(b))$

    i.e. $p_i \circ \mathcal{B}(ab) + p_i \circ \mathcal{B}(\psi(a)\psi(b)) \leq p_i \circ \mathcal{A}(\psi(a)) \wedge p_i \circ \mathcal{A}(\psi(b))$.

    Therefore, for all $ab \in \Omega$, $(i = 1, 2, \ldots, m)$

$$\sum_{a \neq b} p_i \circ \mathcal{B}(ab) + \sum_{a \neq b} p_i \circ \mathcal{B}(\psi(a)\psi(b))$$

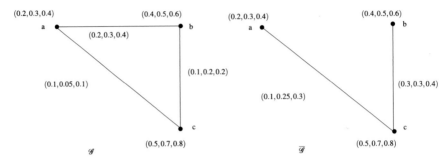

Fig. 8.2 Weak self-complement 3-polar fuzzy graphs

$$\leq \sum_{a\neq b} p_i \circ \mathscr{A}(\psi(a)) \wedge p_i \circ \mathscr{A}(\psi(b))$$

$$= \sum_{a\neq b} p_i \circ \mathscr{A}(a) \wedge p_i \circ \mathscr{A}(b)$$

i.e.

$$2\sum_{a\neq b} p_i \circ \mathscr{B}(ab) \leq \sum_{a\neq b} p_i \circ \mathscr{A}(a) \wedge p_i \circ \mathscr{A}(b)$$

i.e.

$$\sum_{a\neq b} p_i \circ \mathscr{B}(ab) \leq \frac{1}{2}\sum_{a\neq b} p_i \circ \mathscr{A}(a) \wedge p_i \circ \mathscr{A}(b).$$

□

Remark 8.1 The converse of the above theorem is not true in general. For example, consider the 3-polar fuzzy graph $\mathscr{G}$ of Fig. 8.2. Then it satisfies the condition of Theorem 8.4. But, $\mathscr{G}$ is not a weak self-complement as there is no weak isomorphism from $\mathscr{G}$ onto $\overline{\mathscr{G}}$.

Theorem 8.5 *Let $\mathscr{G}$ be an mPFG such that*

$$p_i \circ \mathscr{B}(ab) \leq \frac{1}{2} p_i \circ \mathscr{A}(a) \wedge p_i \circ \mathscr{A}(b) \text{ for all } ab \in \Omega, (i = 1, 2, \ldots, m).$$

Then $\mathscr{G}$ is a weak self-complement.

Proof Let $\overline{\mathscr{G}} = (\mathscr{V}, \overline{\mathscr{A}}, \overline{\mathscr{B}})$ be the complement of $\mathscr{G}$, where $\overline{\mathscr{A}}(a) = \mathscr{A}(a)$ for all $a \in \mathscr{V}$ and $p_i \circ \overline{\mathscr{B}}(ab) = p_i \circ \mathscr{A}(a) \wedge p_i \circ \mathscr{A}(b) - p_i \circ \mathscr{B}(ab)$ for $ab \in \Omega, (i = 1, 2, \ldots, m)$.

Let us now consider the identity map $I : \mathscr{V} \to \mathscr{V}$. Then $\mathscr{A}(a) = \mathscr{A}(I(a)) = \overline{\mathscr{A}}(I(a))$ for all $a \in \mathscr{V}$ and

$$p_i \circ \overline{\mathscr{B}}(I(a)I(b)) = p_i \circ \overline{\mathscr{B}}(ab)$$

$$= p_i \circ \mathscr{A}(a) \wedge p_i \circ \mathscr{A}(b) - p_i \circ \mathscr{B}(ab)$$

$$\geq p_i \circ \mathscr{A}(a) \wedge p_i \circ \mathscr{A}(b) - \frac{1}{2} p_i \circ \mathscr{A}(a) \wedge p_i \circ \mathscr{A}(b)$$

$$= \frac{1}{2} p_i \circ \mathscr{A}(a) \wedge p_i \circ \mathscr{A}(b)$$

$$\geq p_i \circ \mathscr{B}(ab).$$

So, $p_i \circ \mathscr{B}(ab) \leq p_i \circ \overline{\mathscr{B}}(I(a)I(b))$ for $ab \in \Omega$, $(i = 1, 2, \ldots, m)$.

Hence, $I : \mathscr{V} \to \mathscr{V}$ is a weak isomorphism. $\qquad \square$

8.5 Busy Value of Vertices of m-Polar Fuzzy Graphs

In this section, the order, size, and busy value of vertices of an mPFG are defined.

Definition 8.19 The **order of an mPFG** $\mathscr{G}$ is denoted by $|\mathscr{V}|$ (or $O(\mathscr{G})$), where

$$O(\mathscr{G}) = |\mathscr{V}| = \sum_{a \in \mathscr{V}} \frac{1 + \sum\limits_{i=1}^{m} p_i \circ \mathscr{A}(a)}{2}.$$

The **size** of $\mathscr{G}$ is denoted by $|E|$ (or $S(\mathscr{G})$) where

$$S(\mathscr{G}) = |E| = \sum_{ab \in E} \frac{1 + \sum\limits_{i=1}^{m} p_i \circ \mathscr{B}(ab)}{2}.$$

Theorem 8.6 Let $\mathscr{G}_1 = (\mathscr{V}_1, \mathscr{A}_1, \mathscr{B}_1)$ and $\mathscr{G}_2 = (\mathscr{V}_2, \mathscr{A}_2, \mathscr{B}_2)$ be two isomorphic mPFGs. Then they have the same order and size.

Proof Let ψ be an isomorphism from $\mathscr{G}_1$ onto $\mathscr{G}_2$. Then $\mathscr{A}_1(a) = \mathscr{A}_2(\psi(a))$ for all $a \in \mathscr{V}_1$ and $p_i \circ \mathscr{B}_1(ab) = p_i \circ \mathscr{B}_2(\psi(a)\psi(b))$ for $ab \in \Omega_1$, $(i = 1, 2, \ldots, m)$.

Now,

$$O(\mathscr{G}_1) = |\mathscr{V}_1| = \sum_{a \in \mathscr{V}_1} \frac{1 + \sum\limits_{i=1}^{m} p_i \circ \mathscr{A}_1(a)}{2}$$

$$= \sum_{a \in \mathscr{V}_1} \frac{1 + \sum\limits_{i=1}^{m} p_i \circ \mathscr{A}_2(\psi(a))}{2}$$

$$= O(\mathscr{G}_2)$$

and

$$S(\mathscr{G}_1) = |E_1| = \sum_{ab \in E_1} \frac{1 + \sum\limits_{i=1}^{m} p_i \circ \mathscr{B}_1(ab)}{2}$$

$$= \sum_{ab \in E_1} \frac{1 + \sum\limits_{i=1}^{m} p_i \circ \mathscr{B}_2(\psi(a)\psi(b))}{2}$$

$$= S(\mathscr{G}_2).$$

□

Definition 8.20 The **busy value** of a vertex a of an mPFG $\mathscr{G}$ is denoted as $BV(a) = (p_1 \circ BV(a), p_2 \circ BV(a), \ldots, p_m \circ BV(a))$ where $p_i \circ BV(a) = \sum\limits_{k} p_i \circ \mathscr{A}(a) \wedge p_i \circ \mathscr{A}(a_k)$, a_k being the neighbors of a. The busy value of $\mathscr{G}$ is denoted as $BV(\mathscr{G})$, where $BV(\mathscr{G}) = \sum\limits_{k} D(a_k)$, $a_k \in \mathscr{V}$.

Example 8.10 Consider the 3-polar fuzzy graph $\mathscr{G} = (\mathscr{V}, \mathscr{A}, \mathscr{B})$ of $G^* = (\mathscr{V}, E)$ where $\mathscr{V} = \{a, b, c, d\}$, $E = \{ab, bc, cd, ca, bd\}$, $\mathscr{A}(a) = (0.6, 0.3, 0.5)$, $\mathscr{A}(b) = (0.8, 0.4, 0.3)$, $\mathscr{A}(c) = (0.5, 0.6, 0.4)$, $\mathscr{A}(d) = (0.7, 0.5, 0.6)$, $\mathscr{B}(ab) = (0.5, 0.2, 0.2)$, $\mathscr{B}(bc) = (0.1, 0.3, 0.2)$, $\mathscr{B}(cd) = (0.6, 0.2, 0.4)$, $\mathscr{B}(ca) = (0.3, 0.2, 0.3)$, $\mathscr{B}(bd) = ((0.7, 0.4, 0.2)$. Then we have,
$BV(a) = (1.7, 0.9, 1.2)$, $BV(b) = (1.8, 1.1, 0.9)$, $BV(c) = (1, 0.7, 0.7)$, $BV(d) = (1.3, 0.7, 0.8)$.

Definition 8.21 The vertex a of $\mathscr{G}$ is called a **busy vertex** if

$$p_i \circ \mathscr{A}(a) \leq p_i \circ deg(a), \ (i = 1, 2, \ldots, m).$$

Definition 8.22 An edge $a_1b_1 \in E$ is called an **effective edge** of $\mathscr{G}$ if

$$p_i \circ \mathscr{B}(a_1b_1) = p_i \circ \mathscr{A}(a_1) \wedge p_i \circ \mathscr{A}(b_1), \ (i = 1, 2, \ldots, m).$$

Definition 8.23 Let $\mathscr{G}$ be an mPFG and $a \in \mathscr{V}$ be a vertex.
   (i) a is called a **partial free vertex** if it is a **free vertex** of $\mathscr{G}$ and $\overline{\mathscr{G}}$.
   (ii) a is called a **fully free vertex** if it is a free vertex of $\mathscr{G}$ and it is a busy vertex of $\overline{\mathscr{G}}$.
   (iii) a is called a **partial busy vertex** if it is a busy vertex of $\mathscr{G}$ and $\overline{\mathscr{G}}$.
   (iv) a is called a **fully busy vertex** if it is a busy vertex in $\mathscr{G}$ and it is a free vertex of $\overline{\mathscr{G}}$.

Theorem 8.7 *Let ψ be an isomorphism from $\mathscr{G}_1 = (\mathscr{V}_1, \mathscr{A}_1, \mathscr{B}_1)$ onto $\mathscr{G}_2 = (\mathscr{V}_2, \mathscr{A}_2, \mathscr{B}_2)$. Then $deg(a) = deg(\psi(a))$ for all $a \in \mathscr{V}_1$.*

Proof Since ψ is an isomorphism between $\mathcal{G}_1$ and $\mathcal{G}_2$, we have $p_i \circ \mathcal{A}_1(a) = p_i \circ \mathcal{A}_2(\psi(a))$ for all $a \in \mathcal{V}_1$ and $p_i \circ \mathcal{B}_1(a_1 b_1) = p_i \circ \mathcal{B}_2(\psi(a_1)\psi(b_1))$ for all $a_1, b_1 \in \mathcal{V}_1, (i = 1, 2, \ldots, m)$.

Hence, for $a \in \mathcal{V}_1, (i = 1, 2, \ldots, m)$

$$
p_i \circ deg(a) = \sum_{\substack{a \neq b \\ ab \in E_1}} p_i \circ \mathcal{B}_1(ab)
$$

$$
= \sum_{\substack{\psi(a) \neq \psi(b) \\ \psi(a)\psi(b) \in E_2}} p_i \circ \mathcal{B}_2(\psi(a)\psi(b))
$$

$$
= p_i \circ deg(\psi(a)).
$$

So, $deg(a) = deg(\psi(a))$ for all $a \in \mathcal{V}_1$. $\qquad \square$

Theorem 8.8 *If ψ is an isomorphism from $\mathcal{G}_1$ onto $\mathcal{G}_2$ and a is a busy vertex of $\mathcal{G}_1$, then $\psi(a)$ is a busy vertex of $\mathcal{G}_2$.*

Proof We have,
$p_i \circ \mathcal{A}_1(a) = p_i \circ \mathcal{A}_2(\psi(a)) \ a \in \mathcal{V}_1$ and
$p_i \circ \mathcal{B}_1(a_1 b_1) = p_i \circ \mathcal{B}_2(\psi(a_1)\psi(b_1))$ for $a_1, b_1 \in \mathcal{V}_1, i = 1, 2, \ldots, m$.
If a is a busy vertex of $\mathcal{G}_1$, then $p_i \circ \mathcal{A}_1(a) \le p_i \circ deg(a)$ for $i = 1, 2, \ldots, m$.
Then by the above and Theorem 8.7,

$$
p_i \circ \mathcal{A}_2(\psi(a)) = p_i \circ \mathcal{A}_1(a) \le p_i \circ deg(a) = p_i \circ deg(\psi(a)) \text{ for } i = 1, 2, \ldots, m.
$$

Hence, $\psi(a)$ is a busy vertex in $\mathcal{G}_2$. $\qquad \square$

Theorem 8.9 *Let two mPFGs $\mathcal{G}_1$ and $\mathcal{G}_2$ be weak isomorphic. If $a \in \mathcal{V}_1$ is a busy vertex of $\mathcal{G}_1$, then the image of a under the weak isomorphism is also busy in $\mathcal{G}_2$.*

Proof Let $\psi : \mathcal{V}_1 \to \mathcal{V}_2$ be a weak isomorphism between $\mathcal{G}_1$ and $\mathcal{G}_2$.

Then, $p_i \circ \mathcal{A}_1(a) = p_i \circ \mathcal{A}_2(\psi(a))$ for all $a \in \mathcal{V}_1$ and $p_i \circ \mathcal{B}_1(a_1 b_1) \le p_i \circ \mathcal{B}_2(\psi(a_1)\psi(b_1))$ for all $a_1, b_1 \in \mathcal{V}_1, i = 1, 2, \ldots, m$.

Let $a \in \mathcal{V}_1$ be a busy vertex. Then $p_i \circ \mathcal{A}_1(a) \le p_i \circ deg(a), (i = 1, 2, \ldots, m)$.

Now by the above, we have for $i = 1, 2, \ldots, m$

$$
p_i \circ \mathcal{A}_2(a) = p_i \circ \mathcal{A}_1(a) \le p_i \circ deg(a)
$$

$$
= \sum_{\substack{a \neq b \\ ab \in E_1}} p_i \circ \mathcal{B}_1(ab)
$$

$$
\le \sum_{\substack{\psi(a) \neq \psi(b) \\ \psi(a)\psi(b) \in E_2}} p_i \circ \mathcal{B}_2(\psi(a)\psi(b))
$$

$$
= p_i \circ deg(\psi(a)).
$$

Hence, $\psi(a)$ is a busy vertex in $\mathcal{G}_2$. $\qquad \square$

8.6   Products on *m*-Polar Fuzzy Graphs

In this section, some products are defined between mPFGs. Below the **direct product** of two mPFGs is defined.

Definition 8.24 Let $\mathscr{G}_1 = (\mathscr{V}_1, \mathscr{A}_1, \mathscr{B}_1)$ and $\mathscr{G}_2 = (\mathscr{V}_2, \mathscr{A}_2, \mathscr{B}_2)$ be two mPFGs such that $\mathscr{V}_1 \cap \mathscr{V}_2 = \emptyset$. The direct product of $\mathscr{G}_1$ and $\mathscr{G}_2$ is defined to be the mPFG $\mathscr{G}_1 \sqcap \mathscr{G}_2 = (\mathscr{V}_1 \times \mathscr{V}_2, \mathscr{A}_1 \sqcap \mathscr{A}_2, \mathscr{B}_1 \sqcap \mathscr{B}_2)$ of the graph $G^* = (\mathscr{V}_1 \times \mathscr{V}_2, E)$, where $E = \{(u_1, v_1)(u_2, v_2) | u_1 u_2 \in E_1, v_1 v_2 \in E_2\}$ such that for $i = 1, 2, \ldots, m$

1. $p_i \circ (\mathscr{A}_1 \sqcap \mathscr{A}_2)(u, v) = p_i \circ \mathscr{A}_1(u) \wedge p_i \circ \mathscr{A}_2(v)$ for all $(u, v) \in \mathscr{V}_1 \times \mathscr{V}_2$,
2. $p_i \circ (\mathscr{B}_1 \sqcap \mathscr{B}_2)((u_1, v_1)(u_2, v_2)) = p_i \circ \mathscr{B}_1(u_1 u_2) \wedge p_i \circ \mathscr{B}_2(v_1 v_2)$ for all $u_1 u_2 \in E_1$ and $v_1 v_2 \in E_2$,

Theorem 8.10 *The direct product $\mathscr{G}_1 \sqcap \mathscr{G}_2$ of two mPFGs $\mathscr{G}_1$ and $\mathscr{G}_2$ is an mPFG.*

Proof Let $(u_1, v_1)(u_2, v_2) \in E$.
Then $u_1 u_2 \in E_1$ and $v_1 v_2 \in E_2$. We have for $i = 1, 2, \ldots, m$

$$
\begin{aligned}
p_i \circ (\mathscr{B}_1 \sqcap \mathscr{B}_2)((u_1, v_1)(u_2, v_2)) &= p_i \circ \mathscr{B}_1(u_1 u_2) \wedge p_i \circ \mathscr{B}_2(v_1 v_2) \\
&\leq p_i \circ \mathscr{A}_1(u_1) \wedge p_i \circ \mathscr{A}_1(u_2) \wedge p_i \circ \mathscr{A}_2(v_1) \wedge p_i \circ \mathscr{A}_2(v_2) \\
&\quad \text{(since } \mathscr{G}_1 \text{ and } \mathscr{G}_2 \text{ are } mPFs) \\
&= p_i \circ (\mathscr{A}_1 \sqcap \mathscr{A}_2)(u_1, v_1) \wedge p_i \circ (\mathscr{A}_1 \sqcap \mathscr{A}_2)(u_2, v_2)
\end{aligned}
$$

This shows that $\mathscr{G}_1 \sqcap \mathscr{G}_2$ is an mPFG.   □

Theorem 8.11 *Let $\mathscr{G}_1 = (\mathscr{V}_1, \mathscr{A}_1, \mathscr{B}_1)$ and $\mathscr{G}_2 = (\mathscr{V}_2, \mathscr{A}_2, \mathscr{B}_2)$ be two strong mPFGs. Then $\mathscr{G}_1 \sqcap \mathscr{G}_2$ is also strong.*

Proof Let $(u_1, v_1)(u_2, v_2) \in E$.
Since $\mathscr{G}_1$ and $\mathscr{G}_2$ are strong, therefore for $i = 1, 2, \ldots, m$

$$
\begin{aligned}
p_i \circ (\mathscr{B}_1 \sqcap \mathscr{B}_2)((u_1, v_1)(u_2, v_2)) &= p_i \circ \mathscr{B}_1(u_1 u_2) \wedge p_i \circ \mathscr{B}_2(v_1 v_2) \\
&= p_i \circ \mathscr{A}_1(u_1) \wedge p_i \circ \mathscr{A}_1(u_2) \wedge p_i \circ \mathscr{A}_2(v_1) \wedge p_i \circ \mathscr{A}_2(v_2) \\
&= p_i \circ (\mathscr{A}_1 \sqcap \mathscr{A}_2)(u_1, v_1) \wedge p_i \circ (\mathscr{A}_1 \sqcap \mathscr{A}_2)(u_2, v_2).
\end{aligned}
$$

Hence, $\mathscr{G}_1 \sqcap \mathscr{G}_2$ is strong mPFG.   □

Now, **semi-strong** product is defined between two *mPFs* to construct a new *mPF*.

Definition 8.25 The semi-strong product of two *mPFs* $\mathscr{G}_1 = (\mathscr{V}_1, \mathscr{A}_1, \mathscr{B}_1)$ of $G_1^* = (\mathscr{V}_1, E_1)$ and $\mathscr{G}_2 = (\mathscr{V}_2, \mathscr{A}_2, \mathscr{B}_2)$ of $G_2^* = (\mathscr{V}_2, E_2)$, where it is assumed that $\mathscr{V}_1 \cap \mathscr{V}_2 = \emptyset$, is defined to be the *mPF* $\mathscr{G}_1 \bullet \mathscr{G}_2 = (\mathscr{V}_1 \times \mathscr{V}_2, \mathscr{A}_1 \bullet \mathscr{A}_2, \mathscr{B}_1 \bullet \mathscr{B}_2)$ of $G^* = (\mathscr{V}_1 \times \mathscr{V}_2, E)$, where $E = \{(u, v_1)(u, v_2) | u \in \mathscr{V}_1, v_1 v_2 \in E_2\} \cup \{(u_1, v_1)(u_2, v_2) | u_1 u_2 \in E_1, v_1 v_2 \in E_2\}$ satisfying the following: for $i = 1, 2, \ldots, m$

1. $p_i \circ (\mathscr{A}_1 \bullet \mathscr{A}_2)(u, v) = p_i \circ \mathscr{A}_1(u) \wedge p_i \circ \mathscr{A}_2(v)$ for all $(u, v) \in \mathscr{V}_1 \times \mathscr{V}_2$,
2. $p_i \circ (\mathscr{B}_1 \bullet \mathscr{B}_2)((u, v_1)(u, v_2)) = p_i \circ \mathscr{A}_1(u) \wedge p_i \circ \mathscr{B}_2(v_1 v_2)$ for all $u \in \mathscr{V}_1$ and $v_1 v_2 \in E_2$,
3. $p_i \circ (\mathscr{B}_1 \bullet \mathscr{B}_2)((u_1, v_1)(u_2, v_2)) = p_i \circ \mathscr{B}_1(u_1 u_2) \wedge p_i \circ \mathscr{B}_2(v_1 v_2)$ for all $u_1 u_2 \in E_1$ and $v_1 v_2 \in E_2$,

Theorem 8.12 *If* $\mathscr{G}_1 = (\mathscr{V}_1, \mathscr{A}_1, \mathscr{B}_1)$ *and* $\mathscr{G}_2 = (\mathscr{V}_2, \mathscr{A}_2, \mathscr{B}_2)$ *are mPFGs, then* $\mathscr{G}_1 \bullet \mathscr{G}_2$ *is an mPFG.*

Proof Let $(u, v_1)(u, v_2) \in E$. Then $u \in \mathscr{V}_1$ and $v_1 v_2 \in E_2$.
Since $\mathscr{G}_2$ is an mPFG, we have for $i = 1, 2, \ldots, m$

$$
\begin{aligned}
p_i \circ (\mathscr{B}_1 \bullet \mathscr{B}_2)((u, v_1)(u, v_2)) &= p_i \circ \mathscr{A}_1(u) \wedge p_i \circ \mathscr{B}_2(v_1 v_2) \\
&\le p_i \circ \mathscr{A}_1(u) \wedge p_i \circ \mathscr{A}_2(v_1) \wedge p_i \circ \mathscr{A}_2(v_2) \\
&= p_i \circ (\mathscr{A}_1 \bullet \mathscr{A}_2)(u, v_1) \wedge p_i \circ (\mathscr{A}_1 \bullet \mathscr{A}_2)(u, v_2).
\end{aligned}
$$

Let $(u_1, v_1)(u_2, v_2) \in E$.
Then $u_1 u_2 \in E_1$ and $v_1 v_2 \in E_2$.
$\mathscr{G}_1$ and $\mathscr{G}_2$ being mPFGs, we have for $i = 1, 2, \ldots, m$

$$
\begin{aligned}
p_i \circ (\mathscr{B}_1 \bullet \mathscr{B}_2)((u_1, v_1)(u_2, v_2)) &= p_i \circ \mathscr{B}_1(u_1 u_2) \wedge p_i \circ \mathscr{B}_2(v_1 v_2) \\
&\le p_i \circ \mathscr{A}_1(u_1) \wedge p_i \circ \mathscr{A}_1(u_2) \wedge p_i \circ \mathscr{A}_2(v_1) \wedge p_i \circ \mathscr{A}_2(v_2) \\
&= p_i \circ (\mathscr{A}_1 \bullet \mathscr{A}_2)(u_1, v_1) \wedge p_i \circ (\mathscr{A}_1 \bullet \mathscr{A}_2)(u_2, v_2).
\end{aligned}
$$

Hence, the proof. $\square$

Theorem 8.13 *Let* $\mathscr{G}_1 = (\mathscr{V}_1, \mathscr{A}_1, \mathscr{B}_1)$ *and* $\mathscr{G}_2 = (\mathscr{V}_2, \mathscr{A}_2, \mathscr{B}_2)$ *be strong mPFGs. Then* $\mathscr{G}_1 \bullet \mathscr{G}_2$ *is a strong mPFG.*

Proof Let $(u, v_1)(u, v_2) \in E$. Then $u \in \mathscr{V}_1$ and $v_1 v_2 \in E_2$.
$\mathscr{G}_2$ being strong, we have for $i = 1, 2, \ldots, m$

$$
\begin{aligned}
p_i \circ (\mathscr{B}_1 \bullet \mathscr{B}_2)((u, v_1)(u, v_2)) &= p_i \circ \mathscr{A}_1(u) \wedge p_i \circ \mathscr{B}_2(v_1 v_2) \\
&= p_i \circ \mathscr{A}_1(u) \wedge p_i \circ \mathscr{A}_2(v_1) \wedge p_i \circ \mathscr{A}_2(v_2) \\
&= p_i \circ (\mathscr{A}_1 \bullet \mathscr{A}_2)(u, v_1) \wedge p_i \circ (\mathscr{A}_1 \bullet \mathscr{A}_2)(u, v_2).
\end{aligned}
$$

If $(u_1, v_1)(u_2, v_2) \in E$. Then $u_1 u_2 \in E_1$ and $v_1 v_2 \in E_2$.
Now, $\mathscr{G}_1$ and $\mathscr{G}_2$ being strong, we have for $i = 1, 2, \ldots, m$

$$
\begin{aligned}
p_i \circ (\mathscr{B}_1 \bullet \mathscr{B}_2)((u_1, v_1)(u_2, v_2)) &= p_i \circ \mathscr{B}_1(u_1 u_2) \wedge p_i \circ \mathscr{B}_2(v_1 v_2) \\
&= p_i \circ \mathscr{A}_1(u_1) \wedge p_i \circ \mathscr{A}_1(u_2) \wedge p_i \circ \mathscr{A}_2(v_1) \wedge p_i \circ \mathscr{A}_2(v_2) \\
&= p_i \circ (\mathscr{A}_1 \bullet \mathscr{A}_2)(u_1, v_1) \wedge p_i \circ (\mathscr{A}_1 \bullet \mathscr{A}_2)(u_2, v_2).
\end{aligned}
$$

Hence, $\mathscr{G}_1 \bullet \mathscr{G}_2$ is strong mPFG. $\square$

The **strong product** between mPFGs is an important tool to construct new mPFG. It is is defined below.

Definition 8.26 The strong product of two mPFGs $\mathscr{G}_1 = (\mathscr{V}_1, \mathscr{A}_1, \mathscr{B}_1)$ of $\mathscr{G}_1^* = (\mathscr{V}_1, E_1)$ and $\mathscr{G}_2 = (\mathscr{V}_2, \mathscr{A}_2, \mathscr{B}_2)$ of $G_2^* = (\mathscr{V}_2, E_2)$ such that $\mathscr{V}_1 \cap \mathscr{V}_2 = \emptyset$, is defined to be the *mPF* $\mathscr{G}_1 \otimes \mathscr{G}_2 = (\mathscr{A}_1 \otimes \mathscr{A}_2, \mathscr{B}_1 \otimes \mathscr{B}_2)$ of $\mathscr{G}^* = (\mathscr{V}_1 \times \mathscr{V}_2, E)$, where $E = \{(u, v_1)(u, v_2) | u \in \mathscr{V}_1, v_1 v_2 \in E_2\} \cup \{(u_1, w)(u_2, w) | w \in \mathscr{V}_2, u_1 u_2 \in E_1\} \cup \{(u_1, v_1)(u_2, v_2) | u_1 u_2 \in E_1, v_1 v_2 \in E_2\}$ such that the following condition holds: for $i = 1, 2, \ldots, m$

1. $p_i \circ (\mathscr{A}_1 \otimes \mathscr{A}_2)(u, v) = p_i \circ \mathscr{A}_1(u) \wedge p_i \circ \mathscr{A}_2(v)$ for all $(u, v) \in \mathscr{V}_1 \times \mathscr{V}_2$,
2. $p_i \circ (\mathscr{B}_1 \otimes \mathscr{B}_2)((u, v_1)(u, v_2)) = p_i \circ \mathscr{A}_1(u) \wedge p_i \circ \mathscr{B}_2(v_1 v_2)$ for all $u \in \mathscr{V}_1$ and $v_1 v_2 \in E_2$,
3. $p_i \circ (\mathscr{B}_1 \otimes \mathscr{B}_2)((u_1, w)(u_2, w)) = p_i \circ \mathscr{B}_1(u_1 u_2) \wedge p_i \circ \mathscr{A}_2(w)$ for all $w \in \mathscr{V}_2$ and $u_1 u_2 \in E_1$,
4. $p_i \circ (\mathscr{B}_1 \otimes \mathscr{B}_2)((u_1, v_1)(u_2, v_2)) = p_i \circ \mathscr{B}_1(u_1 u_2) \wedge p_i \circ \mathscr{B}_2(v_1 v_2)$ for all $u_1 u_2 \in E_1$ and $v_1 v_2 \in E_2$.

Theorem 8.14 *The strong product $\mathscr{G}_1 \otimes \mathscr{G}_2$ of two mPFGs is an mPFG.*

Theorem 8.15 *If $\mathscr{G}_1 = (\mathscr{V}_1, \mathscr{A}_1, \mathscr{B}_1)$ and $\mathscr{G}_2 = (\mathscr{V}_2, \mathscr{A}_2, \mathscr{B}_2)$ are complete mPFGs, then $\mathscr{G}_1 \otimes \mathscr{G}_2$ is complete.*

Proof We have s strong product of mPFGs is an mPFG by Theorem 8.14. Since $\mathscr{G}_1$ and $\mathscr{G}_2$ are complete, therefore $E = \mathscr{V}_1 \times \mathscr{V}_2$ in the graph $\mathscr{G}_1 \otimes \mathscr{G}_2$.

Let $(u, v_1)(u, v_2) \in E$. Since $\mathscr{G}_2$ is complete, we have for $i = 1, 2, \ldots, m$

$$
\begin{aligned}
p_i \circ (\mathscr{B}_1 \otimes \mathscr{B}_2)((u, v_1)(u, v_2)) &= p_i \circ \mathscr{A}_1(u) \wedge p_i \circ \mathscr{B}_2(v_1 v_2) \\
&= p_i \circ \mathscr{A}_1(u) \wedge p_i \circ \mathscr{A}_2(v_1) \wedge p_i \circ \mathscr{A}_2(v_2) \\
&= p_i \circ (\mathscr{A}_1 \otimes \mathscr{A}_2)(u, v_1) \wedge p_i \circ (\mathscr{A}_1 \otimes \mathscr{A}_2)(u, v_2).
\end{aligned}
$$

Let $(u_1, w)(u_2, w) \in E$. Since $\mathscr{G}_1$ is complete, we have for $i = 1, 2, \ldots, m$

$$
\begin{aligned}
p_i \circ (\mathscr{B}_1 \otimes \mathscr{B}_2)((u_1, w)(u_2, w)) &= p_i \circ \mathscr{B}_1(u_1 u_2) \wedge p_i \circ \mathscr{A}_2(w) \\
&= p_i \circ \mathscr{A}_1(u_1) \wedge p_i \circ \mathscr{A}_1(u_2) \wedge p_i \circ \mathscr{A}_2(w) \\
&= p_i \circ (\mathscr{A}_1 \otimes \mathscr{A}_2)(u_1, w) \wedge p_i \circ (\mathscr{A}_1 \otimes \mathscr{A}_2)(u_2, w).
\end{aligned}
$$

Finally, let $(u_1, v_1)(u_2, v_2) \in E$. Then since $\mathscr{G}_1$ and $\mathscr{G}_2$ are complete, we have for $i = 1, 2, \ldots, m$

$$
\begin{aligned}
p_i \circ (\mathscr{B}_1 \otimes \mathscr{B}_2)((u_1, v_1)(u_2, v_2)) &= p_i \circ \mathscr{B}_1(u_1 u_2) \wedge p_i \circ \mathscr{B}_2(v_1 v_2) \\
&= p_i \circ \mathscr{A}_1(u_1) \wedge p_i \circ \mathscr{A}_1(u_2) \wedge p_i \circ \mathscr{A}_2(v_1) \wedge p_i \circ \mathscr{A}_2(v_2) \\
&= p_i \circ (\mathscr{A}_1 \otimes \mathscr{A}_2)(u_1, v_1) \wedge p_i \circ (\mathscr{A}_1 \otimes \mathscr{A}_2)(u_2, v_2).
\end{aligned}
$$

Hence, $\mathscr{G}_1 \otimes \mathscr{G}_2$ is complete. □

Theorem 8.16 *Let $\mathscr{G}_1 = (\mathscr{V}_1, \mathscr{A}_1, \mathscr{B}_1)$ and $\mathscr{G}_2 = (\mathscr{V}_2, \mathscr{A}_2, \mathscr{B}_2)$ be mPFGs such that $\mathscr{G}_1 \sqcap \mathscr{G}_2$ is strong. Then at least $\mathscr{G}_1$ or $\mathscr{G}_2$ must be strong.*

Proof Let us assume that both $\mathscr{G}_1$ and $\mathscr{G}_2$ are not strong mPFGs. Then there exists at least one $u_1v_1 \in E_1$ and $u_2v_2 \in E_2$ such that for $i = 1, 2, \ldots, m$

$p_i \circ \mathscr{B}_1(u_1v_1) < p_i \circ \mathscr{A}_1(u_1) \wedge p_i \circ \mathscr{A}_1(v_1)$ and
$p_i \circ \mathscr{B}_2(u_2v_2) < p_i \circ \mathscr{A}_2(u_2) \wedge p_i \circ \mathscr{A}_2(v_2)$.
Now, we have for $i = 1, 2, \ldots, m$

$$
\begin{aligned}
p_i \circ (\mathscr{B}_1 \sqcap \mathscr{B}_2)((u_1, v_1)(u_2, v_2)) &= p_i \circ \mathscr{B}_1(u_1u_2) \wedge p_i \circ \mathscr{B}_2(v_1v_2) \\
&< p_i \circ \mathscr{A}_1(u_1) \wedge p_i \circ \mathscr{A}_1(u_2) \wedge p_i \circ \mathscr{A}_2(v_1) \wedge p_i \circ \mathscr{A}_2(v_2) \\
&\quad \text{(from the above assumption)} \\
&= p_i \circ (\mathscr{A}_1 \sqcap \mathscr{A}_2)(u_1, v_1) \wedge p_i \circ (\mathscr{A}_1 \sqcap \mathscr{A}_2)(u_2, v_2).
\end{aligned}
$$

This shows that, $\mathscr{G}_1 \sqcap \mathscr{G}_2$ is not strong, which is a contradiction.

So our assumption is wrong. This means one of $\mathscr{G}_1$ or $\mathscr{G}_2$ is strong. □

The following result follows from the preceding theorem.

Theorem 8.17 *Let $\mathscr{G}_1 = (\mathscr{V}_1, \mathscr{A}_1, \mathscr{B}_1)$ and $\mathscr{G}_2 = (\mathscr{V}_2, \mathscr{A}_2, \mathscr{B}_2)$ be mPFGs such that $\mathscr{G}_1 \bullet \mathscr{G}_2$ or $\mathscr{G}_1 \otimes \mathscr{G}_2$ is strong. Then at least $\mathscr{G}_1$ or $\mathscr{G}_2$ must be strong.*

8.7 Balanced *m*-Polar Fuzzy Graphs

In this section, we define the density of an mPFG and balanced mPFG.

Definition 8.27 The **density** of an mPFG $\mathscr{G} = (\mathscr{V}, \mathscr{A}, \mathscr{B})$ is denoted by $\mathscr{D}(\mathscr{G}) = (p_1 \circ \mathscr{D}(\mathscr{G}), p_2 \circ \mathscr{D}(\mathscr{G}), \ldots, p_m \circ \mathscr{D}(\mathscr{G}))$, where

$$
p_i \circ \mathscr{D}(\mathscr{G}) = \frac{2\left(\sum\limits_{u,v \in \mathscr{V}} p_i \circ \mathscr{B}(uv) \right)}{\sum\limits_{u,v \in \mathscr{V}} \left(p_i \circ \mathscr{A}(u) \wedge p_i \circ \mathscr{A}(v) \right)}, \quad (i = 1, 2, \ldots, m).
$$

$\mathscr{G}$ is said to be **balanced** if

$$
p_i \circ \mathscr{D}(\mathscr{H}) \leq p_i \circ \mathscr{D}(\mathscr{G}), \quad (i = 1, 2, \ldots, m)
$$

for all non-empty subgraphs $\mathscr{H}$ of $\mathscr{G}$.

Example 8.11 Consider the 3-polar fuzzy graph $\mathscr{G} = (\mathscr{V}, \mathscr{A}, \mathscr{B})$ of $\mathscr{G}^* = (\mathscr{V}, E)$, where $\mathscr{V} = \{a, b, c\}$, $E = \{ab, bc, ca\}$, $\mathscr{A}(a) = (0.3, 0.4, 0.5)$, $\mathscr{A}(b) = (0.3, 0.4, 0.5)$, $\mathscr{A}(c) = (0.3, 0.4, 0.5)$, $\mathscr{B}(ab) = (0.1, 0.2, 0.2)$, $\mathscr{B}(bc) = (0.1, 0.2, 0.2)$, $\mathscr{B}(ca) = (0.1, 0.2, 0.2)$.

We have, $p_1 \circ \mathscr{D}(\mathscr{G}) = \dfrac{2\left(p_1 \circ \mathscr{B}(ab) + p_1 \circ \mathscr{B}(bc) + p_1 \circ \mathscr{B}(ca) \right)}{\left(p_1 \circ \mathscr{A}(a) \wedge p_1 \circ \mathscr{A}(b) + p_1 \circ \mathscr{A}(b) \wedge p_1 \circ \mathscr{A}(c) + p_1 \circ \mathscr{A}(c) \wedge p_1 \circ \mathscr{A}(a) \right)} =$

$\dfrac{2(0.1 + 0.1 + 0.1)}{0.3 + 0.3 + 0.3} = 0.67$.

Similarly, $p_2 \circ \mathcal{D}(\mathcal{G}) = 1$ and $p_3 \circ \mathcal{D}(\mathcal{G}) = 0.8$.

Hence, $\mathcal{D}(\mathcal{G}) = (0.67, 1, 0.8)$.

The non-empty subgraphs of $\mathcal{G}$ are $\mathcal{H}_1 = \{a, b\}$, $\mathcal{H}_2 = \{b, c\}$ and $\mathcal{H}_3 = \{c, a\}$.

Then $\mathcal{D}(\mathcal{H}_1) = (\frac{2 \times 0.1}{0.3}, \frac{2 \times 0.2}{0.4}, \frac{2 \times 0.2}{0.5}) = (0.67, 1, 0.8)$,

$\mathcal{D}(\mathcal{H}_2) = (\frac{2 \times 0.1}{0.3}, \frac{2 \times 0.2}{0.4}, \frac{2 \times 0.2}{0.5}) = (0.67, 1, 0.8)$ and

$\mathcal{D}(\mathcal{H}_3) = (\frac{2 \times 0.1}{0.3}, \frac{2 \times 0.2}{0.4}, \frac{2 \times 0.2}{0.5}) = (0.67, 1, 0.8)$.

We see that, $\mathcal{D}(\mathcal{H}_1) = \mathcal{D}(\mathcal{H}_2) = \mathcal{D}(\mathcal{H}_3) = \mathcal{D}(\mathcal{G}) = (0.67, 1, 0.8)$.

Hence, $\mathcal{G}$ is a balanced 3-polar fuzzy graph.

Theorem 8.18 *Any complete mPFG is balanced.*

Proof Let $\mathcal{G} = (\mathcal{V}, \mathcal{A}, \mathcal{B})$ be a complete mPFG and $\mathcal{H}$ be a non-empty subgraph of $\mathcal{G}$.

Then for $i = 1, 2, \ldots, m$

$$
p_i \circ \mathcal{D}(\mathcal{G}) = \frac{2 \left(\sum_{u,v \in \mathcal{V}} p_i \circ \mathcal{B}(uv) \right)}{\sum_{u,v \in \mathcal{V}} \left(p_i \circ \mathcal{A}(u) \wedge p_i \circ \mathcal{A}(v) \right)}
$$

$$
= \frac{2 \left(\sum_{u,v \in \mathcal{V}} p_i \circ \mathcal{A}(u) \wedge p_o \mathcal{A}(v) \right)}{\sum_{u,v \in \mathcal{V}} \left(p_i \circ \mathcal{A}(u) \wedge p_i \circ \mathcal{A}(v) \right)}
$$

$$
= 2
$$

and

$$
p_i \circ \mathcal{D}(\mathcal{H}) = \frac{2 \left(\sum_{u,v \in \mathcal{V}(\mathcal{H})} p_i \circ \mathcal{B}(uv) \right)}{\sum_{u,v \in \mathcal{V}(\mathcal{H})} \left(p_i \circ \mathcal{A}(u) \wedge p_i \circ \mathcal{A}(v) \right)}
$$

$$
\leq \frac{2 \left(\sum_{u,v \in \mathcal{V}(\mathcal{H})} p_i \circ \mathcal{A}(u) \wedge p_i \circ \mathcal{A}(v) \right)}{\sum_{u,v \in \mathcal{V}(\mathcal{H})} \left(p_i \circ \mathcal{A}(u) \wedge p_i \circ \mathcal{A}(v) \right)}
$$

$$
= 2 \text{ (where } \mathcal{V}(\mathcal{H}) \text{ represents the vertex of } \mathcal{H}).
$$

This shows that $\mathcal{G}$ is balanced. □

The converse of the above theorem is not necessarily true always. For example, the 3-polar fuzzy graph in Example 8.11 is balanced but not complete.

Below, we will discuss two types of mPFGs each with density equal to $\mathbf{1} = (1, 1, \ldots, 1)$.

Theorem 8.19 *Every self-complementary strong mPFG has density equal to $\mathbf{1} = (1, 1, \ldots, 1)$.*

Proof Let $\mathscr{G} = (\mathscr{V}, \mathscr{A}, \mathscr{B})$ be a self-complementary strong mPFG of $\mathscr{G}^* = (\mathscr{V}, E)$. Then we have for $xy \in \Omega$, $(i = 1, 2, \ldots, m)$

$$\sum_{x \neq y} p_i \circ \mathscr{B}(xy) = \frac{1}{2} \sum_{x \neq y} (p_i \circ \mathscr{A}(x) \wedge p_i \circ \mathscr{A}(y)).$$

Hence,

$$p_i \circ \mathscr{D}(\mathscr{G}) = \frac{2\left(\sum_{u, v \in \mathscr{V}} p_i \circ \mathscr{B}(uv)\right)}{\sum_{u, v \in \mathscr{V}} (p_i \circ \mathscr{A}(u) \wedge p_i \circ \mathscr{A}(v))} = 1 \ (i = 1, 2, \ldots, m), \ \text{(by the above)}$$

Thus, $\mathscr{D}(\mathscr{G}) = \mathbf{1}$. □

The converse of this theorem is not necessarily true in general.

Example 8.12 Consider the 3-polar fuzzy graph $\mathscr{G} = (\mathscr{V}, \mathscr{A}, \mathscr{B})$ be a 3-polar fuzzy graph of $G^* = (\mathscr{V}, E)$ where $\mathscr{V} = \{a, b, c\}$, $\mathscr{A}(a) = \mathscr{A}(b) = \mathscr{A}(c) = (1, 1, 1)$, $\mathscr{B}(ab) = (0.3, 0.3, 0.3)$, $\mathscr{B}(bc) = (0.2, 0.2, 0.2)$, $\mathscr{B}(ca) = (1, 1, 1)$. Then $\overline{\mathscr{G}} = (\mathscr{V}, \overline{\mathscr{A}}, \overline{\mathscr{B}})$ is an 3-polar fuzzy graph, where $\overline{\mathscr{A}} = \mathscr{A}$ and $\overline{\mathscr{B}}(ab) = (0.7, 0.7, 0.7)$, $\overline{\mathscr{B}}(bc) = (0.8, 0.8, 0.8)$, $\overline{\mathscr{B}}(ab) = (0, 0, 0)$.

Here, we see that $\mathscr{D}(\mathscr{G}) = (1, 1, 1)$ but $\mathscr{G} \ncong \overline{\mathscr{G}}$.

Theorem 8.20 *Let $\mathscr{G} = (\mathscr{V}, \mathscr{A}, \mathscr{B})$ be a strong mPFG such that for $uv \in \Omega$, $(i = 1, 2, \ldots, m)$*

$$p_i \circ \mathscr{B}(uv) = \frac{1}{2}(p_i \circ \mathscr{A}(u) \wedge p_i \circ \mathscr{A}(v)).$$

Then, $\mathscr{D}(\mathscr{G}) = \mathbf{1} = (1, 1, \ldots, 1)$.

Proof Since $\mathscr{G}$ is a strong mPFG such that for $uv \in \Omega$, $(i = 1, 2, \ldots, m)$
$p_i \circ \mathscr{B}(uv) = \frac{1}{2}(p_i \circ \mathscr{A}(u) \wedge p_i \circ \mathscr{A}(v))$, Therefore, by Proposition 8.16, $\mathscr{G}$ is self-complementary.
This implies that $\mathscr{D}(\mathscr{G}) = \mathbf{1}$ by Theorem 8.19. □

Below, necessary and sufficient conditions are given below for different products of two mPFGs to be balanced.

Theorem 8.21 *Let $\mathscr{G}_1 = (\mathscr{V}_1, \mathscr{A}_1, \mathscr{B}_1)$ and $\mathscr{G}_2 = (\mathscr{V}_2, \mathscr{A}_2, \mathscr{B}_2)$ be two mPFGs. Then, $\mathscr{D}(\mathscr{G}_k) \leq \mathscr{D}(\mathscr{G}_1 \sqcap \mathscr{G}_2)$ for $k = 1, 2$ if and only if $\mathscr{D}(\mathscr{G}_1) = \mathscr{D}(\mathscr{G}_2) = \mathscr{D}(\mathscr{G}_1 \sqcap \mathscr{G}_2)$.*

Proof Let $\mathscr{D}(\mathscr{G}_k) \leq \mathscr{D}(\mathscr{G}_1 \sqcap \mathscr{G}_2)$ for $k = 1, 2$.
Then for $i = 1, 2, \ldots, m$

$$p_i \circ \mathscr{D}(\mathscr{G}_1) = \frac{2\left(\sum_{u_1,u_2\in\mathscr{V}_1} p_i \circ \mathscr{B}_1(u_1u_2)\right)}{\sum_{u_1,u_2\in\mathscr{V}_1} \left(p_i \circ \mathscr{A}_1(u_1) \wedge p_i \circ \mathscr{A}_1(u_2)\right)}$$

$$\geq \frac{2\left(\sum_{\substack{u_1,u_2\in\mathscr{V}_1 \\ v_1,v_2\in\mathscr{V}_2}} p_i \circ \mathscr{B}_1(u_1u_2) \wedge \mathscr{A}_2(v_1) \wedge \mathscr{A}_2(v_2)\right)}{\sum_{\substack{u_1,u_2\in\mathscr{V}_1 \\ v_1,v_2\in\mathscr{V}_2}} \left(p_i \circ \mathscr{A}_1(u_1) \wedge p_i \circ \mathscr{A}_1(u_2) \wedge \mathscr{A}_2(v_1) \wedge \mathscr{A}_2(v_2)\right)}$$

$$= \frac{2\left(\sum_{\substack{u_1,u_2\in\mathscr{V}_1 \\ v_1,v_2\in\mathscr{V}_2}} p_i \circ \mathscr{B}_1(u_1u_2) \wedge p_i \circ \mathscr{B}_2(v_1v_2)\right)}{\sum_{\substack{u_1,u_2\in\mathscr{V}_1 \\ v_1,v_2\in\mathscr{V}_2}} \left(p_i \circ \mathscr{A}_1(u_1) \wedge p_i \circ \mathscr{A}_1(u_2) \wedge \mathscr{A}_2(v_1) \wedge \mathscr{A}_2(v_2)\right)}$$

$$= \frac{2\left(\sum_{\substack{u_1,u_2\in\mathscr{V}_1 \\ v_1,v_2\in\mathscr{V}_2}} p_i \circ (\mathscr{B}_1 \sqcap \mathscr{B}_2)(u_1, u_2)(v_1, v_2)\right)}{\sum_{\substack{u_1,u_2\in\mathscr{V}_1 \\ v_1,v_2\in\mathscr{V}_2}} \left(p_i \circ (\mathscr{A}_1 \sqcap \mathscr{A}_2)(u_1, u_2)(v_1, v_2)\right)}$$

$$= p_i \circ \mathscr{D}(\mathscr{G}_1 \sqcap \mathscr{G}_2).$$

Hence, $p_i \circ \mathscr{D}(\mathscr{G}_1) \geq p_i \circ \mathscr{D}(\mathscr{G}_1 \sqcap \mathscr{G}_2)$ for $i = 1, 2, \ldots, m$
i.e. $\mathscr{D}(\mathscr{G}_1) \geq \mathscr{D}(\mathscr{G}_1 \sqcap \mathscr{G}_2)$.
Similarly, $\mathscr{D}(\mathscr{G}_2) \geq \mathscr{D}(\mathscr{G}_1 \sqcap \mathscr{G}_2)$.
Therefore, $\mathscr{D}(\mathscr{G}_1) = \mathscr{D}(\mathscr{G}_2) = \mathscr{D}(\mathscr{G}_1 \sqcap \mathscr{G}_2)$.   □

Theorem 8.22 *Let $\mathscr{G}_1 = (\mathscr{V}_1, \mathscr{A}_1, \mathscr{B}_1)$ and $\mathscr{G}_2 = (\mathscr{V}_2, \mathscr{A}_2, \mathscr{B}_2)$ be two balanced mPFGs. Then $\mathscr{G}_1 \sqcap \mathscr{G}_2$ is balanced if and only if $\mathscr{D}(\mathscr{G}_1) = \mathscr{D}(\mathscr{G}_2) = \mathscr{D}(\mathscr{G}_1 \sqcap \mathscr{G}_2)$.*

Proof Suppose $\mathscr{D}(\mathscr{G}_1 \sqcap \mathscr{G}_2)$ is balanced.
Then $\mathscr{D}(\mathscr{G}_k) \leq \mathscr{D}(\mathscr{G}_1 \sqcap \mathscr{G}_2)$ for $k = 1, 2$ and by Theorem 8.21, $\mathscr{D}(\mathscr{G}_1) = \mathscr{D}(\mathscr{G}_2) = \mathscr{D}(\mathscr{G}_1 \sqcap \mathscr{G}_2)$.
Conversely, let $\mathscr{D}(\mathscr{G}_1) = \mathscr{D}(\mathscr{G}_2) = \mathscr{D}(\mathscr{G}_1 \sqcap \mathscr{G}_2)$ and $\mathscr{H}$ be a non-empty subgraph of $\mathscr{G}_1 \sqcap \mathscr{G}_2$.
Then there exist subgraphs $\mathscr{H}_1$ of $\mathscr{G}_1$ and $\mathscr{H}_2$ of $\mathscr{G}_2$.
Let $p_i \circ \mathscr{D}(\mathscr{G}_1) = p_i \circ \mathscr{D}(\mathscr{G}_2) = \frac{q_i}{r_i}$,
$p_i \circ \mathscr{D}(\mathscr{H}_1) = \frac{s_i}{t_i}$ and
$p_i \circ \mathscr{D}(\mathscr{H}_2) = \frac{a_i}{b_i}$ for $i = 1, 2, \ldots, m$ where $a_i, b_i, q_i, r_i, s_i, t_i \in \mathbb{R}$.
Since $\mathscr{G}_1$ and $\mathscr{G}_2$ are balanced, therefore for $i = 1, 2, \ldots, m$
$p_i \circ \mathscr{D}(\mathscr{H}_1) = \frac{s_i}{t_i} \leq p_i \circ \mathscr{D}(\mathscr{G}_1) = \frac{q_i}{r_i}$ and
$p_i \circ \mathscr{D}(\mathscr{H}_2) = \frac{a_i}{b_i} \leq p_i \circ \mathscr{D}(\mathscr{G}_2) = \frac{q_i}{r_i}$.
Thus, $s_i r_i + a_i r_i \leq t_i q_i + b_i q_i$
i.e. $\frac{s_i + a_i}{t_i + b_i} \leq \frac{q_i}{r_i}$.

Hence, $p_i \circ \mathscr{D}(\mathscr{H}) \leq \frac{s_i + a_i}{t_i + b_i} \leq \frac{q_i}{r_i} = p_i \circ \mathscr{D}(\mathscr{G}_1 \sqcap \mathscr{G}_2)$, $(i = 1, 2, \ldots, m)$.
Therefore, $\mathscr{G}_1 \sqcap \mathscr{G}_2$ is balanced. $\qquad\square$

Similarly, we have the following results.

Theorem 8.23 *Let $\mathscr{G}_1 = (\mathscr{A}_1, \mathscr{B}_1)$ and $\mathscr{G}_2 = (\mathscr{A}_2, \mathscr{B}_2)$ be two balanced mPFs. Then*
(i) $\mathscr{G}_1 \bullet \mathscr{G}_2$ is balanced $\Leftrightarrow \mathscr{D}(\mathscr{G}_1) = \mathscr{D}(\mathscr{G}_2) = \mathscr{D}(\mathscr{G}_1 \bullet \mathscr{G}_2)$.
(ii) $\mathscr{G}_1 \otimes \mathscr{G}_2$ is balanced $\Leftrightarrow \mathscr{D}(\mathscr{G}_1) = \mathscr{D}(\mathscr{G}_2) = \mathscr{D}(\mathscr{G}_1 \otimes \mathscr{G}_2)$.

References

1. M. Akram, D. Saleem, G. Ghorai, Energy of m-polar fuzzy digraphs, in *Handbook of Research on Advanced Applications of Graph Theory in Modern Society* (IGI Global, 2020), pp. 469–491. https://doi.org/10.4018/978-1-5225-9380-5.ch020
2. M. Akram, Bipolar fuzzy graphs. Inf. Sci. **181**(24), 5548–5564 (2011)
3. M. Akram, M. Sarwar, Novel applications of m-polar fuzzy competition graphs in decision support system. Neural Comput. Appl. **30**(10), 3145–3165 (2018)
4. M. Akram, N. Waseem, Certain metrics in m-polar fuzzy graphs. New Math. Nat. Comput. **12**(2), 135–155 (2016)
5. M. Akram, H.R. Younas, Certain types of irregular m-polar fuzzy graphs. J. Appl. Math. Comput. **53**, 365–382 (2017)
6. T. AL-Hawary, Complete fuzzy graphs. Int. J. Math. Combin. **4**, 26–34 (2011)
7. S. Bera, M. Pal, Certain types of m-polar interval-valued fuzzy graph. J. Intell. Fuzzy Syst. (2020). https://doi.org/10.3233/IFS-191587
8. S. Bera, M. Pal, On m-polar interval-valued fuzzy graph and its application. Fuzzy Inf. Eng. (2020). https://doi.org/10.1080/16168658.2020.1785993
9. K.R. Bhutani, On automorphism of fuzzy graphs. Pattern Recogn. Lett. **9**(3), 159–162 (1989)
10. K.R. Bhutani, A. Battou, On M-strong fuzzy graphs. Inf. Sci. **155**(1–2), 103–109 (2003)
11. J. Chen, S. Li, S. Ma, X. Wang, m-polar fuzzy sets: an extension of bipolar fuzzy sets (Hindwai Publishing Corporation). Sci. World J. 8 (2014), Article Id: 416530. https://doi.org/10.1155/2014/416530
12. G. Ghorai, M. Pal, Some properties of m-polar fuzzy graphs. Pacif. Sci. Rev. A Nat. Sci. Eng. **18**(1), 38–46 (2016)
13. G. Ghorai, M. Pal, On some operations and density of m-polar fuzzy graphs. Pac. Sci. Rev. A Nat. Sci. Eng. **17**(1), 14–22 (2015)
14. G. Ghorai, M. Pal, Some isomorphic properties of m-polar fuzzy graphs with applications. SpringerPlus **5**, 1–21 (2016)
15. G. Ghorai, M. Pal, A study on m-polar fuzzy planar graphs. Int. J. Comput. Sci. Math. **7**(3), 283–292 (2016)
16. G. Ghorai, M. Pal, Faces and dual of m-polar fuzzy planar graphs. J. Intell. Fuzzy Syst. **31**(3), 2043–2049 (2016)
17. G. Ghorai, M. Pal, Novel concepts of strongly edge irregular m-polar fuzzy graphs. Int. J. Appl. Comput. Math. **3**(4), 3321–3332 (2017)
18. G. Ghorai, M. Pal, On Degrees of m-polar fuzzy graphs with application. J. Uncertain Syst. **11**(4), 294–305 (2017)
19. G. Ghorai, M. Pal, Results of m-polar fuzzy graphs with application. J. Uncertain Syst. **12**(1), 47–55 (2018)
20. G. Ghorai, M. Pal, H. Rashmanlou, R.A. Borzooei, New concepts of regularity in product m-polar fuzzy graphs. Int. J. Math. Comput. **28**(4), 9–20 (2017)
21. F. Harary, *Graph Theory* (Narosa Publishing House, New Delhi, 2001)

22. K.M. Lee, Bipolar valued fuzzy sets and their basic operations, in *Proceedings of the International conference*, Bangkok, Thailand (2000), pp. 307–317
23. S. Mandal, S. Sahoo, G. Ghorai, M. Pal, Genus value of *m*-polar fuzzy graphs. J. Intell. Fuzzy Syst. **34**(3), 1947–1957 (2018)
24. S. Mandal, S. Sahoo, G. Ghorai, M. Pal, Application of strong arcs in *m*-polar fuzzy graphs. Neural Process. Lett. **50**(1), 771–784 (2018)
25. S. Mandal, S. Sahoo, G. Ghorai, M. Pal, Different types of arcs in *m*-polar fuzzy graphs with application. J. Multiple-valued Logic Soft Comput. **34**(3–4), 263–282 (2020)
26. J.N. Mordeson, P.S. Nair, *Fuzzy Graphs and Hypergraphs* (Physica Verlag, Heidelberg, 2000)
27. J.N. Mordeson, C.S. Peng, Operations on fuzzy graphs. Inf. Sci. **19**, 159–170 (1994)
28. A. Nagoorgani, K. Radha, On regular fuzzy graphs. J. Phys. Sci. **12**, 33–40 (2008)
29. P. S. Nair and S. C. Cheng, Cliques and fuzzy cliques in fuzzy graphs, in *IFSA World Congress and 20th NAFIPS International Conference*, vol. 4 (2001), pp. 2277–2280
30. H. Rashmanlou, S. Samanta, M. Pal, R.A. Borzooei, A study on bipolar fuzzy graphs. J. Intell. Fuzzy Syst. **28**(2), 571–580 (2015)
31. H. Rashmanlou, S. Samanta, M. Pal, R.A. Borzooei, Bipolar fuzzy graphs with categorical properties. Int. J. Comput. Intell. Syst. **8**(5), 808–818 (2015)
32. A. Rosenfeld, Fuzzy graphs, in *Fuzzy Sets and Their Applications*, ed. by L.A. Zadeh, K.S. Fu, M. Shimura (Academic, New York, 1975), pp. 77–95
33. M.S. Sunitha, A. Vijayakumar, Complement of fuzzy graphs. Indian J. Pure Appl. Math. **33**(9), 1451–1464 (2002)
34. A.A. Talebi, H. Rashmanlou, Complement and isomorphism on bipolar fuzzy graphs. Fuzzy Inf. Eng. **6**(4), 505–522 (2014)
35. W.R. Zhang, Bipolar fuzzy sets and relations: a computational framework for cognitive modeling and multiagent decision analysis, in *Proceedings of IEEE Conference* (1994), pp. 305–309

Chapter 9
Intuitionistic Fuzzy Graphs

List of abbreviation

$\overrightarrow{P}\,^m_{a,b}$	A directed Intuitionistic fuzzy path from a to b of length m
$N^+_m(v)$	Intuitionistic fuzzy m-step out-nbd of the node v
$N^-_m(v)$	Intuitionistic fuzzy m-step in-nbd of the node v
$C_m(\overrightarrow{G})$	m-step Intuitionistic fuzzy competition graph of $\overrightarrow{G}$
mS-IFCG	m-step intuitionistic fuzzy competition graph
IFCG	Intuitionistic fuzzy competition graph
UIFG	Undirected Intuitionistic fuzzy graph
DIFG	Directed Intuitionistic fuzzy graph
SNC	Strong node cover
SAC	Strong arc cover
CIFG	Complete intuitionistic fuzzy graphs
IFL	Intuitionistic fuzzy labeling
IFLG	Intuitionistic fuzzy labeling
IFI	Intuitionistic fuzzy interval
IFT	Intuitionistic fuzzy tolerance
IFTGs	Intuitionistic fuzzy tolerance graphs
IF ϕ-TG	Intuitionistic fuzzy ϕ-tolerance graph
IFICG	Intuitionistic fuzzy interval containment graph
IFMTG	Intuitionistic fuzzy min-tolerance graph
IFPG	Intuitionistic fuzzy planar graph
DOP	Degree of planarity

It is well known that in fuzzy set only one characteristic of an element is considered called membership value, i.e. the degree of belongingness (or degree of acceptance) of an element in the set. But, the counter complement part, i.e. the degree of non-belongingness (or degree of rejection) is not considered, though this case occurs naturally in almost all real situations. So to handle both these characteristics, Atanassov [3, 4] introduced a new type of fuzzy set called intuitionistic fuzzy set (IFS) and it became popular within a very short period and used it in many fields of science and

M. Pal et al., *Modern Trends in Fuzzy Graph Theory*, https://doi.org/10.1007/978-981-15-8803-7_9

engineering. Based on the concept of IFS, Parvathi and Karunambigai [19] defined intuitionistic fuzzy graph (IFG) in 2006.

In the first section, different types of products, viz. Cartesian product, composition, normal product, tensor product, modular product, homomorphic product, etc. are defined. Also, the degrees of the IFGs in these products are calculated. The concept of independent sets and coverings are used in modeling several real world problems. In this chapter, covering and matching in IFGs using strong arcs are defined. This area is enhanced by introducing new concepts like strong independent nodes, strong cover nodes, strong arc cover, and strong matching in IFGs. We have come up with new concepts in connectivity in intuitionistic fuzzy (IF) labeling graphs. The concept of strong arc, partial cut node, bridge and block along with IF labeling trees are investigated. In an ecosystem, two species can compete if and only if they have intersecting ecological niches. The competition can be defined indecently by using a food web for the ecosystem and this notion of competition gives rise to a competition graph (CG). But, crisp CG does not model properly the ecological system. Thus, for proper modeling, the intuitionistic fuzzy competition graphs (IFCGs) and their variants, viz. intuitionistic fuzzy k-CG (IFkCG), p-competition IFGs, and $m-$step IFCGs (mS-IFCGs) are introduced. Intuitionistic fuzzy tolerance graph (IFTG) is also discussed. The intuitionistic fuzzy planarity and genus of IFGs are investigated.

This chapter, devoted to IFGs, is written based on the papers [23–30].

For more information on IFGs readers may consult the articles [5, 9, 12, 13, 15–17, 20, 21, 32, 35].

9.1 Definitions and Basic Properties

The IFS defined by Atanasov [3] is cited below.

Definition 9.1 An **intuitionistic fuzzy set** A defined on the universal set U is characterized as follows $A = \{(x, \mu_A(x), \nu_A(x)) : x \in U\}$, where the membership function $\mu_A : U \to [0, 1]$ and non-membership function $\nu_A : U \to [0, 1]$ satisfy the condition $0 \le \mu_A(x) + \nu_A(x) \le 1$, for all $x \in U$.

Definition 9.2 The **support of an IFS** $A = (U, \mu_A, \nu_A)$ is defined as $Supp(A) = \{x \in U : \mu_A(x) \ge 0$ and $\nu_A(x) \le 1\}$. Also, the **support length** is $SL(A) = |Supp(A)|$.

Definition 9.3 The **core of an IFS** $A = (U, \mu_A, \nu_A)$ is defined as $Core(A) = \{x \in U : \mu_A(x) = 1$ and $\nu_A(x) = 0\}$. Also, the core length is $CL(A) = |Core(A)|$.

Now, we define the height of an IFS

Definition 9.4 The **height of an IFS** $A = (U, \mu_A, \nu_A)$ is defined as

$$h(A) = (\sup_{x \in U} \mu_A(x), \inf_{x \in U} \nu_A(x)) = (h_1(A), h_2(A)).$$

Based on the IFS, the IFG is defined. Throughout this chapter, we assume that $G^* = (V, E)$ is the underlying crisp graph of all IFGs.

Definition 9.5 Let $G^* = (V, E)$ be a crisp graph. An IFG is an algebraic structure $\mathscr{G} = (\mathscr{V}, \sigma, \mu)$ whose underlying graph is G^*. σ and μ have two components, viz. $\sigma = (\sigma_1, \sigma_2)$, $\mu = (\mu_1, \mu_2)$ and
(i) $\sigma_1 : \mathscr{V} \to [0, 1]$ and $\sigma_2 : \mathscr{V} \to [0, 1]$, denote the degree of membership and non-membership of the node $a \in \mathscr{V}$ respectively with $0 \le \sigma_1(a) + \sigma_2(a) \le 1$ for every $a \in \mathscr{V}$,
(ii) $\mu_1 : \mathscr{V} \times \mathscr{V} \to [0, 1]$ and $\mu_2 : \mathscr{V} \times \mathscr{V} \to [0, 1]$, where $\mu_1(a, b)$ and $\mu_2(a, b)$ denote the degree of membership and non-membership value of the edge (a, b) respectively such that

$$\mu_1(a, b) \le \min\{\sigma_1(a), \sigma_1(b)\} \quad \text{and} \tag{9.1}$$
$$\mu_2(a, b) \le \max\{\sigma_2(a), \sigma_2(b)\}, \tag{9.2}$$

$0 \le \mu_1(a, b) + \mu_2(a, b) \le 1$ for all $(a, b) \in E$.

Two types of IFGs are defined in literature. Different conditions are imposed on non-membership value of an edge with end vertices. These conditions are stated in Eqs. (9.2) and (9.3).

$$\mu_2(a, b) \ge \max\{\sigma_2(a), \sigma_2(b)\}. \tag{9.3}$$

The above condition is mentioned in [1]. Although the condition on membership value on edge remains same. With both these assumptions, lot of interesting results are presented on IFGs. But, there is a difference in the complement graphs. The **complement of an IFG** $\mathscr{G}$, with condition (9.3) is structurally similar to the crisp graph. But it does not hold the property $(\mathscr{G}^c)^c = \mathscr{G}$, $\mathscr{G}^c$ is the complement of $\mathscr{G}$. While the complement of IFG based on the condition (9.2) is similar to the complement of FG and it satisfies the property $(\mathscr{G}^c)^c = \mathscr{G}$.
An example of IFG is considered below.

Example 9.1 Let $\mathscr{V} = \{a, b, c, d\}$ and $E = \{ab, bc, cd, da, ac, bd\}$. Assume that $\sigma(a) = (0.3, 0.6)$, $\sigma(b) = (0.8, 0.2)$, $\sigma(c) = (0.2, 0.8)$, $\sigma(d) = (0.5, 0.4)$ and μ $(ab) = (0.2, 0.5)$, $\mu(bc) = (0.15, 0.70)$, $\mu(cd) = (0.15, 0.64)$, $\mu(da) = (0.3, 0.4)$, $\mu(ac) = (0.2, 0.7)$, $\mu(bd) = (0.5, 0.4)$.
The corresponding IFG is displayed in Fig. 9.1.

Definition 9.6 Let $\mathscr{F} = \{A_1, A_2, \ldots, A_n\}$ be a finite family of IFSs defined on the set U and consider each IFS as the vertex of the IFG. Let the vertex set be $\mathscr{V} = \{v_1, v_2, \ldots, v_n\}$. Then the **intuitionistic fuzzy intersection graph** of $\mathscr{F}$ is an IFG $Int(\mathscr{F}) = (\mathscr{V}, \sigma, \mu)$. The membership and non-membership value of the vertices are given by $\sigma_1(v_i) = h_1(A_i)$ and $\sigma_2(v_i) = h_2(A_i)$. Also, the membership and non-membership value of the edge (v_i, v_j) in $Int(\mathscr{F})$ are given by

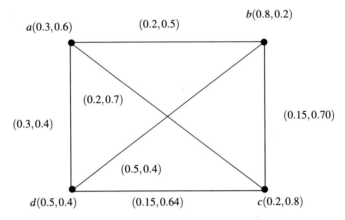

Fig. 9.1 An intuitionistic fuzzy graph

$$\mu_1(v_i, v_j) = \begin{cases} h_1(A_i \cap A_j), & \text{if } i \neq j \\ 0, & \text{if } i = j \end{cases}$$

and

$$\mu_2(v_i, v_j) = \begin{cases} h_2(A_i \cap A_j), & \text{if } i \neq j \\ 0, & \text{if } i = j. \end{cases}$$

If the fuzzy sets $\{A_1, A_2, \ldots, A_n\}$ are intuitionistic fuzzy intervals then the graph $Int(\mathscr{F})$ is called **intuitionistic fuzzy interval graph**.

9.2 Product of Intuitionistic Fuzzy Graphs

Operations on graphs play a very crucial role in graph theory. They are widely used in computer science, combinatorial applications, number theory, geometry, algebra, operations research, etc. Moradi [18] introduced tensor product of graphs, Khelladi and Gravier [11] found domination of the cross-product of graphs, Yang and Xu [34] introduced connectivity concept of the Cartesian product of two graphs.

In this section, five new operations on IFG such as Cartesian Product, composition, tensor product, normal product, modular product, and homomorphic product are defined. The main contribution of this section is from [23, 25].

9.2.1 Cartesian Product on IFGs

Definition 9.7 The **Cartesian product** of two IFGs, $\mathscr{G}' = (\mathscr{V}', E', \sigma', \mu')$ and $\mathscr{G}'' = (\mathscr{V}'', E'', \sigma'', \mu'')$ is defined as $\mathscr{G} = \mathscr{G}' \times \mathscr{G}'' = (\mathscr{V}, E, \sigma' \times \sigma'', \mu' \times \mu'')$, where $\mathscr{V} = \mathscr{V}' \times \mathscr{V}''$ and $E = \{((p_1, q_1), (p_2, q_2)) \mid p_1 = p_2, (q_1, q_2) \in E'' \text{ or } q_1 = q_2, (p_1, p_2) \in E'\}$ with

$$(\sigma_1' \times \sigma_1'')(p_1, q_1) = \sigma_1'(p_1) \wedge \sigma_1''(q_1)$$

$$(\sigma_2' \times \sigma_2'')(p_1, q_1) = \sigma_2'(p_1) \vee \sigma_2''(q_1)$$

for all $(p_1, q_1) \in \mathscr{V}' \times \mathscr{V}''$ and

$$(\mu_1' \times \mu_1'')((p_1, q_1), (p_2, q_2)) = \begin{cases} \sigma_1'(p_1) \wedge \mu_1''(q_1, q_2), & \text{if } p_1 = p_2, (q_1, q_2) \in E'' \\ \mu_1'(p_1, p_2) \wedge \sigma_1''(q_1), & \text{if } q_1 = q_2, (p_1, p_2) \in E' \end{cases}$$

$$(\mu_2' \times \mu_2'')((p_1, q_1), (p_2, q_2)) = \begin{cases} \sigma_2'(p_1) \vee \mu_2''(q_1, q_2), & \text{if } p_1 = q_2, (q_1, q_2) \in E'' \\ \mu_2'(p_1, p_2) \vee \sigma_2''(q_1), & \text{if } q_1 = q_2, (p_1, p_2) \in E'. \end{cases}$$

Definition 9.8 The degree of the node (p_1, q_1) in $\mathscr{G}' \times \mathscr{G}''$ is denoted by $deg_{\mathscr{G}' \times \mathscr{G}''}(p_1, q_1) = (deg_{1\mathscr{G}' \times \mathscr{G}''}(p_1, q_1), deg_{2\mathscr{G}' \times \mathscr{G}''}(p_1, q_1))$ where

$$deg_{1\mathscr{G}' \times \mathscr{G}''}(p_1, q_1) = \sum_{((p_1,q_1)(p_2,q_2)) \in E} (\mu_1' \times \mu_1'')((p_1, q_1)(p_2, q_2))$$

$$= \sum_{p_1 = p_2, (q_1, q_2) \in E''} \sigma_1'(p_1) \wedge \mu_1''(q_1, q_2) + \sum_{q_1 = q_2, (p_1, p_2) \in E'} \mu_1'(p_1, p_2) \wedge \sigma_1''(q_1)$$

$$deg_{2\mathscr{G}' \times \mathscr{G}''}(p_1, q_1) = \sum_{((p_1,q_1)(p_2,q_2)) \in E} (\mu_2' \times \mu_2'')((p_1, q_1)(p_2, q_2))$$

$$= \sum_{p_1 = p_2, (q_1, q_2) \in E''} \sigma_2'(p_1) \vee \mu_2''(q_1, q_2) + \sum_{q_1 = q_2, (p_1, p_2) \in E'} \mu_2'(p_1, p_2) \vee \sigma_2''(q_1).$$

Under certain conditions, the degree of a vertex (p_1, p_2) in $\mathscr{G}' \times \mathscr{G}''$ can be calculated in terms of the degree of p_1 in $\mathscr{G}'$ and that of p_2 in $\mathscr{G}''$.

Theorem 9.1 *Let $\mathscr{G}'$ and $\mathscr{G}''$ be two IFGs. If $\sigma_1' \geq \mu_1'', \sigma_2' \leq \mu_2'', \sigma_1'' \geq \mu_1', \sigma_2'' \leq \mu_2',$ then $deg_{\mathscr{G}' \times \mathscr{G}''}(p_1, q_1) = deg_{\mathscr{G}'}(p_1) + deg_{\mathscr{G}''}(q_1).$*

Proof From the definition of Cartesian product, we have

$$deg_{1\mathscr{G}'\times\mathscr{G}''}(p_1, q_1) = \sum_{((p_1,q_1)(p_2,q_2))\in E} (\mu_1' \times \mu_1'')((p_1, q_1)(p_2, q_2))$$

$$= \sum_{p_1=p_2,(q_1,q_2)\in E''} \sigma_1'(p_1) \wedge \mu_1''(q_1, q_2) + \sum_{q_1=q_2,(p_1,p_2)\in E'} \mu_1'(p_1, p_2) \wedge \sigma_1''(q_1)$$

$$= \sum_{(q_1,q_2)\in E''} \mu_1''(q_1, q_2) + \sum_{(p_1,p_2)\in E'} \mu_1'(p_1, p_2) \qquad [since \ \sigma_1' \geq \mu_1'' \ and \ \sigma_1'' \geq \mu_1']$$

$$= \sum_{(p_1,p_2)\in E'} \mu_1'(p_1, p_2) + \sum_{(q_1,q_2)\in E''} \mu_1''(q_1, q_2)$$

$$= deg_{1\mathscr{G}'}(p_1) + deg_{1\mathscr{G}''}(q_1).$$

and

$$deg_{2\mathscr{G}'\times\mathscr{G}''}(p_1, q_1) = \sum_{((p_1,q_1)(p_2,q_2))\in E} (\mu_2' \times \mu_2'')((p_1, q_1)(p_2, q_2))$$

$$= \sum_{p_1=p_2,(q_1,q_2)\in E''} \sigma_2'(p_1) \vee \mu_2''(q_1, q_2) + \sum_{q_1=q_2,(p_1,p_2)\in E'} \mu_2'(p_1, p_2) \vee \sigma_2''(q_1)$$

$$= \sum_{(q_1,q_2)\in E''} \mu_2''(q_1, q_2) + \sum_{(p_1,p_2)\in E'} \mu_2'(p_1, p_2) \qquad [since \ \sigma_2' \leq \mu_2'' \ and \ \sigma_2'' \leq \mu_2']$$

$$= \sum_{(p_1,p_2)\in E'} \mu_2'(p_1, p_2) + \sum_{(q_1,q_2)\in E''} \mu_2''(q_1, q_2)$$

$$= deg_{2\mathscr{G}'}(p_1) + deg_{2\mathscr{G}''}(q_1).$$

Hence, $deg_{\mathscr{G}'\times\mathscr{G}''}(p_1, q_1) = deg_{\mathscr{G}'}(p_1) + deg_{\mathscr{G}''}(q_1).$ □

9.2.2 Composition on IFGs

Here, **composition of two IFGs** is defined and that the degree of a vertex is determined.

Definition 9.9 The composition of two IFGs $\mathscr{G}'$ and $\mathscr{G}''$ is defined as an IFG $\mathscr{G} = \mathscr{G}'[\mathscr{G}''] = (\mathscr{V}, E, \sigma' \circ \sigma'', \mu' \circ \mu'')$ where $\mathscr{V} = \mathscr{V}' \times \mathscr{V}''$ and $E = \{((p_1, q_1), (p_2, q_2)) \mid p_1 = p_2, (q_1, q_2) \in E'' \ or \ q_1 = q_2, (p_1, p_2) \in E'\} \cup \{((p_1, q_1), (p_2, q_2)) \mid q_1 \neq q_2, (p_1, p_2) \in E'\}$ with

$$(\sigma_1' \circ \sigma_1'')(p_1, q_1) = \sigma_1'(p_1) \wedge \sigma_1''(q_1)$$

$$(\sigma_2' \circ \sigma_2'')(p_1, q_1) = \sigma_2'(p_1) \vee \sigma_2''(q_1)$$

for all $(p_1, q_1) \in \mathcal{V}' \times \mathcal{V}''$ and

$$(\mu_1' \circ \mu_1'')((p_1, q_1), (p_2, q_2)) = \begin{cases} \sigma_1'(p_1) \wedge \mu_1''(q_1, q_2), & \text{if } p_1 = p_2, (q_1, q_2) \in E'' \\ \mu_1'(u_1, p_2) \wedge \sigma_1''(q_1), & \text{if } q_1 = q_2, (p_1, p_2) \in E' \\ \mu_1'(p_1, p_2) \wedge \sigma_1'(q_1) \wedge \sigma_1''(q_2), & \text{if } q_1 \neq q_2, (p_1, p_2) \in E' \end{cases}$$

$$(\mu_2' \circ \mu_2'')((p_1, q_1), (p_2, q_2)) = \begin{cases} \sigma_2'(p_1) \vee \mu_2''(q_1, q_2), & \text{if } p_1 = p_2, (q_1, q_2) \in E'' \\ \mu_2'(p_1, p_2) \vee \sigma_2''(q_1), & \text{if } q_1 = q_2, (p_1, p_2) \in E' \\ \mu_2'(p_1, p_2) \vee \sigma_2'(q_1) \vee \sigma_1''(q_2), & \text{if } q_1 \neq q_2, (p_1, p_2) \in E'. \end{cases}$$

Definition 9.10 The degree of the node (p_1, q_1) in $deg_{\mathcal{G}'[\mathcal{G}'']}$ is denoted by $deg_{\mathcal{G}'[\mathcal{G}'']}(p_1, q_1) = (deg_{1\mathcal{G}'[\mathcal{G}'']}(p_1, q_1), deg_{2\mathcal{G}'[\mathcal{G}'']}(p_1, q_1))$, where

$$deg_{1\mathcal{G}'[\mathcal{G}'']}(p_1, q_1) = \sum_{((p_1, q_1)(p_2, q_2)) \in E} (\mu_1' \circ \mu_1'')((p_1, q_1)(p_2, q_2))$$

$$= \sum_{p_1 = p_2, (q_1, q_2) \in E''} \sigma_1'(p_1) \wedge \mu_1''(q_1, q_2) + \sum_{q_1 = q_2, (p_1, p_2) \in E'} \mu_1'(p_1, p_2) \wedge \sigma_1''(q_1)$$

$$+ \sum_{q_1 \neq q_2, (p_1, p_2) \in E'} \mu_1'(p_1, p_2) \wedge \sigma_1'(q_1) \wedge \sigma_1''(q_2).$$

$$deg_{2\mathcal{G}'[\mathcal{G}'']}(p_1, q_1) = \sum_{((p_1, q_1)(p_2, q_2)) \in E} (\mu_2' \circ \mu_2'')((p_1, q_1)(p_2, q_2))$$

$$= \sum_{p_1 = p_2, (q_1, q_2) \in E''} \sigma_2'(p_1) \vee \mu_2''(q_1, q_2) + \sum_{q_1 = q_2, (p_1, p_2) \in E'} \mu_2'(p_1, p_2) \vee \sigma_2''(q_1)$$

$$+ \sum_{q_1 \neq q_2, (p_1, p_2) \in E'} \mu_2'(p_1, p_2) \vee \sigma_2'(q_1) \vee \sigma_1''(q_2).$$

Theorem 9.2 Let $\mathcal{G}'$ and $\mathcal{G}''$ be two IFGs. If $\sigma_1' \geq \mu_1''$, $\sigma_2' \leq \mu_2''$ and $\sigma_1'' \geq \mu_1'$, $\sigma_2'' \leq \mu_2'$, then $deg_{\mathcal{G}'[\mathcal{G}'']}(p_1, q_1) = |\mathcal{V}''| deg_{\mathcal{G}'}(p_1) + deg_{\mathcal{G}''}(q_1)$.

9.2.3 Tensor Product of Two IFGs

Now, tensor product of two IFGs is defined and the formula to compute the degree of each vertex in the resulted graph is presented.

Definition 9.11 The **tensor product of two IFGs** $\mathcal{G}'$ and $\mathcal{G}''$ is defined as an IFG $\mathcal{G} = \mathcal{G}' \otimes \mathcal{G}'' = (\mathcal{V}, E, \sigma' \otimes \sigma'', \mu' \otimes \mu'')$ where $\mathcal{V} = \mathcal{V}' \times \mathcal{V}''$ and $E = \{((p_1, q_1), (p_2, q_2)) \mid (p_1, p_2) \in E', (q_1, q_2) \in E''\}$ with

$$(\sigma_1' \otimes \sigma_1'')(p_1, q_1) = \sigma_1'(p_1) \wedge \sigma_1''(q_1)$$

$$(\sigma_2' \otimes \sigma_2'')(p_1, q_1) = \sigma_2'(p_1) \vee \sigma_2''(q_1)$$

for all $(p_1, q_1) \in \mathscr{V}' \times \mathscr{V}''$ and

$$(\mu_1' \otimes \mu_1'')((p_1, q_1), (p_2, q_2)) = \mu_1'(p_1, p_2) \wedge \mu_1''(q_1, q_2)$$

$$(\mu_2' \circ \mu_2'')((p_1, q_1), (p_2, q_2)) = \mu_2'(p_1, p_2) \vee \mu_2''(q_1, q_2)$$

for all $(p_1, p_2) \in E'$, $(q_1, q_2) \in E''$.

The degree of a vertex in $\mathscr{G}' \otimes \mathscr{G}''$ is defined below.

Definition 9.12 Let $\mathscr{G} = \mathscr{G}' \otimes \mathscr{G}'' = (\mathscr{V}, E, \sigma' \otimes \sigma'', \mu' \otimes \mu'')$ be the tensor product of two IFGs $\mathscr{G}'$ and $\mathscr{G}''$. Then the degree of the node (p_1, q_1) in $\mathscr{V}$ is denoted by $deg_{\mathscr{G}' \otimes \mathscr{G}''}(p_1, q_1) = (deg_{1\mathscr{G}' \otimes \mathscr{G}''}(p_1, q_1), deg_{2\mathscr{G}' \otimes \mathscr{G}''}(p_1, q_1))$ and defined by

$$deg_{1\mathscr{G}' \otimes \mathscr{G}''}(p_1, q_1) = \sum_{((p_1,q_1)(p_2,q_2)) \in E} (\mu_1' \otimes \mu_1'')((p_1, q_1)(p_2, q_2))$$

$$= \sum_{(p_1,p_2) \in E'} \mu_1'(p_1, p_2) \wedge \mu_1''(q_1, q_2),$$

$$deg_{2\mathscr{G}' \otimes \mathscr{G}''}(p_1, q_1) = \sum_{((p_1,q_1)(p_2,q_2)) \in E} (\mu_2' \otimes \mu_2'')((p_1, q_1)(p_2, q_2))$$

$$= \sum_{(p_1,p_2) \in E'} \mu_2'(p_1, p_2) \vee \mu_2''(q_1, q_2).$$

Theorem 9.3 Let $\mathscr{G}'$ and $\mathscr{G}''$ be two IFGs. If $\mu_1'' \geq \mu_1'$, $\mu_2'' \leq \mu_2'$, then $deg_{\mathscr{G}' \otimes \mathscr{G}''}(p_1, q_1) = deg_{\mathscr{G}'}(p_1)$ and if $\mu_1' \geq \mu_1''$, $\mu_2' \leq \mu_2''$, then $deg_{\mathscr{G}' \otimes \mathscr{G}''}(p_1, q_1) = deg_{\mathscr{G}''}(q_1)$.

9.2.4 Normal Product of Two IFGs

We define the degree of a node in normal product of two IFGs.

Definition 9.13 The **normal product of two IFGs** $\mathscr{G}'$ and $\mathscr{G}''$ is defined as an IFG $\mathscr{G} = \mathscr{G}' \bullet \mathscr{G}'' = (\mathscr{V}, E, \sigma' \bullet \sigma'', \mu' \bullet \mu'')$ where $\mathscr{V} = \mathscr{V}' \times \mathscr{V}''$ and $E = \{((p_1, q_1), (p_2, q_2)) \mid p_1 = p_2, (q_1, q_2) \in E'' \text{ or } q_1 = q_2, (p_1, p_2) \in E'\} \cup \{((p_1, q_1), (p_2, q_2)) \mid (p_1, p_2) \in E', (q_1, q_2) \in E''\}$ with

$$(\sigma_1' \bullet \sigma_1'')(p_1, q_1) = \sigma_1'(p_1) \wedge \sigma_1''(q_1)$$

$$(\sigma'_2 \bullet \sigma''_2)(p_1, q_1) = \sigma'_2(p_1) \vee \sigma''_2(q_1)$$

for all $(p_1, q_1) \in \mathscr{V}' \times \mathscr{V}''$ and

$$(\mu'_1 \bullet \mu''_1)((p_1, q_1), (p_2, q_2)) = \begin{cases} \sigma'_1(p_1) \wedge \mu''_1(q_1, q_2), & \text{if } p_1 = p_2, (q_1, q_2) \in E'' \\ \mu'_1(p_1, p_2) \wedge \sigma''_1(q_1), & \text{if } q_1 = q_2, (p_1, p_2) \in E' \\ \mu'_1(p_1, p_2) \wedge \mu''_1(q_1, q_2), & \text{if } (p_1, p_2) \in E', (q_1, q_2) \in E'' \end{cases}$$

$$(\mu'_2 \bullet \mu''_2)((p_1, q_1), (p_2, q_2)) = \begin{cases} \sigma'_2(p_1) \vee \mu''_2(q_1, q_2), & \text{if } p_1 = p_2, (q_1, q_2) \in E'' \\ \mu'_2(p_1, p_2) \vee \sigma''_2(q_1), & \text{if } q_1 = q_2, (p_1, p_2) \in E' \\ \mu'_2(p_1, p_2) \vee \mu''_2(q_1, q_2), & \text{if } (p_1, p_2) \in E', (q_1, q_2) \in E''. \end{cases}$$

Definition 9.14 Let $\mathscr{G} = \mathscr{G}' \bullet \mathscr{G}'' = (\mathscr{V}, E, \sigma' \bullet \sigma'', \mu' \bullet \mu'')$ be the normal product of two IFGs $\mathscr{G}'$ and $\mathscr{G}''$. Then the degree of the node (p_1, q_1) in $\mathscr{V}$ is symbolised by $deg_{\mathscr{G}' \bullet \mathscr{G}''}(p_1, q_1) = (deg_{1\mathscr{G}' \bullet \mathscr{G}''}(p_1, q_1), deg_{2\mathscr{G}' \bullet \mathscr{G}''}(p_1, q_1))$ and is defined by

$$deg_{1\mathscr{G}' \bullet \mathscr{G}''}(p_1, q_1) = \sum_{((p_1, q_1)(p_2, q_2)) \in E} (\mu'_1 \bullet \mu''_1)((p_1, q_1)(p_2, q_2))$$

$$= \sum_{p_1 = p_2, (q_1, q_2) \in E''} \sigma'_1(p_1) \wedge \mu''_1(q_1, q_2) + \sum_{q_1 = q_2, (p_1, p_2) \in E'} \mu'_1(p_1, p_2) \wedge \sigma''_1(q_1)$$

$$+ \sum_{(p_1, p_2) \in E', (q_1, q_2) \in E''} \mu'_1(p_1, p_2) \wedge \mu''_1(q_1, q_2)$$

$$deg_{2\mathscr{G}' \bullet \mathscr{G}''}(p_1, q_1) = \sum_{((p_1, q_1)(p_2, q_2)) \in E} (\mu'_2 \bullet \mu''_2)((p_1, q_1)(p_2, q_2))$$

$$= \sum_{p_1 = p_2, (q_1, q_2) \in E''} \sigma'_2(p_1) \vee \mu''_2(q_1, q_2) + \sum_{q_1 = q_2, (p_1, p_2) \in E'} \mu'_2(p_1, p_2) \vee \sigma''_2(q_1)$$

$$+ \sum_{(p_1, p_2) \in E', (q_1, q_2) \in E''} \mu'_2(p_1, p_2) \vee \mu''_2(q_1, q_2).$$

Theorem 9.4 Let $\mathscr{G}'$ and $\mathscr{G}''$ be two IFGs. If $\sigma'_1 \geq \mu''_1, \sigma'_2 \leq \mu''_2, \sigma''_1 \geq \mu'_1, \sigma''_2 \leq \mu'_2, \mu'_1 \leq \mu''_1$ and $\mu'_2 \geq \mu''_2$, then $deg_{\mathscr{G}' \bullet \mathscr{G}''}(p_1, q_1) = |\mathscr{V}''| deg_{\mathscr{G}'}(p_1) + deg_{\mathscr{G}''}(q_1)$.

9.2.5 Modular Product of Two IFGs

Definition 9.15 The **modular product of two IFGs** $\mathscr{G}'$ and $\mathscr{G}''$ is defined as an IFG $\mathscr{G} = \mathscr{G}' \odot \mathscr{G}'' = (\mathscr{V}, E, \sigma' \odot \sigma'', \mu' \odot \mu'')$, where $\mathscr{V} = \mathscr{V}' \odot \mathscr{V}'' = \{(p_1, q_1) \mid p_1 \in \mathscr{V}' \text{ and } q_1 \in \mathscr{V}''\}$ and $E = E' \odot E'' = \{((p_1, q_1), (p_2, q_2)) \mid (p_1, p_2) \in E', (q_1, q_2) \in E'' \text{ or } (p_1, p_2) \notin E', (q_1, q_2) \notin E''\}$ with

$$(\sigma_1' \odot \sigma_1'')(p_1, q_1) = \sigma_1'(p_1) \wedge \sigma_1''(q_1)$$

$$(\sigma_2' \odot \sigma_2'')(p_1, q_1) = \sigma_2'(p_1) \vee \sigma_2''(q_1)$$

for all $(p_1, q_1) \in \mathcal{V}$ and

$(\mu_1' \odot \mu_1'')((p_1, q_1), (p_2, q_2))$
$$= \begin{cases} \mu_1'(p_1, p_2) \wedge \mu_1''(q_1, q_2), & \text{if } (p_1, p_2) \in E', (q_1, q_2) \in E'' \\ \sigma_1'(p_1) \wedge \sigma_1'(p_2) \wedge \sigma_1''(q_1) \wedge \sigma_1''(q_2), & \text{if } (p_1, p_2) \notin E', (q_1, q_2) \notin E'' \end{cases}$$

$(\mu_2' \odot \mu_2'')((p_1, q_1), (p_2, q_2))$
$$= \begin{cases} \mu_2'(p_1, p_2) \vee \mu_2''(q_1, q_2), & \text{if } (p_1, p_2) \in E', (q_1, q_2) \in E'' \\ \sigma_2'(p_1) \vee \sigma_2'(p_2) \vee \sigma_2''(q_1) \vee \sigma_2''(q_2), & \text{if } (p_1, p_2) \notin E', (q_1, q_2) \notin E''. \end{cases}$$

Definition 9.16 Let $\mathcal{G} = \mathcal{G}' \odot \mathcal{G}'' = (\mathcal{V}, E, \sigma' \odot \sigma'', \mu' \odot \mu'')$ be the modular product of two IFGs $\mathcal{G}'$ and $\mathcal{G}''$. Then the degree of the node (p_1, q_1) in $\mathcal{V}$ is denoted by $deg_{\mathcal{G}' \odot \mathcal{G}''}(p_1, q_1) = (deg_{1\mathcal{G}' \odot \mathcal{G}''}(p_1, q_1), deg_{2\mathcal{G}' \odot \mathcal{G}''}(p_1, q_1))$ and defined by

$$deg_{1\mathcal{G}' \odot \mathcal{G}''}(p_1, q_1) = \sum_{((p_1,q_1)(p_2,q_2)) \in E} (\mu_1' \odot \mu_1'')((p_1, q_1)(p_2, q_2))$$

$$= \sum_{(p_1,p_2) \in E', (q_1,q_2) \in E''} \mu_1'(p_1, p_2) \wedge \mu_1''(q_1, q_2)$$

$$+ \sum_{(p_1,p_2) \notin E', (q_1,q_2) \notin E''} \sigma_1'(p_1) \wedge \sigma_1'(p_2) \wedge \sigma_1''(q_1) \wedge \sigma_1''(q_2)$$

and $$deg_{2\mathcal{G}' \odot \mathcal{G}''}(p_1, q_1) = \sum_{((p_1,q_1)(p_2,q_2)) \in E} (\mu_2' \odot \mu_2'')((p_1, q_1)(p_2, q_2))$$

$$= \sum_{(p_1,p_2) \in E', (q_1,q_2) \in E''} \mu_2'(p_1, p_2) \vee \mu_2''(q_1, q_2)$$

$$+ \sum_{(p_1,p_2) \notin E', (q_1,q_2) \notin E''} \sigma_2'(p_1) \vee \sigma_2'(p_2) \vee \sigma_2''(q_1) \vee \sigma_2''(q_2).$$

Theorem 9.5 Let $\mathcal{G}'$ and $\mathcal{G}''$ be two IFGs. If both IFGs are complete and $\mu_1' \leq \mu_1''$, $\mu_2' \geq \mu_2''$, then $deg_{\mathcal{G}' \odot \mathcal{G}''}(p_1, q_1) = deg_{\mathcal{G}'}(p_1)$.

9.2.6 Homomorphic Product of Two IFGs

Definition 9.17 The **homomorphic product of two IFGs** $\mathcal{G}'$ and $\mathcal{G}''$ is defined as an IFG $\mathcal{G} = \mathcal{G}' \diamond \mathcal{G}'' = (\mathcal{V}, E, \sigma' \diamond \sigma'', \mu' \diamond \mu'')$ where $\mathcal{V} = \mathcal{V}' \diamond \mathcal{V}'' = \{(p_1, q_1) \mid p_1 \in \mathcal{V}' \text{ and } q_1 \in \mathcal{V}''\}$ and $E = E' \diamond E'' = \{((p_1, q_1), (p_2, q_2)) \mid p_1 = p_2, (q_1, q_2) \in E'' \text{ or } (p_1, p_2) \in E', (q_1, q_2) \notin E''\}$ with

$$(\sigma_1' \diamond \sigma_1'')(p_1, q_1) = \sigma_1'(p_1) \wedge \sigma_1''(q_1)$$

$$(\sigma_2' \diamond \sigma_2'')(p_1, q_1) = \sigma_2'(p_1) \vee \sigma_2''(q_1)$$

for all $(p_1, q_1) \in \mathcal{V}$ and

$$(\mu_1' \diamond \mu_1'')((p_1, q_1), (p_2, q_2))$$
$$= \begin{cases} \sigma_1'(p_1) \wedge \mu_1''(q_1, q_2), & \text{if } p_1 = p_2, (q_1, q_2) \in E'' \\ \mu_1'(p_1, p_2) \wedge \sigma_1''(q_1) \wedge \sigma_1''(q_2), & \text{if } (p_1, p_2) \in E', (q_1, q_2) \notin E''. \end{cases}$$
$$(\mu_2' \diamond \mu_2'')((p_1, q_1), (p_2, q_2))$$
$$= \begin{cases} \sigma_2'(p_1) \vee \mu_2''(q_1, q_2), & \text{if } p_1 = p_2, (q_1, q_2) \in E'' \\ \mu_2'(p_1, p_2) \vee \sigma_2''(q_1) \vee \sigma_2''(q_2), & \text{if } (p_1, p_2) \in E', (q_1, q_2) \notin E''. \end{cases}$$

Definition 9.18 Let $\mathscr{G} = \mathscr{G}' \diamond \mathscr{G}'' = (\mathscr{V}, E, \sigma' \diamond \sigma'', \mu' \diamond \mu'')$ be the homomorphic product of two IFGs $\mathscr{G}'$ and $\mathscr{G}''$. Then the degree of the node (p_1, q_1) in $\mathscr{V}$ is denoted by $deg_{\mathscr{G}' \diamond \mathscr{G}''}(p_1, q_1) = (deg_{1\mathscr{G}' \diamond \mathscr{G}''}(p_1, q_1), deg_{2\mathscr{G}' \diamond \mathscr{G}''}(p_1, q_1))$ and defined by

$$deg_{1\mathscr{G}' \diamond \mathscr{G}''}(p_1, q_1) = \sum_{((p_1,q_1)(p_2,q_2)) \in E} (\mu_1' \diamond \mu_1'')((p_1, q_1)(p_2, q_2))$$
$$= \sum_{p_1=p_2, (q_1,q_2) \in E''} \sigma_1'(p_1) \wedge \mu_1''(q_1, q_2) + \sum_{(p_1,p_2) \in E', (q_1,q_2) \notin E''} \mu_1'(p_1, p_2) \wedge \sigma_1''(q_1) \wedge \sigma_1''(q_2)$$

and $deg_{2\mathscr{G}' \diamond \mathscr{G}''}(p_1, q_1) = \sum_{((p_1,q_1)(p_2,q_2)) \in E} (\mu_2' \diamond \mu_2'')((p_1, q_1)(p_2, q_2))$
$$= \sum_{p_1=p_2, (q_1,q_2) \in E''} \sigma_2'(p_1) \vee \mu_2''(q_1, q_2) + \sum_{(p_1,p_2) \in E', (q_1,q_2) \notin E''} \mu_2'(p_1, p_2) \vee \sigma_2''(q_1) \vee \sigma_2''(q_2).$$

Theorem 9.6 *Let $\mathscr{G}'$ and $\mathscr{G}''$ be two IFGs. If $\mathscr{G}''$ is a complete intuitionistic fuzzy graph (CIFG) and $\sigma_1' \leq \mu_1''$, $\sigma_2' \geq \mu_2''$, then $deg_{\mathscr{G}' \diamond \mathscr{G}''}(p_1, q_1) = (|\mathscr{V}''| - 1)\sigma'(p_1)$.*

Proof From the expression of the degree of node in homomorphic product, one can write

$$deg_{1\mathscr{G}' \diamond \mathscr{G}''}(p_1, q_1) = \sum_{((p_1,q_1)(p_2,q_2)) \in E} (\mu_1' \diamond \mu_1'')((p_1, q_1)(p_2, q_2))$$
$$= \sum_{p_1=p_2, (q_1,q_2) \in E''} \sigma_1'(p_1) \wedge \mu_1''(q_1, q_2) + \sum_{(p_1,p_2) \in E', (q_1,q_2) \notin E''} \mu_1'(p_1, p_2) \wedge \sigma_1''(q_1) \wedge \sigma_1''(q_2)$$
$$= \sum_{p_1=p_2, (q_1,q_2) \in E''} \sigma_1'(p_1) \wedge \mu_1''(q_1, q_2) \quad [\text{Since } \mathscr{G}'' \text{ is an CIFG}]$$
$$= \sum_{p_1=p_2} \sigma_1'(p_1) \quad [\text{Since } \sigma_1' \leq \mu_1'']$$
$$= (|\mathscr{V}''| - 1)\sigma_1'(p_1).$$

and $deg_{2\mathscr{G}' \diamond \mathscr{G}''}(p_1, q_1) = \sum_{((p_1,q_1)(p_2,q_2)) \in E} (\mu_2' \diamond \mu_2'')((p_1, q_1)(p_2, q_2))$
$$= \sum_{p_1=p_2, (q_1,q_2) \in E''} \sigma_2'(p_1) \vee \mu_2''(q_1, q_2) + \sum_{(p_1,p_2) \in E', (q_1,q_2) \notin E''} \mu_2'(p_1, p_2) \vee \sigma_2''(q_1) \vee \sigma_2''(q_2)$$
$$= \sum_{p_1=p_2, (q_1,q_2) \in E''} \sigma_2'(p_1) \vee \mu_2''(q_1, q_2) \quad [\text{Since } \mathscr{G}'' \text{ is an CIFG}]$$
$$= \sum_{p_1=p_2} \sigma_2'(p_1) \quad [\text{Since } \sigma_2' \geq \mu_2'']$$
$$= (|\mathscr{V}''| - 1)\sigma_2'(p_1).$$

Hence, $deg_{\mathscr{G}' \diamond \mathscr{G}''}(p_1, q_1) = (|\mathscr{V}''| - 1)\sigma'(p_1)$. $\qquad \square$

9.2.7 Direct Product of Two IFGs

Now, direct product of two IFGs is defined.

Definition 9.19 The **direct product of two IFGs** $\mathcal{G}'$ and $\mathcal{G}''$ such that $\mathcal{V}' \cap \mathcal{V}'' = \phi$, is defined to be the IFG $\mathcal{G}' \sqcap \mathcal{G}'' = (\mathcal{V}, E, \sigma' \sqcap \sigma'', \mu' \sqcap \mu'')$, where $\mathcal{V} = \mathcal{V}' \times \mathcal{V}''$, $E = \{((p_1, q_1), (p_2, q_2)) \mid (p_1, p_2) \in E', (q_1, q_2) \in E''\}$. The value of membership and value of non-membership of the node (p, q) in $\mathcal{G}' \sqcap \mathcal{G}''$ are given by

$$(\sigma_1' \sqcap \sigma_1'')(p, q) = \sigma_1'(p) \wedge \sigma_1''(q)$$

and

$$(\sigma_2' \sqcap \sigma_2'')(p, q) = \sigma_2'(p) \vee \sigma_2''(q).$$

Also, the values of membership and value non-membership of the arc $((p_1, q_1), (p_2, q_2))$ in $\mathcal{G}' \sqcap \mathcal{G}''$ are given by

$$(\mu_1' \sqcap \mu_1'')((p_1, q_1), (p_2, q_2)) = \mu_1'(p_1, p_2) \wedge \mu_1''(q_1, q_2)$$

$$(\mu_2' \sqcap \mu_2'')((p_1, q_1), (p_2, q_2)) = \mu_2'(p_1, p_2) \vee \mu_2''(q_1, q_2).$$

Theorem 9.7 *If $\mathcal{G}'$ and $\mathcal{G}''$ are strong IFGs, then $\mathcal{G}' \sqcap \mathcal{G}''$ is also strong.*

Proof Since $\mathcal{G}'$ and $\mathcal{G}''$ are strong IFGs, $\mu_1'(p_1, p_2) = \sigma_1'(p_1) \wedge \sigma_1''(p_2)$, $\mu_2'(p_1, p_2) = \sigma_2'(p_1) \vee \sigma_2''(p_2)$, $\mu_1''(q_1, q_2) = \sigma_1''(q_1) \wedge \sigma_1''(q_2)$ and $\mu_2''(q_1, q_2) = \sigma_2''(q_1) \vee \sigma_2''(q_2)$ for all $(p_1, p_2) \in E'$ and $(q_1, q_2) \in E''$.
Now,

$$\begin{aligned}
(\mu_1' \sqcap \mu_1'')((p_1, q_1), (p_2, q_2)) &= \mu_1'(p_1, p_2) \wedge \mu_1''(q_1, q_2) \\
&= [\sigma_1'(p_1) \wedge \sigma_1''(p_2)] \wedge [\sigma_1''(q_1) \wedge \sigma_1''(q_2)] \\
&= [\sigma_1'(p_1) \wedge \sigma_1''(q_1)] \wedge [\sigma_1'(p_2) \wedge \sigma_1''(q_2)] \\
&= (\sigma_1' \sqcap \sigma_1'')(p_1, q_1) \wedge (\sigma_1' \sqcap \sigma_1'')(p_2, q_2).
\end{aligned}$$

Also,

$$\begin{aligned}
(\mu_2' \sqcap \mu_2'')((p_1, q_1), (p_2, q_2)) &= \mu_2'(p_1, p_2) \vee \mu_2''(q_1, q_2) \\
&= [\sigma_2'(p_1) \vee \sigma_2''(p_2)] \vee [\sigma_2''(q_1) \vee \sigma_2''(q_2)] \\
&= [\sigma_2'(p_1) \vee \sigma_2''(q_1)] \vee [\sigma_2'(p_2) \vee \sigma_2''(q_2)] \\
&= (\sigma_2' \sqcap \sigma_2'')(p_1, q_1) \vee (\sigma_2' \sqcap \sigma_2'')(p_2, q_2).
\end{aligned}$$

Hence, $\mathcal{G}' \sqcap \mathcal{G}''$ is a strong IFG. $\square$

Theorem 9.8 *If $\mathcal{G}'$ and $\mathcal{G}''$ are two IFGs such that $\mathcal{G}' \sqcap \mathcal{G}''$ is strong, then either one or both of them must be strong.*

9.2.8 Semi-strong Product of Two IFGs

Next, **semi-strong product** of two IFGs is defined.

Definition 9.20 The **semi-strong product** of two IFGs $\mathscr{G}'$ and $\mathscr{G}''$ such that $\mathscr{V}' \cap \mathscr{V}'' = \phi$, is defined to be the IFG $\mathscr{G}' \bullet \mathscr{G}'' = (\mathscr{V}, E, \sigma' \bullet \sigma'', \mu' \bullet \mu'')$, where $\mathscr{V} = \mathscr{V}' \times \mathscr{V}''$, $E = \{((p, q_1), (p, q_2)) \mid p \in E', (q_1, q_2) \in E''\} \cap \{((p_1, q_1), (p_2, q_2)) \mid (p_1, p_2) \in E', (q_1, q_2) \in E''\}$. The values of membership and that of non-membership of the node (p, q) in $\mathscr{G}' \bullet \mathscr{G}''$ are given by

$$(\sigma_1' \bullet \sigma_1'')(p, q) = \sigma_1'(p) \wedge \sigma_1''(q)$$

$$(\sigma_2' \bullet \sigma_2'')(p, q) = \sigma_2'(p) \vee \sigma_2''(q).$$

Also, the values of membership and value of non-membership of the arc in $\mathscr{G}' \bullet \mathscr{G}''$ are given by

$$\langle i \rangle \begin{cases} (\mu_1' \bullet \mu_1'')((p, q_1), (p, q_2)) = \sigma_1'(p) \wedge \mu_1''(q_1, q_2) \\ (\mu_1' \bullet \mu_1'')((p_1, q_1), (p_2, q_2)) = \mu_1'(p_1, p_2) \wedge \mu_1''(q_1, q_2) \end{cases}$$

$$\langle ii \rangle \begin{cases} (\mu_2' \bullet \mu_2'')((p, q_1), (p, q_2)) = \sigma_2'(p) \vee \mu_2''(q_1, q_2) \\ (\mu_2' \bullet \mu_2'')((p_1, q_1), (p_2, q_2)) = \mu_2'(p_1, p_2) \vee \mu_2''(q_1, q_2). \end{cases}$$

Example 9.2 Consider $\mathscr{G}'$ and $\mathscr{G}''$ be two IFGs as shown in Fig. 9.2. Then associated $\mathscr{G}' \bullet \mathscr{G}''$ is shown in Fig. 9.2.

Theorem 9.9 If $\mathscr{G}'$ and $\mathscr{G}''$ are strong IFGs, then $\mathscr{G}' \bullet \mathscr{G}''$ is also strong.

Theorem 9.10 If $\mathscr{G}'$ and $\mathscr{G}''$ are two IFGs such that $\mathscr{G}' \bullet \mathscr{G}''$ is strong, then either one or both of them must be strong.

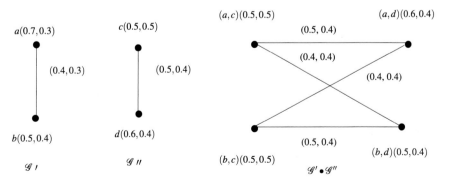

Fig. 9.2 Semi-strong product of two IFGs

9.2.9 Strong Product of Two IFGs

Next, we define **strong product** of two IFGs.

Definition 9.21 The **strong product of two IFGs** $\mathscr{G}'$ and $\mathscr{G}''$ such that $\mathscr{V}' \cap \mathscr{V}'' = \phi$, is defined to be the IFG $\mathscr{G}' \otimes \mathscr{G}'' = (\mathscr{V}, E, \sigma' \otimes \sigma'', \mu' \otimes \mu'')$, where $\mathscr{V} = \mathscr{V}' \times \mathscr{V}''$, $E = \{((p, q_1), (p, q_2)) \mid p \in E', (q_1, q_2) \in E''\} \cup \{((p_1, q), (p_2, q)) \mid (p_1, p_2) \in E', q \in E''\} \cup \{((p_1, q_1), (p_2, q_2)) \mid (p_1, p_2) \in E', (q_1, q_2) \in E''\}$. The value of membership and value of non-membership of the node (p, q) in $\mathscr{G}' \otimes \mathscr{G}''$ are given by

$$(\sigma_1' \otimes \sigma_1'')(p, q) = \sigma_1'(p) \wedge \sigma_1''(q)$$

$$(\sigma_2' \otimes \sigma_2'')(p, q) = \sigma_2'(p) \vee \sigma_2''(q).$$

Also, the value of membership and value of non-membership of the arc in $\mathscr{G}' \otimes \mathscr{G}''$ are given by

$$\langle i \rangle \begin{cases} (\mu_1' \otimes \mu_1'')((p, q_1), (p, q_2)) = \sigma_1'(p) \wedge \mu_1''(q_1, q_2) \\ (\mu_1' \otimes \mu_1'')((p_1, q), (p_2, q)) = \mu_1'(p_1, p_2) \wedge \sigma_1''(q) \\ (\mu_1' \otimes \mu_1'')((p_1, q_1), (p_2, q_2)) = \mu_1'(p_1, p_2) \wedge \mu_1''(q_1, q_2) \end{cases}$$

$$\langle ii \rangle \begin{cases} (\mu_2' \otimes \mu_2'')((p, q_1), (p, q_2)) = \sigma_2'(p) \vee \mu_2''(q_1, q_2) \\ (\mu_2' \otimes \mu_2'')((p_1, q), (p_2, q)) = \mu_2'(p_1, p_2)(p) \vee \sigma_2''(q) \\ (\mu_2' \otimes \mu_2'')((p_1, q_1), (p_2, q_2)) = \mu_2'(p_1, p_2) \vee \mu_2''(q_1, b_2). \end{cases}$$

Example 9.3 Let $\mathscr{G}'$ and $\mathscr{G}''$ be two IFGs depicted in Fig. 9.3. Then the associated $\mathscr{G}' \otimes \mathscr{G}''$ are shown in Fig. 9.3.

Theorem 9.11 If $\mathscr{G}'$ and $\mathscr{G}''$ are CIFGs, then $\mathscr{G}' \otimes \mathscr{G}''$ is complete.

Theorem 9.12 If $\mathscr{G}'$ and $\mathscr{G}''$ are two IFGs such that $\mathscr{G}' \otimes \mathscr{G}''$ is strong, then at least one of $\mathscr{G}'$ or $\mathscr{G}''$ must be strong.

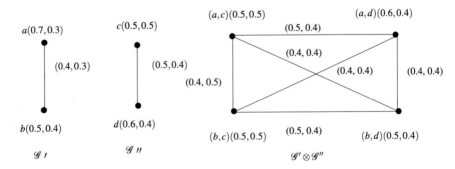

Fig. 9.3 Strong product of two IFGs

9.2.10 Product Intuitionistic Fuzzy Graphs

Now, we define product IFG as below.

Definition 9.22 Let $\mathscr{G}$ be an IFG. If $\mu_1(p_i, p_j) \leq \sigma_1(p_i) \times \sigma_1(p_j)$ and $\mu_2(p_i, p_j) \leq \sigma_2(p_i) \times \sigma_2(p_j)$ for all $(p_i, p_j) \in \mathscr{V}$, where " $\times$ " represents ordinary multiplication, then the IFG $\mathscr{G}$ is said to be the **product IFG**.

Remark 9.1 If $\mathscr{G}$ be an IFG and since σ_1 and σ_2 are less than or equal to 1, it follows that

$$\mu_1(p_i, p_j) \leq \sigma_1(p_i) \times \sigma_1(p_j) \leq \min\{\sigma_1(p_i), \sigma_1(p_j)\}$$

and

$$\mu_2(p_i, p_j) \leq \sigma_2(p_i) \times \sigma_2(p_j) \leq \max\{\sigma_2(p_i), \sigma_2(p_j)\} \, for \, all \, (p_i, p_j) \in \mathscr{V}.$$

Thus every product IFG is an IFG.

Definition 9.23 A product IFG $\mathscr{G}$ is said to be complete if $\mu_1(p_i, p_j) = \sigma_1(p_i) \times \sigma_1(p_j)$ and $\mu_2(p_i, p_j) = \sigma_2(p_i) \times \sigma_2(p_j)$ for all $(p_i, p_j) \in \mathscr{V}$.

Definition 9.24 The complement of product IFG $\mathscr{G}$ is an IFG $\bar{\mathscr{G}} = (\mathscr{V}, E, \bar{\sigma}, \bar{\mu})$, where $\sigma = \bar{\sigma}, \bar{\mu}_1(p_i, p_j) = \sigma_1(p_i) \times \sigma_1(p_j) - \mu_1(p_i, p_j)$ and $\bar{\mu}_2(p_i, p_j) = \sigma_2(p_i) \times \sigma_2(p_j) - \mu_2(p_i, p_j)$ for all $(p_i, p_j) \in \mathscr{V}$.

Remark 9.2 It follows that, if $\mathscr{G}$ is a product IFG, then $\bar{\bar{\mathscr{G}}} = \mathscr{G}$.

Definition 9.25 Consider the product IFGs $\mathscr{G}'$ and $\mathscr{G}''$. The isomorphism between two product IFGs $\mathscr{G}', \mathscr{G}''$ is a bijection $h : \mathscr{V}' \rightarrow \mathscr{V}''$ such that $\sigma_1'(a_1) = \sigma_1''(h(a_1))$, $\sigma_2'(a_1) = \sigma_2''(h(a_1))$ for all $a_1 \in \mathscr{V}'$ and $\mu_1'(a_1, b_1) = \mu_1''(h(a_1), h(b_1))$, $\mu_2'(a_1, b_1) = \mu_2''(h(a_1), h(b_1))$ for all $a_1, b_1 \in \mathscr{V}'$.
This is written as $\mathscr{G}' \cong \mathscr{G}''$.

Definition 9.26 The union of two product IFGs $\mathscr{G}'$ and $\mathscr{G}''$ is determined by $\mathscr{G} = \mathscr{G}' \cup \mathscr{G}'' = (\mathscr{V}, E, \sigma, \mu)$, where $\mathscr{V} = \mathscr{V}' \cup \mathscr{V}''$ and $E = E' \cup E''$. The value of membership and value of non-membership of the node $p_1 \in \mathscr{G}$ are given by

$$\sigma_1(p_1) = (\sigma_1' \cup \sigma_1'')(p_1) = \begin{cases} \sigma_1'(p_1), & \text{if } p_1 \in \mathscr{V}' - \mathscr{V}'' \\ \sigma_1''(p_1), & \text{if } p_1 \in \mathscr{V}'' - \mathscr{V}' \\ \sigma_1'(p_1) \vee \sigma_1''(p_1), & \text{if } p_1 \in \mathscr{V}' \cap \mathscr{V}'' \end{cases}$$

$$\sigma_2(p_1) = (\sigma_2' \cup \sigma_2'')(p_1) = \begin{cases} \sigma_2'(p_1), & \text{if } p_1 \in \mathscr{V}' - \mathscr{V}'' \\ \sigma_2''(p_1), & \text{if } p_1 \in \mathscr{V}'' - \mathscr{V}' \\ \sigma_2'(p_1) \wedge \sigma_2''(p_1), & \text{if } p_1 \in \mathscr{V}' \cap \mathscr{V}''. \end{cases}$$

Also, the value of membership and value of non-membership of the arc $(p_1, q_1) \in \mathscr{G}$ are given by

$$\mu_1(p_1, q_1) = (\mu_1' \cup \mu_1'')(p_1, q_1) = \begin{cases} \mu_1'(p_1, q_1), & \text{if } (p_1, q_1) \in E' - E'' \\ \mu_1''(p_1, q_1), & \text{if } (p_1, q_1) \in E'' - E' \\ \mu_1'(p_1, q_1) \vee \mu_1''(p_1, q_1), & \text{if } (p_1, q_1) \in E' \cap E'' \end{cases}$$

$$\mu_2(p_1, q_1) = (\mu_2' \cup \mu_2'')(p_1, q_1) = \begin{cases} \mu_2'(p_1, q_1), & \text{if } (p_1, q_1) \in E' - E'' \\ \mu_2''(p_1, q_1), & \text{if } (p_1, q_1) \in E'' - E' \\ \mu_2'(p_1, q_1) \wedge \mu_2''(p_1, q_1), & \text{if } (p_1, q_1) \in E' \cap E''. \end{cases}$$

Definition 9.27 The join $\mathscr{G}' + \mathscr{G}'' = (\mathscr{V}, E, \sigma' + \sigma'', \mu' + \mu'')$ of two product IFGs $\mathscr{G}'$ and $\mathscr{G}''$ is defined as follows:

$(\sigma_1' + \sigma_1'')(p_1) = (\sigma_1' \cup \sigma_1'')(p_1)$, $(\sigma_2' + \sigma_2'')(p_1) = (\sigma_2' \cap \sigma_2'')(p_1)$, if $p_1 \in \mathscr{V}' \cup \mathscr{V}''$;

$(\mu_1' + \mu_1'')(p_1, q_1) = (\mu_1' \cup \mu_1'')(p_1, q_1)$, $(\mu_2' + \mu_2'')(p_1, q_1) = (\mu_2' \cup \mu_2'')(p_1, q_1)$, if $(p_1, q_1) \in E' \cup E''$ and

$(\mu_1' + \mu_1'')(p_1, q_1) = \sigma_1'(p_1) \vee \sigma_1''(q_1)$ $(\mu_2' + \mu_2'')(p_1, q_1) = \sigma_2'(p_1) \wedge \sigma_2''(q_1)$, if $(p_1, q_1) \in E_1$, where E_1 is the set of all arcs joining the nodes of $\mathscr{V}', \mathscr{V}''$.

Theorem 9.13 *If $\mathscr{G}'$ and $\mathscr{G}''$ are the product IFGs, then*

(i) $(\mathscr{G}' \cup \mathscr{G}'')^c \cong (\mathscr{G}')^c + (\mathscr{G}'')^c$

(ii) $(\mathscr{G}' + \mathscr{G}'')^c \cong (\mathscr{G}')^c \cup (\mathscr{G}'')^c$.

Theorem 9.14 *The direct product of two IFGs is a product IFG.*

Definition 9.28 Let $\mathscr{G}'$ and $\mathscr{G}''$ be the product IFG. Then, the ring sum of two product IFGs $\mathscr{G}'$ and $\mathscr{G}''$ is $\mathscr{G}' \oplus \mathscr{G}'' = (\mathscr{V}, E, \sigma' \oplus \sigma'', \mu' \oplus \mu'')$ where $(\sigma_1' \oplus \sigma_1'')(p_1) = \sigma_1'(p_1) \cup \sigma_1''(p_1)$, $(\sigma_2' \oplus \sigma_2'')(p_1) = \sigma_2'(p_1) \cup \sigma_2''(p_1)$ for all $p_1 \in \mathscr{V} = \mathscr{V}' \cup \mathscr{V}''$ and

$$(\mu_1' \oplus \mu_1'')(p_1, q_1) = \begin{cases} \mu_1'(p_1, q_1), & \text{if } (p_1, q_1) \in E' - E'' \\ \mu_1''(p_1, q_1), & \text{if } (p_1, q_1) \in E'' - E' \\ 0, & otherwise \end{cases}$$

$$(\mu_2' \oplus \mu_2'')(p_1, q_1) = \begin{cases} \mu_2'(p_1, q_1), & \text{if } (p_1, q_1) \in E' - E'' \\ \mu_2''(p_1, q_1), & \text{if } (p_1, q_1) \in E'' - E' \\ 0, & otherwise. \end{cases}$$

Theorem 9.15 *If $\mathscr{G}'$ and $\mathscr{G}''$ are product IFGs, $\mathscr{G}' \oplus \mathscr{G}''$ is also a product IFG.*

Theorem 9.16 *Let $\mathscr{G}'$ and $\mathscr{G}''$ be two product IFGs and $E' \cap E'' = \phi$, then*

(i) $(\mathscr{G}' + \mathscr{G}'')^c \cong (\mathscr{G}')^c \oplus (\mathscr{G}'')^c$,

(ii) $(\mathscr{G}' \oplus \mathscr{G}'')^c \cong (\mathscr{G}')^c + (\mathscr{G}'')^c$.

9.3 Covering and Matching in Intuitionistic Fuzzy Graphs

In a crisp graph G, a node (vertex) cover is a set of vertices such that each edge of the graph is incident to at least one vertex of the set. The minimum number of vertices in a vertex cover is called the vertex covering a number of the graph G and it is denoted by $\alpha_0(G)$. In a graph G, a set of nodes is independent if no two vertices in the set are adjacent. The maximum cardinality of an independent set is called the independent number and it is denoted by $\beta_0(G)$.

An edge cover in a graph G without isolated nodes is the set of edges of G that covers all nodes of G. The edge covering number is denoted by $\alpha_1(G)$, which is the minimum cardinality among all other edge cover of G. If two edges are not adjacent then they are called independent. A matching M in a graph is a set of independent edges.

9.3.1 Covering in Intuitionistic Fuzzy Graphs

Here, the concept of covering and matching in IFGs using strong arcs are discussed and many interesting results are established. The fuzzy strong node cover (SNC) for IFG is now defined.

Definition 9.29 Let $\mathscr{G} = (\mathscr{V}, \sigma, \mu)$ be an IFG on $G^* = (V, E)$. A node and an incident strong arc are said to be **strong cover** to each other. A SNC in $\mathscr{G}$ is the set D of nodes that cover all the strong arcs of $\mathscr{G}$. The membership and non-membership values of the SNC D is explained as

$$W_1(D) = \sum_{a \in D} \mu_1(a, b) \quad \text{and} \quad W_2(D) = \sum_{a \in D} \mu_2(a, b),$$

where $\mu_1(a, b)$ (resp. $\mu_2(a, b)$) is the minimum (maximum) membership (non-membership) value of the strong arc incident to a.

The strong node covering number of $\mathscr{G}$ is defined by

$$\alpha_{s_0}(\mathscr{G}) = \alpha_{s_0} = (\alpha_{s_{10}}, \alpha_{s_{20}}),$$

where $\alpha_{s_{10}}$ and $\alpha_{s_{20}}$ are the minimum membership value and maximum non-membership value of the SNC of $\mathscr{G}$.

A minimum SNC in $\mathscr{G}$ is a SNC among all other SNCs of $\mathscr{G}$ such that $\alpha_{s_{10}}$ is minimum and $\alpha_{s_{20}}$ is maximum.

Theorem 9.17 *If $\mathscr{G}$ is a CIFG, then $\alpha_{s_{10}} = (n-1)\mu_1(a, b)$ and $\alpha_{s_{20}} = (n-1)\mu_2(a, b)$, where n is the number of nodes in $\mathscr{G}$ and $\mu_1(a, b), \mu_2(a, b)$ are the membership and non-membership values of the weakest arc in $\mathscr{G}$ respectively.*

Proof Given that $\mathscr{G}$ is CIFG. Then every arc is strong and every node is adjacent to every other node. Therefore, any set containing $(n-1)$ nodes forms a SNC of $\mathscr{G}$. Let a be a node in $\mathscr{G}$ having minimum membership value and maximum non-membership value. Let $b_1, b_2, \ldots, b_{n-1}$ be the nodes adjacent to a. Then $(a, b_1), (a, b_2), \ldots, (a, b_{n-1})$ are the weakest arcs of $\mathscr{G}$. Now, for some $b \in \{b_1, b_2, \ldots, b_{n-1}\}$ the membership and non-membership strengths of each arc are $\mu_1(a, b)$ and $\mu_2(a, b)$.

Thus, the set $D = \{b_1, b_2, \ldots, b_{n-1}\}$ forms a SNC of $\mathscr{G}$ with

$$W_1(D) = \sum_{b_i \in D} \mu_1(a, b_i).$$

Thus,

$$\alpha_{s_{10}} = \sum_{i=1}^{n-1} \mu_1(a, b_i) = (n-1)\mu_1(a, b).$$

Similarly,

$$\alpha_{s_{20}} = (n-1)\mu_2(a, b).$$

$\square$

The above theorem is extended to the complete bipartite IFG (CBIFG).

Theorem 9.18 *Let* $K_{\sigma', \sigma''}$ *be a CBIFG with partite sets* V' *and* V''. *Then* $\alpha_{s_{10}}(K_{\sigma', \sigma''}) = \min\{W_1(V'), W_1(V'')\}$ *and* $\alpha_{s_{20}}(K_{\sigma', \sigma''}) = \max\{W_2(V'), W_2(V'')\}$.

Theorem 9.19 *Let* G^* *be a crisp cycle and* $\mathscr{G}$ *be an IF cycle. Then* $\alpha_{s_{10}}(G) = \min\{W_1(D) : D \text{ is a SNC in } \mathscr{G} \text{ with } |D| \geq [\frac{n}{2}]\}$ *and* $\alpha_{s_{20}}(\mathscr{G}) = \max\{W_2(D) : D \text{ is a SNC in } \mathscr{G} \text{ with } |D| \geq [\frac{n}{2}]\}$.

Proof By definition of IF cycle, all arcs are strong. Also, the number of nodes in the SNC of both the graphs $\mathscr{G}$ and G^* are same, as each arc in both graphs is strong. The strong node covering number of the crisp graph G^* is $[\frac{n}{2}]$ [6]. Therefore, the least number of nodes in SNC of $\mathscr{G}$ is $[\frac{n}{2}]$.

Hence, $\alpha_{s_{10}}(\mathscr{G}) = \min\{W_1(D)\}$ and $\alpha_{s_{20}}(\mathscr{G}) = \max\{W_2(D)\}$. $\square$

Definition 9.30 Two nodes are called **strongly independent** if there is no strong arc between them. A set of nodes in $\mathscr{G}$ is said to be strongly independent if every pair of nodes in the set are strongly independent.

Definition 9.31 The membership and non-membership values of a strong independent set D of an IFG $\mathscr{G} = (\mathscr{V}, \sigma, \mu)$ are $W_1(D) = \sum_{a \in D} \mu_1(a, b)$ and $W_2(D) = \sum_{a \in D} \mu_2(a, b)$, where $\mu_1(a, b)$ $(\mu_2(a, b))$ is the least (greatest) membership (non-membership) value of strong arcs incident on a.

The strong independent number of an IFG $\mathscr{G}$ is denoted by $\beta_{s_0}(G) = \beta_{s_0} = (\beta_{s_{10}}, \beta_{s_{20}})$, where $\beta_{s_{10}}$ (resp. $\beta_{s_{20}}$) is greatest (least) membership (non-membership) value of the nodes of strong independent set. A maximum strong independent set is a strong independent set with maximum membership value and minimum non-membership value.

The results for CIFG, CBIFG and intuitionistic fuzzy cycle are stated below.

Theorem 9.20 *For a CIFG $\mathscr{G}$, $\beta_{s_{10}} = \mu_1(u, v)$ and $\beta_{s_{20}} = \mu_2(u, v)$, where $\mu_1(u, v)$, $\mu_2(u, v)$ are the membership and non-membership values of the weakest arc in $\mathscr{G}$.*

Theorem 9.21 *For a CBIFG, $K_{\sigma',\sigma''}$ with partite sets V' and V'', $\beta_{s_{10}}(K_{\sigma',\sigma''}) = \max\{W_1(V'), W_1(V'')\}$ and $\beta_{s_{20}}(K_{\sigma',\sigma''}) = \min\{W_2(V'), W_2(V'')\}$.*

In the case of cycle, the result is stated as below.

Theorem 9.22 *Let D be a SNC in $\mathscr{G}$ such that $|D| \leq [\frac{n}{2}]$. For an IF cycle $\mathscr{G}$, $\beta_{s_{10}}(\mathscr{G}) = \max\limits_{D}\{W_1(D)\}$ and $\beta_{s_{20}}(G) = \min\limits_{D}\{W_2(D)\}$.*

Now, **strong arc cover** (SAC) and strong covering number in IFGs are defined as follows.

Definition 9.32 Let $\mathscr{G} = (\mathscr{V}, \sigma, \mu)$ be an IFG without any isolated node. A SAC of $\mathscr{G}$ is a set $\mathscr{C}$ of strong arcs that cover all the nodes of $\mathscr{G}$. The membership and non-membership values of the strong arc cover $\mathscr{C}$ is defined by

$$W_1(\mathscr{C}) = \sum_{(u,v) \in \mathscr{C}} \mu_1(u, v) \quad \text{and} \quad W_2(\mathscr{C}) = \sum_{(u,v) \in \mathscr{C}} \mu_2(u, v).$$

The strong arc covering number of $\mathscr{G}$ is defined by $\alpha_{s_1}(\mathscr{G}) = \alpha_{s_1} = (\alpha_{s_{11}}, \alpha_{s_{21}})$, where $\alpha_{s_{11}}$ and $\alpha_{s_{21}}$ represent the minimum membership and maximum non-membership values of the SAC of $\mathscr{G}$.

The minimum SAC of $\mathscr{G}$ is a SAC of minimum membership value and maximum non-membership value among all SACs in $\mathscr{G}$.

For a CIFG, every arc is strong and every node is adjacent to every other node. Also, the number of arcs in SAC of both $\mathscr{G}$ and G^* are same because each arc in both graphs is strong. Thus, one can conclude the following result.

Theorem 9.23 *Let $\mathscr{C}$ be a SAC in $\mathscr{G}$ such that $|\mathscr{C}| \geq [\frac{n}{2}]$.*
If $\mathscr{G} = (\mathscr{V}, \sigma, \mu)$ is a CIFG or a cycle, then $\alpha_{s_{11}}(\mathscr{G}) = \min\limits_{\mathscr{C}}\{W_1(\mathscr{C})\}$ and $\alpha_{s_{21}}(\mathscr{G}) = \max\limits_{\mathscr{C}}\{W_2(\mathscr{C})\}$.

Similar result also holds for CBIFG.

Theorem 9.24 *Let $\mathscr{G} = K_{\sigma',\sigma''}$ be a CBIFG and $\mathscr{C}$ be a SAC in $K_{\sigma',\sigma''}$ such that $|\mathscr{C}| \geq \max\{|V'|, |V''|\}$, where V' and V'' are the bipartition of nodes. Then $\alpha_{s_{11}}(K_{\sigma',\sigma''}) = \min\limits_{\mathscr{C}}\{W_1(\mathscr{C})\}$ and $\alpha_{s_{21}}(K_{\sigma',\sigma''}) = \max\limits_{\mathscr{C}}\{W_1(\mathscr{C})\}$.*

Definition 9.33 Let $\mathcal{M}$ be a set of strong arcs of an IFG $\mathcal{G}$ such that no two arcs in $\mathcal{M}$ have a common node. This set $\mathcal{M}$ is called a **strong independent set** of arcs or a **strong matching** in $\mathcal{G}$.

Definition 9.34 Let $\mathcal{M}$ be a strong matching in $\mathcal{G} = (\mathcal{V}, \sigma, \mu)$. If $(a, b) \in \mathcal{M}$, then $\mathcal{M}$ strongly matches a with b. The membership and non-membership values of strong matching are defined as

$$W_1(\mathcal{M}) = \sum_{(a,b)\in\mathcal{M}} \mu_1(a, b) \quad \text{and} \quad W_2(\mathcal{M}) = \sum_{(a,b)\in\mathcal{M}} \mu_2(a, b).$$

Definition 9.35 The strong arc independent number or strong matching number of $\mathcal{G}$ is denoted by $\beta_{s_1}(\mathcal{G}) = (\beta_{s_{11}}, \beta_{s_{21}})$, where $\beta_{s_{11}}$ and $\beta_{s_{21}}$ are the maximum membership value and minimum non-membership value of the strong matchings of $\mathcal{G}$.

A maximum strong matching of $\mathcal{G}$ is the strong matching of maximum membership value and minimum non-membership value among all such matchings.

The following theorems determine β_{s_1} of CIFG, CBIFG and cycle.

Theorem 9.25 *Let $\mathcal{M}$ be a strong matching of $\mathcal{G}$ such that $|\mathcal{M}| \leq [\frac{n}{2}]$. If $\mathcal{G}$ is a CIFG or an IF cycle, then $\beta_{s_{11}}(\mathcal{G}) = \max_{\mathcal{M}}\{W_1(\mathcal{M}) :\}$ and $\beta_{s_{21}}(\mathcal{G}) = \min_{\mathcal{M}}\{W_2(\mathcal{M})\}$.*

A similar result for CBIFG is stated below.

Theorem 9.26 *Let $\mathcal{G} = K_{\sigma',\sigma''}$ be a CBIFG with partite sets V' and V'' and $\mathcal{M}$ is a strong matching in $K_{\sigma',\sigma''}$ with $|\mathcal{M}| \leq \min\{|V'|, |V''|\}\}$. Then $\beta_{s_{11}}(K_{\sigma',\sigma''}) = \max_{\mathcal{M}}\{W_1(\mathcal{M})\}$ and $\beta_{s_{21}}(K_{\sigma',\sigma''}) = \min_{\mathcal{M}}\{W_1(\mathcal{M})\}$.*

Example 9.4 Let $\mathcal{G} = (\mathcal{V}, \sigma, \mu)$ be an IFG depicted in Fig. 9.4, where $\mathcal{V} = \{a, b, c, d\}$ and $(a, b), (b, c), (c, d), (d, a)$ are the edges. The vertex and edge membership and non-membership values are shown in Fig. 9.4.

For this graph, the strong arc are $(a, c), (c, d)$ and (d, b).

Fig. 9.4 Illustration of strong covering in IFG

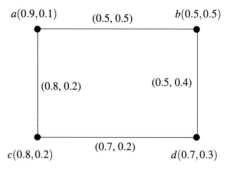

There are several strong node covers for this graph and they are listed below:
$D_1 = \{c, d\}$, $D_2 = \{b, c\}$, $D_3 = \{a, d\}$, $D_4 = \{a, b, c\}$, $D_5 = \{a, b, d\}$, $D_6 = \{b, c, d\}$, $D_7 = \{a, c, d\}$ and $D_8 = \{a, b, c, d\}$.

Now, $W(D_1) = (0.7 + 0.5, 0.2 + 0.4) = (1.2, 0.6)$, $W(D_2) = (0.5 + 0.7, 0.4 + 0.2) = (1.2, 0.6)$,
$W(D_3) = (0.8 + 0.5, 0.2 + 0.4) = (1.3, 0.6)$, $W(D_4) = (0.8 + 0.5 + 0.7, 0.2 + 0.4 + 0.2) = (2.0, 0.8)$,
$W(D_5) = (0.8 + 0.5 + 0.5, 0.2 + 0.4 + 0.4) = (1.8, 1.0)$,
$W(D_6) = (0.5 + 0.7 + 0.5, 0.4 + 0.2 + 0.4) = (1.7, 1.0)$,
$W(D_7) = (0.8 + 0.7 + 0.5, 0.2 + 0.2 + 0.4) = (2.0, 0.8)$ and
$W(D_8) = (0.8 + 0.5 + 0.7 + 0.5, 0.2 + 0.4 + 0.2 + 0.4) = (2.5, 1.2)$.

Among these the minimum value is $(1.2, 0.6)$. Hence, $\alpha_{s_0} = (1.2, 0.6)$.

The strong independent sets are $D_1 = \{a, b\}$, $D_2 = \{a, d\}$ and $D_3 = \{b, d\}$.

Now, $W(D_1) = (0.8 + 0.5, 0.2 + 0.4) = (1.3, 0.6)$, $W(D_2) = (0.8 + 0.5, 0.2 + 0.4) = (1.3, 0.6)$ and $W(D_3) = (0.5 + 0.7, 0.4 + 0.2) = (1.2, 0.6)$.

The maximum value among these is $(1.3, 0.6)$ and hence $\beta_{s_0} = (1.3, 0.6)$.

The set $\mathscr{C} = \{(a, c), (b, d)\}$ is the only strong arc cover and strong matching in G and $W(\mathscr{C}) = (0.8 + 0.5, 0.2 + 0.4) = (1.3, 0.6)$.

Hence, $\alpha_{s_1} = \beta_{s_1} = (1.3, 0.6)$.

Theorem 9.27 *Let $\mathscr{G} = (\mathscr{V}, \sigma, \mu)$ be any IFG of order m without any isolated vertex. Then*
(i) $\alpha_{s_0} + \beta_{s_0} = W(\mathscr{V}) \leq m$
(ii) $\alpha_{s_1} + \beta_{s_1} \leq m$.

Proof Let N_{s_0} be the minimum SNC of $\mathscr{G}$ and $\alpha_{s_0} = W(N_{s_0})$. Then $\mathscr{V} - N_{s_0}$ is a strong independent set. Thus, the nodes in $\mathscr{V} - N_{s_0}$ are incident on strong arcs of $\mathscr{G}$. Therefore, $\beta_{s_0} \geq W(\mathscr{V} - N_{s_0}) = W(\mathscr{V}) - \alpha_{s_0}$.

Hence,

$$\alpha_{s_0} + \beta_{s_0} \geq W(\mathscr{V}). \tag{9.4}$$

Let M_{s_0} be a maximum strong independent set of nodes in $\mathscr{G}$ and $\beta_{s_0} = W(M_{s_0})$. Therefore, no two nodes in M_{s_0} are adjacent to each other by a strong arc. Hence, $\mathscr{V} - M_{s_0}$ strong covers all the strong arcs of $\mathscr{G}$. So, $\mathscr{V} - M_{s_0}$ is a strong node cover of $\mathscr{G}$. Hence, $\alpha_{s_0} \leq W(\mathscr{V} - M_{s_0}) = W(\mathscr{V}) - \beta_{s_0}$.

$$\alpha_{s_0} + \beta_{s_0} \leq W(\mathscr{V}). \tag{9.5}$$

From Eqs. (9.4) and (9.5), $\alpha_{s_0} + \beta_{s_0} = W(\mathscr{V})$.
By definition, $W(\mathscr{V}) \leq m$. Hence, $\alpha_{s_0} + \beta_{s_0} = W(\mathscr{V}) \leq m$.
The second inequality follows directly. $\qquad\square$

9.4 Intuitionistic Fuzzy Labeling Graphs

Intuitionistic fuzzy labeling (IFL) is a very flexible and compatible model for classical and intuitionistic fuzzy setup. Also, IFL has several applications in physics, chemistry, computer science, etc. Gallian [10] prepared a good review on (crisp) graph labeling. Square difference labeling for some graphs is available in [33]. Lakshmiprasanna et al. [14] presented an application of graph labeling in computer science, whereas Saha et al. [22] presented a very appropriate application of graph labeling in scheduling.

Here, the concept of connectivity in intuitionistic fuzzy labeling (IFLG) is introduced and some useful results are investigated. Apart from this, partial cut node, strong arc, bridge and block are introduced with several examples.

First of all, we define IFLGs.

Definition 9.36 An IFG $\mathcal{G} = (\mathcal{V}, \sigma, \mu)$, $\sigma = (\sigma_1, \sigma_2)$, $\mu = (\mu_1, \mu_2)$ is said to be an IFLG if $\sigma_i : \mathcal{V} \to [0, 1]$ and $\mu_i : \mathcal{V} \times \mathcal{V} \to [0, 1]$ for $i = 1, 2$ are bijective and have distinct membership and non-membership values of nodes and arcs. Also, $\mu_1(v_i, v_j) < \min\{\sigma_1(v_i), \sigma_1(v_j)\}$, $\mu_2(v_i, v_j) < \max\{\sigma_2(v_i), \sigma_2(v_j)\}$, $0 \leq \mu_1(v_i, v_j) + \mu_2(v_i, v_j) \leq 1$ for each edge (v_i, v_j) of $\mathcal{G}$.

Definition 9.37 Let $\mathcal{G} = (\mathcal{V}, \sigma, \mu)$ be an IFG. A IF subgraph $H = (\mathcal{V}, \tau, \rho)$ of $\mathcal{G}$ is said to be an IFL subgraph of $\mathcal{G}$ if $\tau_1(p) \leq \sigma_1(p)$, $\tau_2(p) \geq \sigma_2(p)$ for all $p \in \mathcal{V}$ and $\rho_1(p, q) \leq \mu_1(p, q)$, $\rho_2(p, q) \geq \mu_2(p, q)$ for all arcs (p, q).

We denote $CONN_{1\mathcal{G}}(a, b)$ and $CONN_{2\mathcal{G}}(a, b)$ for the connectedness of the arc (a, b) in $\mathcal{G}$ for membership and non-membership values respectively.

Theorem 9.28 If $H = (\mathcal{V}, \tau, \rho)$ is an IFL subgraph of $\mathcal{G} = (\mathcal{V}, \sigma, \mu)$, then $CONN_{1H}(p, q) \leq CONN_{1\mathcal{G}}(p, q)$ and $CONN_{2H}(p, q) \geq CONN_{2\mathcal{G}}(p, q)$ for all $p, q \in \mathcal{V}$.

Proof Since H is a subgraph of $\mathcal{G}$, then $\tau_1(p) \leq \sigma_1(p)$, $\rho_1(p, q) \leq \mu_1(p, q)$ and $\tau_2(p) \geq \sigma_2(p)$, $\rho_2(p, q) \geq \mu_2(p, q)$.

Thus, $CONN_{1H}(p, q) \leq CONN_{1\mathcal{G}}(p, q)$ and $CONN_{2H}(p, q) \geq CONN_{2\mathcal{G}}(p, q)$ for all $p, q \in \mathcal{V}$. $\square$

The following result follows from the definition of the union of IFGs.

Theorem 9.29 *The union of two IFLGs $\mathcal{G}'$ and $\mathcal{G}''$ is also an IFLG, provided membership and non-membership values of the arcs of $\mathcal{G}'$ and $\mathcal{G}''$ are distinct.*

Let us consider the strength of an arc in IFG.

Definition 9.38 Let $\mathcal{G}$ be an IFLG. The **strength of a path** P of m edges e_i for $i = 1, 2, \ldots, m$ is defined $S(P) = (S_1(P), S_2(P))$, where $S_1(P) = \min_{1 \leq i \leq m} \mu_1(e_i)$ and $S_2(P) = \max_{1 \leq i \leq m} \mu_2(e_i)$.

Fig. 9.5 An IFLG in which p_2 is a partial cut node

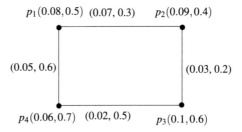

$p_1(0.08, 0.5)$ $(0.07, 0.3)$ $p_2(0.09, 0.4)$

$(0.05, 0.6)$ $(0.03, 0.2)$

$p_4(0.06, 0.7)$ $(0.02, 0.5)$ $p_3(0.1, 0.6)$

Definition 9.39 Let $\mathscr{G} = (\mathscr{V}, \sigma, \mu)$ be an $IFLG$. The **connectedness strength** of a pair of nodes $p, q \in \mathscr{V}$ is defined as $CONN_{\mathscr{G}}(p, q) = (CONN_{1\mathscr{G}}(p, q), CONN_{2\mathscr{G}}(p, q))$, where $CONN_{1\mathscr{G}}(p, q) = \max\{S_1(Q) : Q$ is a p-q path in $\mathscr{G}\}$ and $CONN_{2\mathscr{G}}(p, q) = \min\{S_2(Q) : Q$ is a p-q path in $\mathscr{G}\}$.

If p and q are isolated nodes of $\mathscr{G}$, then $CONN_{\mathscr{G}}(p, q) = (0, 0)$.

The following result follows from definition of connectedness.

Lemma 9.1 *Let H be an IFL subgraph of an IFLG $\mathscr{G} = (\mathscr{V}, \sigma, \mu)$. Then for each pair of nodes $p, q \in \mathscr{V}$, $CONN_{1H}(p, q) \leq CONN_{1\mathscr{G}}(p, q)$ and $CONN_{2H}(p, q) \geq CONN_{2\mathscr{G}}(p, q)$.*

Definition 9.40 A p-q path in an IFLG $\mathscr{G}$ is said to be a **strongest p-q path** if $S_1(P) = CONN_{1\mathscr{G}}(p, q)$ and $S_2(P) = CONN_{2\mathscr{G}}(p, q)$.

Definition 9.41 Let $\mathscr{G} = (\mathscr{V}, \sigma, \mu)$ be an IFLG. A node x in $\mathscr{G}$ is called a **partial cut node** or p-cut node of $\mathscr{G}$ if there exits a pair of nodes $p, q \in \mathscr{V}, p \neq q \neq x$ such that $CONN_{1(\mathscr{G}-x)}(p, q) < CONN_{1\mathscr{G}}(p, q), CONN_{2(\mathscr{G}-x)}(p, q) > CONN_{2\mathscr{G}}(p, q)$.

A connected IFLG without any p-cut node is called a **partial block** or p-block.

Definition 9.42 Let $\mathscr{G}$ be an IFLG. An arc $e = (p, q)$ of $\mathscr{G}$ is said to be a **partial bridge** (p- bridge) if $CONN_{1(\mathscr{G}-e)}(a, b) < CONN_{1\mathscr{G}}(p, q), CONN_{2(\mathscr{G}-e)}(p, q) > CONN_{2\mathscr{G}}(p, q)$.

Example 9.5 Let $\mathscr{G}$ be an IFLG, which is depicted in Fig. 9.5. In this graph, node p_2 is a partial cut node, since $CONN_{1(\mathscr{G}-p_2)}(p_1, p_3) = 0.02 < 0.03 = CONN_{1\mathscr{G}}(p_1, p_3)$ and $CONN_{2(\mathscr{G}-p_2)}(p_1, p_3) = 0.6 > 0.3 = CONN_{2\mathscr{G}}(p_1, p_3)$.

Definition 9.42 Let $\mathscr{G}$ be an IFLG. An arc $e = (p, q)$ of $\mathscr{G}$ is said to be a **partial bridge** (p- bridge) if $CONN_{1(\mathscr{G}-e)}(a, b) < CONN_{1\mathscr{G}}(p, q), CONN_{2(\mathscr{G}-e)}(p, q) > CONN_{2\mathscr{G}}(p, q)$.

Definition 9.43 Let $\mathscr{G}$ be an IFLG on $G^* = (V, E)$ and $\mathscr{C}$ is a cycle in $\mathscr{G}$. Then, (i) If all arcs in $\mathscr{C}$ are strong, $\mathscr{C}$ is called a **strong cycle**. (ii) An arc $e = (a, b) \in E$ is called α**-strong** if $CONN_{1(\mathscr{G}-e)}(a, b) < \mu_1(c, d)$ and $CONN_{2(\mathscr{G}-e)}(a, b) > \mu_2(c, d)$; e is called a δ**-strong arc** if $CONN_{1(\mathscr{G}-e)}(a, b) > \mu_1(c, d)$ and $CONN_{2(\mathscr{G}-e)}(a, b) < \mu_2(c, d)$ for all $(c, d) \in E$. (iii) A c-d path P in $\mathscr{G}$ is called a strong c-d path if all the arcs of P are strong. In particular, if all the arcs of P are α-strong !α**-strong path**, then P is said to be an α**-strong path**.

Clearly, an arc $e = (a, b)$ is called strong if it is α-strong. If (a, b) is a strong arc, then a and b are said to be in strong nbd of each other.

Fig. 9.6 α-strong path

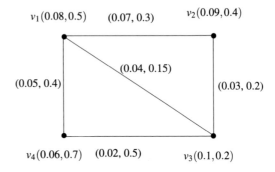

Let us illustrates α-strong arc and α-strong path of an IFG.

Example 9.6 In Fig. 9.6, the arcs (v_1, v_2), (v_1, v_3), (v_1, v_4) are α-strong and the arc (v_3, v_4) is a δ-strong arc. Also $P = v_4 v_1 v_2$ is an α-strong path. But, there is no strong cycle.

Theorem 9.30 *If $\mathcal{G}$ is a connected IFLG, then there is a strong path any two nodes.*

Proof Let $\mathcal{G} = (\mathcal{V}, \sigma, \mu)$ be a connected IFLG and a and b be any two nodes of $\mathcal{G}$. If the arc (a, b) is strong, then there is nothing to prove. Otherwise, there are two cases, either (i) (a, b) is a δ-arc or (ii) there exists a path from a to b of length more than one.

In case (i), there is a path P (say) such that $S_1(P) > \mu_1(a, b)$ and $S_2(P) < \mu_2(a, b)$.

In case (ii), if some arcs on the path is not strong, replace it by a path having more strength. This argument can not be repeated infinitely because the graph is finite; therefore, one can find a path from a to b for which all arcs are strong. Hence, there is a strong path between a and b. $\square$

9.4.1 Intuitionistic Fuzzy Labeling Tree

Here, we define intuitionistic fuzzy labeling tree (IFLT) as follows.

Definition 9.44 An IFG $\mathcal{G} = (\mathcal{V}, \sigma, \mu)$ is called a **IFLT**, if it has IFL and an IF spanning subgraph $F = (\mathcal{V}, \sigma, \rho)$ that is a tree, where for every edge (p, q) not in F, $\mu_1(p, q) < CONN_{1F}(p, q)$ and $\mu_2(p, q) > CONN_{2F}(p, q)$.

Theorem 9.31 *If $\mathcal{G}$ is an IFLT, then the arcs of IF spanning subgraph F are IF bridges of $\mathcal{G}$.*

Proof Let $\mathcal{G} = (\mathcal{V}, \sigma, \mu)$ be an IFLT and $F = (\mathcal{V}, \sigma, \rho)$ be its spanning subgraph. Let (p, q) be an arc in F. Then $CONN_{1F}(p, q) < \mu_1(p, q) \leq CONN_{1\mathcal{G}}(p, q)$ and

$CONN_{2F}(p, q) > \mu_2(p, q) \geq CONN_{2\mathcal{G}}(p, q)$, which indicates that the arc (p, q) is an IF bridge of $\mathcal{G}$. Since the arc (p, q) is arbitrary, then every arc of F are IF bridges of $\mathcal{G}$.   □

Theorem 9.32 *Every IFLG is an IFLT.*

Proof Let $\mathcal{G} = (\mathcal{V}, \sigma, \mu)$ be an IFLG. Since μ is bijective, each and every vertex of $\mathcal{G}$ have at least one arc as IF bridge. Therefore, there exists a spanning subgraph F, whose arcs are IF bridges. Hence, every IFLG is an IFLT.   □

9.4.2 Partial Intuitionistic Fuzzy Labeling Tree

Partial IFLT is another type of labeling tree.

Definition 9.45 A connected IFLG $\mathcal{G}$ is said to be a **partial IFLT** if $\mathcal{G}$ has a spanning subgraph F that is a tree and for every arc $(p, q)(\notin F)$ of $\mathcal{G}$, $CONN_{1\mathcal{G}}(p, q) > \mu_1(p, q)$ and $CONN_{2\mathcal{G}}(p, q) < \mu_2(p, q)$.

When a graph $\mathcal{G}$ is disconnected and the above condition holds for all components of $\mathcal{G}$, then $\mathcal{G}$ is called a **partial fuzzy forest**.

Following is a characterization of partial IFLT.

Theorem 9.33 *A connected IFLG $\mathcal{G}$ is a partial IFLT if and only if for any cycle $\mathcal{C}$ in $\mathcal{G}$, there is an arc $e = (p, q)$ with $\mu_1(e) < CONN_{1(\mathcal{G}-e)}(p, q)$ and $\mu_2(e) > CONN_{2(\mathcal{G}-e)}(p, q)$.*

Proof Suppose $\mathcal{G}$ is a connected IFLG. If there is no cycle, then $\mathcal{G}$ is obviously a tree and also a partial fuzzy tree. If there is cycle in $\mathcal{G}$, let (p, q) be an arc of $\mathcal{C}$ with the least membership and greatest non-membership values in $\mathcal{G}$. Remove the arc (p, q) from $\mathcal{G}$. If $\mathcal{G}$ has another cycle, repeat the process. Not at each step no previously removed arc is strongest the arc being presently deleted. When the graph $\mathcal{G}$ does not contain any cycle, then the subgraph is a tree F. Suppose the arc (p, q) is not in F. Then (p, q) is one of the arcs removed in the process to construct F. Since F is a tree and (p, q) is the arc having least membership and greatest non-membership values among all the arcs of a cycle in $\mathcal{G}$, it follows that there is a path between a and b whose membership value is greater than $\mu_1(p, q)$ and non-membership value is less than $\mu_2(p, q)$ and this does not cover (p, q) or any other edges removed before. If that path covers arcs that were removed later, then the path can be further detached and so on. This process gives a path belonging entirely the arcs of F. Thus $\mathcal{G}$ is a partial IFLT.

Conversely, suppose $\mathcal{G}$ is a partial IFLT and $\mathcal{P}$ is cycle, then for some arc $e = (p, q)$ of $\mathcal{P}$ does not lie on F. Thus, by definition, $\mu_1(e) < CONN_{1(\mathcal{G}-e)}(p, q) < CONN_{1\mathcal{G}}(p, q)$ and $\mu_2(e) > CONN_{2(\mathcal{G}-e)}(p, q) > CONN_{2\mathcal{G}}(p, q)$.   □

Theorem 9.34 *If between any two vertices of $\mathcal{G}$, there is at most one strongest path, then $\mathcal{G}$ is a partial tree.*

Proof Suppose $\mathcal{G}$ is not a partial tree. Then there exists a cycle $\mathcal{C}$ in $\mathcal{G}$ with $\mu_1(p,q) \geq CONN_{1\mathcal{G}}(p,q)$ and $\mu_2(p,q) \leq CONN_{2\mathcal{G}}(p,q)$, for all arcs $(p,q) \in \mathcal{C}$. Thus, (p,q) is the strongest path between a and b. If we choose (p,q) as the weakest arc of $\mathcal{C}$, then the remaining of $\mathcal{C}$ forms the strongest path between a and b, a contradiction. Hence, $\mathcal{G}$ must be a partial tree. $\qquad\square$

The above theorem is true for a tree, but it can be proved for a forest also. The following theorem follows from above one.

Theorem 9.35 *If $\mathcal{G}$ is not a tree but a partial tree, then there is at least one arc (p,q) for which $\mu_1(p,q) < CONN_{1\mathcal{G}}(p,q)$ and $\mu_2(p,q) > CONN_{2\mathcal{G}}(p,q)$.*

Following theorem is a characterization of a bridge in IFG.

Theorem 9.36 *Let $\mathcal{G}$ be a partial tree and F be a spanning tree of $\mathcal{G}$. Then all arcs of F are partial bridges of $\mathcal{G}$.*

Proof Let (p,q) be an arc in F. Then the arc (p,q) is the one and only path between the nodes a and b in F. Since there is no other path between a and b in $\mathcal{G}$, then (p,q) is a bridge of $\mathcal{G}$ and hence a partial bridge of $\mathcal{G}$. If there is a path P (say) from a to b in $\mathcal{G}$, then P must have an arc $(x,y) \notin F$ such that $\mu_1(x,y) < CONN_{1G}(p,q)$ and $\mu_2(x,y) > CONN_{2G}(p,q)$. Therefore, the arc (p,q) is the strongest arc of any cycle of $\mathcal{G}$.

Hence, (p,q) is a partial bridge. $\qquad\square$

9.5 Intuitionistic Fuzzy Competition Graphs

In 1968, Cohen [7] first developed the concept of competition graph (CG) to solve the problem of the food web in ecology. The problem of a food web is to describe the predator-prey relationship among species in the community. Let $\overrightarrow{G} = (V, \overrightarrow{E})$ be a digraph, which corresponds to the food web. A node $a \in V(\overrightarrow{G})$ characterizes a species in the food web, an edge $\overrightarrow{(a,c)} \in \overrightarrow{E}(\overrightarrow{G})$ represents that a preys on the species c. If c is a common prey of two species a and b, then they compete for prey c.

In this section, the strength of competition, i.e. the capability (membership) and non-capability (non-membership) of competition between species are considered. Hence, naturally, IFG is the best tool to explain the CGs. The intuitionistic fuzzy competition graph (IFCG) and its several variants like IF k-CGs (IFkCGs) and p-CGs (p-IFCGs), m-steps IFCG are discussed and several relevant results are investigated. An application of IFCG in an ecosystem is described.

Definition 9.46 The **IF out-nbd (neighborhood)** of a node b in a directed intuitionistic fuzzy graph (DIFG) $\overrightarrow{\mathcal{G}}$ is an IFS $N^+(b) = (X_b^+, m_b^+, n_b^+)$, where $X_b^+ = \{a \mid \mu_1\overrightarrow{(b,a)} > 0$ and $\mu_2\overrightarrow{(b,a)} > 0\}$ and $m_b^+ : X_b^+ \to [0,1]$ expressed by $m_b^+(a) = \mu_1\overrightarrow{(b,a)}$ and $n_b^+ : X_b^+ \to [0,1]$ expressed by $n_b^+(a) = \mu_2\overrightarrow{(b,a)}$.

Likewise, the IF in-nbd of a node is defined as follows:

Definition 9.47 The **IF in-nbd** of a node b of directed IFG $\overrightarrow{\mathscr{G}} = (\mathscr{V}, \sigma, \mu)$ is an IFS $N^-(b) = (X_b^-, m_b^-, n_b^-)$, where $X_b^- = \{a \mid \mu_1\overrightarrow{(a, b)} > 0 \text{ and } \mu_2\overrightarrow{(a, b)} > 0\}$ and $m_b^- : X_b^+ \to [0, 1]$ explained by $m_b^-(a) = \mu_1\overrightarrow{(a, b)}$ and $n_b^- : X_b^+ \to [0, 1]$ expressed by $n_b^-(a) = \mu_2\overrightarrow{(a, b)}$.

The following example illustrates IF out- and in-nbd of a node.

Example 9.7 Let $\overrightarrow{\mathscr{G}}$ be a DIFG and $\{b_1, b_2, b_3, b_4\}$ be the set of nodes. Let $\sigma(b_1) = (0.3, 0.6), \sigma(b_2) = (0.8, 0.2)$ $\sigma(b_3) = (0.5, 0.4)$, $\sigma(b_4) = (0.2, 0.8)$ and $\mu\overrightarrow{(b_1, b_2)} = (0.2, 0.6)$, $\mu\overrightarrow{(b_3, b_2)} = (0.4, 0.3)$, $\mu\overrightarrow{(b_3, b_4)} = (0.15, 0.52)$, $\mu\overrightarrow{(b_4, b_2)}$ $= (0.2, 0.75)$, $\mu\overrightarrow{(b_1, b_3)} = (0.3, 0.5)$.

The out-nbd and in-nbd of all nodes are shown below:
$N^+(b_1) = \{(b_2, 0.2, 0.6), (b_3, 0.3, 0.5)\}$, $N^+(b_3) = \{(b_2, 0.4, 0.3), (b_4, 0.15, 0.52)\}$, $N^+(b_4) = \{(b_2, 0.2, 0.75)\}$ and $N^-(b_3) = \{(b_1, 0.3, 0.5)\}$.

Now, we are in a position to define IFCG.

Definition 9.48 The **IFCG** for a DIFG $\overrightarrow{\mathscr{G}}$ denoted by $\mathscr{C}(\overrightarrow{\mathscr{G}})$ is an UIFG $\mathscr{G} = (\mathscr{V}, \sigma, \mu)$ which has same set of IF nodes as in $\overrightarrow{\mathscr{G}}$. There is an IF arc between the nodes $p, q \in \mathscr{V}$ in $\mathscr{C}(\overrightarrow{\mathscr{G}})$ if and only if $N^+(p) \cap N^+(q) (\neq \emptyset)$ is an IFS in $\overrightarrow{\mathscr{G}}$. The membership and non-membership values of the arc (p, q) in IFCG are $\mu_1(p, q) = (\sigma_1(p) \wedge \sigma_1(q)) h_1(N^+(p) \cap N^+(q))$ and $\mu_2(p, q) = (\sigma_2(p) \vee \sigma_2(q)) h_2(N^+(p) \cap N^+(q))$.

The following example illustrates IFCG.

Example 9.8 Let $\overrightarrow{\mathscr{G}}$ be a DIFG and the set of nodes be $\{b_1, b_2, b_3, b_4\}$. Let the membership and non-membership values of the nodes and arcs be $\sigma(b_1) = (0.3, 0.6)$, $\sigma(b_2) = (0.8, 0.2)$, $\sigma(b_3) = (0.5, 0.4)$, $\sigma(b_4) = (0.2, 0.8)$ and $\mu\overrightarrow{(b_1, b_2)} = (0.2, 0.6)$, $\mu\overrightarrow{(b_1, b_3)} = (0.3, 0.5)$, $\mu\overrightarrow{(b_3, b_4)} = (0.15, 0.52)$, $\mu\overrightarrow{(b_4, b_2)} = (0.2, 0.75)$.

Then $N^+(b_1) = \{(b_2, 0.2, 0.6), (b_3, 0.3, 0.5)\}$, $N^+(b_2) = \phi$, $N^+(b_3) = \{(b_4, 0.15, 0.52)\}$, $N^+(b_4) = \{(b_2, 0.2, 0.75)\}$. Therefore, $N^+(b_1) \cap N^+(b_4) = \{(b_2, 0.2, 0.75)\}$ (shown in Fig. 9.7a).

Now, the set of vertices in IFCG $\mathscr{C}(\overrightarrow{G})$ is $\{b_1, b_2, b_3, b_4\}$. By definition of CG, there is only one arc between the nodes b_1 and b_4 in $\mathscr{C}(\overrightarrow{\mathscr{G}})$. The membership and non-membership values are given by $\mu_1(b_1, b_4) = 0.04$ and $\mu_2(b_1, b_4) = 0.6$. For all other edges, $\mu_1 = \mu_2 = 0$. The corresponding IFCG is shown in Fig. 9.7b.

Theorem 9.37 Let $\overrightarrow{\mathscr{G}}$ be a DIFG. If $N^+(c) \cap N^+(d)$ contains only one element of $\overrightarrow{\mathscr{G}}$, then the arc (c, d) of $C(\overrightarrow{\mathscr{G}})$ is independently strong (IS) if and only if $\|[N^+(c) \cap N^+(d)]\|_m > 0.5$ and $\|[N^+(c) \cap N^+(d)]\|_n < 0.5$.

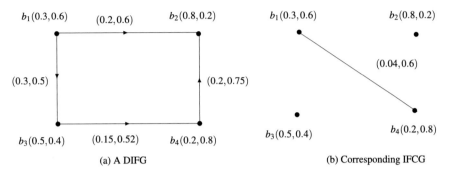

Fig. 9.7 Construction of IFCG

Proof Without loss of generality, let $N^+(c) \cap N^+(d) = (z, \sigma_1, \sigma_2)$ in which σ_1 and σ_2 are the value of membership and the value of non-membership of either the arc (c, z) or the arc (d, z). Here, $|[N^+(c) \cap N^+(d)]|_m = \sigma_1 = h_1(N^+(c) \cap N^+(d))$ and $|[N^+(c) \cap N^+(d)]|_n = \sigma_2 = h_2(N^+(c) \cap N^+(d))$. So $\mu_1(c, d) = \sigma_1 \times [\sigma_1(c) \wedge \sigma_1(d)]$ and $\mu_2(c, d) = \sigma_2 \times [\sigma_2(c) \vee \sigma_2(d)]$. Therefore, the arc (c, d) in $C(\overrightarrow{\mathcal{G}})$ is IS if and only if $\sigma_1 > 0.5$ and $\sigma_2 < 0.5$. Hence, the result. □

9.5.1 An Application of IFCGs

The IFGs are applied in many areas, one of them is ecosystem. To illustrate the IFCG, let us consider a small ecosystem containing six species, viz. kelp, small fish, sea urchin, sea star, sea otter, and shark. As per ecosystem, sharks eat sea otters, sea otters eat sea stars and sea urchins, sea stars eat sea urchins and small fishes, both small fishes and sea urchins eat kelp. These six species are considered as the nodes. Suppose, the degree of existence of kelp in the ecosystem is 90% and degree of non-existence is 10%, i.e. the value of membership and value of non-membership of the vertex kelp is $(0.9, 0.1)$. Similarly, we assume the membership and non-membership values of the other (species) vertices be $(0.7, 0.3)$, $(0.8, 0.2)$, $(0.9, 0.1)$, $(0.8, 0.2)$, and $(0.9, 0.1)$. Suppose, sharks like to eat sea otter, say 70% and dislike to eat, say 20%. The likeness and dislikeness of preys for predators are shown in Table 9.1.

Then the membership and non-membership values of the edge between shark and sea otter are $(0.7, 0.2)$ and similar for other edges. The corresponding food web is shown in Fig. 9.8.

It may be observed that if sea otters are removed from the food web, then the shark must become extinct and in this case, sea stars and sea urchins will grow uncontrollable. Thus, one can analyze the food web with the help of IFGs.

For this food web, $N^+(\text{shark}) = \phi$, $N^+(\text{sea otter}) = \{(\text{shark}, 0.7, 0.2)\}$, $N^+(\text{sea urchin}) = \{(\text{sea otter}, 0.7, 0.2), (\text{sea star}, 0.8, 0.1)\}$, $N^+(\text{sea star}) = \{(\text{sea otter}, 0.8, 0.2)\}$,

Table 9.1 Likeness and dislikeness of preys and predators

Name of predator	Shark	Sea otter	Sea otter	Sea star	Sea star	Small fish	Sea urchin
Name of prey	Sea otter	Sea urchin	Sea star	Sea urchin	Small fish	Kelp	Kelp
Like to eat (%)	70	70	80	80	70	70	80
Dislike to eat (%)	20	20	20	10	20	25	10

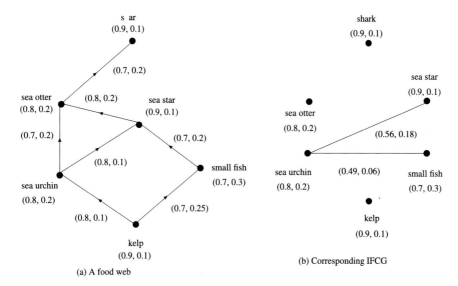

Fig. 9.8 An intuitionistic fuzzy food web

N^+(small fish) = {(sea star, 0.7, 0.2)} and N^+(kelp) = {(sea urchin, 0.8, 0.1), (small fish, 0.7, 0.25)}.

Therefore, $(N^+($ sea urchin $) \cap N^+($ sea star $)) = \{$(sea otter , 0.7, 0.2)$\}$ and $(N^+($ sea urchin$) \cap N^+($small fish$)) = \{$(sea star, 0.7, 0.2)$\}$.

Now, there is an edge between sea urchin and sea star, sea urchin and small fish in the IFCGs and for other pairs of vertices, there is no edge in the IFCG. The value of membership and value of non-membership of the edges between two vertices sea urchin and sea star is (0.56, 0.18) and sea urchin and small fish are (0.49, 0.06), calculated by the definition of IFCG. Hence, there is a competition between sea urchin and sea star, sea urchin and small fish in the ecosystem.

9.5.2 Intuitionistic Fuzzy k-Competition Graph

An extension of IFCG, **Intuitionistic fuzzy k-competition graph** (IFkCG) is defined below.

Definition 9.49 Let k be a given positive number. The IFkCG $\mathscr{C}_k(\overrightarrow{G})$ of a DIFG $\overrightarrow{\mathscr{G}}$ is a UIFG $\mathscr{G} = (\mathscr{V}, \sigma, \mu)$ which has same set of nodes as in $\overrightarrow{\mathscr{G}}$ and there is an arc between two nodes $p, q \in V$ if and only if $|(N^+(p) \cap N^+(q))|_m > k$ and $|(N^+(p) \cap N^+(q))|_n > k$. The membership and non-membership values of the arc (p, q) in $\mathscr{C}_k(\overrightarrow{\mathscr{G}})$ are given by
$\mu_1(p, q) = \frac{k_1 - k}{k_1}[\sigma_1(p) \wedge \sigma_1(q)]h_1(N^+(p) \cap N^+(q))$, where $k_1 = |(N^+(p) \cap N^+(q))|_m$
and $\mu_2(p, q) = \frac{k_2 - k}{k_2}[\sigma_2(p) \vee \sigma_2(q)]h_2(N^+(p) \cap N^+(q))$, where $k_2 = |(N^+(p) \cap N^+(q))|_n$.

Example 9.9 Let $\overrightarrow{\mathscr{G}}$ be a DIFG and the node set be $\mathscr{V} = \{x, y, p_1, p_2, p_3, p_4\}$. The node and arc membership and non-membership values are taken as $\sigma(x) = (0.5, 0.5)$, $\sigma(y) = (0.4, 0.5)$, $\sigma(p_1) = (0.5, 0.4)$, $\sigma(p_2) = (0.6, 0.3)$, $\sigma(p_3) = (0.7, 0.2)$, $\sigma(p_4) = (0.6, 0.4)$ and $\mu(\overrightarrow{x, p_1}) = (0.4, 0.4)$, $\mu(\overrightarrow{x, p_2}) = (0.5, 0.4)$, $\mu(\overrightarrow{x, p_3}) = (0.4, 0.3)$, $\mu(\overrightarrow{x, p_4}) = (0.3, 0.5)$, $\mu(\overrightarrow{y, p_1}) = (0.3, 0.4)$, $\mu(\overrightarrow{y, p_2}) = (0.4, 0.4)$, $\mu(\overrightarrow{y, p_3}) = (0.4, 0.5)$, $\mu(\overrightarrow{y, p_4}) = (0.3, 0.3)$.

The set of nodes of k-CG is $\mathscr{V}$. By definition of k-CG, there is only one arc (x, y) in IFkCG. The data for this arc is calculated below.

$N^+(x) = \{(p_1, 0.4, 0.4), (p_2, 0.5, 0.4), (p_3, 0.4, 0.3), (p_4, 0.3, 0.5)\}$ and
$N^+(y) = \{(p_1, 0.3, 0.4), (p_2, 0.4, 0.4), (p_3, 0.4, 0.5), (p_4, 0.3, 0.3)\}$.

Therefore, $N^+(x) \cap N^+(y) = \{(p_1, 0.3, 0.4), (p_2, 0.4, 0.4), (p_3, 0.4, 0.5), (p_4, 0.3, 0.5)\}$.

Thus, $k_1 = 0.3 + 0.4 + 0.4 + 0.3 = 1.4$ and $k_2 = 0.4 + 0.4 + 0.5 + 0.5 = 1.8$. Let $k = 0.5$.

Then $\mu_1(x, y) = 0.08$ and $\mu_2(x, y) = 0.18$. This graph is shown in the Fig. 9.9.

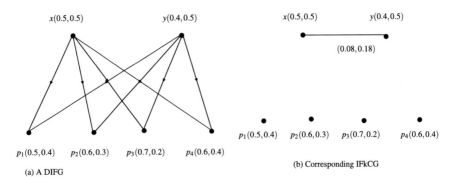

(a) A DIFG

(b) Corresponding IFkCG

Fig. 9.9 An IFkCG for $k = 0.5$

Theorem 9.38 *Let $\overrightarrow{\mathscr{G}}$ be a DIFG and (p, q) be an arc. If $|(N^+(p) \cap N^+(q))|_m >$ $2k$, $|(N^+(p) \cap N^+(q))|_n < 2k$ and heights of both the set $(N^+(p) \cap N^+(q))$ and $(N^+(p) \cap N^+(q))$ are 1, the arc (p, q) is independently strong in $\mathscr{C}_k(\overrightarrow{\mathscr{G}})$.*

Proof Let $\mathscr{C}_k(\overrightarrow{\mathscr{G}})$ be the IFkCG for the DIFG $\overrightarrow{\mathscr{G}}$. Now, since $h_1(N^+(p) \cap N^+(q)) = 1$ and $|(N^+(p) \cap N^+(q))|_m > 2k$, then $k_1 > 2k$ and hence,

$$\mu_1(p, q) = \frac{k_1 - k}{k_1}[\sigma_1(p) \wedge \sigma_1(q)]h_1(N^+(p) \cap N^+(q)) = \frac{k_1 - k}{k_1}[\sigma_1(p) \wedge \sigma_1(q)]$$

or, $\quad \dfrac{\mu_1(p, q)}{[\sigma_1(p) \wedge \sigma_1(q)]} = \dfrac{k_1 - k}{k_1} > 0.5$

Similarly,

$$\frac{\mu_2(p, q)}{[\sigma_2(p) \vee \sigma_2(q)]} = \frac{k_2 - k}{k_2} < 0.5$$

Hence, the arc (p, q) in $\mathscr{C}_k(\overrightarrow{G})$ is independently strong. □

Another variation of competition graph called p-competition graph is described below.

9.5.3 Intuitionistic Fuzzy p-Competition Graph

In IFkCG, k is a positive real number and p is any non-negative integer.

Definition 9.50 Let p be a non-negative integer. Then **IFG p-competition graph** (IFpCG) $\mathscr{C}^p(\overrightarrow{\mathscr{G}})$ of a DIFG $\overrightarrow{\mathscr{G}}$ is an UIFG $\mathscr{G}$ which has the same set of nodes as in $\overrightarrow{\mathscr{G}}$. There is an IF arc between two nodes $a, b \in \mathcal{V}$ in $\mathscr{C}^p(\overrightarrow{\mathscr{G}})$ if and only if $|supp(N^+(a) \cap N^+(b))| \geq p$. The membership and non-membership values of the arc (a, b) in $\mathscr{C}^p(\overrightarrow{\mathscr{G}})$ are given by $\mu_1(a, b) = \frac{(\lambda - p) + 1}{\lambda}[\sigma_1(a) \wedge \sigma_1(b)]h_1(N^+(a) \cap N^+(b))$ and $\mu_2(a, b) = \frac{(\lambda - p) + 1}{\lambda}[\sigma_2(a) \vee \sigma_2(b)]h_2(N^+(a) \cap N^+(b))$ where $\lambda = |supp(N^+(a) \cap N^+(b))|$.

Let us construct a 3-competition IFG.

Example 9.10 Let $\overrightarrow{\mathscr{G}}$ be a DIFG and its node set be $\mathcal{V} = \{x, y, z, p, q, r, \}$ (see Fig. 9.10). The membership and non-membership values of nodes and arcs are taken as
$\sigma(x) = (0.6, 0.4), \sigma(y) = (0.7, 0.3), \sigma(z) = (0.4, 0.5), \sigma(p) = (0.7, 0.3), \sigma(q) = (0.7, 0.2), \sigma(r) = (0.6, 0.3)$ and $\mu\overrightarrow{(x, p)} = (0.5, 0.4), \mu\overrightarrow{(x, q)} = (0.6, 0.2), \mu\overrightarrow{(x, r)} = (0.6, 0.3), \quad \mu\overrightarrow{(y, p)} = (0.6, 0.3), \quad \mu\overrightarrow{(y, q)} = (0.7, 0.2), \quad \mu\overrightarrow{(y, r)} = (0.6, 0.3), \mu\overrightarrow{(z, p)} = (0.4, 0.4), \mu\overrightarrow{(z, q)} = (0.3, 0.4)$.

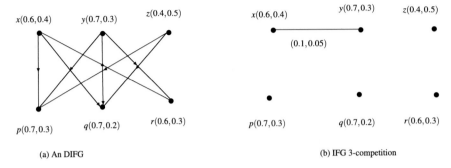

(a) An DIFG   (b) IFG 3-competition

Fig. 9.10 IGpCG for $p = 3$

Now, the out-nbd of x, y, z are
$N^+(x) = \{(p, 0.5, 0.4), (q, 0.6, 0.2), (r, 0.6, 0.3)\}$,
$N^+(y) = \{(p, 0.6, 0.3), (q, 0.7, 0.2), (r, 0.6, 0.3)\}$ and $N^+(z) = \{(p, 0.4, 0.4), (q, 0.3, 0.4)\}$.

Therefore, $N^+(x) \cap N^+(y) = \{(p, 0.5, 0.4), (q, 0.6, 0.2), (r, 0.6, 0.3)\}$,
$N^+(x) \cap N^+(z) = \{(p, 0.4, 0.4), (q, 0.3, 0.4)\}$ and $N^+(y) \cap N^+(z) = \{(p, 0.4, 0.4), (q, 0.3, 0.4)\}$.

Hence, $\lambda = |supp(N^+(x) \cap N^+(y))| = 3$.

If we consider $p = 3$, then $\mu_1(x, y) = 0.1$ and $\mu_2(x, y) = 0.05$. Hence, $\mu(x, y) = (0.1, 0.05)$

Theorem 9.39 *Let $\overrightarrow{\mathscr{G}}$ be a DIFG. If, for the arc (p, q), $h_1(N^+(p) \cap N^+(q)) = 1$ and $h_2(N^+(p) \cap N^+(q)) = 0$ in $C^{\lceil \frac{\lambda}{2} \rceil}(\overrightarrow{\mathscr{G}})$, then (p, q) is independently strong, where $\lambda = |supp(N^+(p) \cap N^+(q))|$.*

Proof Let $\mathscr{G} = (\mathscr{V}, \sigma, \mu)$ be an IGpCG for $p = [\frac{\lambda}{2}]$, where $\lambda = |supp(N^+(p) \cap N^+(q))|$. Suppose, $h_1(N^+(p) \cap N^+(q)) = 1$ and $h_1(N^+(p) \cap N^+(q)) = 0$, $p, q \in \mathscr{V}$.

Now, $\mu_1(p, q) = \frac{\lambda - [\frac{\lambda}{2}] + 1}{\lambda}[\sigma_1(p) \wedge \sigma_1(q)] \times 1$ and $\mu_2(p, q) = \frac{\lambda - [\frac{\lambda}{2}] + 1}{\lambda}[\sigma_2(p) \vee \sigma_2(q)] \times 0$.

Hence, $\frac{\mu_1(p,q)}{[\sigma_1(p) \wedge \sigma_1(q)]} = \frac{\lambda - [\frac{\lambda}{2}] + 1}{\lambda} > 0.5$ and $\frac{\mu_2(p,q)}{[\sigma_2(p) \vee \sigma_2(q)]} = 0 < 0.5$.
Thus, (p, q) is an independently strong arc.   □

9.5.4 Intuitionistic Fuzzy Open-Nbd and Closed-Nbd Graphs

The IF open-nbd and closed-nbd of a node for an IFG are defined below.

Definition 9.51 The **IF open-nbd** of a node of an IFG $\mathscr{G} = (\mathscr{V}, \sigma, \mu)$ is an IFS. For the node u, the open-nbd is defined as $N(u) = (X_u, m_u, n_u)$, where $X_u = \{z :$

$\mu_1(u, z) > 0$ and $\mu_2(u, z) > 0$}, $m_u : X_u \to [0, 1]$ and $n_u : X_u \to [0, 1]$ explained by $m_u(z) = \mu_1(u, z)$ and $n_u(z) = \mu_2(u, z)$.

The **IF closed-nbd** of the node u is $N[u] = N(u) \cup \{u\}$.

Definition 9.52 Let $\mathscr{G} = (\mathscr{V}, \sigma, \mu)$ be an IFG. **IF open-nbd graph** of $\mathscr{G}$ is an IFG denoted by $N(\mathscr{G})$ and defined as $N(\mathscr{G}) = (\mathscr{V}, \sigma, \mu')$, where node sets of both the graph $\mathscr{G}$ and $N(\mathscr{G})$ are same. Two nodes p and q are connected by an arc in $N(\mathscr{G})$ if and only if $N(p) \cap N(q)$ is a non-empty IFS. The membership and non-membership values of the arc (p, q) are $[\sigma_1(p) \wedge \sigma_1(q)]h_1(N(p) \cap N(q))$ and $[\sigma_2(p) \vee \sigma_2(q)]h_2(N(p) \cap N(q))$.

Like open-nbd IF graph, the closed-nbd graph can be defined.

Definition 9.53 **IF closed-nbd graph** of an IFG $\mathscr{G}$ is another IFG defined as $N[\mathscr{G}] = (\mathscr{V}, \sigma, \mu')$ with same node set as $\mathscr{G}$. The nodes p and q are connected by an arc in $N[\mathscr{G}]$ if and only if $N[p] \cap N[q]$ is a non-empty IFS and $\mu(p, q) = ([\sigma_1(p) \wedge \sigma_1(q)]h_1(N[p] \cap N[q]), [\sigma_2(p) \vee \sigma_2(q)]h_2(N[p] \cap N[q]))$.

By definition of the closed-nbd graph, the following result can easily be proved.

Theorem 9.40 *For each arc of an IFG $\mathscr{G}$, there is an arc in $N[\mathscr{G}]$.*

9.5.5 m-Step IFCG

Another extension of IFCG, called m-step IFCG (mS-IFCG), is considered here. The mS-IFCG is a DIFG over a DIFG. It is used in many real applications.

Definition 9.54 The m-step CG of a DIFG $\overrightarrow{\mathscr{G}}$ is denoted by $\overrightarrow{\mathscr{G}}_m$, where the set of nodes of $\overrightarrow{\mathscr{G}}_m$ and $\overrightarrow{\mathscr{G}}$ are same. Two nodes p and q are connected by a directed IF arc in $\overrightarrow{\mathscr{G}}_m$ if there is an IF directed path $\overrightarrow{P}^{\,m}_{(p,q)}$ in $\overrightarrow{\mathscr{G}}$.

The m-step out-nbd and in-nbd graphs are defined likewise.

Definition 9.55 IF **m-step out-nbd** of a node u of a DIFG $\overrightarrow{\mathscr{G}}$ is the IFS $N^+_m(u) = (X^+_u, \rho^+_1(u), \rho^+_2(u))$, where $X^+_u = \{z :$ there exists a directed path from u to z of length m, $\overrightarrow{P}^{\,m}_{u,z}\}$, $\rho^+_1(u) : X^+_u \to [0, 1]$ and $\rho^+_2(u) : X^+_u \to [0, 1]$ and explained by $\rho^+_1(u) = \min\{\mu_1\overrightarrow{(p, q)}, (p, q)$ is an arc of $\overrightarrow{P}^{\,m}_{u,z}\}$, $\rho^+_2(u) = \max\{\mu_2\overrightarrow{(p, q)}, (p, q)$ is an arc of $\overrightarrow{P}^{\,m}_{u,z}\}$.

Definition 9.56 IF **m-step in-nbd** of a node u of a DIFG $\overrightarrow{\mathscr{G}}$ is an IFS defined as $N^-_m(u) = (X^-_u, \rho^-_1(u), \rho^-_2(u))$, where $X^-_u = \{z :$ there exists a directed path from z to u of length m, $\overrightarrow{P}^{\,m}_{z,u}\}$, $\rho^-_1(u) : X^-_u \to [0, 1]$ and $\rho^-_2(u) : X^-_u \to [0, 1]$ and explained by $\rho^-_1(u) = \min\{\mu_1\overrightarrow{(p, q)}, (p, q)$ is an arc of $\overrightarrow{P}^{\,m}_{z,u}\}$, $\rho^-_2(u) = \max\{\mu_2\overrightarrow{(p, q)}, (p, q)$ is an arc of $\overrightarrow{P}^{\,m}_{z,u}\}$.

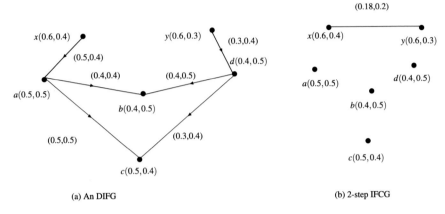

(a) An DIFG   (b) 2-step IFCG

Fig. 9.11 Two-step IFCG

Definition 9.57 The *m*-**step IFCG** of a DIFG $\overrightarrow{\mathcal{G}}$ is denoted by $C_m(\overrightarrow{\mathcal{G}}) = (\mathcal{V}, \sigma, \mu)$ whose node set is the node set of $\overrightarrow{\mathcal{G}}$. Two nodes $p, q \in \mathcal{V}$ are connected by an arc in $C_m(\overrightarrow{\mathcal{G}})$ if and only if $(N_m^+(p) \cap N_m^+(q))$ is a non-empty IFS. The $\mu(p, q)$ in $C_m(\overrightarrow{\mathcal{G}})$ is $(\{\sigma_1(p) \wedge \sigma_1(q)\} h_1(N_m^+(p) \cap N_m^+(q)), \{\sigma_2(p) \vee \sigma_2(q)\} h_2(N_m^+(p) \cap N_m^+(q)))$.

An example of two-step IFCG is considered below.

Example 9.11 Let $\overrightarrow{\mathcal{G}} = (\mathcal{V}, \sigma, \mu)$ be a DIFG with node set $\mathcal{V} = \{x, y, a, b, c, d\}$. Let $\sigma(x) = (0.6, 0.4)$, $\sigma(y) = (0.6, 0.3)$, $\sigma(a) = (0.5, 0.5)$, $\sigma(b) = (0.4, 0.5)$, $\sigma(c) = (0.5, 0.4), \sigma(d) = (0.4, 0.5)$ and $\mu(\overrightarrow{x, a}) = (0.5, 0.4), \mu(\overrightarrow{a, b}) = (0.4, 0.4)$, $\mu(\overrightarrow{a, c}) = (0.5, 0.5)$, $\mu(\overrightarrow{y, d}) = (0.3, 0.4)$, $\mu(\overrightarrow{d, b}) = (0.4, 0.5)$, $\mu(\overrightarrow{d, c}) = (0.3, 0.4)$.
Then $N_2^+(x) = \{(b, 0.4, 0.4), (c, 0.5, 0.5)\}$ and $N_2^+(y) = \{(b, 0.4, 0.5), (c, 0.3, 0.4)\}$. So $N_2^+(x) \cap N_2^+(y) = \{(b, 0.4, 0.5), (c, 0.3, 0.5)\}$. Then $\mu(x, y) = 0.18$ and $\mu_2(x, y) = 0.2$. This two-step IFCG is shown in Fig. 9.11.

The *m*-step nbd of a node in an IFG is defined below.

Definition 9.58 IF *m*-**step nbd** of a node u in an IFG $\mathcal{G}$ is defined as $N_m(u) = (X_u, \rho_1(u), \rho_2(u))$,   where $X_u = \{z :$ there exists a directed path from u to z of length m, $P_{u,z}^m\}$, $\rho_1(u)$ and $\rho_2(u)$ both maps from X_u to $[0, 1]$ and also $\rho_1(b) = \min\{\mu_1(p, q), (p, q)$ is an arc of $P_{b,a}^m\}$, $\rho_2(b) = \max\{\mu_2(p, q), (p, q)$ is an arc of $P_{b,a}^m\}$. Note that $N_m(u)$ is an IFS.

Now, one can define *m*-step IF nbd graph.

Definition 9.59 The *m*-**step IF nbd graph** for an IFG $\mathcal{G}$ is defined by $N_m(\mathcal{G}) = (\mathcal{V}, \sigma, \eta)$, where node set of $N_m(\mathcal{G})$ and $\mathcal{G}$ are same, $\sigma = (\sigma_1, \sigma_2)$, $\eta = (\eta_1, \eta_2)$. Both η_1 and η_2 maps from $\mathcal{V} \times \mathcal{V}$ to $[0, 1]$ such that $\eta_1(p, q) = \sigma_1(a) \wedge \sigma_1(q) h_1(N_m(p) \cap N_m(q))$ and $\eta_2(p, q) = \sigma_2(p) \vee \sigma_2(q) h_2(N_m(p) \cap N_m(q))$.

Theorem 9.41 *Let $\overrightarrow{\mathscr{G}}$ be a DIFG and all of its arcs are independently strong. Then all arcs of $C_m(\overrightarrow{\mathscr{G}})$ are also independently strong.*

Theorem 9.42 *Let $\overrightarrow{\mathscr{G}}$ be a DIFG having n number of vertices. If $m > n$ then $C_m(\overrightarrow{\mathscr{G}})$ has no arcs.*

Proof Let $C_m(\overrightarrow{\mathscr{G}})$ be an m-step IFCG over $\mathscr{G}$ and the membership and non-membership values of an arc (p, q) are given by $\mu_1(p, q) = (\sigma_1(p) \wedge \sigma_1(q)) h_1 (N_m^+(p) \cap N_m^+(q))$ and $\mu_2(p, q) = (\sigma_2(p) \vee \sigma_2(q)) h_2(N_m^+(p) \cap N_m^+(q))$.

If $m > n$, there is no directed IF path of length m in $\overrightarrow{\mathscr{G}}$. Therefore, $N_m^+(p) \cap N_m^+(q)$ is a null set and so $\mu_1(p, q) = 0$ and $\mu_2(p, q) = 0$ for all $(p, q) \in C_m(\overrightarrow{\mathscr{G}})$. Hence, $C_m(\overrightarrow{\mathscr{G}})$ has no arcs. $\qquad\square$

Definition 9.60 Let $\overrightarrow{\mathscr{G}}$ be a DIFG. Assume that p be the common node of m-step out-nbds of the nodes $x_1, x_2, \ldots, x_k$. Also, let $\overrightarrow{(y_i, z_i)}$ be an arc on the path $\overrightarrow{P}^m_{(x_i, p)}$ with the least membership value $\mu_1\overrightarrow{(y_i, z_i)}$ and the greatest non-membership value $\mu_2\overrightarrow{(y_i, z_i)}$ for $i = 1, 2, \ldots, k$. The m-step vertex $p \in \mathscr{V}$ is **independently strong vertex** if $\mu_1\overrightarrow{(y_i, z_i)} > 0.5$ and $\mu_2\overrightarrow{(y_i, z_i)} < 0.5$ for all $i = 1, 2, \ldots, k$.

The strength of the vertex p is denoted by $(s_1(p), s_2(p))$ where $s_1 : \mathscr{V} \to [0, 1]$ and $s_2 : \mathscr{V} \to [0, 1]$ such that

$$s_1(p) = \frac{1}{k} \sum_{i=1}^{k} \mu_1\overrightarrow{(y_i, z_i)} \quad \text{and} \quad s_2(p) = \frac{1}{k} \sum_{i=1}^{k} \mu_2\overrightarrow{(y_i, z_i)}.$$

Example 9.12 In Fig. 9.11, the strength of the vertex b is $(\frac{0.4+0.4}{2}, \frac{0.4+0.5}{2}) = (0.4, 0.45)$. Therefore b is not independently strong two-step vertex.

Theorem 9.43 *If a vertex p of $\overrightarrow{\mathscr{G}}$ is independently strong, then $s_1(p) > 0.5$ and $s_2(p) < 0.5$.*

Proof Let p be a common node of m-step out-nbd of the nodes $x_1, x_2, \ldots, x_k$, i.e. there exists the k paths $\overrightarrow{P}^m_{(x_1, p)}, \overrightarrow{P}^m_{(x_2, p)}, \ldots, \overrightarrow{P}^m_{(x_k, p)}$ in $\overrightarrow{\mathscr{G}}$. Also, let $\mu_1\overrightarrow{(y_i, z_i)}$ be the least membership value of an arc on the path $\overrightarrow{P}^m_{(x_i, p)}$ and $\mu_2\overrightarrow{(y_i, z_i)}$ be the greatest non-membership value of an arc on the path $\overrightarrow{P}^m_{(x_i, p)}$ for $i = 1, 2, \ldots, k$.

If p is independently strong, each arc $\overrightarrow{(y_i, z_i)}, i = 1, 2, \ldots, k$ is independently strong. Therefore, $\mu_1\overrightarrow{(y_i, z_i)} > 0.5$ and $\mu_2\overrightarrow{(y_i, z_i)} < 0.5, i = 1, 2, \ldots, k$.

Hence, by definition of $s_1(p)$ and $s_2(p)$, $s_1(p) > 0.5$ and $s_2(p) < 0.5$. $\qquad\square$

9.6   Intuitionistic Fuzzy Tolerance Graphs

In this section, intuitionistic fuzzy tolerance graphs (IFTGs) are defined. Also, its several variations, viz. intuitionistic fuzzy ϕ-tolerance graph (IF ϕ-TG) and its particular cases—max, min, sum; IF proper TG and unit TG are considered along with several properties. A very useful application of IFTG is provided.

Before going to define tolerance graph, let us define intuitionistic fuzzy intervals.

Definition 9.61 Let V be a linearly ordered set. An **intuitionistic fuzzy interval** (IFI) $\mathscr{I}$ on V is a convex, normal IF subset of V and is denoted by $\mathscr{I} = (V, \mu_1, \mu_2)$. That is, there is an element $a \in V$ with $h(a) = (1, 0)$ and the order $d \leq b \leq c$ means $\mu_1(b) \geq \min\{\mu_1(d), \mu_1(c)\}$ and $\mu_2(b) \leq \max\{\mu_2(d), \mu_2(c)\}$.

Definition 9.62 Intuitionistic fuzzy tolerance (IFT) of an IFI is denoted by $\mathscr{T}$ and described by an arbitrary IFI whose length of core is a non-negative real number. If L is the real number and $\mid i_k - i_{k-1} \mid = L$ where $i_{k-1}, i_k \in R$ (set of real), then the IFT is the IF set of the interval $[i_{k-1}, i_k]$.

The support and core of an IFT are called tolerance support and tolerance core respectively. Length of tolerance support and core of an IFT $\mathscr{T}$ are denoted by $SL(\mathscr{T})$ and $CL(\mathscr{T})$ respectively. IFT may be an IF number.

The IFT is shown in Fig. 9.12. Here, the dotted line and thick line represent the non-membership and membership functions respectively.

Intuitionistic fuzzy ϕ-TG is the general form of IFI graph that is introduced below.

Definition 9.63 Let $\phi : R^+ \times R^+ \to R^+$ (R^+ be a positive real number) be a real-valued function. Let $\mathscr{I} = \{\mathscr{I}_1, \mathscr{I}_2, \ldots, \mathscr{I}_n\}$ be a finite family of IFIs on real line and the corresponding IFTs be $\mathscr{T} = \{\mathscr{T}_1, \mathscr{T}_2, \ldots, \mathscr{T}_n\}$. For these IFIs and IFTs a IFG $\mathscr{G}$ is constructed as follows:

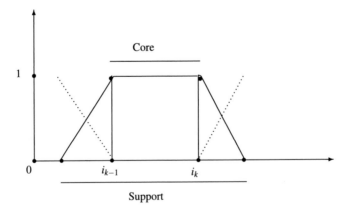

Fig. 9.12 An intuitionistic fuzzy tolerance

For each IFI $\mathscr{I}_i$, there is a node v_i, $i = 1, 2, \ldots, n$. The membership and non-membership values of the node v_i are given by $\sigma_1(v_i) = h_1(\mathscr{I}_i)$ and $\sigma_2(v_i) = h_2(\mathscr{I}_i)$ for all v_i. The membership and non-membership values of the arc (v_i, v_j) in $\mathscr{G}$ are

$$
\mu_1(v_i, v_j) = \begin{cases} 1, & \text{if } CL(\mathscr{I}_i \cap \mathscr{I}_j) \geq \phi\{CL(\mathscr{T}_i), CL(\mathscr{T}_j)\} \\ \frac{SL(\mathscr{I}_i \cap \mathscr{I}_j) - \phi\{SL(T_i), SL(\mathscr{T}_j)\}}{SL(\mathscr{I}_i \cap \mathscr{I}_j)} h_1(\mathscr{I}_i \cap \mathscr{I}_j), & \text{if } SL(\mathscr{I}_i \cap \mathscr{I}_j) \geq \phi\{SL(\mathscr{T}_i), SL(\mathscr{T}_j)\} \\ 0, & \text{otherwise.} \end{cases}
$$

and

$$
\mu_2(v_i, v_j) = \begin{cases} 0, & \text{if } CL(\mathscr{I}_i \cap \mathscr{I}_j) \geq \phi\{CL(\mathscr{T}_i), CL(\mathscr{T}_j)\} \\ \frac{SL(\mathscr{I}_i \cap \mathscr{I}_j) - \phi\{SL(\mathscr{T}_i), SL(\mathscr{T}_j)\}}{SL(\mathscr{I}_i \cap \mathscr{I}_j)} h_2(\mathscr{I}_i \cap \mathscr{I}_j), & \text{if } SL(\mathscr{I}_i \cap \mathscr{I}_j) \geq \phi\{SL(\mathscr{T}_i), SL(\mathscr{T}_j)\} \\ 1, & \text{otherwise.} \end{cases}
$$

where $CL(\mathscr{I}_i \cap \mathscr{I}_j)$ and $SL(\mathscr{I}_i \cap \mathscr{I}_j)$ are the lengths of core and support of the intersection of intervals $\mathscr{I}_i$ and $\mathscr{I}_j$ respectively $h_1(\mathscr{I}_i \cap \mathscr{I}_j)$ is the degree of membership and $h_2(\mathscr{I}_i \cap \mathscr{I}_j)$ is the degree of non-membership of the height $(\mathscr{I}_i \cap \mathscr{I}_j)$.

From the definition, it is clear that for each IF ϕ-TG there are two IFSs, viz. IFIs $(\mathscr{I} = \{\mathscr{I}_1, \mathscr{I}_2, \ldots, \mathscr{I}_n\})$ and IFTs $(\mathscr{T} = \{\mathscr{T}_1, \mathscr{T}_2, \ldots, \mathscr{T}_n\})$. So, for an IF ϕ-TG, it means that these sets are associated with it. Note that the number of nodes of IF ϕ-TG is n.

Theorem 9.44 *Let $\mathscr{G}$ be an IF ϕ-TG with IFIs $\mathscr{I}$ and IFTs $\mathscr{T}$. If $h(\mathscr{I}_i \cap \mathscr{I}_j) = (1, 0)$ and $SL(\mathscr{I}_i \cap \mathscr{I}_j) \geq 2\phi\{SL(\mathscr{T}_i), SL(\mathscr{T}_j)\}$, then $\mu_1(v_i, v_j) \leq \frac{1}{2}$ and $\mu_2(v_i, v_j) \leq \frac{1}{2}$ for all (v_i, v_j) in $\mathscr{G}$.*

Proof Let $\mathscr{G}$ be an IF ϕ-TG with IFIs $\mathscr{I}$ and IFTs $\mathscr{T}$. Since $h(\mathscr{I}_i \cap \mathscr{I}_j) = (1, 0)$, then $h_1(\mathscr{I}_i \cap \mathscr{I}_j) = 1$ and $h_2(\mathscr{I}_i \cap \mathscr{I}_j) = 0$. Also, since $SL(\mathscr{I}_i \cap \mathscr{I}_j) \geq 2\phi\{SL(\mathscr{T}_i), SL(\mathscr{T}_j)\}$, then $SL(\mathscr{I}_i \cap \mathscr{I}_j) > \phi\{SL(\mathscr{T}_i), SL(\mathscr{T}_j)\}$. Then,

$$
\begin{aligned}
\mu_1(v_i, v_j) &= \frac{SL(\mathscr{I}_i \cap \mathscr{I}_j) - \phi\{SL(\mathscr{T}_i), SL(\mathscr{T}_j)\}}{SL(\mathscr{I}_i \cap \mathscr{I}_j)} h_1(\mathscr{I}_i \cap \mathscr{I}_j) \\
&= 1 - \frac{\phi\{SL(\mathscr{T}_i), SL(\mathscr{T}_j)\}}{SL(\mathscr{I}_i \cap \mathscr{I}_j)} \\
&\leq 1 - \frac{1}{2} = \frac{1}{2}
\end{aligned}
$$

and

$$
\begin{aligned}
\mu_2(v_i, v_j) &= \frac{SL(\mathscr{I}_i \cap \mathscr{I}_j) - \phi\{SL(\mathscr{T}_i), SL(\mathscr{T}_j)\}}{SL(\mathscr{I}_i \cap \mathscr{I}_j)} h_2(\mathscr{I}_i \cap \mathscr{I}_j) \\
&= [1 - \frac{\phi\{SL(\mathscr{T}_i), SL(\mathscr{T}_j)\}}{SL(\mathscr{I}_i \cap \mathscr{I}_j)}] \times 0 = 0 < \frac{1}{2}.
\end{aligned}
$$

Hence, the result obtained. $\qquad\square$

9.6.1 Intuitionistic Fuzzy Min-Tolerance Graphs

The function ϕ is a generic function and it can be chosen in many different ways. Let us consider $\phi(x, y) = \min\{x, y\}$. Then the IF ϕ-TG is said to be intuitionistic fuzzy min-tolerance graph (IFMTG).

Definition 9.64 An IF ϕ-TG is said to be **intuitionistic fuzzy min-tolerance graph** if $\phi(x, y) = \min\{x, y\}$ for any x, y.

Now, an IFMTG is illustrated below.

Example 9.13 Let $\mathscr{I}_1$, $\mathscr{I}_2$, $\mathscr{I}_3$ and $\mathscr{I}_4$ be four IFIs and the corresponding IFTs be $\mathscr{T}_1$, $\mathscr{T}_2$, $\mathscr{T}_3$ and $\mathscr{T}_4$. Let $[2, 6]$, $[4.75, 13.25]$, $[10.25, 13.75]$, $[15, 18.25]$ be the cores and $[0.75, 6.5]$, $[3, 14]$, $[9.25, 16.25]$, $[13.5, 20]$ be the supports of these IFIs respectively.
    Then $CL(\mathscr{T}_1) = 2, CL(\mathscr{T}_2) = 2.5, CL(\mathscr{T}_3) = 4, CL(\mathscr{T}_4) = 0.75$ and $SL(\mathscr{T}_1) = 3.25$, $SL(\mathscr{T}_2) = 6.75$, $SL(\mathscr{T}_3) = 5.5$, $SL(\mathscr{T}_4) = 2.5$. The IFIs and IFTs are shown in Fig. 9.13.
    Then $h(\mathscr{I}_1 \cap \mathscr{I}_2) = (1, 0)$, $h(\mathscr{I}_2 \cap \mathscr{I}_3) = (1, 0)$, $h(\mathscr{I}_2 \cap \mathscr{I}_4) = (0.25, 0.75)$ and $h(\mathscr{I}_3 \cap \mathscr{I}_4) = (0.69, 0.31)$. Therefore, $\mu(v_1, v_2) = (0.07, 0.00)$, $\mu(v_2, v_3) = (1.00, 0.00)$, $\mu(v_2, v_4) = (0.00, 1.00)$ and $\mu(v_3, v_4) = (0.06, 0.03)$. The corresponding IFMTG is shown in Fig. 9.14.

Definition 9.65 Let $\mathscr{G}$ be an IFTG with IFIs $\mathscr{I}$ and IFTs $\mathscr{T}$. If $CL(\mathscr{I}_i) \geq CL(\mathscr{T}_i)$ and $SL(\mathscr{I}_i) \geq SL(\mathscr{T}_i)$ of the IFTG, then the IFTG is called an **IF bounded TG**.

Theorem 9.45 *If $\mathscr{G}$ is an IFTG having constant length of core and support, then $\mathscr{G}$ bounded.*

Proof Let $\mathscr{G}$ be an IFTG with IFI $\mathscr{I}$ and IFT $\mathscr{T}$. Let k_1, k_2 be two natural numbers with $CL(\mathscr{T}_i) = k_1$ and $SL(\mathscr{T}_i) = k_2$ for $i = 1, 2, \ldots, n$.
    If $CL(\mathscr{I}_i) \geq k_1$ and $SL(\mathscr{I}_i) \geq k_2$, then $CL(\mathscr{I}_i) \geq CL(\mathscr{T}_i)$ and $SL(\mathscr{I}_i) \geq SL(\mathscr{T}_i)$ for all $i = 1, 2, \ldots, n$. Hence, $\mathscr{G}$ bounded.
    If $CL(\mathscr{I}_i) < k_1$ and $SL(\mathscr{I}_j) < k_2$ for some i and j, then we take $CL(\mathscr{I}_i) = k_1$ and $SL(\mathscr{I}_j) = k_2$ to make $\mathscr{G}$ bounded.
    Hence, $\mathscr{G}$ is bounded. $\qquad\square$

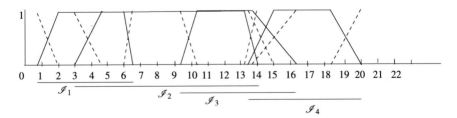

Fig. 9.13 Intuitionistic fuzzy intervals

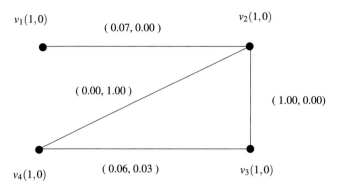

Fig. 9.14 An IFMTG

9.6.2 Application of if Min-Tolerance Graph

Consider a group of employees each scheduled to work for a fixed interval of time in a company. Each employee is each assigned to a single work for their entire work interval. A conflict arises if two employees are assigned to work on the same work station at the same time. Then a tolerance arises in a natural way. Let three employees p_1, p_2, p_3 are assigned for the same work and their time intervals be $\mathscr{I}_1 = [2, 4], \mathscr{I}_2 = [6, 9], \mathscr{I}_3 = [11, 14]$. These are taken as core of the intervals. The time interval $[2, 4]$ means the working time of an employee from 2 a.m. to 4 a.m. (say). Also, assume that they are working inactively as well as actively in the time intervals $[1, 6], [3, 10], [8, 15]$, which are taken as the support of the intervals.

Then $h(\mathscr{I}_1 \cap \mathscr{I}_2) = (0.6, 0.4), h(\mathscr{I}_2 \cap \mathscr{I}_3) = (0.5, 0.5), h(\mathscr{I}_1 \cap \mathscr{I}_3) = (0, 0)$, which is shown in Fig. 9.15.

Since the time intervals intersect, the tolerance needs to be considered. Let $\mathscr{T}_1, \mathscr{T}_2, \mathscr{T}_3$ be the tolerances. Therefore, $CL(\mathscr{T}_1) = 1, CL(\mathscr{T}_2) = 2, CL(\mathscr{T}_3) = 3$ and $SL(\mathscr{T}_1) = 2, SL(\mathscr{T}_2) = 1, SL(\mathscr{T}_3) = 1$, and, $\mu(p_1, p_2) = (0.4, 0.27), \mu(p_2, p_3) = (0.25, 0.25)$ and $\mu(p_1, p_3) = (0, 0)$. Then there is a relation between the employees p_1 and p_2; p_2 and p_3. But there is no relation between the employees p_1 and p_3. Then the corresponding IF min-tolerance graph is shown in Fig. 9.16.

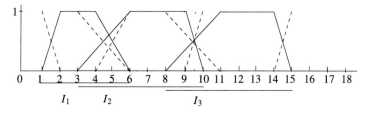

Fig. 9.15 Intuitionistic fuzzy intervals for the problem

Fig. 9.16 An IF
min-tolerance graph for the
intervals of Fig. 9.15

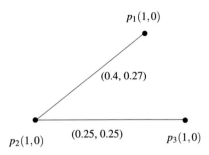

Hence, there is a tolerance between the employees p_1 and p_2 whose membership and non-membership degrees are 0.4 and 0.27. Also, there is a tolerance between the employees p_2 and p_3 whose membership and non-membership degrees are 0.25 and 0.25 respectively. But, there is no tolerance between the employees p_1 and p_3. Thus, the company can schedule two employees to work on the same work station with minimum tolerance.

9.6.3   Intuitionistic Fuzzy Max-Tolerance Graphs

Like min-tolerance graph one can define max-tolerance graph by defining the function ϕ as $\phi(x, y) = \max\{x, y\}$. Then the IF ϕ-TG is said to be intuitionistic fuzzy max-tolerance graph (IFMxTG).

Definition 9.66 An IF ϕ-TG is said to be **intuitionistic fuzzy max-tolerance graph** if $\phi(x, y) = \max\{x, y\}$ for any x, y.

The following example describes an IFMxTG.

Example 9.14 Let us consider three IFIs $\mathscr{I}_1$, $\mathscr{I}_2$, $\mathscr{I}_3$ with corresponding IFTs $\mathscr{T}_1$, $\mathscr{T}_2$, $\mathscr{T}_3$. Let [2, 4], [6, 9], [11, 14] be the cores and [0, 6], [5.5, 12], [10.5, 16] be the supports of the IFIs respectively. For these three IFIs, we consider three nodes p_1, p_2, p_3.

Therefore, $CL(\mathscr{T}_1) = 2, CL(\mathscr{T}_2) = 1, CL(\mathscr{T}_3) = 3$ and $SL(\mathscr{T}_1) = 0.2, SL(\mathscr{T}_2) = 0.4, SL(\mathscr{T}_3) = 0.7$.

Then $h(\mathscr{I}_1 \cap \mathscr{I}_2) = (0.2, 0.8), h(\mathscr{I}_2 \cap \mathscr{I}_3) = (0.43, 0.56)$.

The graph has three nodes for three IFIs and two arcs. The membership and non-membership values of the arcs are $\mu(p_1, p_2) = (0.04, 0.16), \mu(p_2, p_3) = (0.23, 0.30)$.

Definition 9.67 An IFTG is called a unit IFMxTG if all tolerances and all IFIs have same core and same support lengths.

The following result follows from the definition.

Lemma 9.2 *The chordless IF cycle C_k is a unit IFMxTG.*

9.6.4 Intuitionistic Fuzzy Sum-Tolerance Graphs

Let us consider a value of the function ϕ. Let $\phi(x, y) = x + y$. The corresponding IF ϕ-TG is called intuitionistic fuzzy sum-tolerance graphs (IFSTGs).

Definition 9.68 An IF ϕ-TG is said to be **intuitionistic fuzzy sum-tolerance graph** if $\phi(x, y) = x + y = \max\{x, y\}$ for any x, y.

Following is an example of IFSTG.

Example 9.15 Let $\mathscr{I}_1, \mathscr{I}_2, \mathscr{I}_3$ be three IFIs and the corresponding IFTs be $\mathscr{T}_1,$ $\mathscr{T}_2, \mathscr{T}_3$.

Let the cores of IFIs be $[2, 7], [4, 8], [9, 13]$ and the supports be $[1, 8], [2.5, 9.5],$ $[6.5, 15.5]$ respectively.

and $CL(\mathscr{T}_1) = 2.5$, $CL(\mathscr{T}_2) = 3$, $CL(\mathscr{T}_3) = 7.5$ and $SL(\mathscr{T}_1) = 1$, $SL(\mathscr{T}_2) = 2.5$, $SL(\mathscr{T}_2) = 0.4$.

Then $h(\mathscr{I}_1 \cap \mathscr{I}_2) = (1, 0)$, $h(\mathscr{I}_1 \cap \mathscr{I}_3) = (0.43, 0.54)$, $h(\mathscr{I}_2 \cap \mathscr{I}_3) = (0.75,$ $0.25)$.

The membership and non-membership values of the nodes p_1, p_2, p_3 are $(1, 0)$.

Therefore, the membership and non-membership values of the arcs are $\mu(p_1, p_2)$ $= (0.36, 0.00)$, $\mu(p_2, p_3) = (0.03, 0.01)$, $\mu(p_2, p_1) = (0.03, 0.04)$.

Now, we introduce a kind of IFTG, called proper IFTG.

Definition 9.69 A **proper IFTG** is one which has tolerance representation in which no IFI support and core properly contain another IFI support and core.

Theorem 9.46 *Every IFMTG $\mathscr{G}$ with $CL(\mathscr{T}_i) = CL(\mathscr{I}_i)$ and $SL(\mathscr{T}_i) = SL(\mathscr{I}_i)$ for all i is an IFICG.*

Proof Let $\mathscr{G}$ be an IFMTG with IFIs $\mathscr{I}$ and IFTs $\mathscr{T}$ such that $CL(\mathscr{T}_i) = CL(\mathscr{I}_i)$ and $SL(\mathscr{T}_i) = SL(\mathscr{I}_i)$ for every $i = 1, 2, \ldots, n$.

Now, the inequalities $CL(\mathscr{I}_i \cap \mathscr{I}_j) \geq \min\{CL(\mathscr{T}_i), CL(\mathscr{T}_j)\} = \min\{CL(\mathscr{I}_i),$ $CL(\mathscr{I}_j)\}$ are true if and only if the cores of $\mathscr{I}_i$, $\mathscr{I}_j$ contains one another.

Again, $SL(\mathscr{I}_i \cap \mathscr{I}_j) \geq \min\{SL(\mathscr{T}_i), SL(\mathscr{T}_j)\} = \min\{SL(\mathscr{I}_i), SL(\mathscr{I}_j)\}$ are true if and only if the supports of $\mathscr{I}_i$, $\mathscr{I}_j$ contains one other. These imply that $\mu_1(v_i, v_j) \geq 0$ and $\mu_2(v_i, v_j) \geq 0$. Therefore, one of $\mathscr{I}_i$, $\mathscr{I}_j$ contains another. Hence, $\mathscr{G}$ is an IFICG. □

The following result trivially holds.

Theorem 9.47 *Any proper or unit IFT has a bounded IFT.*

Theorem 9.48 *Let $\mathscr{G} = (\mathscr{V}, \sigma, \mu)$ be an IFSTG, which is an unbounded IFTG. Then $\mathscr{G}'$ obtained by adding a new pendant node to every node of $\mathscr{G}$ is not an IFTG.*

Proof Let if possible, $\mathscr{G}'$ be an IFSTG with certain IFT representation. As $\mathscr{G}'$ is an IFSTG with arbitrary IFT, then we assume that $CL(\mathscr{I}_a) \geq CL(T_a)$ and $SL(\mathscr{I}_a) \geq SL(T_a)$ for all $a \in \mathscr{V}$. This implies that $\mathscr{G}$ is a bounded IFTG, a contradiction. □

9.6.5 Intuitionistic Fuzzy Interval Containment Graphs

Let us consider another variation of IFTG known as **intuitionistic fuzzy interval containment graph** (IFICG).

Definition 9.70 Let $\mathscr{I}$ and $\mathscr{T}$ be the IFIs and IFTs respectively. Let v_i be the node for the interval $\mathscr{I}_i \in \mathscr{I}$, $i = 1, 2, \ldots, n$, of the graph $\mathscr{G}$, where n is the number of intervals of $\mathscr{I}$. The membership and non-membership values of the nodes are given by $\sigma_1(v_i) = h_1(\mathscr{I}_i)$ and $\sigma_2(v_i) = h_2(\mathscr{I}_i)$ for all nodes v_i.

The membership and non-membership values of the arc (v_i, v_j) in $\mathscr{G}$ are determined as

$$\mu_1(v_i, v_j) = \begin{cases} 1, \text{ if support and core of one of } \mathscr{I}_i, \mathscr{I}_j \text{ comprise the other} \\ \frac{1}{2}[\frac{CL(\mathscr{I}_i \cap \mathscr{I}_j)}{min\{CL(\mathscr{I}_i), CL(\mathscr{I}_j)\}} + \frac{SL(\mathscr{I}_i \cap \mathscr{I}_j)}{min\{SL(\mathscr{I}_i), SL(\mathscr{I}_j)\}}]h_1(\mathscr{I}_i \cap \mathscr{I}_j), \text{ otherwise.} \end{cases}$$

and

$$\mu_2(v_i, v_j) = \begin{cases} 0, \text{ if support and core of one of } \mathscr{I}_i, \mathscr{I}_j \text{ comprise the other} \\ \frac{1}{2}[\frac{CL(\mathscr{I}_i \cap \mathscr{I}_j)}{min\{CL(\mathscr{I}_i), CL(\mathscr{I}_j)\}} + \frac{SL(\mathscr{I}_i \cap \mathscr{I}_j)}{min\{SL(\mathscr{I}_i), SL(\mathscr{I}_j)\}}]h_2(\mathscr{I}_i \cap \mathscr{I}_j), \text{ otherwise.} \end{cases}$$

The graph IFICG is illustrated by the following example.

Example 9.16 Let us consider four IFIs $\mathscr{I}_1, \mathscr{I}_2, \mathscr{I}_3$ and $\mathscr{I}_4$ with corresponding IFTs $\mathscr{T}_1, \mathscr{T}_2, \mathscr{T}_3$ and $\mathscr{T}_4$. Let the cores and supports of the intervals be $[2, 6], [4.75, 13.25], [10.25, 13.75], [15, 18.25]$ and $[0.75, 6.5], [3, 14], [9.25, 16.25], [13.5, 20]$ respectively. Let p_1, p_2, p_3, p_4 be the nodes of the graph.

Now, the height of the intersecting sets are $h(\mathscr{I}_1 \cap \mathscr{I}_2) = (1, 0), h(\mathscr{I}_2 \cap \mathscr{I}_3) = (1, 0), h(\mathscr{I}_2 \cap \mathscr{I}_4) = (0.25, 0.75)$ and $h(\mathscr{I}_3 \cap \mathscr{I}_4) = (0.69, 0.31)$. Therefore, the membership and non-membership values of the arcs are $\mu(p_1, p_2) = (0.46, 0.00)$, $\mu(p_2, p_3) = (0.78, 0.00)$, $\mu(p_2, p_4) = (0.02, 0.03)$ and $\mu(p_3, p_4) = (0.14, 0.07)$. Finally, the resultant IFICG is shown in Fig. 9.17.

Fig. 9.17 An IF interval containment graph

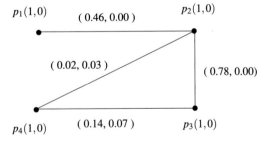

9.7 Genus and Planarity of Intuitionistic Fuzzy Graphs

The **embedding of a graph** into a surface is a drawing of the graph on the surface in such a way that its edges may intersect only at the endpoints. It is proved that any finite graph can be embedded in three-dimensional Euclidean space $\mathbb{R}^3$ [8], whereas a planar graph can be embedded in two-dimensional Euclidean space $\mathbb{R}^2$.

More specifically, an embedding of a graph G on a surface S is a representation of G on S in which points of S are associated with vertices and simple edge are associated with arcs which follow the conditions stated below:
(iii) Two arcs never intersect at a point which is interior to either of the arcs,
(ii) No arcs include points associated with other vertices,
(i) The endpoints of the arc associated with an edge e are the points associated with the end vertices of e.

Definition 9.71 Let S be any surface and, it an IFG $\mathscr{G}$ can be drawn in S where no edges of IFG are intersected, then the graph is said to be an S-embedding.

The embedding of any graph, maybe IFG, in plane and sphere are equal. Because a sphere can be deformed by pressing or stretching, but its embedding property will remain same. An IFG drawn in the plane surface can be drawn on the surface of the sphere because it posseses the embedding property like a plane surface. The drawing approach of crisp graph is used for IFG.

Theorem 9.49 *Every IFG embedded on the surface of a plane can be embedded on the surface of the sphere.*

9.7.1 Genus of Crisp Graphs

A planar graph can be drawn in a plane surface (2D surface) without crossing over of its edges. But, it is not possible for a non-planar graph. A non-planar graph can be drawn on a sphere by introducing a number of handles or holes so that it can be drawn without crossing of edges. The number of handles or holes is referred to as the genus of the surface and is denoted by $g(G)$ or $gen(G)$. The genus of a graph G means the smallest genus of all surfaces on which G can be embedded. Trivially, a graph with m edges can be embedded on a surface of genus m. The genus of a planar graph is 0. Thus, the genus $g(G)$ of a finite graph G of size m, satisfies the inequality $0 \le g(G) \le m$.

If a graph is embedded in a surface with n handles, it is called S_n embeddable.

The first four members of an infinite family of surfaces i.e. sphere, one-hole torus, two-hole torus, and three-hole torus are shown in Fig. 9.18. The genus of a sphere (Fig. 9.18a), one-hole torus (Fig. 9.18b), two-hole torus (Fig. 9.18c), and three-hole torus (Fig. 9.18d) are respectively 0, 1, 2, 3 as number of holes are 0, 1, 2, 3 respectively.

The graph of Fig. 9.18d can also be drawn as in Fig. 9.19.

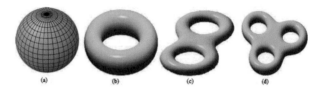

Fig. 9.18 First four torus. Courtesy. https://en.wikipedia.org

Fig. 9.19 A sphere with
three handles

Example 9.17 (i) Every planar graph has genus 0.
(ii) Genus of K_3, K_4, C_n and tree are 0 as they all are planar.
(iii) Genus of K_5 is 1.
(iv) Genus of $K_{3,3}$ is 1.
(v) Genus of K_n is $\lceil \frac{(n-3)(n-4)}{12} \rceil$.

If we draw K_5 and $K_{3,3}$ in a plane surface then at least one edge, say e, must cross other edge(s), and it can not be avoided. For these two graphs, if we draw such edge e on a surface (torus) with one hole, then no edge crosses with another. So, the genus of K_5 and $K_{3,3}$ is 1.

The following result is obvious for a crisp graph.

Theorem 9.50 *(i) Every graph has a genus and its value is a non-negative integer.*
(ii) Let the genus of the graph G be g, then G can be drawn without crossings of edges on every surface S_n for which $n \geq g$.

9.7.2   Planarity of Intuitionistic Fuzzy Graphs

Fuzzy planar graph is defined by Samanta and Pal [31]. Using the same concept, Alshehri and Akram [2] defined **intuitionistic fuzzy planar graph** (IFPG).

Definition 9.72 **Strength of the intuitionistic fuzzy edge** xy can be defined as

$$(M_{xy}, N_{xy}) = \left(\frac{\mu_B(xy)}{\min\{\mu_A(x), \mu_A(y)\}}, \frac{\nu_B(xy)}{\max\{\nu_A(x), \nu_A(y)\}} \right).$$

An edge xy is said to be **intuitionistic fuzzy strong** if $M_{xy} \geq 0.5$ or $N_{xy} \leq 0.5$, otherwise weak.

Let the edges ab and cd intersect at the point P in the graph. For this point P, we define intersecting value $\mathscr{I}_P$ as

$$\mathscr{I}_P = (M_P, N_P) = \left(\frac{M_{ab} + M_{cd}}{2}, \frac{N_{ab} + N_{cd}}{2} \right).$$

By definition of the planar graph, there is no point of intersection among the edges. But, if there is an intersecting point between very weak edges (say, its membership value is very low, such that it tends to 0) then one can say that this intersection is insignificant and we can neglect the presence of this intersection in the graph. And this graph may be considered as a planar graph with certain doubt. So, in general, it is assumed that in an IFG the number of points of intersections be k and let these points be $P_1, P_2, \ldots, P_k$ for a certain geometrical representation. In our new concept, we say that any IFG is an IFPG with some degrees of planarity (DOP). Such DOP is defined below.

Definition 9.73 The DOP of an IFG is defined as

$$f = (f_1, f_2) = \left(\frac{1}{1 + \{M_{P_1} + M_{P_2} + \cdots + M_{P_k}\}}, \frac{1}{1 + \{N_{P_1} + N_{P_2} + \cdots + N_{P_k}\}} \right).$$

By the definition of DOP, it is clear that $0 < f_1 \leq 1$ and $0 < f_2 \leq 1$. If there is no point of intersection for a certain geometrical representation of an IFPG, then its DOP is $(1, 1)$. In this case, G^*, the underlying crisp graph of this IFG is a planar graph (crisp sense). If f_1 goes to 0 and f_2 goes to 1, then the number of points of intersection between the edges increases. In this case, the nature of planarity decreases. Thus, we can say that every IFG is an IFPG with a certain DOP.

From the above analogy, one can define strong and weak IFPGs.

Definition 9.74 An IFG $\mathscr{G}$ is called a **strong IFPG** if $f_1 \geq 0.5$ and $f_2 \leq 0.5$, weak otherwise.

9.7.3 Genus of an Intuitionistic Fuzzy Graph

There is a connection between genus of an IFG and the degree of planarity of an IFG.

Let us define the genus of an IFG.

The genus of a crisp graph is a non-negative integer, but for a FG or an IFG, it is a non-negative real number. For IFG, genus is defined below.

Definition 9.75 An IFG $G = (V, \sigma, \mu)$ has a genus if it is S_n-**embeddable** for some least positive integer n. In other words, if $P_1, P_2, \ldots, P_n$ be the points of intersections between the edges for a certain geometrical representation, then the genus of the IFG $\mathscr{G}$ is given by $f^*(\mathscr{G})$ or simply $f^* = (f_1^*, f_2^*)$, where

$$f_1^* = \frac{\sum\limits_{i=1}^{n} M_{Pi}}{1 + \sum\limits_{i=1}^{n} M_{Pi}} \quad \text{and} \quad f_2^* = \frac{\sum\limits_{i=1}^{n} N_{Pi}}{1 + \sum\limits_{i=1}^{n} N_{Pi}}.$$

Remark 9.3 Similar to the DOP of an IFG, the genus of the IFG is bounded, i.e. $0 < f_1^* \le 1$ and $0 < f_2^* \le 1$.

Note 1 From definitions of genus and DOP of an IFG, it is obvious that

$$f = (f_1, f_2) = (1 - f_1^*, 1 - f_2^*).$$

Example 9.18 Let us consider an IFG $\mathscr{G}$ whose underlying graph is K_5 (see Fig. 9.20a). In this graph, there are five points of intersection. But, this graph is redrawn and seen that there is only one point of intersection (see Fig. 9.20b). Now, we calculate the genus of the IFG $\mathscr{G}$. The membership and non-membership values of the vertices and edges are shown in Fig. 9.20a.

For this graph, $M_{(a,c)} = \frac{0.4}{\min\{0.7, 0.8\}} = \frac{0.4}{0.7} = 0.57$, $N_{(a,c)} = \frac{0.1}{\max\{0.3, 0.2\}} = \frac{0.1}{0.3} = 0.33$, $M_{(b,d)} = \frac{0.3}{\min\{0.9, 0.7\}} = \frac{0.3}{0.7} = 0.43$ and $N_{(b,d)} = \frac{0.1}{\max\{0.2, 0.2\}} = \frac{0.1}{0.2} = 0.50$. Therefore, $M_P = \frac{M_{(a,c)} + M_{(b,d)}}{2} = \frac{0.57 + 0.43}{2} = 0.5$ and $N_P = \frac{N_{(a,c)} + N_{(b,d)}}{2} = \frac{0.33 + 0.5}{2} = 0.42$. Then,

$$f_1^* = \frac{\sum\limits_{i=1}^{n} M_{Pi}}{1 + \sum\limits_{i=1}^{n} M_{Pi}} = \frac{M_P}{1 + M_P} = \frac{0.5}{1 + 0.5} = 0.33$$

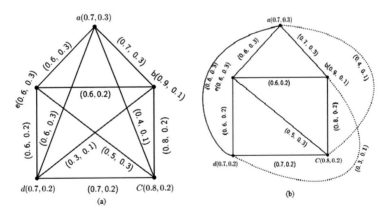

Fig. 9.20 **a** IFG K_5 with five crossings, **b** IFG K_5 with 1 crossing

$$and \ f_2^* = \frac{\sum\limits_{i=1}^{n} N_{Pi}}{1 + \sum\limits_{i=1}^{n} N_{Pi}} = \frac{N_P}{1 + N_P} = \frac{0.42}{1 + 0.42} = 0.30.$$

Hence, the genus of $\mathscr{G}$ is $(0.33, 0.30)$.

Also, the DOP of IFG $\mathscr{G}$ is $(1 - 0.33, 1 - 0.30) = (0.67, 0.70)$.

Theorem 9.51 *Every IFG has a genus.*

Proof Let $\mathscr{G}$ be an IF multigraph. The geometric representation of the IF multigraph is drawn so that the number of intersections of IF edges are determined.

Case 1: Let $\sum\limits_{i=1}^{n} M_{Pi} = 0$ and $\sum\limits_{i=1}^{n} N_{Pi} = 0$. Then the number of intersections of IF edges in $\mathscr{G}$ is zero. Hence, the DOP of IFG is $(1, 1)$. Then the given IFG is obviously planar IFG.

Now, $f_1^* = 1 - f_1 = 0$ and $f_2^* = 1 - f_2 = 0$.

Therefore, the genus of $\mathscr{G}$ is $(0, 0)$.

Case 2: Let $\sum_{i=1}^{n} M_{Pi} \neq 0$ and $\sum_{i=1}^{n} N_{Pi} \neq 0$.

Then there exist an intersection of edges in the IFG $\mathscr{G}$. The genus of an IFG is obtained by the number of intersection of the IF edges in $\mathscr{G}$.

Hence, every IFG has a genus. $\square$

The following result is obtained from the definition.

Lemma 9.3 *Every IFG $\mathscr{G}$ has genus $(0, 0)$ if and only if the DOP of IFG is $(1, 1)$.*

Now, we define strong and weak genus of a graph.

Definition 9.76 For an IFG $\mathscr{G}$, if $f_1^* \geq 0.5$ and $f_2^* \leq 0.5$, then the genus is called strong; otherwise, it is weak.

An IFG $\mathscr{G}$ with strong genus is said to be an **IF strong genus graph**, otherwise a weak genus graph.

Theorem 9.52 *(i) If $\mathscr{G}$ is a weak IFPG, $\mathscr{G}$ is an IF strong genus graph.*
(ii) If $\mathscr{G}$ is an IF weak genus graph, then it is a strong IFPG.

The proof follows from the definitions of weak IFPG and IF strong genus graphs.

Theorem 9.53 *Let G_1 and G_2 be two isomorphic IFGs with IF genus (f_1^*, f_2^*) and (g_1^*, g_2^*) respectively. Then $f_1^* = g_1^*$ and $f_2^* = g_2^*$.*

9.7.4   Euler Polyhedral Equation for Intuitionistic Fuzzy Graph

Euler presented a no. of concepts in the discipline of graph theory. Euler polyhedral equation of graphs helps in finding genus of a graph and the bounds for genus. Here, **Euler polyhedral equation** for IFG is presented.

The following theorem is called Euler's second formula.

Theorem 9.54 *If G is a connected (crisp) graph then $n - m + f = 2 - 2g$, where n, m, f, g represent the number of vertices, edges, faces, and genus respectively of G.*

If $g = 0$, then the above equation reduces to $n - m + f = 2$. It is valid for any planar graph and this equation is known as **Euler's first formula**.

Theorem 9.55 *Let $\mathcal{G} = (\mathcal{V}, \sigma, \mu)$ be a connected IFG with set of vertices $\mathcal{V}$, set of edges E and F be the set of faces. Let $\sigma = (\sigma_1, \sigma_2)$, $\mu = (\mu_1, \mu_2)$ and number of vertices, edges, and faces of $\mathcal{G}$ be n, m and r respectively. Then the IF genus of $\mathcal{G}$ follows the following inequalities: $V_1^* - E_1^* + F_1^* \leq 2 - f_1^*$ and $V_2^* - E_2^* + F_2^* \leq 2 - f_2^*$, where*

$$V_1^* = \frac{1}{n} \sum_{v_i \in \mathcal{V}} \sigma_1(v_i),$$

$$V_2^* = \frac{1}{n} \sum_{v_i \in \mathcal{V}} \sigma_2(v_i),$$

$$E_1^* = \frac{1}{m} \sum_{(v_i, v_j) \in E} \mu_1(v_i, v_j),$$

$$E_2^* = \frac{1}{m} \sum_{(v_i, v_j) \in E} \mu_2(v_i, v_j),$$

$$F_1^* = \frac{\text{The sum of the membership values of the IF faces in } G}{r} \text{ and}$$

$$F_2^* = \frac{\text{The sum of the non-membership values of the IF faces in } G}{r}.$$

Proof We prove this theorem by the method of contradiction. Let us assume that $V_1^* - E_1^* + F_1^* > 2 - f_1^*$ and $V_2^* - E_2^* + F_2^* > 2 - f_2^*$.
Case 1: Let $f_1^* = 0$ and $f_2^* = 0$.

Then obviously IF genus is $(0, 0)$. Then the above inequalities are strictly greater than 2. But, from the definitions of V_1^*, V_2^*, E_1^*, E_2^*, F_1^* and F_2^*, all of them lies between 0 and 1. Then the values of $V_1^* - E_1^* + F_1^*$ and $V_2^* - E_2^* + F_2^*$ lie in $(0, 1)$. Hence, the above inequalities do not hold. So, it is a contradiction and therefore, $V_1^* - E_1^* + F_1^* \leq 2 - f_1^*$ and $V_2^* - E_2^* + F_2^* \leq 2 - f_2^*$.
Case 2: Let $0 < f_1^* < 1$ and $0 < f_2^* < 1$.

Then the IF genus lies between $(0, 1)$. Then by similar argument the above inequalities do not hold—a contradiction and therefore, $V_1^* - E_1^* + F_1^* \leq 2 - f_1^*$ and $V_2^* - E_2^* + F_2^* \leq 2 - f_2^*$.

Hence, the result follows. $\qquad\qquad\qquad\qquad\qquad\qquad\qquad\qquad\qquad\qquad\qquad\square$

References

1. M. Akram, B. Davvaz, Strong intuitionistic fuzzy graphs. Filomat **26**(1), 177–196 (2012)
2. N. Alshehri, M. Akram, Intuitionistic fuzzy planar graphs. Discrete Dyn. Nature Soc. **2014**, 9 (2014). Article ID 397823. https://doi.org/10.1155/2014/397823
3. K.T. Atanassov, Intuitionistic fuzzy sets, in *VII ITKR's Seession, Deposed in Central for Science-Technical Library of Bulgarian Academy of Science, 1697/84* (Sofia, Bulgaria, 1983)
4. K.T. Atanassov, Intuitionistic fuzzy sets. Fuzzy Sets Syst. **20**(1), 87–96 (1986)
5. A. Bozhenyuk, M. Knyazeva, I. Rozenberg, Algorithm for finding domination set in intuitionistic fuzzy graph, in *11th Conference of the European Society for Fuzzy Logic and Technology (EUSFLAT 2019)*, Atlantis Studies in Uncertainty Modelling, August 2019, pp. 72–76. https://doi.org/10.2991/eusflat-19.2019.11
6. G. Chartrand, P. Zang, *Introduction to Graph Theory* (Tata McGraw-Hill Edition, 2005)
7. J.E. Cohen, *Interval Graphs and Food Webs: A Finding and a Problem, Document 17696-PR* (RAND Corporation, Santa Monica, 1968)
8. R.F. Cohen, P. Eades, T. Lin, F. Ruskey, Three-dimensional graph drawing, in *Graph Drawing: DIMACS International Workshop, GD '94 Princeton, New Jersey, USA, October 1012, 1994*, eds. by R. Tamassia, I.G. Tollis. Proceedings, Lecture Notes in Computer Science, vol. 894 (Springer, 1995), p. 111. https://doi.org/10.1007/3-540-58950-3_351
9. B. Davvaz, N. Jan, T. Mahmood, K. Ullah, Intuitionistic fuzzy graphs of nth type with applications. J. Intell. Fuzzy Syst. **36**(4), 3923–3932 (2019)
10. J.A. Gallian, A dynamic survey of graph labeling. Electron. J. Comb. **17**, 61–150 (2014)
11. S. Gravier, A. Khelladi, On the domination number of cross products of graphs. Discrete Math. **145**(1), 273–277 (1995)
12. M.G. Karunambigai, M. Akram, S. Sivasankar, K. Palanivel, Balanced intuitionistic fuzzy graphs. Appl. Math. Sci. **7**(51), 2501–2514 (2013)
13. M.G. Karunambigai, M. Akram, S. Sivasankar, K. Palanivel, Clustering algorithm for intuitionistic fuzzy graphs. Int. J. Uncertainty Fuzziness Knowl.-Based Syst. **25**(3), 367–383 (2017)
14. N. Lakshmiprasanna, K. Sravanthi, N. Sudhakar, Applications of graph labeling in major areas of computer science. Int. J. Res. Comput. Commun. Technol. **3**(8), 819–823 (2014)
15. T. Mahapatra, M. Pal, Fuzzy colouring of m-polar fuzzy graph and its application. J. Intell. Fuzzy Syst. **35**, 6379–6391 (2018)
16. R. Mahapatra, S. Samanta, M. Pal, Applications of edge colouring of fuzzy graphs. Informatica **31**(2), 118 (2020). https://doi.org/10.15388/20-INFOR403
17. S. Mandal, M. Pal, Product of bipolar intuitionistic fuzzy graphs and their degree. TWMS J. App. Eng. Math. **9**(2), 327–338 (2019)
18. S. Moradi, A note on tensor product of graphs. Iranian J. Math. Sci. Inf. **7**(1), 73–81 (2012)
19. R. Parvathi, M.G. Karunambigai, Intuitionistic fuzzy graphs. Comput. Intell. Theory Appl. **38**, 139–150 (2006)
20. H. Rashmanlou, S. Samanta, M. Pal, R.A. Borzooei, Intuitionistic fuzzy graphs with categorical properties. Fuzzy Inf. Eng. **7**(3), 317–334 (2015)
21. H. Rashmanlou, M. Pal, S. Raut, F. Mofidnakhaei, B. Sarkar, Novel concepts in intuitionistic fuzzy graphs with application. J. Intell. Fuzzy Syst. **37**, 37433749 (2019). https://doi.org/10.3233/JIFS-182961

22. A. Saha, M. Pal, T.K. Pal, Selection of programme slots of television channels for giving advertisement: a graph theoretic approach. Inf. Sci. **177**(12), 2480–2492 (2007)
23. S. Sahoo, M. Pal, Different types of products on intuitionistic fuzzy graphs. Pac. Sci. Rev. A. Natural Sci. Eng. **17**(3), 87–96 (2015)
24. S. Sahoo, M. Pal, Intuitionistic fuzzy competition graphs. J. Appl. Math. Comput. **52**(1), 37–57 (2016). https://doi.org/10.1007/s12190-015-0928-0
25. S. Sahoo, M. Pal, Product of intuitionistic fuzzy graphs and their degree. J. Intell. Fuzzy Syst. **32**(1), 1059–1067 (2017)
26. S. Sahoo, M. Pal, Intuitionistic fuzzy tolerance graphs with application. J. Appl. Math. Comput. **55**, 495511 (2017)
27. S. Sahoo, M. Pal, H. Rashmanlou, R.A. Borzooei, Covering and paired domination in intuitionistic fuzzy graphs. J. Intell. Fuzzy Syst. **33**(6), 4007–4015 (2017)
28. S. Sahoo, G. Ghorai, M. Pal, Embedding and genus of intuitionistic fuzzy graphs on spheres. J. Mult.-Valued Logic Soft Comput. **31**(1–2), 139–154 (2018)
29. S. Sahoo, M. Pal, Certain types of edge irregular intuitionistic fuzzy graphs. J. Intell. Fuzzy Syst. **34**, 295–305 (2018)
30. S. Sahoo, M. Pal, Intuitionistic fuzzy labeling graphs. TWMS J. Appl. Eng. Math. **8**(2), 466–476 (2018)
31. S. Samanta, M. Pal, Fuzzy planar graphs. IEEE Trans. Fuzzy Syst. **23**(6), 1936–1942 (2016)
32. S. Samanta, T. Pramanik, M. Pal, Fuzzy colouring of fuzzy graphs. Afrika Matematika **27**(1–2), 37–50 (2016). https://doi.org/10.1007/s13370-015-0317-8
33. Z. Sherman, Square difference labeling for some graph. Cybern. Syst. Anal. **52**(4), 636–640 (2016)
34. J. Xu, C. Yang, Connectivity of Cartesian product graphs. Discrete Math. **306**(1), 159–165 (2006)
35. O. Zihni, Y. elik, Irregular intuitionistic fuzzy graphs, in *AIP Conference Proceedings*, American Institute of Physics, vol. 1726 (2016), p. 020042. https://doi.org/10.1063/1.4945868

Chapter 10
Few Applications of Fuzzy Graphs

Like crisp graphs, **fuzzy graphs** (FGs) also have useful many applications. But, a very few number of applications of FGs are available in the literature. In this chapter, some applications of FGs are incorporated. The main use of fuzzy graph theory is to represent the network with ambiguity. An ecological problem is modeled as a fuzzy digraph and

Some problems are considered here which are modeled as FGs, viz. representation of ecological problem, representation of social network, representation of telecommunication system, link prediction in fuzzy social networks, competition in manufacturing industries, patrolling of bus network, image contraction, installation of cell phone tower, traffic signaling, job selection, etc. The ecological system and the churn prediction of a fuzzy telecommunication network are analyzed.

10.1 Application of Fuzzy Graph in Ecology

The **ecological system** can be successfully modeled using crisp graph as well as fuzzy graph. In this section, we explain how an ecological problem can be modeled using FG.

Different types of organisms are present in an ecosystem. Each organism depends for food on one or more organisms for food, except primary producers (plants and some chemicals). The organisms are divided into two groups â Preys and Predators. A predator can be a prey for another predator as well. The **prey-predator relationship** can be explained easily with the help of a food web, i.e. a directed graph (digraph).

For example, the food chilli-chicken is made by chicken, but hen is not a primary producer. Hen eats grass which is a primary producer. So we eat chicken and hen eats grass. This is a very simple food web with three species—grass, hen, and human.

Let us consider a small ecosystem.

© The Editor(s) (if applicable) and The Author(s), under exclusive license
to Springer Nature Singapore Pte Ltd. 2020
M. Pal et al., *Modern Trends in Fuzzy Graph Theory*,
https://doi.org/10.1007/978-981-15-8803-7_10

Fig. 10.1 A small food web

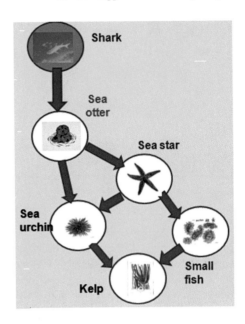

(a) sharks eat sea otters;
(b) sea otters eat sea stars and sea urchins;
(c) sea stars eat sea urchins and small fishes;
(d) sea urchins eat kelp;
(e) small fishes eat kelp.

These relationships can be represented by a digraph, called **food web**, depicted in Fig. 10.1.

For this food web, one can construct a digraph ($\overrightarrow{\mathscr{G}}$) using the following principal: The six species, viz. kelp, sea urchin, small fish, sea star, sea otter, and shark are taken as vertices. There is a directed edge from the species A to the species B if A is a prey for the predator B. Using this principle the food web of Fig. 10.1 can be represented as a digraph $\overrightarrow{\mathscr{G}}$ shown in Fig. 10.2. For simplicity, we rename the species kelp, sea urchin, small fish, sea star, sea otter, shark by $a_1, a_2, a_3, a_4, a_5, a_6$ respectively.

In a food web, it may happen that one particular species may eat more than one species and the amount of eating may also vary. For example, sea star eats 'sea urchins' and 'small fishes', but the amount of consumption of these two species are different. It is assumed that sea stars intake diet 40% of sea urchins and 60% of small fishes as daily. These values (percentage of food) can be assigned to the corresponding edges as weights and hence the resultant graph becomes a **weighted digraph**. Let the weight of the directed edge (i, j) is denoted by w_{ij} (Fig. 10.3).

The numbers associated with the edges represent the amount of food (in [0, 1] scale) is consumed by a species. Note that consumption of a particular prey by a predator is not fixed, it varies time to time, depending on the availability of such

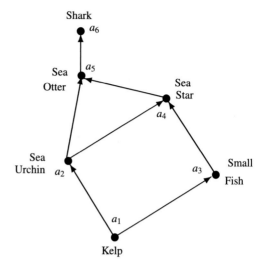

Fig. 10.2 The digraph corresponding to the food web of Fig. 10.1

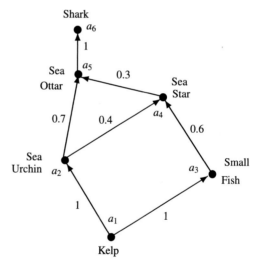

Fig. 10.3 The weighted digraph for the food web of Fig. 10.1

prey, and other factors. Thus, the weights of the edges are not fixed quantities, so it is better to say these are membership values of the edges.

Advantage to represent a food web as a digraph is that one can analyze the food web regarding the extinction and growth of species. For example, if the small fishes get extinct due to some physical phenomenon from the food web then food habit of the sea star will change. In this case, sea star will eat only sea urchins as no other food will be available for them. Then there is a chance of extinction of sea urchins as well. Similarly, if sea otters become extinct, then shark will also become extinct in a matter of time and then the sea stars and sea urchins will grow uncontrolled.

10.1.1 Competition Graph

In the food web, there is also a competition among the species. In the food web of Fig. 10.1, the small fishes and sea urchins both eat kelp. Thus, these two species are competitors to each other. Similarly, sea stars and sea otters are competitors as sea urchins are their common prey. The entire competition of a food web can be represented by another type of graph called **competition graph** (discussed in detail in Chap. 4).

The competition graph is constructed as follows:

The vertices in the competition graph ($\mathscr{C}(\overrightarrow{\mathscr{G}})$) are the species and there exists an edge between the species a and b iff a and b have a common prey, say x, i.e. there is an undirected edge (a, b) in the competition graph if there are directed edges (x, a) and (x, b) in the food web.

Many results are available for competition graphs in [16, 17, 25–28].

For the digraph of Fig. 10.2, sea urchin (a_2) and small fish (a_3) are competitors for kelp (a_1). So we draw an (undirected) edge between the vertices (a_2) and (a_3). Similarly, the vertices (a_4) and (a_5) are competitors and hence there is an edge between them in $\mathscr{C}(\overrightarrow{\mathscr{G}})$.

The weight (W_{ij}) of the edge (i, j) in $\mathscr{C}(\overrightarrow{\mathscr{G}})$ is calculated by

$$(W_{ij}) = \frac{|prey_{a_i} \cap prey_{a_j}|}{|prey_{a_i} \cup prey_{a_j}|},$$

where $|\cdot|$ represents the number of elements in the set.

In the present example, $prey_{a_2} = prey_{a_3} = \{a_1\}$ and $prey_{a_4} = \{a_2, a_3\}$ and $prey_{a_5} = \{a_2, a_4\}$.

Therefore,

$$W_{a_2 a_3} = \frac{|\{a_1\}|}{|\{a_1\}|} = 1$$

and

$$W_{a_4 a_5} = \frac{|\{a_2\}|}{|\{a_2, a_3, a_4\}|} = \frac{1}{3}.$$

Thus, the competition graph of the food web of Fig. 10.2 is shown in Fig. 10.4.

10.1.2 Energy of the Food Web

Any digraph $\overrightarrow{\mathscr{G}}$ can be represented by a matrix called adjacency matrix and it is denoted by $A(\overrightarrow{\mathscr{G}}) = [x_{ij}]$, where $x_{ij} = $ weight of the directed edge (i, j).

The rows and columns are numbered by $1, 2, 3, 4, 5, 6$ corresponding to the species $a_1, a_2, a_3, a_4, a_5, a_6$ respectively. The adjacency matrix of the graph of Fig. 10.2 is

Fig. 10.4 The competition graph of food web of Fig. 10.2

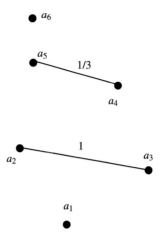

$$A(\overrightarrow{\mathscr{G}}) = \begin{bmatrix} 0 & 1 & 1 & 0 & 0 & 0 \\ 0 & 0 & 0 & 0.4 & 0.7 & 0 \\ 0 & 0 & 0 & 0.6 & 0 & 0 \\ 0 & 0 & 0 & 0 & 0.3 & 0 \\ 0 & 0 & 0 & 0 & 0 & 1 \\ 0 & 0 & 0 & 0 & 0 & 0 \end{bmatrix}$$

Note that all the entries for the column 1 (kelp) and the row 6 (shark) are zero, as a_1 has no incoming edge and a_6 has no outgoing edge. Remove these row and column from the adjacency matrix and the reduced matrix is shown below.

$$A(\overrightarrow{\mathscr{G}}) = \begin{bmatrix} 1 & 1 & 0 & 0 & 0 \\ 0 & 0 & 0.4 & 0.7 & 0 \\ 0 & 0 & 0.6 & 0 & 0 \\ 0 & 0 & 0 & 0.3 & 0 \\ 0 & 0 & 0 & 0 & 1 \end{bmatrix}$$

Now, we use the conventional method to find the eigenvalues of the matrix $A(\overrightarrow{\mathscr{G}})$ and the eigenvalues are 1, 0, 0.6, 0.3 and 1.

The energy of a graph is the sum of absolute eigenvalues, i.e. if λ_i, $i = 1, 2, \ldots, n$ are the eigenvalues of a matrix of order $n \times n$, then the energy of the corresponding graph is

$$E(\overrightarrow{\mathscr{G}}) = \sum_{i=1}^{n} |\lambda_i|.$$

For this problem the energy of the graph, i.e. the energy of the food web of Fig. 10.2 is $1 + 0 + 0.6 + 0.3 + 1 = 2.9$.

But, what this particular value of energy represent for food web is still not known for sure.

10.2   Application of Fuzzy Graph in Social Network

The **Social networks** (SNs) are the platforms for staying in touch with friends in the twenty-first century. These serve as inter focus to come across information on various issues and concerns essential tools for public awareness by inability quick spread of messages to a wide number of user at a time.

Indeed, SNs are essential for the dissemination of information, e-commerce and e-business, influential players (scientists, innovators, employees, etc.).

There are a lot of advantages are of SNs, even though some drawbacks are also there in modern SNs. These are listed below.

(i) Sometimes, it may be observed that the existence of nodes and edges (links) are not certain. For example, if two nodes represent two friends and the edge between them portrays the strength of the friendship (very strong, strong, very weak, etc.), then the corresponding graph is not the classical graph.

(ii) In all these SNs, all social units (individuals, groups, etc.) are presumed to have the same relevance. In reality, not every individual has the same influential power.

(iii) In the existing SNs, it is assumed that the strength of connectedness between each pair of individuals is same, but in reality that is not happens.

All these limitations are removed in fuzzy social network by introducing membership values to the individual units and also in the links.

A new SN called **fuzzy SN** (FSN) is described below.

10.2.1   Construction of Fuzzy Social Network

We assumed all social actors as fuzzy vertices and all linkages between them as fuzzy edges. Thus, a SN can be modeled as a fuzzy multigraph $(\mathcal{V}, \sigma, \mu)$, where $\mathcal{V} = \{P_1, P_2, \ldots, P_\lambda\}$ be a set of social units, λ be large positive integer represent the total number of vertices. $\sigma(P_i)$ represents the membership value of the vertex P_i. The set of edges $E = \{((P_i, P_j), \mu^r(P_i, P_j)), r = 1, 2, \ldots, g | (P_i, P_j) \in \mathcal{V} \times \mathcal{V}\}$ is a fuzzy multiset of $\mathcal{V} \times \mathcal{V}$ such that $\mu^r(P_i, P_j) \leq \sigma(P_i) \wedge \sigma(P_j)$ for all $P_i, P_j \in \mathcal{V}$ and for all $r = 1, 2, \ldots, g$, where $g = max\{r | \mu^r(P_i, P_j) \neq 0\}$.

The **fuzzy social network** (FSN) can be represented by a FG $FSN = (\mathcal{V}, \sigma, \mu)$, where $\sigma : \mathcal{V} \to [0, 1]$ and $\mu : \mathcal{V} \times \mathcal{V} \to [0, 1]$ are mappings such that $\mu(P_i, P_j) \leq \sigma(P_i) \wedge \sigma(P_j)$ for all $P_i, P_j \in \mathcal{V}$.

The membership values of the vertices and edges can be determined by many different ways. The membership value of a vertex may be considered as its performance in SN. Similarly, the membership value of an edge is taking as the frequency of interaction between end vertices.

10.2.2 Centrality of a Social Unit in Fuzzy Social Network

In the network analysis, indicators of centrality identify the most important vertices within a graph. It is used to identify the most influential person(s) in a social network, key infrastructure nodes in the Internet, super-spreaders of disease and many others. **Centrality** is the one of the most studied topics for evaluation of SN. Various methodologies have been proposed, including **degree centrality**, **betweenness centrality**, **core index**, etc. Recently, Das et al. [4] described a new type of centrality.

Degree centrality measures the direct linkage of a social unit to others. Betweenness centrality measures the number of paths between any pair of social units through the social unit. In all types of networks, the **central persons** are more valuable than others. They can send messages to more people, can collect more information compared to non-central persons. Degree centrality measures the number of direct friends (or linkages). So it does not measure the number of connected people by a path. It may be noted that friend of a friend in Facebook shares information with the person. So betweenness centrality is also useful. But, friend of friend does not share so much information like the direct friend. So importance of the linkage will gradually decrease from a person to another person by a connected path.

In FSN, if a unit X_f is connected with the unit X by an edge, then X_f is called **distance-1 friend** of X. $d_1(X)$ denotes the set of all distance-1 friends of X. That is,

$$d_1(X) = \{X_i \in \mathcal{V} : X_i \text{ is a distance-1 friend of } X\}.$$

Similarly, if there is a shortest path (i.e. minimum number of edges) between X and X_f of length k, then X_f is called a **distance-k friend** of X. That is,

$$d_k(X) = \{X_i \in \mathcal{V} : X_i \text{ is a distance-} k \text{ friend of } X\}.$$

Naturally, distance-1 friends are more influential than distance-2 friends; distance-2 friends are more influential than distance-3 friends; and so on. The linguistic term "more significant" can be represented mathematically by a suitable weight. Let w_k, $0 \leq w_k \leq 1$ be the weight which signifies the importance between the distance-k friends. The weights gradually decrease if the distance between friends increase. Thus, $w_1 \geq w_2 \geq \cdots \geq w_k \geq \cdots$

Let $u_1(= X_i), u_2, u_3, \ldots, u_k(= X_j)$ be the vertices on a path between X_i and X_j. We define fuzzy distance $D_f(X_i, X_j)$ between X_i and X_j along this path as

$$D_f(X_i, X_j) = \sum_{l=1}^{k-1} \mu(u_l, u_{l+1}).$$

In a network, it may be observed that there are multiple paths between two vertices. In FSN, we consider such path (may be more than one) whose fuzzy distance D_f is maximum and this path is called the **longest path between the vertices**. If there are k edges in this longest path, then we denote the length of the longest path by D_f^k, i.e. $D_f^k(X_i, X_j)$ represents the fuzzy distance between the vertices X_i and X_j in FSN along a specific path including exactly k edges. It is assumed that a social unit X has at most distance-p friends $(p \geq k)$.

Let us define $d_k'(X) = d_k(X) - d_{k-1}'(X)$, where $k = 2, 3, \ldots$ and $d_1'(X) = d_1(X)$. The centrality $C(X)$ of a social unit X of FSN is given as follows:

$$C(X) = \sum_{u_1 \in d_1'(X)} w_1 D_f^1(X, u_1) + \sum_{u_2 \in d_2'(X)} w_2 D_f^2(X, u_2) + \cdots + \sum_{u_p \in d_p'(X)} w_p D_f^p(X, u_p).$$

In this estimation, the influence of a close friend is more than the next to close friend and gradually decrease the most distant friend. The whole process is carried on by including the weight w_i, for distance-i friend, $i = 1, 2, 3, \ldots$

10.3 Application of Fuzzy Graphs in Telecommunication System

The **telecommunications system** is a platform for sharing information over long distance by means of a telecommunications route. The complete, single telecommunication circuit consists of two stations, each equipped with a transmitter and a receiver. The communication network may be an electrical wire or a cable, fiber optic, or magnetic waves. The free-space transmitting and processing of data using magnetic waves is called wireless.

The **churn prediction** is currently a subject in data mining and has been applied in the field of banking [15], mobile telecommunication [6, 13], life insurances [12], and other fields to predict persons are going to leave the system.

A telecommunication network is a SN and it can be represented appropriately with the help of FG. Here, a method to represent a telecommunication network by FG is described.

Let $V_1 = \{P_1, P_2, \ldots, P_\lambda\}$, where λ represents the number of registered consumers in a particular telecom circle or a particular telephone service provider, say $\mathcal{T}_1$ and $V_2 = \{P_{\lambda+1}, P_{\lambda+2}, \ldots, P_\varphi\}$ be another set of valid registered customers in other telecom circle or telephone service provider, say $\mathcal{T}_2$. Let $V = V_1 \cup V_2$ be the set of all customers. The customers are taken as the vertices of a FG and if two customers are connected over telephone, whatever may be the frequency of communication, then an edge is consider between such vertices and the resultant graph is called a **fuzzy telecommunication network** (FTN). The membership values of the consumers are determined by a given function ϕ, where $\phi : V \to [0, 1]$ and the membership values of the edges is given by $\overrightarrow{\mu} : V \times V \to [0, 1]$. Then, the **FTN** is represented by a directed FG $\overrightarrow{\xi} = (V, \phi, \overrightarrow{\mu})$.

A service provider gives more attention to those consumers who have lots of connected consumers. In every society, the recognition of a personality is measured by some segments of society. In FTN, the value of such members is called a **recognition number**, denoted by n. The recognition number of a person may be different for other network. The telecom companies roughly calculate this number for each customer. But, for simplicity n is taking a fixed number of a FTN.

The membership value $\sigma(P)$ of a customer P, i.e. a vertex P is determined by

$$\sigma(P) = \begin{cases} \frac{i}{n}, & \text{if } i = 1, 2, \ldots, n \\ 1, & \text{if } i > n. \end{cases},$$

where i is the number of distinct customers connected with $P \in V_1$ per unit interval of time.

Every group of people are connected with many other customers of other telecommunication network. In this study, we are interested with those customers who are communicated with the customers of other FTN. It is hard to collect all the data of such customers, but the number of calls (with duration) of the customers of other networks are available for a particular FTN.

All service providers fix their plans in such a way that customer talks to others within the same network by low call charges. If a customer of the network $\mathscr{T}_1$ (say) is talking with a customer of another network $\mathscr{T}_2$ during a significant amount of time, then it can be concluded that the customer of $\mathscr{T}_1$ has many friends in $\mathscr{T}_2$ and hence there is a change to chance the service provider to $\mathscr{T}_2$.

If a customer of $\mathscr{T}_2$ calls for a significant amount of time, say T, to the customers of $\mathscr{T}_1$, then such customer of $\mathscr{T}_2$ is also valuable for $\mathscr{T}_1$. This time T, in general, a fixed amount of time for a particular FTN. If the call duration of a customer more than T, then such customer is considered as valuable one. This fixed amount of time T is called the **satisfied time of calling**.

Then the membership value of a customer $P \in \mathscr{T}_2$ is

$$\sigma(P) = \begin{cases} \frac{t}{T}, & \text{if } t \leq T \\ 1, & \text{if } t > T. \end{cases}$$

Now, we calculate the membership value of an edge between two customers. So strength between two friends depends on how much time they are talking to each other over phones per unit interval of time. If P_i calls P_j, then there is a directed edge from P_i to P_j and it is denoted by $\overrightarrow{(P_i, P_j)}$, $P_i, P_j \in V$. Let $\overrightarrow{\mu} : V \times V \to [0, 1]$ be a mapping, which represents the membership value of an edge, defined by

$$\overrightarrow{\mu}(P_i, P_j) = \begin{cases} \frac{t}{T}(\sigma(P_i) \wedge \sigma(P_j)), & \text{if } t \in [0, T] \\ \sigma(P_i) \wedge \sigma(P_j), & \text{if } t > T. \end{cases},$$

where t is the calling duration per unit interval of time.

The edges and membership values play a crucial role to predict subscriber churn. Let P^* be a customer of $\mathscr{T}_1$ and we assume that P^* is communicating with k_1 number of customers $P_1^1, P_2^1, \ldots, P_{k_1}^1$ of the FTN $\mathscr{T}_1$ and k_2 number of customers $P_1^2, P_2^2, \ldots, P_{k_2}^2$ of the FTN $\mathscr{T}_2,$, in general, k_1 and k_2 are different. Let us define

$$W_1 = \sum_{i=1}^{k_1} \overrightarrow{\mu}\,(P^*, P_i^1)) \quad \text{and} \quad W_2 = \sum_{i=1}^{k_2} \overrightarrow{\mu}\,(P^*, P_i^2)).$$

If W_1 is significantly less than W_2, then there is no chance of changing the FTN by P^*. If W_2 is much more that W_1 then there is a high change to change FTN. But, if these two values are same or almost same then prediction of churn is difficult and in this case more investigation is required.

10.4   Application of Fuzzy Graph in Link Prediction

Link prediction (LP) is one of the most important research topics in graph theory and social networks. The aim of link prediction problem is to identify whether a pair of nodes will form a link in the future or not. The link prediction problem is stated below.

Given a social network (SN) $G = (V, E)$ and, in general, G is a multigraph. The set of edges E is not fixed, it changes from time to time. An edge $e = (a, b) \in E$ represents some sort of interaction between the end points a and b at a particular time t. Let $G[t_1, t_2]$ be the (sub)graph during the time interval $[t_1, t_2]$. Again, let $[t_1, t_2]$ and $[t_3, t_4]$, $t_2 < t_3$ be two time slots. In link prediction problem, find a list of edges that are not present in $G[t_1, t_2]$, but are predicted to appear in $G[t_3, t_4]$.

Link prediction problem has many applications in real life problems. Some of them are listed below.

(i) Predict which customers are likely to be a member of an online marketplace, viz. Amazon, flifkart, etc. It can help in making better product recommendations

(ii) Suggest collaborations or interactions between employees in an organization.

(iii) Extract vital insights from terrorist networks.

(iv) Protein-protein interaction (PPI) prediction or to annotate the PPI graph.

(v) Identify hidden groups of terrorists and criminals.

All available methods for link prediction are defined based on the neighbors. The score of link prediction depends on the number of neighbors. If the number of neighbors increases, then the chance of link between the vertices will increase. Several methods are available for link prediction, but no one is superior than the other. Some of them are presented below.

We denote the degree of a vertex a by $deg(a)$ for all vertex of a graph. Also, let $N(a)$ be the neighbor of the vertex a and score of LP between the vertices a and b is denoted by $\mathscr{L}$.

10.4.1 Common Neighbors

Name indicates that this method is based on common neighbors. The score of LP of this method is based on the idea that two nodes a and b have a high score if they have a common neighbor c. With an increasing number of common neighbors c score grows even higher.

The score of LP for this method is defined as

$$\mathscr{L}(a, b) = |N(a) \bigcap N(b)|.$$

10.4.2 Salton Index

This index is also called Salton Cosine index. It is used to find the similarity index based on cosine angle between rows of adjacency matrix having nodes a and b. The score of LP is defined by

$$\mathscr{L}(a, b) = \frac{|N(a) \bigcap N(b)|}{\sqrt{deg(a) \times deg(b)}}.$$

10.4.3 Jaccard Index

Jaccard proposed a statistic to compare similarity and diversity of sample sets. It is the ratio of common neighbors and all neighbors of nodes x and y. So, the score of link prediction by Jaccard index prevents higher degree nodes to have high similarity index with other nodes. This method was introduced by Jaccard and is defined as

$$\mathscr{L}(a, b) = \frac{|N(a) \bigcap N(b)|}{|N(a) \bigcup N(b)|}.$$

10.4.4 Sorensen Index

This method is similar to Jaccard and was proposed by Sorenson to find LP. This index has calculated the score as a ratio of twice the common neighbors and the sum of degrees of nodes a and b. This method is mainly used in ecological community. Soresen index is defined as

$$\mathscr{L}(a, b) = \frac{2|N(a) \bigcap N(b)|}{deg(a) + deg(b)}.$$

10.4.5 Hub Promoted Index

Hub Promoted index is another index used for LP which is the ratio of common neighbors and the minimum of degrees of nodes a and b. This method gives the high score because the denominator is the minimum degree of nodes a and b.

Hub Promoted index is defined as

$$\mathscr{L}(a, b) = \frac{|N(a) \cap N(b)|}{\min(deg(a), deg(b))}.$$

10.4.6 Hub Depressed Index

This index is similar to Hub Promoted index. This index is defined as the ratio of common neighbors and the maximum of degrees of nodes a or b. The Hub Depressed index is always less than the Hub Promoted index because the denominator is the maximum of degrees of nodes a and b.

The Hub Depressed index is defined as

$$\mathscr{L}(a, b) = \frac{|N(a) \cap N(b)|}{\max(deg(a), deg(b)}.$$

10.4.7 Leicht–Holme–Newman Index

This index was proposed by Leicht, Holme, and Newman. It is the ratio of common neighbors and the product of degrees of the nodes a and b.

This index is defined as

$$\mathscr{L}(a, b) = \frac{|N(a) \cap N(b)|}{(deg(a) \times deg(b))}.$$

This index and Salton index are almost similar, differ only in their denominator. As a result, for a same pair of nodes, a and b Salton index always assign a higher score compared to Leicht–Holme–Newman index.

10.4.8 Preferential Attachment Index

Preferential Attachment index is a similarity score calculated independently of the neighborhood of each node. Social networks increase as and when new nodes join and the new nodes join with the existing nodes having a higher degree compared to lower degree nodes. The chance of LP of two nodes a and b by this method is proportional to $deg(a) \times deg(b)$ and is defined as

$$\mathscr{L}(a, b) = deg(a) \times deg(b).$$

10.4.9 Adamic-Aar Index

Adamic-Adar index of LP is proposed by Adamic and Adar. This is calculated by adding weights to the nodes which are connected to both nodes a and b. This index depends on common neighbor and is defined as

$$\mathscr{L}(a, b) = \sum_{c \in N(a) \cap N(b)} \frac{1}{(\log deg(c))},$$

where c is a common neighbor of the nodes a and b.

10.4.10 Resource Allocation Index

In this index, a pairs of nodes a and b have no edges, then the node a can send some information to the node b by their neighbor. This index is defined as

$$\mathscr{L}(a, b) = \sum_{c \in N(a) \cap N(b)} \frac{1}{deg(c)}.$$

All these methods are defined based on the neighbors and the number of neighbors. If the number of neighbors increase then the chance of LP will also increase. In these methods, the quality, familiarity, acquaintances, etc. of neighbors are not considered. But, practically it is observed that the quality, familiarity, acquaintances, etc. of neighbors are required to make a link between two persons in the network. For example, a notable person is used to buy some products from a particular company say Amazon, then a person who have no idea about Amazon may buy some product from this company based only on the faith that a notable person is a customer of Amazon. And a link is established between that person and Amazon. Notice that the linguistic terms quality, familiarity, acquaintances, etc. can not be measured in any method, but these terms can be mathematized by using the concept of fuzzy logic. By incorporating these linguistic terms, one new method is described for link prediction called RSM method and the parameter associated with this method is called **RSM index**.

10.4.11   RSM Index for Link Prediction

Let $\mathscr{G} = (\mathscr{V}, \sigma, \mu)$ be a connected FG, i.e. a SN which is represented by a FG. Let u and v be any two members of $\mathscr{G}$ and their common neighbors be $w_1, w_2, \ldots, w_n$. Also, $\mu(u, w_i)$ is the membership value of the edge (u, w_i), where $i = 1, 2, 3, \ldots, n$. This membership value may be measured based on the frequency of interaction or familiarity of the neighbor w_i. We have mentioned that, if a notable person is a common neighbor of the persons u and v, then there is a high chance of a link between them.

To incorporate this feature in LP, familiarity index or 'fam' index is defined for a neighbor, defined below

$$fam(i) = min\{\mu(u, w_i), \mu(v, w_i)\},$$

where $i = 1, 2, 3, \ldots, n$.

Now, the score of LP between the vertices u and v is denoted by S_{uv} and determined by

$$S_{uv} = \sum_{i=1}^{n} \frac{fam(i)}{n}.$$

Thus, a link is established between the vertices u and v and the membership value of the link (u, v) is S_{uv}.

10.5   Application of Fuzzy Competition Graph in Manufacturing Industries

In manufacturing industry, several production companies and markets are there to sell products. Any production company produces the products as per market demands. They are also liable to transport the products to the market so that the end user can use their product within a reasonable time. They wish to deliver with minimum cost as much as they can. Market has the time-bound factor to get the production from company within a reasonable cost. Market has various opportunities to choose the company as well as company can choose market for their sake. So, there is a fair competition between companies. The problem is to find out which companies are in competition and the strengths of their competition to achieve markets that they serve, considering all the cases of production, demands and the time that they can spare. This problem can be modeled as a fuzzy competition graph by considering the following correspondences [19]:

- Companies and markets are treated as vertices.
- The membership value of a vertex that is taken as a company represents the power of production capability.

- Similarly, the membership value of a vertex corresponding to a market represents the grade of demand of the market.
- The companies and markets are connected, that is, they have an edge if they both have the same time tenure to transport or receive the product. A grade is assigned to each time within the tenure.

Assuming the companies and markets have higher membership values than that of their shared time, i.e. membership value of each edge is less than the minimum of membership values of all the vertices, the problem is well defined for a FG.

For simplicity let us consider three companies, namely C_1, C_2 and C_3 to produce a certain product. Each company has a fixed capacity for production and suppose the companies produce 20%, 87%, and 90% of market demands respectively. These production capacities are considered as the membership values of the vertices, i.e. these are 0.20, 0.87, 0.90. Assume that there are two markets M_1 and M_2 and the demands of these markets be 90% and 85% respectively. Thus, the membership values of the markets are 0.90 and 0.85. Since the transportation time is an uncertain quantity, so the transportation time from companies to markets (C_1, M_1), (C_1, M_2), (C_2, M_1), (C_2, M_2), (C_3, M_1), and (C_3, M_2) are considered as 0.1, 0.2, 0.85, 0.75, 0.8, and 0.8 respectively. The entire representation is depicted in Fig. 10.5. Note that this is a directed complete bipartite fuzzy graph and this graph is denoted as $\overrightarrow{\mathscr{G}}$.

Based on the competitions among the manufacturing companies a fuzzy competition graph $\mathscr{C}(\overrightarrow{\mathscr{G}})$ is constructed. In the fuzzy competition graph $\mathscr{C}(\overrightarrow{\mathscr{G}})$, the vertices and their membership values are same as $\overrightarrow{\mathscr{G}}$. Only the edge membership values are to be determined.

Computation of edge membership values:
Now, the out-neighborhood of the vertices C_1, C_2, C_3 are calculated as follows:

$$\Delta^+(C_1) = \{M_1(0.1), M_2(0.2)\}$$
$$\Delta^+(C_2) = \{M_1(0.85), M_2(0.75)\}$$
$$\Delta^+(C_3) = \{M_1(0.8), M_2(0.8)\}$$

Fig. 10.5 The relationship between companies and markets

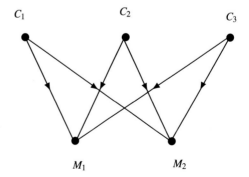

Fig. 10.6 Fuzzy competition
graph of Fig. 10.5

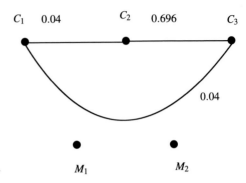

Table 10.1 Degree of competition among the companies

Companies	Degree of competition	Competition in %
C_1, C_2	0.04	4
C_2, C_3	0.696	69.6
C_3, C_1	0.04	4

The heights of the intersecting sets are

$$h(\Delta^+(C_1) \cap \Delta^+(C_2)) = h(\{M_1(0.1), M_2(0.2)\}) = 0.2$$
$$h(\Delta^+(C_1) \cap \Delta^+(C_3)) = h(\{M_1(0.1), M_2(0.2)\}) = 0.2$$
$$h(\Delta^+(C_2) \cap \Delta^+(C_3)) = h(\{M_1(0.8), M_2(0.75)\}) = 0.8$$

The membership value of the edge (a, b) (denoted by $\mu'(a, b)$) in $\mathscr{C}(\overrightarrow{\mathscr{G}})$ is computed as $\mu'(a, b) = \sigma(a) \wedge \sigma(b) h(\Delta^+(a) \cap \Delta^+(a))$

$$\mu'(C_1, C_2) = 0.2 \times 0.2 = 0.04, \quad \mu'(C_1, C_3) = 0.2 \times 0.2 = 0.04, \quad \mu'(C_2, C_3) = 0.87 \times 0.8 = 0.696.$$

The corresponding fuzzy competition graph of Fig. 10.5 is shown in Fig. 10.6. The membership value (degree) of competition among the companies is shown in Table 10.1.

Highest degree of competition among companies is 0.696 and it is in between C_2 and C_3.

10.6 Application of *m*-Polar Fuzzy Graph in Multi Attribute Decision-Making Problem

In this section, an application of *m*-polar fuzzy graph in Decision-making problem is presented. Suppose Ms. B want to buy the 'best' mobile from a branded company [2, 3]. In nearby market only five mobile companies are available, viz. Samsung (x_1), Apple (x_2), Huawei (x_3), Oppo (x_4), and Xiaomi (x_5). Each mobile has some common features (f_i) such as operating system (f_1), internal memory/RAM (f_2), camera quality (f_3), display (f_4), high demand/rating (f_5). Notice that no feature can be measured as a specific way, as measurement of these features depend on the decision-maker.

Comparing the features of mobiles among these companies one may assign membership value $(\sigma^{x_i}(f_j))$ to the feature (f_j) for the company x_i. Thus, each vertex has multiple attributes comprising the features of the mobile. We consider the membership values of the features for each company as per Table 10.2.

Now, for these five companies and five features an *m*-polar FG is constructed as follows:

For each company a vertex is considered and the membership value of each vertex is a 5-tuple vector $(f_1, f_2, f_3, f_4, f_5)$. An edge is drawn between the vertices if they have at least one common feature. In this problem, all features are available for all kinds of mobiles. So, each vertex is adjacent to all other vertices. Now, the membership value of the edge (x_i, x_j) for the feature f_j is determined as

$$\mu^{f_j}(x_i, x_k) = \min\{\sigma^{x_i}(f_j), \sigma^{x_k}(f_j)\}.$$

Using Table 10.2 and above formula, the membership values of all edges for all features are calculated and tabulated in Table 10.3.

Now the *m*-polar FG for the above tabulated data is shown in Fig. 10.7.

To select the best mobile, we define a score function $(w(x_i))$ for the company x_i as,

$$w(x_i) = \frac{1 + \sum_j \sigma^{x_i}(f_j)}{2},$$

for $i = 1, 2, 3, 4, 5$.

Table 10.2 Membership value assigned to each node

σ	x_1	x_2	x_3	x_4	x_5
$\sigma(f_1)$	0.75	0.85	0.70	0.72	0.70
$\sigma(f_2)$	0.76	0.72	0.75	0.78	0.82
$\sigma(f_3)$	0.70	0.82	0.72	0.72	0.68
$\sigma(f_4)$	0.72	0.75	0.66	0.65	0.62
$\sigma(f_5)$	0.76	0.80	0.70	0.72	0.65

Table 10.3   Membership values of the edges

Edges	f_1	f_2	f_3	f_4	f_5
$x_1 x_2$	0.75	0.72	0.70	0.72	0.76
$x_1 x_3$	0.70	0.75	0.70	0.66	0.70
$x_1 x_4$	0.72	0.76	0.70	0.65	0.72
$x_1 x_5$	0.70	0.76	0.68	0.62	0.65
$x_2 x_3$	0.70	0.72	0.72	0.66	0.70
$x_2 x_4$	0.72	0.72	0.72	0.65	0.72
$x_2 x_5$	0.70	0.72	0.68	0.62	0.65
$x_3 x_4$	0.70	0.75	0.72	0.65	0.70
$x_3 x_5$	0.70	0.75	0.68	0.62	0.65
$x_4 x_5$	0.70	0.78	0.68	0.62	0.65

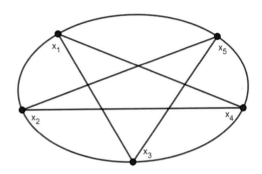

Fig. 10.7   The m-polar fuzzy graph for the problem

Table 10.4   Preferences of mobiles

Candidate	Score	Rank
Samsung	2.345	2
Apple	2.47	1
Huawei	2.295	4
Oppo	2.265	3
Xiaomi	2.235	5

Thus, the score for all vertices are $w(x_1) = 2.345$, $w(x_2) = 2.47$, $w(x_3) = 2.265$, $w(x_4) = 2.295$, $w(x_5) = 2.235$.

Depending on the score of each vertex, the preference (rank) of all mobile companies are calculated and shown in Table 10.4.

Hence, Ms. B can buy a mobile from 'Apple' company as its score is maximum among all other companies. It is very necessary for a person to take an appropriate decision to select a 'good' mobile with all these features.

10.7 Applications of Fuzzy Graph Coloring

There are several applications on coloring of FGs. Two of them are mentioned here, viz. traffic light signaling and job selection among several companies. For other application, readers may consult [14]. Fuzzy coloring of FG is discussed in Chap. 7 and more results are available in [9–11, 29]. A very good application of coloring of graph is discuss in [24].

10.7.1 Traffic Light Signaling Using Fuzzy Graph Coloring

In traffic light signaling, red color is used to indicate 'stop', green is used to indicate 'go', and the third color (usually amber) is used to indicate that the route is open, but needs attention. Consequently, this third light does not indicate how much risk is there on the route. It creates a dilemma as it is hard to figure out which route is more crowded than the other. Large number of traffic needs to be controlled using these lights at all time. Also, some paths collide precariously as compared to different routes. The terms "crowder", "caution" can not be represented by a crisp graph or FGs with crisp vertices and hence a crisp graph or a FG with crisp vertices does not represent the traffic light signaling properly.

In traffic light signaling, red and green colors are used. Here, a modified approach is used for traffic light management.

In the proposed approach, the end vertices of a strong edge are colored by red and green colors and the end vertices of a weak edge are colored by fuzzy colors.

The traffic light signaling system can be modeled using a FG as described below: For each route a vertex is considered and the membership value of the vertex is the degree of crowdedness of the corresponding route. An edge is drawn between two (vertices) routes if they intersect. The membership value of an edge denotes the probability of accident when the corresponding routes are opened. Obviously, the resultant graph is a FG, i.e. any traffic light signaling system can be modeled by a FG. Now, the red signal indicates full stoppage of the route and green color means the route is fully open. Fuzzy red symbolizes some danger rather than full closing of the route and fuzzy green will symbolize little less safety than full green signal. In such FG, strong edges express that two routes corresponding to the end vertices can not be opened simultaneously. Therefore, when one route is opened, the other is closed. A weak edge in the FG represents little less danger between the routes of the corresponding vertices. In this case, if one end vertex is colored with red (or green), then another end vertex must be colored with fuzzy green (or fuzzy red).

Let us explain how this works. Let the north, south, east, and west routes at a particular junction are denoted as N, S, E, and W. For simplicity, it is assumed that E is a one way route, i.e. traveling is possible from the route E to other routes but traveling is not allowed from any route to the route E. Traveling from north to south

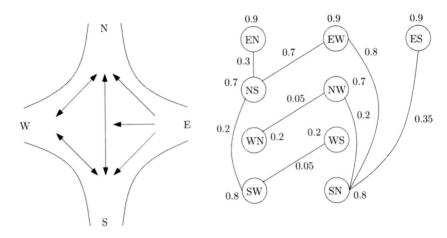

(a) Junction of some routes   (b) The corresponding fuzzy graph with fuzzy colour

Fig. 10.8 FG corresponding to a traffic signal system

direction is denoted by NS and similar notations are used for other directions. So, the possible routes are $NS, NW, WN, WS, SN, SW, EN, ES, EW$ (see Fig. 10.8a).

Suppose in a particular time, the route W is less crowded as compared to other routes. Depending on the traffic on the routes, let the membership values of the vertices be $(NS, 0.7), (NW, 0.7), (WN, 0.2), (WS, 0.2), (SN, 0.8), (SW, 0.8), (EN, 0.9), (ES, 0.9), (EW, 0.9)$. Considering the probability of accidents at the crossings, let the membership values of the edges be $((EN, NS), 0.3), ((EW, NS), 0.7), ((EW, SN), 0.8), ((ES, SN), 0.35), ((NS, SW), 0.2), ((NW, SN), 0.2), ((NW, WN), 0.05), ((WS, SW), 0.05)$. Now, the strengths of the roads are $0.43, 1, 1, 0.44, 0.29, 0.29, 0.25, 0.25$ respectively. Thus, the routes (EW, NS) and (EW, SN) are only strong edges. The corresponding FG is displayed in Fig. 10.8b.

Now, we discuss about coloring this FG. Since there are only two strong edges in the graph, so only two basic colors, viz. red and green are required. Let EW be colored with basic color green. Therefore, the route EW is open. We need to color NS and SN. Here, (EW, NS) and (EW, SN) are strong edges. So, NS and SN are colored by basic color red. As EN is adjacent to NS which is colored by red, therefore, EN will be colored by a fuzzy color whose basic color is green. Here, strength of the edge (EN, NS) is 0.43. Therefore, EN is colored by the fuzzy color (green, $1 - 0.43$) = (green, 0.57). Similarly, the other vertices SW, NW, ES are colored by the fuzzy colors (green, 0.71), (green, 0.71), (green, 0.56) respectively. Finally, WN gets fuzzy color (red, 0.75) and WS gets fuzzy color (red, 0.75).

10.7.2 Selection of Job

The edge coloring of FG can also be used to represent the job selection problem efficiently. Nowadays, organizations/companies are designing good web sites to advertise for their vacant posts with detailed information such as post name, salary structure, eligibility criteria, terms & conditions, etc. Potential candidates can understand which companies are appropriate as per his/her expertise.

Suppose M web sites are available from M organizations mentioning the above details. Again, we assume that N number of candidates have registered for jobs with their curriculum vitae and other details. For the entire processes, the organizations and candidates are considered as vertices. If the minimum eligibility criteria of a company is fulfilled by a candidate then an edge is drawn between them (see Fig. 10.9).

Now the membership values of the vertices corresponding to the companies may depend on the following issues.

Salary, company brand value, product value, job security, medical facility, car facility, insurance facility, accommodation facility, service rule, service hours, job responsibility, etc.

The membership values of vertices corresponding to the candidates depend on following parameters—academic qualification, experience, languages known, communication skill, age, salary requirement, compensation, behavior, etc.

The membership values of edges are computed based on the matching of criteria of the companies with the profile of the applicants. The relationship between companies and applicants is constructed as a fuzzy bipartite graph $\mathscr{G}_B$.

Till now, we talked about the formation of the FG for the companies and applicants. Now, we give a detailed explanation of the selection process.

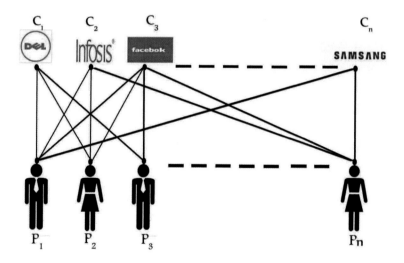

Fig. 10.9 Relationship between companies and applicants

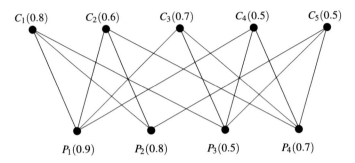

Fig. 10.10 Fuzzy bipartite graph for five companies and four applicants

From the entire FG, any company can find a list of all applicants along with applicant's score (i.e. the edge membership value). This list of applicants can be found out by finding the subgraph of the original graph $\mathcal{G}_B$. From this subgraph, a company can prepare a short list from the complete list of applicants by using the concept of strong CI.

For detailed illustration, let us consider a job oriented web site (i.e. the FG $\mathcal{G}_B$) containing five companies and four applicants (Fig. 10.10).

For this case, the companies are denoted by C_1, C_2, C_3C_5, C_5 and the applicants by P_1, P_2, P_3, P_4. The edges are drawn as per the method described earlier. Thus, the edges are (C_1, P_1), (C_1, P_2),(C_1, P_3), (C_2, P_1), (C_2, P_2), (C_2, P_4), (C_3, P_1), (C_3, P_3), (C_3, P_4), (C_4, P_1), (C_4, P_3), (C_4, P_4), (C_5, P_2), (C_5, P_3), (C_5, P_4). The membership values of vertices C_1, C_2, C_3, C_4, C_5 depend on the salary, company's brand value, product value, etc. as stated above. The membership values of the vertices P_1, P_2, P_3, P_4 are determined based on the applicant's qualification, experience, etc. as stated above. The membership values of edges are calculated based on matching between the companies and applicants criteria.

For illustration, in a particular case the membership values of vertices C_1, C_2, C_3, C_4, C_5 are taken as 0.8, 0.6, 0.7, 0.5, 0.5. The membership values of vertices P_1, P_2, P_3, P_4 are considered as 0.9, 0.8, 0.5, 0.7. Suppose the membership values of the edges (C_1, P_1), (C_1, P_2), (C_1, P_3), (C_2, P_1), (C_2, P_2), (C_2, P_4), (C_3, P_1), (C_3, P_3), (C_3, P_4), (C_4, P_1), (C_4, P_3), (C_4, P_4), (C_5, P_2), (C_5, P_3), (C_5, P_4) are determined as 0.7, 0.6, 0.5, 0.6, 0.5, 0.4, 0.7, 0.4, 0.6, 0.4, 0.4, 0.5, 0.2, 0.3, 0.2 respectively.

By signing in to the site, the company as well as applicant can see the fuzzy subgraph along with the membership values of all vertices and edges. For example, if the applicant P_2 logs in to the site, then he/she can see the fuzzy subgraph depicted as in Fig. 10.11a and same for the company C_3 as is shown in Fig. 10.11b.

If these two fuzzy subgraphs are colored by fuzzy colors, then the applicant P_2 gets a colored fuzzy subgraph (see Fig. 10.12a. Here, three basic colors red, green, and brown are used. The edge (P_2, C_1) is colored by (red, 0.75). Similarly, other two edges (P_2, C_2) and (P_2, C_5) are colored by (green, 0.83) and (brown, 0.4) respectively. And the weight is $(0.75 + 0.83 + 0.4) = 1.98$ and strong weight is $(0.75 + 0.83) = 1.58$.

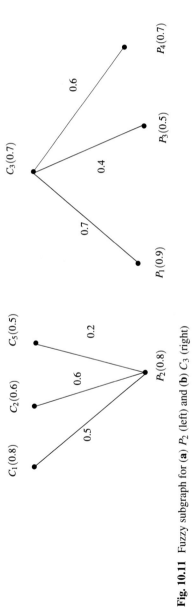

Fig. 10.11 Fuzzy subgraph for (**a**) P_2 (left) and (**b**) C_3 (right)

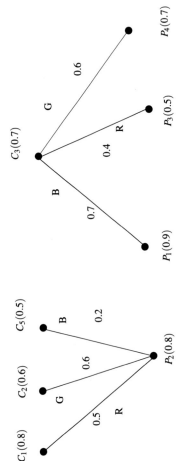

Fig. 10.12 Edge coloring of the fuzzy subgraphs

Thus, the CI is (3, 1.98) and strong CI is (2, 1.58). Therefore, the company will select best two applicants among these four applicants.

From CI, an applicant can realize how many companies are suitable for him/her and from the weight he/she can understand the chances of selection by the companies. And from the intensity of color, he/she can understand which company is best for him/her.

Similar idea is applicable to a company. From the CI and strong CI, the company can find out the number of suitable candidates for the company and from the colors assigned to the edges the company can identify the most deserving candidates.

10.8 Application of Fuzzy Graph in Patrolling of Bus Network

Fuzzy graph theory can be used in the **patrolling of bus network** of a city or a district or any bigger region. The bus network can be modeled as a FG, where the bus depots are the vertices and the route connecting two bus depots is an edge. During patrolling some bus inspectors are deployed and the objective of the patrolling problem is to find the minimum number of inspectors required who will inspect the bus routes for a particular time interval T. Let $V = \{B_1, B_2, \ldots, B_n\}$ be the set of bus depots in a particular bus network and c_i be the capacity (maximum number of buses can stand at a time) of the depot B_i. Suppose the number of buses entering into the depots are $b_1, b_2, \ldots, b_n$ in time T. The node membership value of the depot B_i is determined as b_i/c_i, for $i = 1, 2, \ldots, n$.

Now, define a term "satisfied passenger number" [23], which is the minimum number of passengers for a particular route so that we can say that the route is profitable or valuable and this number is denoted by P. The membership value of the route (B_i, B_j) is determined from the following definition.

$$\mu(B_i, B_j) = \left\{ \begin{array}{ll} \frac{s}{P}\{\sigma(B_i) \wedge \sigma(B_j)\}, & \text{if } 0 \leq s \leq P \\ \sigma(B_i) \wedge \sigma(B_j), & \text{if } s > P \end{array} \right\},$$

where s is the number of passengers traveling the route (B_i, B_j) during T.

For illustration, let the capacity of five depots, number of passengers arriving into the depots are tabulated in Table 10.5.

Let the routes (edges) for this problem be (B_1, B_2), (B_1, B_3), (B_1, B_5), (B_2, B_3), (B_2, B_4), (B_2, B_5), (B_3, B_4), (B_4, B_5).

The number of passengers for all routes and the membership values of them are listed in Table 10.6. Let the 'satisfied passenger number' P be 30.

The entire bus route is shown in Fig. 10.13.

By imposing different conditions/rules, the patrolling problems have been solved in [23].

Table 10.5 Bus depots along with their capacities and membership values

Bus depot	Capacity	Number of buses arrived	Vertex membership value (σ)
B_1	60	30	0.5
B_2	10	10	1.0
B_3	40	30	0.75
B_4	20	25	1.0
B_5	10	40	0.25

Table 10.6 Membership values of the edges

Edges	Number of passengers traveling (s)	Edge membership value
(B_1, B_2)	08	0.133
(B_1, B_3)	40	1
(B_1, B_5)	35	1
(B_2, B_3)	40	1
(B_2, B_4)	05	0.167
(B_2, B_5)	30	1
(B_3, B_4)	25	0.625
(B_4, B_5)	06	0.050

Fig. 10.13 Fuzzy graph representation of bus network

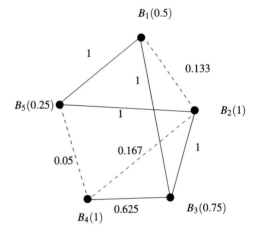

10.9 Application of Fuzzy Graph in Image Compression

Image processing is a very important topic in computer vision. Image is a collection of pixels. For an image, one can construct a graph. The pixels can be considered as nodes and edges connect two pixels if they share a common boundary. Also, a group of similar pixels are considered as a node, and if two groups share a common boundary, then there is an edge between them. An image is considered in Fig. 10.14. The entire image is divided into four quadrants and these are referred to as $a, b, c,$ and d. Depending on pixel density or depth of the color or amount of pixels in a group, membership values of vertices are determined. The membership value of an edge is considered as the minimum of the membership values of end vertices of the edge.

Now, the original image can be compressed using the following method.

In level one, let the membership values of the four vertices a, b, c, d be $\sigma(a), \sigma(b),$ $\sigma(c),$ and $\sigma(d)$. Now, based on the desired image (this step can be executed by considering a constant number), similar nodes are merged. In level two, such a group of pixels are considered where the number of pixels is less in groups. In this level, the vertex membership values are determined by some specific method depending on the problem. The membership values of the segmented images are denoted by $\sigma_1(a), \sigma_1(b), \sigma_1(c),$ and $\sigma_1(d)$, where $\sigma_1(x) \leq \sigma(x)$ for all $x \in \{a, b, c, d\}$. A similar technique is used to reduce the number of pixels in groups in the next levels. This assumption of membership values is purely based on the counts of pixels in the selected groups. If the counts are not possible, then the density of the groups are assumed to define the membership values. The entire segmented steps are illustrated in Fig. 10.14.

10.10 Application of Fuzzy Graph in Installation of Cell Phone Towers

One of the most widely studied graphs in the literature of graph theory is perfect graphs. Be it for their usefulness in solving different kinds of problems or their specific features, they have always attained a higher position in the study of different types of graphs. Raut and Pal [20–22] made an attempt to extend this study into fuzzy graph theory. They introduced perfect fuzzy graphs which is a generalization of the well-known perfect graphs (crisp). Their study highlighted the reason for introducing perfectness in fuzzy graphs not only for theoretical reason but also on applicative grounds.

They showed in [22] how using **perfect fuzzy graphs** and obviously, fuzzy graph coloring, a much needed yet an overlooked problem of how living world, that gets hugely affected due to electromagnetic radiation emitted from cell phone towers can be saved to a great extent. Usually, cell-phone tower installation company nowadays set up towers almost arbitrary anywhere within a city owing to meet up huge demands

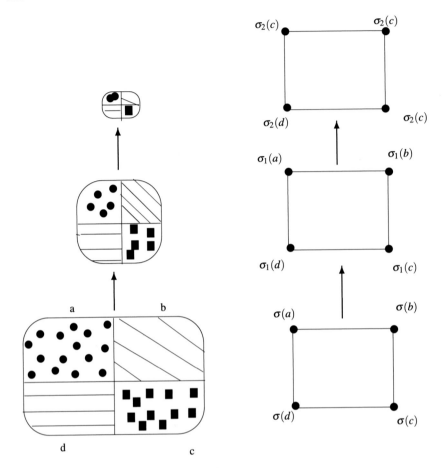

Fig. 10.14 Demonstration of image compression

as well as to prosper their own business. While doing so, they hardly maintain the protocols that are to be followed while installing a tower. As a result, there are plenty of towers in any city including the residential areas as well. As a consequence, the living world is getting badly affected day by day owing to the nearness of these towers. Reports symptoms include sleeplessness, headaches, anxiety, depression, fatigue, loss of libido, and even cancan.

The aim was to find out with the help of graphical modeling which residential units such as schools, college, universities, hospitals, buildings, apartments, etc. are in a comparatively safer zone and which are not. So, small sections are to be picked at a time and consider the mobile towers installed in that area. Consider these towers as vertices of a fuzzy graph joining edges with hereby residential units (other vertices of the graph).

A sample figure is given (see Fig. 10.15).

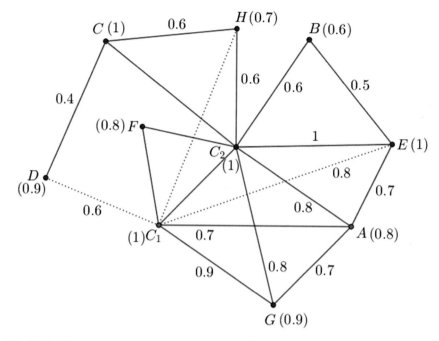

Fig. 10.15 Figure showing relative positions of residential units (*A, B, C, D, E, F, G and H*) with respect to cell-phone towers C_1 and C_2 in a certain graphical location

The solid lines (as edges) represent strong connectivity between vertices, whereas the dotted lines represent weak connectivity between vertices.

The membership values of the towers are taken as to be 1 because of their usually high EMR. The membership values of the residential units vary due to the varying number of users.

Due to varying strengths of signals and varying number of users of mobile networks, fuzziness will come into play which will be expressed in terms of membership values of vertices and edges. A set of vertices that will form a partial fuzzy clique will indicate a wrong placement of the respective mobile tower.

The strong and weak connectivity between the vertices is determined mainly on the basis of distances between them—higher distance, lower the membership value, and vice versa.

To find out which among these residential units (vertices) form partial fuzzy cliques with the cell-phone towers (C_1 or C_2), we need to find out the strengths of the edges based on the formula

$$I_{(p,q)} = \frac{\mu(p,q)}{\sigma(p) \wedge \sigma(q)} \quad \text{for an edge } (p,q).$$

Table 10.7 Table to find out colors of adjacent vertices to C_1 by calculating the strengths of edges

Edge	Strength of edge	Color assigned to residential unit (vertex)
(C_1, A)	$\frac{0.7}{0.8} = 0.875$	Different color from C_1
(C_1, F)	$\frac{0.8}{0.8} = 1$	Different color from C_1
(C_1, E)	$\frac{0.5}{1} = 0.5$	Fuzzy color of C_1
(C_1, D)	$\frac{0.6}{0.9} = 0.67$	Different color from C_1
(C_1, G)	$\frac{0.9}{0.9} = 1$	Different color from C_1

A set of vertices forming a partial fuzzy clique with definitely indicate distinct coloring of vertices using distinct solid colors while those vertices (residential units) that will have fuzzy colors sharing the same solid color so that of the vertex for the mobile tower will ensure lesser threat to the people.

Table 10.7 finds the colors of adjacent vertices to C_1 by calculating the strengths of edges.

Similarly, for C_2, it can be calculated following the same method as in Table 10.7.

Constructing this graphical model on a larger scale (consider the whole city) will help us to figure out which towers are to be kept and which of them need to be shifted to some place possessing less harm to the living beings nearby. For other applications of fuzzy graphs reader may look in [1, 5, 7, 8, 18, 30, 31].

References

1. M. Akram, A. Bashir, S. Samanta, Complex pythagorean fuzzy planar graphs. Int. J. Appl. Comput. Math. **6**(5) (2020)
2. S. Bera, M. Pal, Certain types of m-polar interval-valued fuzzy graph. J. Intell. Fuzzy Syst. (2020). https://doi.org/10.3233/IFS-191587
3. S. Bera, M. Pal, On m-polar interval-valued fuzzy graph and its application. Fuzzy Inf. Eng. (2020). https://doi.org/10.1080/16168658.2020.1785993
4. K. Das, S. Samanta, M. Pal, Study on centrality measures in social networks: a survey. Soc. Netw. Anal. Mining **8**(1), 13 (2018)
5. K. Das, S. Samanta, K. De, Generalized neutrosophic competition graphs. Neutrosophic Sets Syst. **31**, 156–171 (2020)
6. M.P.J. Ferreira, M. Vellasco, C. Barbosa, Data mining techniques on the evaluation of wireless churn, in *ESANN* (2004), pp. 483–488
7. R. Mahapatra, S. Samanta, M. Pal, X. Qin, RSM index: a new way of link prediction in social networks. J. Intell. Fuzzy Syst. **37**(2), 2137–2151 (2019)
8. R. Mahapatra, S. Samanta, M. Pal, Applications of edge colouring of fuzzy graphs. Informatica **31**(2), 313–330 (2020)
9. R. Mahapatra, S. Samanta, T. Allahviranloo, M. Pal, Radio fuzzy graphs and assignment of frequency in radio stations. Comput. Appl. Math. **38**(3), 1–20 (2019)
10. T. Mahapatra, M. Pal, Fuzzy colouring of m-polar fuzzy graph and its application. J. Intell. Fuzzy Syst. **35**(6), 6379–6391 (2018)

11. T. Mahapatra, G. Ghorai, M. Pal, Fuzzy fractional coloring of fuzzy graph with its application. J. Ambient Intell. Humanized Comput. (2020). https://doi.org/10.1007/s12652-020-01953-9

12. K. Morik, H. Kopcke, Analysing customer churn in insurance data a case study, in *Proceedings of the 8th European Conference on Principles and Practice of Knowledge Discovery in Databases*, New York, USA (2004), pp. 325–336

13. D.G.M. Mozer, R. Wolniewicz, H. Kaushansky, Predicting subscriber dissatisfaction and improving retention in the wireless telecommunications industry. IEEE Trans. Neural Netw. **11**, 690–696 (2000)

14. S. Munoz, M.T. Ortuo, J. Ramrez, J. Yez, Coloring fuzzy graphs. Omega **33**(3), 211–221 (2005)

15. D.V.D. Poel, B. Larivire, Customer attrition analysis for financial services using proportional hazard models. Eur. J. Oper. Res. **157**(1), 196–217 (2004)

16. T. Pramanik, S. Samanta, M. Pal, S. Mondal, B. Sarkar, Interval-valued fuzzy ϕ-tolerance competition graphs. SpringerPlus **5**, 1981 (2016). https://doi.org/10.1186/s40064-016-3463-z

17. T. Pramanik, S. Samanta, B. Sarkar, M. Pal, Fuzzy ϕ-tolerance competition graphs. Soft Comput. **21**, 3723–3734 (2017). https://doi.org/10.1007/s00500-015-2026-5 (2016)

18. T. Pramanik, S. Mondal, S. Samanta, M. Pal, A study on bipolar fuzzy planar graph and its application in image shrinking. J. Intell. Fuzzy Syst. **34**(3), 1863–1874 (2018)

19. T. Pramanik, G. Muhiuddin, A.M. Alanazi, M. Pal, An extension of fuzzy competition graph and its uses in manufacturing industries. Mathematics **8**, 1008 (2020)

20. S. Raut, M. Pal, G. Ghorai, Fuzzy permutation graph and its complements. Intell. Fuzzy Syst. **35**(2), 2199–2213 (2018)

21. S. Raut, M. Pal, Generation of maximal fuzzy cliques of fuzzy permutation graph and applications. Artif. Intell. Rev. **53**, 1585–1614 (2020)

22. S. Raut, M. Pal, On perfectness of fuzzy graph, manuscript

23. S. Rehmani, M.S. Sunitha, Edge geodesic number of a fuzzy graph. J. Intell. Fuzzy Syst. **37**(3), 4273–4286 (2019)

24. A. Saha, M. Pal, T.K. Pal, Selection of programme slots of television channels for giving advertisement: a graph theoretic approach. Inf. Sci. **177**(12), 2480–2492 (2007)

25. S. Sahoo, M. Pal, Intuitionistic fuzzy competition graphs. J. Appl. Math. Comput. **52**, 37–57 (2016). https://doi.org/10.1007/s12190-015-0928-0

26. S. Samanta, M. Pal, Fuzzy k-competition graphs and p-competition fuzzy graphs. Fuzzy Eng. Inf. **5**(2), 191–204 (2013)

27. S. Samanta, M. Akram, M. Pal, m-step fuzzy competition graphs. J. Appl. Math. Comput. **47**(1–2), 461–472 (2015)

28. S. Samanta, M. Pal, A. Pal, Some more results on fuzzy k-competition graphs. Intern. J. Adv. Res. Artif. Intell. **3**(1), 60–67 (2014). https://doi.org/10.14569/IJARAI.2014.030109

29. S. Samanta, T. Pramanik, M. Pal, Fuzzy colouring of fuzzy graphs. Afrika Matematica **27**(1–2), 37–50 (2016)

30. S. Samanta, B. Sarkar, A study on generalized fuzzy Euler graphs and Hamiltonian graphs. J. Intell. Fuzzy Syst. **35**(3), 3413–3419 (2018)

31. S. Samanta, B. Sarkar, Representation of generalized fuzzy competition graphs. Int. J. Comput. Intell. Syst. **11**(1), 1005–1015 (2018)

Index

© The Editor(s) (if applicable) and The Author(s), under exclusive license
to Springer Nature Singapore Pte Ltd. 2020
M. Pal et al., *Modern Trends in Fuzzy Graph Theory*,
https://doi.org/10.1007/978-981-15-8803-7

Printed in the United States
by Baker & Taylor Publisher Services